France

| Vintage | Red Bordeaux | | White Bordeaux | | Alsace |
	Médoc/Graves	Pom/St-Ém	Sauternes & sw	Graves & dry	
2019	7–9	7–9	7–8	7–9	6–8
2018	8–9	8–9	7–9	7–8	7–9
2017	6–8	6–7	8–9	7–8	9–10
2016	8–9	8–9	8–10	7–9	7–8
2015	7–9	8–10	8–10	7–9	7–9
2014	7–8	6–8	8–9	8–9	7–8
2013	4–7	4–7	8–9	7–8	8–9
2012	6–8	6–8	5–6	7–9	8–9
2011	7–8	7–8	8–10	7–8	5–7
2010	8–10	7–10	7–8	7–9	8–9
2009	7–10	7–10	8–10	7–9	8–9
2008	6–8	6–9	6–7	7–8	7–8
2007	5–7	6–7	8–9	8–9	6–8
2006	7–8	7–8	7–8	8–9	6–8
2005	9–10	8–9	7–9	8–10	8–9
2004	7–8	7–9	5–7	6–7	6–8
2003	5–9	5–8	7–8	6–7	6–7
2002	6–8	5–8	7–8	7–8	7–8
2001	6–8	7–8	8–10	7–9	6–8

France, continued

| Vintage | Burgundy | | | Rhône | |
	Côte d'Or red	Côte d'Or white	Chablis	North	South
2019	7–9	7–10	7–9	8–9	6–8
2018	6–9	7–8	7–9	7–9	7–8
2017	6–9	8–9	8–9	7–9	7–9
2016	7–8	7–8	5–7	7–9	7–9
2015	7–9	7–8	7–8	8–9	8–9
2014	6–8	7–9	7–9	7–8	6–8
2013	5–7	7–8	6–8	7–9	7–8
2012	8–9	7–8	7–8	7–9	7–9
2011	7–8	7–8	7–8	7–8	6–8
2010	8–10	8–10	8–10	8–10	8–9
2009	7–10	7–8	7–8	7–9	7–8
2008	7–9	7–9	7–9	6–7	5–7
2007	7–8	8–9	8–9	6–8	7–8
2006	7–8	8–10	8–9	7–8	7–9

Beaujolais 19 18 17 15 14 11. Crus will keep. **Mâcon-Villages** (white). Drink 19 18 17 15 14. **Loire** (sweet Anjou and Touraine) best recent vintages: 19 18 15 10 09 07 05 02 97 96 93 90 89; Bourgueil Chinon Saumur-Champigny: 19 18 17 15 14 10 09 06 05 04 02. **Upper Loire** (Sancerre Pouilly-Fumé): 19 18 17 15 14 12. **Muscadet:** DYA.

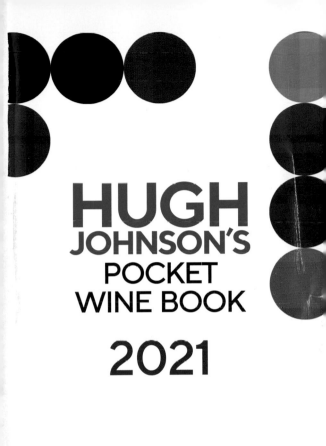

HUGH
JOHNSON'S
POCKET
WINE BOOK

2021

Hugh Johnson's Pocket Wine Book 2021

Edited and designed by Mitchell Beazley,
an imprint of Octopus Publishing Group Limited,
Carmelite House, 50 Victoria Embankment
London EC4Y 0DZ
www.octopusbooks.co.uk

An Hachette UK Company
www.hachette.co.uk

Distributed in the US by Hachette Book Group
1290 Avenue of the Americas
4th and 5th Floors
New York, NY 10020
www.octopusbooksusa.com

General Editor **Margaret Rand**
Commissioning Editor **Hilary Lumsden**
Senior Editor **Pauline Bache**
Proofreader **Jamie Ambrose**
Art Director **Yasia Williams-Leedham**
Designer **Jeremy Tilston**
Picture Researchers **Giulia Hetherington** and **Jennifer Veall**
Senior Production Manager **Katherine Hockley**

Printed and bound in China

Mitchell Beazley would like to acknowledge and thank the following
for supplying photographs for use in this book:

Alamy Stock Photo Gusty 326 left; Montager Productions Ltd 336; Yves
Marcoux/Destinations/Design Pics Inc 330. **Cephas Picture Library Ltd.**
© Andy Christodolo 327; © Matt Wilson 329. **Château Guiraud** 333;
Getty Images Bloomberg/Contributor 331; Jorg Greuel 321; William
Campbell 326 right. **iStock** Ekaterina79 7; georgeclerk 324; gruizza 4;
jacquesvandinteren 325; MarioGuti 10; Morsa Images 6, MvH 14; Paul_June
1, 3; zozzzzo 12. **Shutterstock** FreeProd33 334. ©**Weingut Bründlmayer
Photo:** Robert Herbst 322–23.

HUGH
JOHNSON'S
POCKET
WINE BOOK

2021

GENERAL EDITOR
MARGARET RAND

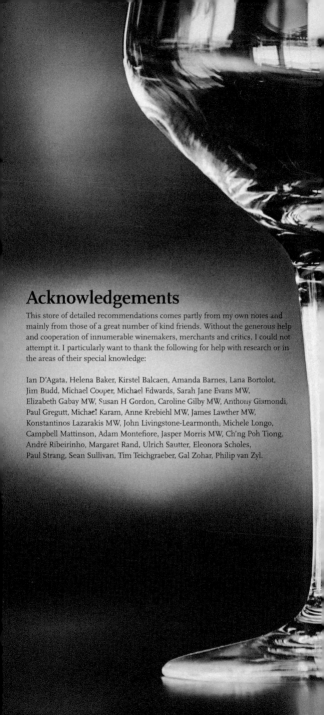

Acknowledgements

This store of detailed recommendations comes partly from my own notes and mainly from those of a great number of kind friends. Without the generous help and cooperation of innumerable winemakers, merchants and critics, I could not attempt it. I particularly want to thank the following for help with research or in the areas of their special knowledge:

Ian D'Agata, Helena Baker, Kirstel Balcaen, Amanda Barnes, Lana Bortolot, Jim Budd, Michael Cooper, Michael Edwards, Sarah Jane Evans MW, Elizabeth Gabay MW, Susan H Gordon, Caroline Gilby MW, Anthony Gismondi, Paul Gregutt, Michael Karam, Anne Krebiehl MW, James Lawther MW, Konstantinos Lazarakis MW, John Livingstone-Learmonth, Michele Longo, Campbell Mattinson, Adam Montefiore, Jasper Morris MW, Ch'ng Poh Tiong, André Ribeirinho, Margaret Rand, Ulrich Sautter, Eleonora Scholes, Paul Strang, Sean Sullivan, Tim Teichgraeber, Gal Zohar, Philip van Zyl.

Contents

The top line of most entries consists of the following information:

1. Aglianico del Vulture Bas
2. r dr (s/sw sp)
3. ★★★
4. 12' 13 14 15 16' 17 (18)

1. Aglianico del Vulture Bas

Wine name and the region the wine comes from, abbreviations of regions are listed in each section.

2. r dr (s/sw sp)

Whether it is red, rosé or white (or brown), dry, sweet or sparkling, or several of these (and which is most important):

r	red
p	rosé
w	white
br	brown
dr	dry*
sw	sweet
s/sw	semi-sweet
sp	sparkling

( ) brackets here denote a less important wine
* assume wine is dry when dr or sw are not indicated

3. ★★★

Its general standing as to quality – a necessarily rough-and-ready guide based on its current reputation as reflected in its prices:

★	plain, everyday quality
★★	above average
★★★	well known, highly reputed
★★★★	grand, prestigious, expensive

So much is more or less objective. Additionally there is a subjective rating:

★ etc.   Stars are coloured for any wine that, in my experience, is usually especially good within its price range. There are good everyday wines as well as good luxury wines. This system helps you find them.

4. 12' 13 14 15 16' 17 (18)

Vintage information: those recent vintages that are outstanding,
and of these, which are ready to drink this year, and which
will probably improve with keeping. Your choice for current
drinking should be one of the vintage years printed in **bold** type.
Buy light-type years for further maturing.

17 etc.	recommended years that may be currently available
16' etc.	vintage regarded as particularly successful for the property in question
13 etc.	years in bold should be ready for drinking (those not in bold will benefit from keeping)
15 etc.	vintages in colour are those recommended as first choice for drinking in 2021. (*See also* Bordeaux introduction, p.100)
(18) etc.	provisional rating

The German vintages work on a different principle again: *see* p.171.

Other abbreviations

DYA	drink the youngest available
NV	vintage not normally shown on label; in Champagne this means a blend of several vintages for continuity
CHABLIS	properties, areas or terms cross-referred within the section; all grapes cross-ref to Grape Varieties chapter on pp.14–24
Foradori	entries styled this way indicate wine especially enjoyed by Hugh Johnson (mid-2019–20)

"You know, Hugh," said James Mitchell, my publisher, in 1977. It was at Easter, and I can remember every detail. "All anyone really needs to know about wine would go in a little book like..." (and he produced his pocket diary). We had just revised my *World Atlas of Wine* (a long job). It took me three seconds to agree: a little job I could polish off in a few months.

James wanted to call it a pocket encyclopaedia. I said no: that would make it sound comprehensive. "Call it my Pocket Wine Book," I said. Then it was up to me what went in and what didn't. The slim volume duly appeared, elegantly laid out with wide margins and plenty of space between entries.

I should have seen what was coming: the 1978 edition, the 1979 and every year since. The wide margins were the first to go, then any spare white space; eventually most of my few precious jokes, and the spoof entries no one ever found. Over the years we've added more pages and squeezed every page. From 188 pages it has swollen to 336. It's not my fault. The wine world has been leading in all sorts of new directions, geographically of course, in adding country after country to the roster; less obviously in a great corolla of spin-offs as wine becomes a near-universal commodity. The Islamic countries are the one great gap – and for how long?

Agenda 2021

In 44 editions of this book we have seen fashions come and go: crazes for this variety or that region, perverse neglect of such classics as Sherry and German wines and absurd adulation of oak, Sauvignon and rosé. Robert Parker built an empire; its traces remain in the statistical nonsense of the 100-point score. Now "orange" wine is riding high; the search for novelty never dies. Once or twice I have been tempted to say we must have reached Peak Wine – only for the crescendo to continue, and the news items multiply. Whoever thought we should be discussing vineyards as firebreaks, or the effects of smoke on wine? There's no avoiding the topic of the large animal in a small space: Global Warming.

All farmers will be affected, but of all farmers wine-growers are the most sensitive, the most finely poised. The great wines of the world are the result of fine equations of land, weather and vines chosen to ripen grapes at the right speed and the right moment. Because the equation is precise it is inevitably marginal, and marginal means fragile.

Great vintages have always been exceptional – and always associated with fine weather. A wine-and-cricket enthusiast once proved that he could predict the quality of a (Bordeaux) vintage by the number of runs scored in first-class cricket in the season. (A dry wicket favours the batsman.)

So far there have been more winners than losers from rising temperatures and longer growing seasons. It is a long time since Germany – in the old days plagued by cold, wet, late harvests – had a really bad one. England has become a serious prospect for fine wine, and even Scandinavia now has a few vineyards. As does Belgium, which we include for the first time this year.

The public, besides, is looking for new things in wine. For years – too long, some think – there has been a consensus about what "good" wine means. Country after country, region after region, has taken Bordeaux as its model. Bordeaux has learned the technology to make its vintages easier to drink younger, rounder, stronger and smoother. "Mouthfeel" is a vogue word. Technology can be replicated – and has been, endlessly, especially by industrial producers aiming for a plausible consistent crowd-pleaser. There are signs that wine-drinkers are getting bored – and cynical. Is this the magic of a sacred beverage? And is this industrial approach good for the planet?

The answer, or one of them, is "natural" wine. What are people looking for in wines sold as "natural" (there is no agreed definition) – apart from the glow of virtue? It may be an alternative to the glossy consistency, or the plain stolid predictability of the commercial ideal. The idea of "fault-free" risks good wines being thrown out for a perceived "fault" that non-technical palates could have enjoyed as "character". True, some faults would have condemned a wine to a short life – and a smelly one. In the world of natural wine wayward tastes can be considered a virtue. Orange wines were perhaps the first to break the rules. Fermenting white grapes on their skins as you would red adds colour, tannin and often unfamiliar acidity, perhaps not making a wine you would sip with pleasure, more of a bracing drink that calls for food.

Try a pét-nat. The name stands for pétillant-naturel, or "naturally fizzy" – which all wines are at the stage before they finish fermenting. They are coltish at this stage, unpredictable, funky, bubbly, yeasty, cloudy and wild. Champagne, when it was first picked out as something special, was like this. The process of disgorgement invented by the Widow Clicquot aimed to remove the yeast and the cloudiness and stabilize the wine. Now you have a brilliantly polished deluxe product, but very far from natural. Pét-nats are breaking out everywhere; wine is joining hands with cider. Let's hope it's just a generational thing.

Never was there so much commentary, so many opinions, websites and blogs, courses, competitions, clubs, wine tourism, speculation, hardware from glasses to storage cabinets. Sommeliers are on a roll. The wine writer has been upgraded to connoisseur, critic, almost to oenotherapist. And wine discussion has blossomed from the latest vintage to terroir, to questions of DNA, of origins, of the first principles of winemaking, to what, if anything, is "natural".

Vintage report 2019

It wasn't long after the 2019 vintage was in the cellars of European producers that Australia started burning – and went on, and on. It's hard to be triumphant about another good year in the wine world when such a loved wine producer as Australia is suffering such tragedies.

Smoke taint was also expected to make a lot of the 2020 vintage unusable. By the 19 vintage, records for high temperatures and low rainfall in **Australia** were already being broken, though 16 and 18 were often warmer than 19. But sugar levels in 19 didn't go through the roof, on the whole, and neither did acidity disappear. Since initial predictions were of a small and problematic vintage, most people were pretty happy with both balance and quantity – and with concentration and colour. In **NZ**, where it was a small vintage, coming after two other small vintages, they're cheering about quality. Everywhere, it seems, was good. Gimblett Gravels has made splendid Syrah. and Martinborough terrific Pinot and Sauvignon. Pinot in Central Otago is bright and aromatic. Drought was the order of the year in **S Africa**, with the harvest the smallest since 05. Stellenbosch had a particularly small harvest; but the Chenin Blanc is particularly good. Swartland too is reporting concentrated wines of great flavour. Pinot Noir in the cooler, maritime regions of Elgin and Walker Bay also looks good; cool nights helped here.

As summer reached the northern hemisphere the pattern was repeated: dry and warm, warm and dry. The difference between balance and too much alcohol and not enough freshness was whether rain arrived at crucial points in the summer. In some places it did, and in others it didn't.

Austria? Good structure, lovely balance. There were unpredictables, as ever: the country's warmest, sunniest and driest June ever, but then compensation from occasional downpours – though happily no hail. Burgenland growers in particular are pleased, though they had to fight off flocks of starlings. In **Germany**, everything is looking balanced, fine and fresh, reds and whites alike, after a year of local drought and sunburn. Some BA and TBA was made. But it was a capricious, uneven year, with some growers hit hard by drought and heat. Baden had some hail as well as drought and sunburn; Franken had the biggest losses overall. In **Italy**, Tuscany managed to combine heat and drought with cold and rain: a cold April slowed everything down, then heat speeded things up, then some late July rain freshened up the vines ready for ripening. Barolo is reporting a "classic" vintage, whatever that means these days. There was no drought here, and high temperatures alternated with rain. In **Spain**, water was an issue: not enough of it in Rioja, so low volumes but lots of concentration; Ramón Bilbao is describing it as a Mediterranean vintage, compared to the Atlantic vintage in 18. Rueda was dry, and the grapes healthy; Ribera del Duero was hot with intermittent storms, and has made big, rich wines with decent acidity. Decent for Ribera del Duero, that is. There's rather more acidity in Rías Baixas, in spite of a warm year and a shorter growing season than the previous year. Priorat had a serious heatwave in June, which dried out grape clusters and burned the leaves; some vineyards lost 40%. But then it rained in the nick of time.

France was hot too in 19. Temperatures rose to 48°C (118°F) in the Hérault in late June, and even heat-resistant vines like Carignan suffered from sunburn and dried-out leaves. West- and southwest-facing vineyards suffered most. The Mâconnais had frost in early April, and there was a bit of *millerandage* in a cool June, but then Burgundy had a remarkably dry summer, which concentrated grapes but reduced volumes. Quality seems to be very good, with wines ticking all the boxes, with plenty of tannins; only light extraction was required for the reds. Whites are described as "rich". Volumes? Yes, down. As they were too in Champagne: some frost, some hail, some mildew, and a humdinger of a hot summer. Aromatic wines, they say, and particularly powerful Pinot Noir. But not a homogeneous year. The Rhône is happy: again, a hot, dry summer seems to have produced unexpectedly balanced wines. There were rainstorms in the north, which would have helped, but the south had a real scorcher of a year. Alsace's heatwave reduced quantities, especially for Gewurztraminer. But what is left is concentrated and powerful. Bordeaux breaks the pattern of low volumes with a generous vintage that is ripe, concentrated, quite high in alcohol and tannins, but fine tannins. Olivier Bernard of Domaine de Chevalier reckons they were saved twice by rain, once at the end of July and again at the end of September; otherwise drought would have taken its toll. The dry whites look better than 18, and the Sauternes reminded Philippe de Lur-Saluces of de Fargues of 03. Another vintage of the century? Why not? We're surely due for another.

And of course Australia was not the only place to have suffered bush fires. Tens of thousands of acres of **California** burned as well, particularly in Sonoma, but damage to vineyards and wineries was contained. Smoke taint on reds might be an issue in some places. In fact one wine group was reported to be suing its insurers for $19 million after the latter refused to pay out for smoke taint from the 17 fires. In northern California, energy supplier PG&E turned the power off to prevent sparks from faulty equipment starting yet more fires – but this was in October, when the grapes were in the presses and power was badly needed. In the end the wines turned out structured, balanced – excellent, in short. **NY State**'s Rieslings are looking good, after a year of less drama than some regions endured. Acidity looks good, alcohol looks moderate.

Moderation is rare this year. Welcome to the new normal.

What do we recommend that you try this year? The world is your oyster; but the following ten wines are ones that we currently love. Some are new discoveries, some are rediscoveries and some are old friends. All are pearls.

10 wines to try in 2021

Red wines

Reunion, Kracher, Burgenland, Austria
Beautiful Blaufränkisch: savoury, complex and accomplished, full of pure, tense, red-cherry fruit, layered and fresh. Blaufränkisch is a terrific grape, in the Gamay-Pinot-Malbec flavour spectrum, and it's a bit of an Austrian speciality.

Las Uvas de la Ira, Vino del Pueblo, El Real de San Vicente, Daniel Landi, Méntrida, Spain
Thyme, incense, cherries, medicinal notes, tension and freshness conjured from old Garnacha vines in the cult region of Gredos, near Madrid. Perfumed and sinuous. This is new-wave Garnacha at its best, showing all the perfume and delicacy of which the grape is capable in certain places.

Bodegas Pintia, Toro, Spain
Toro used to be the home of huge tannins and far too much oak, but when Vega Sicilia moved in it showed that things could be done differently. This is from Vega Sicilia's estate: it smells of cocoa, plums, spice and roses, and is fresh, tense and nicely grippy on the palate. "Precise" and "detailed" are words I never thought I'd use for Toro. Bravo.

Henry's Seven SGV, Henschke, Eden Valley, Australia
We should all drink Australian wine this year, to support a disaster-struck industry and country. This is a beaut from a great grower: a Grenache/Shiraz/Mourvèdre blend, savoury, herbal, complex, perfumed and distinctive.

Terrenus Vinha da Ammaia Amphorae, Alentejo, Portugal
If you fancy tasting an amphora wine, this is a good place to start. The grapes are Trincadeira, Castelão and Moreto, and the vines and the amphorae are both old. The wine is as modern as they come: supple, subtle, poised, full of blackberry and sloe fruit, with a nice fresh earthy touch.

Geyserville, Ridge Vineyards, Sonoma County, California, USA
At some point in your life you have to try Ridge. Geyserville
is not its most expensive, but it's particularly good value.
It's mostly Zinfandel, plus Carignane, Petite Sirah and a
pinch of Mataro: all hot-climate grapes that can be rustic
in the wrong hands. Instead this is taut and subtle, all earth
and spice and lots of tarry black fruit, and even flowers. It's
a traditional field blend, fermented with native yeasts – and
it's unmissable.

White wines

Cuvée Reynolds Stone Brut, Breaky Bottom, Sussex, England
Reynolds Stone was a wood engraver and letter cutter who
designed the original Breaky Bottom label, as well as the royal
coat of arms on every British passport. All Peter Hall's wines
are named after artists and musicians, and they come from
a tiny vineyard hidden in the South Downs, where Peter Hall
has been making wine since 1974. The 10 vintage is saline,
tense and compelling.

White Blend, AA Badenhorst, Swartland, South Africa
I do love a wine name that tells you exactly what it is. This
mixes Chenin Blanc with Roussanne, Marsanne, Grenache
Blanc, Viognier, Verdelho, Grenache Gris, Clairette, Semillon
and Palomino from Swartland's old bush vines, fermented
and aged in very old casks. It's a wonderfully fresh, mineral,
complex wine, with notes of citrus, salt, stone fruit, tea, orange
blossom and wild herbs, with just a touch of tannic grip.

**Ried Lamm 1ÖTW, Grüner Veltliner, Schloss Gobelsburg,
Kamptal, Austria**
Lamm is one of the best vineyards in Austria for Grüner
Veltliner, and Schloss Gobelsburg never puts a foot wrong.
This is silky, graceful, deep and delicate, filigree and firm,
full of quince and salt. 1ÖTW, by the way, indicates that the
vineyard is a "first growth" in the vineyard classification
of the Österreichische Traditionsweingüter association.

**Dafni, Psarades Vineyard, Domaine Lyrarakis, Crete,
Greece**
This tastes and smells fascinatingly of bay – which is why,
as gardeners will have guessed, the grape is called Dafni.
It seems to be an ancient variety, is made without oak, and
is a pungent, delicate wine, very moreish, with a refreshingly
bitter edge.

Grape varieties

In the past two decades a radical change has come about in all except the most long-established wine countries: the names of a handful of grape varieties have become the ready-reference to wine. In senior wine countries, above all France and Italy, more complex traditions prevail. All wine of old prestige is known by its origin, more or less narrowly defined – not just by the particular fruit juice that fermented. For the present the two notions are in rivalry. Eventually the primacy of place over fruit will become obvious, at least for wines of quality. But for now, for most people, grape tastes are the easy reference point – despite the fact that they are often confused by the added taste of oak. If grape flavours were really all that mattered, this would be a very short book. But of course they *do* matter, and a knowledge of them both guides you to flavours you enjoy and helps comparisons between regions. Hence the originally Californian term "varietal wine", meaning, in principle, made from one grape variety. At least seven varieties – Cabernet Sauvignon, Pinot Noir, Riesling, Sauvignon Blanc, Chardonnay, Gewurztraminer and Muscat – taste and smell distinct and memorable enough to form international wine categories. To these add Merlot, Malbec, Syrah, Sémillon, Chenin Blanc, Pinots Blanc and Gris, Sylvaner, Viognier, Nebbiolo, Sangiovese, Tempranillo. The following are the best and/or most popular wine grapes.

All grapes and synonyms are cross-referenced in SMALL CAPITALS throughout every section of this book.

Grapes for red wine

Agiorgitiko Greek; the grape of Nemea, now planted all over Greece. Versatile and delicious, from soft and charming to dense and age-worthy. A must-try.

Aglianico S Italy's best red, the grape of Taurasi; dark, deep and fashionable.

Alicante Bouschet Used to be shunned, now stylish in Alentejo, Chile, esp old vines.

Aragonez *See* TEMPRANILLO.

Auxerrois *See* MALBEC, if red. White Auxerrois has its own entry in White Grapes.

Băbească Neagră Traditional "black grandmother grape" of Moldova; light body and ruby-red colour.

Babić Dark grape from Dalmatia, grown in stony seaside v'yds round Šibenik. Exceptional quality potential.

Baga Portugal. Bairrada grape. Dark and tannic. Great potential but hard to grow.

Barbera Widely grown in Italy, best in Piedmont: high acidity, low tannin, cherry fruit. Ranges from barriqued and serious to semi-sweet and frothy. Fashionable in California and Australia; promising in Argentina.

Blauburger Austrian cross between BLAUER PORTUGIESER and BLAUFRÄNKISCH. Makes simple wines.

Blauburgunder *See* PINOT N.

Blauer Portugieser Central European, esp Germany (Rheinhessen, Pfalz, mostly for rosé), Austria, Hungary. Light, fruity reds: drink young, slightly chilled.

Blaufränkisch (Kékfrankos, Lemberger, Modra Frankinja) Widely planted in Austria's Mittelburgenland: medium-bodied, peppery acidity, a characteristic salty note, berry aromas, eucalyptus. Often lovely, lively. Often blended with CAB SAUV or ZWEIGELT. Lemberger in Germany (speciality of Württemberg), Kékfrankos in Hungary, Modra Frankinja in Slovenia.

Boğaskere Tannic and Turkish. Produces full-bodied wines.

Bonarda Ambiguous name. In Oltrepò Pavese, an alias for Croatina, soft fresh *frizzante* and still red. In Lombardy and Emilia-Romagna an alias for Uva Rara. Different in Piedmont. Argentina's Bonarda can be any of these, or something else. None are great.

Bouchet St-Émilion alias for CAB FR.

Brunello SANGIOVESE, splendid at Montalcino.

Cabernet Franc [Cab Fr] The lesser of two sorts of Cab grown in B'x, but dominant in St-Émilion. Outperforms CAB SAUV in Loire (Chinon, Saumur-Champigny, rosé), in Hungary (depth and complexity in Villány and Szekszárd) and often in Italy. Much of na Italy's Cab Fr turned out to be CARMENÈRE. Used in B'x blends of Cab Sauv/MERLOT across the world.

Cabernet Sauvignon [Cab Sauv] Grape of great character: slow-ripening, spicy, herby, tannic, with blackcurrant aroma. Main grape of the Médoc; also makes some of the best California, S American, E European reds. Vies with SHIRAZ in Australia. Grown almost everywhere, and led vinous renaissance in eg. Italy. Top wines need ageing; usually benefits from blending with eg. MERLOT, CAB FR, SYRAH, TEMPRANILLO, SANGIOVESE, etc. Makes aromatic rosé.

Cannonau GRENACHE in its Sardinian manifestation; can be v. fine, potent.

Carignan (Carignane, Carignano, Cariñena) Low-yielding old vines now fashionable everywhere from s of France to Chile; best: Corbières. Lots of depth and vibrancy, but must never be overcropped. Common in N Africa, Spain (as Cariñena) and California.

Carignano *See* CARIGNAN.

Cariñena *See* CARIGNAN.

Carmenère An old B'x variety now a star, rich and deep, in Chile (where it's pronounced "carmeneary"). B'x is looking at it again.

Castelão *See* PERIQUITA.

Cencibel *See* TEMPRANILLO.

Chiavennasca *See* NEBBIOLO.

Cinsault (Cinsaut) A staple of s France, v.gd if low-yielding, hopeless if not. Makes gd rosé. One of parents of PINOTAGE.

Cornalin du Valais Swiss speciality with high potential, esp in Valais.

Corvina Dark and spicy; one of best grapes in Valpolicella blend. Corvinone, even darker, is a separate variety.

Côt *See* MALBEC.

Dolcetto Source of soft, seductive dry red in Piedmont. Now high fashion.

Dornfelder Gives deliciously light reds, straightforward, often rustic, and well-coloured in Germany, parts of the US, even England. German plantings have doubled since 2000.

Duras Spicy, peppery, structured; exclusive to Gaillac and parts of Tarn Valley, Southwest France.

Fer Servadou Exclusive to Southwest France, aka Mansois in Marcillac, Braucol in Gaillac and Pinenc in St-Mont. Redolent of red summer fruits and spice.

Fetească Neagră Romania: "black maiden grape" with potential as showpiece variety; can give deep, full-bodied wines with character. Acreage increasing.

Frühburgunder An ancient German mutation of PINOT N, mostly in Ahr but also in Franken and Württemberg, where it is confusingly known as Clevner. Lower acidity than Pinot N.

Gamay The Beaujolais grape: light, fragrant wines, best young, except in Beaujolais crus (*see* France) where quality can be high, wines for 2–10 yrs. Grown in the Loire Valley, in central France, in Switzerland and Savoie. California's Napa Gamay is Valdiguié.

Gamza *See* KADARKA.

Garnacha (Cannonau, Garnatxa, Grenache) Important pale, potent grape for warm climates, fashionable with *terroiristes* because it expresses its site. The base of Châteauneuf-du-Pape. Also gd for rosé and *vin doux naturel* – esp in the s of France, Spain and California – but also the mainstay of beefy Priorat. Old-vine versions prized in S Australia. Usually blended with other varieties. Cannonau in Sardinia, Grenache in France.

Garnatxa *See* GARNACHA.

Graciano Spanish; part of Rioja blend. Aroma of violets, tannic, lean structure, a bit like PETIT VERDOT. Difficult to grow but increasingly fashionable.

Grenache *See* GARNACHA.

Grignolino Italy: gd everyday table wine in Piedmont.

Kadarka (Gamza) Makes spicy, light reds in E Europe. In Hungary revived, esp for Bikavér.

Kalecik Karasi Turkish: sour-cherry fruit, fresh, supple. Bit like GAMAY. Drink young.

Kékfrankos Hungarian BLAUFRÄNKISCH.

Lagrein N Italian, dark, bitter finish, rich, plummy. DOC in Alto Adige (*see* Italy).

Lambrusco Productive grape of the lower Po Valley; quintessentially Italian, cheerful, sweet and fizzy red.

Lefkada Rediscovered Cypriot variety, higher quality than MAVRO. Usually blended as tannins can be aggressive.

Lemberger *See* BLAUFRÄNKISCH.

Malbec (Auxerrois, Côt) Minor in B'x, major in Cahors (alias Auxerrois) and the star in Argentina. Dark, dense, tannic but fleshy wine capable of real quality. High-altitude versions in Argentina best. Bringing Cahors back into fashion.

Maratheftiko Deep-coloured Cypriot grape with quality potential.

Marselan CAB SAUV X GRENACHE, 1961. Gd colour, structure, supple tannins, ages well. A success in China.

Mataro *See* MOURVÈDRE.

Mavro Most planted black grape of Cyprus but only moderate quality. Best for rosé.

Mavrodaphne Greek; means "black laurel". Sweet fortifieds, speciality of Patras, also in Cephalonia. Dry versions too, great promise.

Mavrotragano Greek, almost extinct, now revived; found on Santorini. Top quality.

Mavrud Probably Bulgaria's best. Spicy, dark, plummy late-ripener native to Thrace. Ages well.

Melnik Bulgarian grape from the region of the same name. Dark colour and a nice dense, tart-cherry character. Ages well.

Mencía Making waves in Bierzo, n Spain. Aromatic with steely tannins and lots of acidity.

Merlot The grape behind the great fragrant and plummy wines of Pomerol and (with CAB FR) St-Émilion, a vital element in the Médoc, soft and strong in California, Washington, Chile, Australia. Lighter, often gd in n Italy (can be world-class in Tuscany), Italian Switzerland, Slovenia, Argentina, S Africa, NZ, etc. Perhaps too adaptable for own gd: can be v. dull; less than ripe it tastes green. Much planted in E Europe, esp Romania.

Modra Frankinja *See.* BLAUFRÄNKISCH.

Modri Pinot *See* PINOT N.

Monastrell *See* MOURVÈDRE.

Mondeuse Found in Savoie; skier's red; deep-coloured, gd acidity. Related to SYRAH.

Montepulciano Deep-coloured grape dominant in Italy's Abruzzo, important along Adriatic coast from Marches to s Puglia. Also name of a Tuscan town, unrelated.

Morellino SANGIOVESE in Maremma, s Tuscany. Esp Scansano.

Mourvèdre (Mataro, Monastrell) A star of s France (eg. Bandol, growing influence Châteauneuf-du-Pape), Australia (aka Mataro) and Spain (aka Monastrell). Excellent dark, aromatic, tannic grape, gd for blending. Enjoying new interest in eg. S Australia and California.

Napa Gamay Identical to Valdiguié (s of France). Nothing to get excited about.

Nebbiolo (Chiavennasca, Spanna) One of Italy's best red grapes; makes Barolo, Barbaresco, Gattinara and Valtellina. Intense, nobly fruity, perfumed wine with steely tannin: improves for yrs.

Negroamaro Puglian "black bitter" red grape with potential for either high quality or high volume.

Nerello Mascalese Characterful Sicilian red grape, esp on Etna; potentially elegant.

Nero d'Avola Dark-red grape of Sicily, quality levels from sublime to industrial.

Nielluccio Corsican; plenty of acidity and tannin. Gd for rosé.

Öküzgözü Soft, fruity Turkish grape, usually blended with BOĞASKERE, rather as MERLOT in B'x is blended with CAB SAUV.

País Pioneer Spanish grape in Americas. Rustic; some producers now trying harder.

Pamid Bulgarian: light, soft, everyday red.

Periquita (Castelão) Common in Portugal, esp round Setúbal. Originally nicknamed Periquita after Fonseca's popular (trademarked) brand. Firm-flavoured, raspberryish reds develop a figgish, tar-like quality.

Petite Sirah Nothing to do with SYRAH; gives rustic, tannic, dark wine. Brilliant blended with ZIN in California; also found in S America, Mexico, Australia.

Petit Verdot Excellent but awkward Médoc grape, now increasingly planted in CAB areas worldwide for extra fragrance. Mostly blended but some gd varietals, esp in Virginia.

Pinotage Singular S African cross (PINOT N x CINSAULT). Has had a rocky ride, getting better from top producers. Gd rosé too. "Coffee Pinotage" is espresso-flavoured, sweetish, aimed at youth.

Pinot Crni *See* PINOT N.

Pinot Meunier (Schwarzriesling) [Pinot M] 3rd grape of Champagne, better known as Meunier, great for blending but can be fine in its own right; a bridge between PINOT N and CHARD. Best on chalky sites (Damery, Leuvigny, Festigny) nr Épernay.

Pinot Noir (Blauburgunder, Modri Pinot, Pinot Crni, Spätburgunder) [Pinot N] Glory of Burgundy's Côte d'Or. Fine in Alsace and now Germany. V.gd in Austria, esp in Kamptal, Burgenland, Thermenregion. Light in Hungary; mainstream, light to weightier in Switzerland (aka Clevner). Splendid in Sonoma, Carneros, Central Coast, also Oregon, Ontario, Yarra Valley, Adelaide Hills, Tasmania, NZ's Central Otago, S Africa's Walker Bay. Some v. pretty Chileans. New French clones promise improvement in Romania. Modri Pinot in Slovenia; probably country's best red. In Italy, best in ne, gets worse as you go s. PINOTS BL and GR mutations of Pinot N.

Plavac Mali (Crljenak) Croatian, and related to ZIN, like so much round there. Lots of quality potential, can age well, though can also be alcoholic and dull.

Primitivo S Italian grape, originally from Croatia, making big, dark, rustic wines, now fashionable because genetically identical to ZIN. Early ripening, hence the name. The original name for both seems to be Tribidrag.

Refosco (Refošk) Various DOCs in Italy, esp Colli Orientali. Deep, flavoursome and age-worthy wines, particularly in warmer climates. Dark, high acidity. Refošk in Slovenia and points e, genetically different, tastes similar.

Refošk *See* REFOSCO.

Roter Veltliner Austrian; unrelated to GRÜNER V. There is also a Frühroter and a Brauner Veltliner.

Rubin Bulgarian cross, NEBBIOLO X SYRAH. Peppery, full-bodied.

Sagrantino Italian grape grown in Umbria for powerful, cherry-flavoured wines.

St-Laurent Dark, smooth, full-flavoured Austrian speciality. Can be light and juicy or deep and structured. Also in Pfalz.

Sangiovese (Brunello, Morellino, Sangioveto) Principal red grape of Tuscany and central Italy. Its characteristic astringency is hard to get right, but sublime and long-lasting when it is. Dominant in Chianti, Vino Nobile, Brunello di Montalcino, Morellino di Scansano and various fine IGT offerings. Also in Umbria (eg. Montefalco and Torgiano) and across the Apennines in Romagna and Marches. Not so clever in the warmer, lower-altitude v'yds of the Tuscan coast, nor in other parts of Italy despite its nr-ubiquity. Interesting in Australia.

Sangioveto *See* SANGIOVESE.

Saperavi The main red of Georgia, Ukraine, etc. Blends well with CAB SAUV (eg. in Moldova). Huge potential, seldom gd winemaking.

Schiava *See* TROLLINGER.

Schioppettino NE Italian, high acidity, high quality. Elegant, refined, can age.

Schwarzriesling PINOT M in Württemberg.

Sciacarello Corsican, herby and peppery. Not v. tannic.

Shiraz *See* SYRAH.

Spanna *See* NEBBIOLO.

Spätburgunder German for PINOT N.

Syrah (Shiraz) The great Rhône red grape: tannic, purple, peppery, matures superbly. Important as Shiraz in Australia, increasingly gd under either name in Chile, S Africa, terrific in NZ (esp Hawke's Bay). Widely grown: a gd traveller.

Tannat Raspberry-perfumed, highly tannic force behind Madiran, Tursan and other firm reds from Southwest France. Also rosé. The star of Uruguay.

Tempranillo (Aragonez, Cecibel, Tinto Fino, Tinta del País, Tinta Roriz, Ull de Llebre) Aromatic, fine Rioja grape, called Ull de Llebre in Catalonia, Cencibel in La Mancha, Tinto Fino in Ribera del Duero, Tinta Roriz in Douro, Tinta del País in Castile, Aragonez in s Portugal. Now Australia too. V. fashionable; elegant in cool climates, beefy in warm. Early-ripening, long-maturing.

Teran (Terrano) Close cousin of REFOSCO, esp on limestone (karst) in Slovenia.

Teroldego Rotaliano Trentino's best indigenous variety; serious, full-flavoured wine, esp on the flat Campo Rotaliano.

Tinta Amarela *See* TRINCADEIRA.

Tinta del País *See* TEMPRANILLO.

Tinta Negra (Negramoll) Until recently called Tinta Negra Mole. Easily Madeira's most planted grape and the mainstay of cheaper Madeira. Now coming into its own in Colheita wines (*see* Portugal).

Tinta Roriz *See* TEMPRANILLO.

Tinto Fino *See* TEMPRANILLO.

Touriga Nacional [Touriga N] The top Port grape, now widely used in the Douro for floral, stylish table wines. Australian Touriga is usually this; California's Touriga can be either this or Touriga Franca.

Trincadeira (Tinta Amarela) Portuguese; v.gd in Alentejo for spicy wines. Tinta Amarela in the Douro.

Trollinger (Schiava, Vernatsch) Popular pale red in Württemberg; aka Vernatsch and Schiava. Covers group of vines, not necessarily related. In Italy, snappy, brisk.

Vernatsch *See* TROLLINGER.

Xinomavro Greece's answer to NEBBIOLO. "Sharp-black"; the basis for Naoussa, Rapsani, Goumenissa, Amindeo. Some rosé, still or sparkling. Top quality, can age for decades. Being tried in China.

Zinfandel [Zin] Fruity, adaptable grape of California with blackberry-like, and sometimes metallic, flavour. Can be structured and gloriously lush, ageing for decades, but also makes "blush" pink, usually sweet, jammy. Genetically the same as s Italian PRIMITIVO.

Zweigelt (Blauer Zweigelt) BLAUFRÄNKISCH X ST-LAURENT, popular in Austria for aromatic, dark, supple, velvety wines. Also found in Hungary, Germany.

Grapes for white wine

Airén Bland workhorse of La Mancha, Spain: fresh if made well.

Albariño (Alvarinho) Fashionable, expensive in Spain: apricot-scented, gd acidity. Superb in Rías Baixas; shaping up elsewhere, but not all live up to the hype. Alvarinho in Portugal just as gd: aromatic Vinho Verde, esp in Monção, Melgaço.

Aligoté Burgundy's 2nd-rank white grape. Sharp wine for young drinking, perfect for mixing with cassis to make Kir. Widely planted in E Europe, Russia.

Alvarinho *See* ALBARIÑO.

Amigne One of Switzerland's speciality grapes, traditional in Valais, esp Vétroz. Total planted: 43 ha. Full-bodied, tasty, often sweet but also bone-dry.

Ansonica *See* INSOLIA.

Arinto Portuguese; the mainstay of aromatic, citrus wines in Bucelas; also adds welcome zip to blends, esp in Alentejo.

Arneis Nw Italian. Fine, aromatic, appley-peachy, high-priced grape, DOCG in Roero, DOC in Langhe, Piedmont.

Arvine Rare but excellent Swiss *spécialité*, from Valais. Also Petite Arvine. Dry or sweet, fresh, long-lasting wines with salty finish.

Assyrtiko From Santorini; one of the best grapes of the Mediterranean, balancing power, minerality, extract and high acid. Built to age. Could conquer the world...

Auxerrois Red Auxerrois is a synonym for MALBEC, but white Auxerrois is like a fatter, spicier PINOT BL. Found in Alsace and much used in Crémant; also Germany.

Bacchus German-bred crossing, early-ripening so grown in England, v. aromatic, can be coarse.

Beli Pinot *See* PINOT BL.

Blanc Fumé *See* SAUV BL.

Boal *See* BUAL.

Bourboulenc This and the rare Rolle make some of the Midi's best wines.

Bouvier Indigenous aromatic Austrian grape, esp gd for Beerenauslese and Trockenbeerenauslese, rarely for dry wines.

Bual (Boal) Makes top-quality sweet Madeira wines, not quite so rich as MALMSEY.

Carricante Italian. Principal grape of Etna Bianco, regaining ground.

Catarratto Prolific white grape found all over Sicily, esp in w in DOC Alcamo.

Cerceal *See* SERCIAL.

Chardonnay (Morillon) [Chard] The white grape of Burgundy and Champagne, now ubiquitous worldwide, partly because it is one of the easiest to grow and vinify. Also the name of a Mâcon-Villages commune. The fashion for overoaked butterscotch versions now thankfully over. Morillon in Styria, Austria.

Chasselas (Fendant, Gutedel) Swiss (originated in Vaud). Neutral flavour, takes on local character: elegant (Geneva); refined, full (Vaud); exotic, racy (Valais). Fendant in Valais. Makes almost 3rd of Swiss wines but giving way, esp to red. Gutedel in Germany; grown esp in s Baden. Elsewhere usually a table grape.

Chenin Blanc [Chenin Bl] Wonderful white grape of the middle Loire (Vouvray, Layon, etc). Wine can be dry or sweet (or v. sweet), but with plenty of acidity. Superb old-vine versions in S Africa, esp Swartland.

Cirfandl *See* ZIERFANDLER.

Clairette Important Midi grape, low-acid, part of many blends. Improved winemaking helps.

Colombard Slightly fruity, nicely sharp grape, makes everyday wine in S Africa, California and Southwest France. Often blended.

Dimiat Perfumed Bulgarian grape, made dry or off-dry, or distilled. Far more synonyms than any grape needs.

Encruzado Portuguese, serious; fresh, versatile, ages well. Esp gd in Dão.

Ermitage Swiss for MARSANNE.

Ezerjó Hungarian, with sharp acidity. Name means "thousand blessings".

Falanghina Italian: ancient grape of Campanian hills. Gd dense, aromatic dry.

Fendant *See* CHASSELAS.

Fernão Pires *See* MARIA GOMES.

Fetească Albă / Regală Romania has two Fetească grapes, both with slight MUSCAT aroma. F. Regală is a cross of F. Albă and Frâncuşă; more finesse, gd for late-harvest wines. F. NEAGRĂ is dark-skinned.

Fiano High-quality grape giving peachy, spicy wine in Campania, s Italy.

Folle Blanche (Gros Plant) High acid/little flavour make this ideal for brandy. Gros Plant in Brittany, Picpoul in Armagnac, but unrelated to true PICPOUL. Also respectable in California.

Friulano (Sauvignonasse, Sauvignon Vert) N Italian: fresh, pungent, subtly floral. Used to be called Tocai Friulano. Best in Collio, Isonzo, Colli Orientali. Found in nearby Slovenia as Sauvignonasse; also in Chile, where it was long confused with SAUV BL. Ex-Tocai in Veneto now known as Tai.

Fumé Blanc *See* SAUV BL.

Furmint (Šipon) Superb, characterful. The trademark of Hungary, both as the principal grape in Tokaji and as vivid, vigorous table wine, sometimes mineral, sometimes apricot-flavoured, sometimes both. Šipon in Slovenia. Some grown in Rust, Austria for sweet and dry.

Garganega Best grape in Soave blend; also in Gambellara. Top, esp sweet, age well.

Garnacha Blanca (Grenache Blanc) The white version of GARNACHA/Grenache, much used in Spain and s France. Low acidity. Can be innocuous, or surprisingly gd.

Gewurztraminer (Traminac, Traminec, Traminer, Tramini) [Gewurz] One of the most pungent grapes, spicy with aromas of rose petals, face cream, lychees, grapefruit. Wines are often rich and soft, even when fully dry. Best in Alsace; also gd in Germany (Baden, Pfalz, Sachsen), Eastern Europe, Australia, California, Pacific Northwest and NZ. Can be relatively unaromatic if just labelled Traminer (or variants). Italy uses the name Traminer Aromatico for its (dry) "Gewurz" versions. (The name takes an Umlaut in German.) Identical to SAVAGNIN.

Glera Uncharismatic new name for Prosecco vine: Prosecco is now wine only in EU, but still a grape name in Australia.

Godello Top quality (intense, mineral) in nw Spain. Called Verdelho in Dão, Portugal, but unrelated to true VERDELHO.

Grasă (Kövérszőlö) Romanian; name means "fat". Prone to botrytis; important in Cotnari, potentially superb sweet wines. Kövérszőlő in Hungary's Tokaj region.

Graševina See WELSCHRIESLING.

Grauburgunder See PINOT GR.

Grechetto Ancient grape of central and s Italy noted for the vitality and stylishness of its wine. Blended, or used solo in Orvieto.

Greco S Italian: there are various Grecos, probably unrelated, perhaps of Greek origin. Brisk, peachy flavour, most famous as Greco di Tufo. Greco di Bianco is from semi-dried grapes. Greco Nero is a black version.

Grenache Blanc See GARNACHA BLANCA.

Grillo Italy: main grape of Marsala. Also v.gd full-bodied dry table wine.

Gros Plant See FOLLE BLANCHE.

Grüner Veltliner [Grüner V] Austria's fashionable flagship white grape. V. diverse: from simple, peppery, everyday to great complexity, ageing potential. Found elsewhere in Central Europe and outside.

Gutedel See CHASSELAS.

Hárslevelű Other main grape of Tokaji, but softer, peachier than FURMINT. Name means "linden-leaved". Gd in Somló, Eger as well.

Heida Swiss for SAVAGNIN.

Humagne Swiss speciality, older than CHASSELAS. Fresh, plump, not v. aromatic. Humagne Rouge is not related but increasingly popular: same as Cornalin du Aosta. Cornalin du Valais is different. (Keep up at the back, there.)

Insolia (Ansonica, Inzolia) Sicilian; Ansonica on Tuscan coast. Fresh, racy wine at best. May be semi-dried for sweet wine.

Irsai Olivér Hungarian cross; aromatic, MUSCAT-like wine for drinking young.

Johannisberg Swiss for SILVANER.

Kéknyelű Low-yielding, flavourful grape giving one of Hungary's best whites. Has the potential for fieriness and spice. To be watched.

Kerner Quite successful German cross. Early-ripening, flowery (but often too blatant) wine with gd acidity.

Királyleanyka Hungarian; gentle, fresh wines (eg. in Eger).

Koshu More-or-less indigenous Japanese table-turned-wine grape, much hyped. Fresh, tannic. Orange versions gd.

Kövérszőlő See GRASĂ.

Laški Rizling See WELSCHRIESLING.

Leányka Hungarian. Soft, floral wines.

Listán See PALOMINO.

Longyan (Dragon Eye) Chinese original; gd substantial, aromatic wine.

Loureiro Best Vinho Verde grape after ALVARINHO: delicate, floral. Also in Spain.

Macabeo See VIURA.

Maccabeu See VIURA.

Malagousia Rediscovered Greek grape for gloriously perfumed wines.

Malmsey See MALVASIA. The sweetest style of Madeira.

Malvasia (Malmsey, Malvazija, Malvoisie, Marastina) Italy, France and Iberia. Not a single variety but a whole stable, not necessarily related or even alike. Can be white or red, sparkling or still, strong or mild, sweet or dry, aromatic or neutral. Slovenia's and Croatia's version is Malvazija Istarka, crisp and light, or rich, oak-aged. Sometimes called Marastina in Croatia. "Malmsey" (as in the sweetest style of Madeira) is a corruption of Malvasia.

Malvoisie See MALVASIA. A name used for several varieties in France, incl BOURBOULENC, Torbato, VERMENTINO. Also PINOT GR in Switzerland's Valais.

Manseng, Gros / Petit Gloriously spicy, floral whites from Southwest France. The key to Jurançon. Superb late-harvest and sweet wines too.

Maria Gomes (Fernão Pires) Portuguese; aromatic, ripe-flavoured, slightly spicy whites in Barraida and Tejo.

Marsanne (Ermitage) Principal white grape (with ROUSSANNE) of the N Rhône (Hermitage, St-Joseph, St-Péray). Also gd in Australia, California and (as Ermitage Blanc) the Valais. Soft, full wines that age v. well.

Melon de Bourgogne See MUSCADET.

Misket Bulgarian. Mildly aromatic; the basis of most country whites.

Morillon CHARD in parts of Austria.

Moscatel See MUSCAT.

Moscato See MUSCAT.

Moschofilero Pink-skinned, rose-scented, high-quality, high-acid, low-alcohol Greek grape. Makes white, some pink, some sparkling.

Müller-Thurgau [Müller-T] Aromatic wines to drink young. Makes gd sweet wines but usually dull, often coarse, dry ones. In Germany, most common in Pfalz, Rheinhessen, Nahe, Baden, Franken. Has some merit in Italy's Trentino-Alto Adige, Friuli. Sometimes called RIES X SYLVANER (incorrectly) in Switzerland.

Muscadelle Adds aroma to white B'x, esp Sauternes. In Victoria used (with MUSCAT, to which it is unrelated) for Rutherglen Muscat.

Muscadet (Melon de Bourgogne) Makes light, refreshing, v. dry wines with a seaside tang around Nantes in Brittany. Also found (as Melon) in parts of Burgundy.

Muscat (Moscatel, Moscato, Muskateller) Many varieties; the best is Muscat Blanc à Petits Grains (alias Gelber Muskateller, Rumeni Muškat, Sarga Muskotály, Yellow Muscat). Widely grown, easily recognized, pungent grapes, mostly made into perfumed sweet wines, often fortified, as in France's *vin doux naturel*. Superb, dark, sweet in Australia. Sweet, sometimes v.gd in Spain. Most Hungarian Muskotály is Muscat Ottonel except in Tokaj, where Sarga Muskotály rules, adding perfume (in small amounts) to blends. Occasionally (eg. Alsace, Austria, parts of s Germany) made dry. Sweet Cap Corse Muscats often superb. Light Moscato fizz in n Italy.

Muskateller See MUSCAT.

Narince Turkish; fresh and fruity wines.

Neuburger Austrian, rather neglected; mainly in the Wachau (elegant, flowery), Thermenregion (mellow, ample-bodied) and n Burgenland (strong, full).

Olaszrizling See WELSCHRIESLING.

Païen See SAVAGNIN.

Palomino (Listán) The great grape of Sherry; with little intrinsic character, it gains all from production method. As Listán, makes dry white in Canaries.

Pansa Blanca See XAREL·LO.

Pecorino Italian: not a cheese but alluring dry white from a recently nr-extinct variety. IGT in Colli Pescaresi.

Pedro Ximénez [PX] Makes sweet brown Sherry under its own name; used in Montilla, Málaga. Grown in Argentina, Canaries, Australia, California, S Africa.

Picpoul (Piquepoul) S French, best known in Picpoul de Pinet. Should have high acidity. Picpoul Noir is black-skinned.

Pinela Local to Slovenia. Subtle, lowish acidity; drink young.

Pinot Bianco See PINOT BL.

Pinot Blanc (Beli Pinot, Pinot Bianco, Weißburgunder) [Pinot Bl] A cousin of PINOT N, similar to but milder than CHARD. Light, fresh, fruity, not aromatic, to drink young. Gd for Italian *spumante*, and potentially excellent in the ne, esp high sites in Alto Adige. Widely grown. Weißburgunder in Germany and best in s: often racier than Chard.

Pinot Gris (Pinot Grigio, Grauburgunder, Ruländer, Sivi Pinot, Szürkebarát) [Pinot Gr] Ultra-popular as Pinot Grigio in n Italy, even for rosé, but top, characterful versions can be excellent (from Alto Adige, Friuli). Cheap versions are just that. Terrific in Alsace for full-bodied, spicy whites. Once important in Champagne. In Germany can be alias Ruländer (sw) or Grauburgunder (dr): best in Baden (esp Kaiserstuhl) and s Pfalz. Szürkebarát in Hungary, Sivi P in Slovenia (characterful, aromatic).

Pošip Croatian; mostly on Korčula. Quite characterful and citrus; high-yielding.

Prosecco Old name for grape that makes Prosecco. Now you have to call it GLERA.

Renski Rizling Rhine RIES.

Rèze Super-rare ancestral Valais grape used for *vin de glacier*.

Ribolla Gialla / Rebula Acidic but characterful. In Italy, best in Collio. In Slovenia, traditional in Brda. Can be v.gd, even made in eccentric ways.

Rieslaner German cross (SILVANER x RIES); low yields, difficult ripening, now v. rare (less than 50 ha). Makes fine Auslesen in Franken and Pfalz.

Riesling Italico See WELSCHRIESLING.

Riesling (Renski Rizling, Rhine Riesling) [Ries] The greatest, most versatile white grape, diametrically opposite in style to CHARD. Offers a range from steely to voluptuous, always positively perfumed, with far more ageing potential than Chard. Great in all styles in Germany; forceful and steely in Austria; lime cordial and toast fruit in S Australia; rich and spicy in Alsace; Germanic and promising in NZ, NY State, Pacific Northwest; has potential in Ontario, S Africa. In warmer climates soon smells of petrol.

Rkatsiteli Found widely in E Europe, Russia, Georgia. Can stand cold winters and has high acidity; protects to a degree from poor winemaking. Also in ne US.

Robola In Greece (Cephalonia) a top-quality, floral grape, unrelated to RIBOLLA GIALLA.

Roditis Pink grape, all over Greece, usually making whites. Gd when yields low.

Roter Veltliner Austrian; unrelated to GRÜNER V. There is also a Frühroter and an (unrelated) Brauner Veltliner.

Rotgipfler Austrian; indigenous to Thermenregion. With ZIERFANDLER, makes lively, lush, aromatic blend.

Roussanne Rhône grape of real finesse, now popping up in California and Australia. Can age many yrs.

Ruländer See PINOT GR.

Sauvignonasse See FRIULANO.

Sauvignon Blanc [Sauv Bl] Distinctive aromatic, grassy-to-tropical wines, pungent in NZ, often mineral in Sancerre, riper in Australia. V.gd in Rueda, Austria, n Italy (Isonzo, Piedmont, Alto Adige), Chile's Casablanca Valley and S Africa. Blended with SÉM in B'x. Can be austere or buxom (or indeed nauseating). Sauv Gris is a pink-skinned, less aromatic version of Sauv Bl with untapped potential.

Sauvignon Vert See FRIULANO.

Savagnin (Heida, Païen) Grape for *vin jaune* from Jura: aromatic form is GEWURZ. In Switzerland known as Heida, Païen or Traminer. Full-bodied, high acidity.

Scheurebe (Sämling) Grapefruit-scented German RIES x SILVANER (possibly), v. successful in Pfalz, esp Auslese and up. Can be weedy: must be v. ripe to be gd.

Sémillon [Sém] Contributes lusciousness to Sauternes, but decreasingly important for Graves and other dry white B'x. Grassy if not fully ripe; can make soft dry wine of great ageing potential. Superb in Australia; NZ, S Africa promising.

Sercial (Cerceal) Portuguese: makes the driest Madeira. Cerceal, also Portuguese, seems to be this plus any of several others.

Seyval Blanc [Seyval Bl] French-made hybrid of French and American vines. V. hardy and attractively fruity. Popular and reasonably successful in e US and England but dogmatically banned by EU from "quality" wines.

Silvaner (Johannisberg, Sylvaner) Can be excellent in Germany's Rheinhessen, Pfalz, esp Franken, with plant/earth flavours, mineral notes. V.gd (and powerful) as Johannisberg in the Valais, Switzerland. Lightest of the Alsace grapes.

Šipon See FURMINT.

Sivi Pinot See PINOT GR.

Spätrot See ZIERFANDLER.

Sylvaner See SILVANER.

Tămâioasă Românească Romanian "frankincense" grape, with exotic aroma and taste. Belongs to MUSCAT family.

Torrontés Name given to a number of grapes, mostly with an aromatic, floral character, sometimes soapy. A speciality of Argentina; also in Spain. DYA.

Traminac Or Traminec. See GEWURZ.

Traminer Or Tramini (Hungary). See GEWURZ.

Trebbiano (Ugni Blanc) Principal white grape of Tuscany, found all over Italy in many different guises. Rarely rises above the plebeian except in Tuscany's Vin Santo. Some gd dry whites under DOCs Romagna or Abruzzo. Trebbiano di Soave, aka VERDICCHIO, only distantly related. T di Lugana now called Turbiana. Grown in s France as Ugni Blanc, and Cognac as St-Émilion. Mostly thin, bland wine; needs blending (and more careful growing).

Ugni Blanc [Ugni Bl] See TREBBIANO.

Ull de Llebre See TEMPRANILLO.

Verdejo The grape of Rueda in Castile, potentially fine and long-lived.

Verdelho Great quality in Australia (pungent, full-bodied); rare but gd (and medium-sweet) in Madeira.

Verdicchio Potentially gd, muscular, dry; central-e Italy. Wine of same name.

Vermentino Italian, sprightly; satisfying texture and ageing capacity. Potential here.

Vernaccia Name given to many unrelated grapes in Italy. Vernaccia di San Gimignano is crisp, lively; Vernaccia di Oristano is Sherry-like.

Vidal French hybrid much grown in Canada for Icewine.

Vidiano Most Cretan producers love this. Powerful, stylish. Lime/apricot, gd acidity.

Viognier Ultra-fashionable Rhône grape, finest in Condrieu, less fine but still aromatic in the Midi. Gd examples from California, Virginia, Uruguay, Australia.

Viura (Macabeo, Maccabéo, Maccabeu) Workhorse white grape of n Spain, widespread in Rioja and Catalan Cava country. Also found over border in Southwest France. Gd quality potential.

Weißburgunder PINOT BL in Germany.

Welschriesling (Graševina, Laški Rizling, Olaszrizling, Riesling Italico) Not related to RIES. Light and fresh to sweet and rich in Austria; ubiquitous in Central Europe, where it can be remarkably gd for dry and sweet wines.

Xarel·lo (Pansa Blanca) Traditional Catalan grape, used for Cava (with Parellada, MACABEO). Neutral but clean. More character (lime cordial) in Alella (Pansa Blanca).

Xynisteri Cyprus's most planted white grape. Can be simple and is usually DYA; but when grown at altitude makes appealing, minerally whites.

Zéta Hungarian; BOUVIER X FURMINT used by some in Tokaji Aszú production.

Zierfandler (Spätrot, Cirfandl) Found in Austria's Thermenregion; often blended with ROTGIPFLER for aromatic, orange-peel-scented, weighty wines.

Wine & food

Matching wine with food matters, but don't get hung up on it. Red is still best with red meat, white (or red with little tannin) with fish; but contemporary food is rarely that clear-cut. Think of the body, tannin and acidity of the wine when making the match, and think about all the flavours of the dish: spices in eastern food, soy in oriental. Root vegetables are sweetish, chicory is bitter, chilli exaggerates tannins, tomatoes need acidity in the wine. Vinegar is complicated.

On p.37 you'll find a box of can't-go-wrong favourites with food; the entries below give lots of specific matches enjoyed over the years.

Before the meal – apéritifs

The pop of a Champagne cork is the classic appetite sharpener; Fino Sherry is the other. A glass of dry white or pink is good but don't go for the cheapest, most neutral option; Riesling is more interesting and Provence rosé from magnum looks wonderful. You could even try orange wine. Take a risk. Cocktails are not a good base for fine wine, but don't ignore vermouth; there are some great ones out there.

First courses

Aïoli More about mood than matching. Cold Provence rosé, PECORINO, ALIGOTÉ. Beer, marc or grappa... you'll hardly notice.

Antipasti / tapas / mezze In Italy, Spain, Greece or elsewhere, a selection of savoury, salty, meaty, cheesy, fishy, veggie bits and pieces works perfectly with Fino Sherry. Or whatever crisp white or light red is to hand. Gd Prosecco in emergencies.

Burrata Forget mozzarella; this is the crème de la crème. So a top Italian white, FIANO or Cusumano's GRILLO. I'll try Sauternes one day.

Carpaccio, beef or fish The beef version works well with most wines, incl reds. Tuscan is appropriate, but fine CHARDS are gd. So are vintage and pink Champagnes. Give Amontillado a try. **Salmon** Chard or Champagne. **Tuna** VIOGNIER, California Chard, Marlborough SAUV BL. Or sake.

Charcuterie / prosciutto / salami High-acid, unoaked red works better than white. Simple Beaujolais, Valpolicella, REFOSCO, SCHIOPPETINO, TEROLDEGO, BARBERA. If you must have white, it needs acidity. Chorizo makes wines taste metallic. Prosciutto with melon or figs needs full dry or medium white: CHENIN BL, FIANO, MUSCAT, VIOGNIER.

Dim sum Classically, China tea. PINOT GR or classic German dry RIES; light PINOT N. For reds, soft tannins are key. Bardolino, Rioja; Côtes du Rhône. Also NV Champagne or English fizz.

Eggs See also SOUFFLÉS. Not easy: eggs have a way of coating your palate. Omelettes: follow the other ingredients; mushrooms suggest red; Côtes du Rhone is a safe bet. With a truffle omelette, vintage Champagne. As a last resort I can bring myself to drink Champagne with scrambled eggs or eggs Benedict. Florentine, with spinach, is not a winey dish.

    quails' eggs Blanc de blancs Champagne; VIOGNIER.

    gulls' eggs Push the luxury: mature white burgundy or vintage Champagne.

    oeufs en meurette Burgundian genius: eggs in red wine with a glass of the same.

Fish terrine Calls for something fine. Pfalz RIES Spätlese Trocken, top GRÜNER V, Premier Cru Chablis, Clare Valley Ries, Sonoma CHARD; or Manzanilla.

Mozzarella with tomatoes, basil Fresh Italian white, eg. Soave, Alto Adige. VERMENTINO from Liguria or Rolle from the Midi. See also AVOCADO.

Oysters, raw NV Champagne, Chablis, MUSCADET, white Graves, Sancerre, English
Bacchus, or Guinness. Experiment with Sauternes. Manzanilla is my favourite.
Flat oysters worth gd wine; Pacific ones drown it in brine.

   stewed, grilled or otherwise cooked Puligny-Montrachet or gd NZ CHARD.
Champagne is gd with either.

Pasta Red or white according to the sauce:

   cream sauce (eg. carbonara) Orvieto, GRECO di Tufo. Young SANGIOVESE.

   meat sauce MONTEPULCIANO d'Abruzzo, Salice Salentino, MALBEC.

   pesto (basil) sauce BARBERA, Ligurian VERMENTINO, NZ SAUV BL, Hungarian FURMINT.

   seafood sauce (eg. vongole) VERDICCHIO, Lugana, Soave, GRILLO, Cirò, unoaked CHARD.

   tomato sauce Chianti, Barbera, Sicilian red, ZIN, S Australian GRENACHE.

Risotto Follow the flavour:

   with vegetables (eg. Primavera) PINOT GR from Friuli, Gavi, youngish SÉM,
DOLCETTO or BARBERA d'Alba.

   with fungi porcini Finest mature Barolo or Barbaresco.

   nero A rich dry white: VIOGNIER or even Corton-Charlemagne.

   seafood A favourite dry white.

Soufflés As show dishes these deserve ★★★ wines:

   with cheese Mature red burgundy or B'x, CAB SAUV (not Chilean or Australian),
etc. Or fine mature white burgundy.

   with fish (esp smoked haddock & chive cream sauce) Dry white: ★★★ burgundy,
B'x, Alsace, CHARD, etc.

   with spinach (which is tough on wine) Mâcon-Villages, St-Véran, or Valpolicella.
Champagne (esp vintage) can also spark things with the texture of a soufflé.

Fish

Abalone Dry or medium white: SAUV BL, unoaked CHARD. A touch of oak works with
soy sauce, oyster sauce, etc. In Hong Kong: Dom Pérignon (at least).

Anchovies Fino, obviously. Franciacorta or Cava.

   salade Niçoise Provence rosé.

   bocquerones VERDEJO, unoaked SÉM.

Bacalão Salt cod needs acidity: young Portuguese red or white. Or Italian ditto.
Cider or cold beer work well.

Bass, sea Fine white, eg. Clare RIES, Chablis, white Châteauneuf, VERMENTINO from
Sardinia, WEISSBURGUNDER from Baden or Pfalz. Rev the wine up for more
seasoning, eg. ginger, spring onions; more powerful Ries, not necessarily dry.

Beurre blanc, fish (or veg) A top-notch MUSCADET *sur lie*, a SAUV BL/SÉM blend,
Premier Cru Chablis, Vouvray, ALBARIÑO or Rheingau RIES.

Brill More delicate than turbot: hence a top fish for fine old Puligny and the like.
With the richness of hollandaise you could go up to Montrachet.

Caviar Iced vodka (and) full-bodied Champagne (eg. Bollinger, Krug). Don't (ever)
add raw onion.

Ceviche Can be applied to anything now, but here, it's fish. Australian RIES or
VERDELHO, Chilean SAUV BL, TORRONTÉS. Manzanilla.

Cod, roast Gd neutral background for fine dry/medium whites: Chablis, Meursault,
Corton-Charlemagne, Cru Classé Graves, GRÜNER V, German Kabinett or Grosses
Gewächs, or gd lightish PINOT N.

   black cod with miso sauce NZ or Oregon Pinot N, Meursault Premier Cru or
Rheingau RIES Spätlese. Vintage Champagne.

Crab (esp Dungeness) and RIES together are part of the Creator's plan. But He also
created Champagne.

   Chinese, with ginger & onion German RIES Kabinett or Spätlese Halbtrocken.
Tokaji FURMINT, GEWURZ.

cioppino SAUV BL; but West Coast friends say ZIN. Also California sparkling.

cold, dressed Top Mosel Ries, dry Alsace or Australian Ries or Assyrtiko.

softshell Unoaked CHARD, ALBARIÑO or top-quality German Ries Spätlese.

crabcakes Pungent Sauv Bl (Loire, S Africa, Australia, NZ) or Ries (German Spätlese or Australian).

with black bean sauce A big Barossa SHIRAZ or SYRAH. Even a tumbler of Cognac.

with chilli & garlic Quite powerful Ries, perhaps German Grosses Gewächs or Wachau Austrian.

Cured fish Salmon can have a whisky cure, a beetroot cure; all have sweetness and pungency. With gravadlax, sweet mustard sauce is a complication. SERCIAL Madeira (eg. 10-yr-old Henriques), Amontillado, Tokaji Szamarodni, orange wine. Or NV Champagne.

Curry S African CHENIN BL, Alsace PINOT BL, Franciacorta, fruity rosé, not too pale and anodyne; look at other flavours. Prawn and mango needs more sweetness, tomato needs acidity. Fino can be remarkably gd and can handle heat. So can IPA or Pilsner.

Fish pie (with creamy sauce) ALBARIÑO, Soave Classico, RIES Erstes Gewächs, Mâcon Blanc, Spanish GODELLO.

Grilled or fried fish also applies to **fish & chips**, **tempura**, **fritto misto**...

 Dover sole Perfect with fine wines: white burgundy or equivalent. **Plaice**, **flounder** Light, fresh whites. **Cod**, **haddock** CHARD, PINOT BL. Oily fish like **herrings**, **mackerel**, **sardines** More acidity, weight: ASSYRTIKO, VERDELHO, FURMINT. **In saor** Try top Prosecco, and I mean top. Red fish like **salmon**, **red mullet** PINOT N. For salmon, also best Chard, Grand Cru Chablis, top RIES. **Trout** Gd Chard, Ries, Pinot N. **Halibut**, **turbot** Best rich, dry white; top Chard, mature Ries Spätlese. **Swordfish** Full-bodied, dry white (or why not red?) of the country. Nothing grand. **Tuna** Best served rare (or raw) with light red: young Loire CAB FR or red burgundy. Young Rioja is a possibility. **Whitebait** Crisp dry whites, eg. FURMINT, Greek, Touraine SAUV BL, VERDICCHIO, white Dão, Fino Sherry. Or beer.

Herrings Need a sharp white to cut their richness. Rully, Chablis, MUSCADET, Bourgogne ALIGOTÉ, Greek, dry SAUV BL. Or Indian tea. Or cider.

 pickled/raw Challenging: beer or akvavit; but old Mosel RIES Spätlese (c.15 yrs) is brilliant with sweet vinegar marinade. Try Alsace GEWURZ too.

Ikan bakar This classic Indonesian/Malay grilled fish works well with GRÜNER V.

Kedgeree Full white, still or sparkling: Mâcon Villages, S African CHARD, GRÜNER V, German Grosses Gewächs or (at breakfast) Champagne.

Lobster with a rich sauce Eg. Thermidor: Vintage Champagne, fine white burgundy, Cru Classé Graves, Roussanne, top Australian CHARD. Alternatively, for its inherent sweetness, Sauternes, Pfalz Spätlese, even Auslese.

 plain grilled, or cold with mayonnaise NV Champagne, Alsace RIES, Premier Cru Chablis, Condrieu, Mosel Spätlese, GRÜNER V, Hunter SEM, or local fizz.

Monkfish Meaty but neutral; full-flavoured white or red, according to sauce.

Mullet, grey VERDICCHIO, Rully or unoaked CHARD.

Mussels marinière MUSCADET *sur lie*, Premier Cru Chablis, unoaked CHARD.

 curried Something semi-sweet; Alsace RIES.

Paella, shellfish Full-bodied white or rosé, unoaked CHARD, ALBARIÑO, or GODELLO. Or local Spanish red.

Perch, sandre Exquisite freshwater fish for finest wines: top white burgundy, Grand Cru Alsace RIES or noble Mosels. Or try top Swiss CHASSELAS (eg. Dézaley, St-Saphorin). Or Franciacorta.

Prawns, crayfish with mayonnaise Menetou-Salon or Reuilly.

 with garlic Keep the wine light, white, or rosé, and dry.

with spices Up to and incl chilli, go for a bit more body, but not oak: dry RIES or Italian, eg. FIANO, Grillo. *See also* FISH/CURRY.

Sashimi The Japanese preference is for white wine with body (Chablis Premier Cru, Alsace RIES) with white fish, PINOT N with red. Both need acidity: low-acidity wines don't work. Simple Chablis can be too thin. If soy is involved, then low-tannin red (again, Pinot). Remember sake (or Fino). As though you'd forget Champagne. Try Japanese KOSHU.

Scallops An inherently slightly sweet dish, best with medium-dry whites.

   in cream sauces German Spätlese, Montrachets or top Australian CHARD.

   grilled or seared Hermitage Blanc, GRÜNER V, Pessac-Léognan Blanc, vintage Champagne or PINOT N.

   with Asian seasoning NZ Chard, CHENIN BL, GODELLO, Grüner V, GEWURZ.

Scandi fish dishes Scandinavian dishes often have flavours of dill, caraway and cardamom, plus they combine sweet and sharp flavours. Go for acidity and some weight in the wine: FALANGHINA, GODELLO, VERDELHO, Australian, Alsace, or Austrian RIES.

Shellfish Dry white with plain boiled shellfish, richer wines with richer sauces. RIES.

   with plateaux de fruits de mer Chablis, MUSCADET de Sèvre et Maine, PICPOUL de Pinet, Alto Adige PINOT BL.

Skate / raie with brown butter White with some pungency (eg. Alsace PINOT GR or ROUSSANNE) or a clean, straightforward one, ie. MUSCADET, VERDICCHIO.

Smoked fish All need freshness and some pungency; Fino Sherry works with all.

   eel Often with beetroot, crème fraiche: Fino again, or Mosel Ries.

   haddock Gd Chablis, MARSANNE, GRÜNER V. *See also* SOUFFLÉS.

   kippers Try Oloroso Sherry or Speyside malt.

   mackerel Not wine-friendly. Try Fino.

   salmon Condrieu, Alsace PINOT GR, Grand Cru Chablis, German RIES Spätlese, vintage Champagne, vodka, schnapps, or akvavit.

   trout More delicate: Mosel Ries.

Snapper SAUV BL if cooked with oriental flavours; white Rhône or Provence rosé with Med flavours.

Squid / octopus Fresh white: ALBARIÑO, MUSCADET, sparkling, esp with salt and pepper squid. Squid ink (risotto, pasta) needs Soave.

Sushi Hot wasabi is usually hidden in every piece. German QbA Trocken wines, simple Chablis, ALVARINHO or NV Brut Champagne or KOSHU. Obvious fruit doesn't work. Or, of course, sake, or beer.

Tagine N African flavours need substantial whites to balance – Austrian, Rhône – or crisp, neutral whites that won't compete. Go easy on the oak. VIOGNIER or ALBARIÑO can work well.

Taramasalata A Med white with personality, Greek if possible. Fino Sherry works well. Try Rhône MARSANNE.

Teriyaki A way of cooking, and a sauce, used for meat as well as fish. Germans favour off-dry RIES with weight: Kabinett can be too light.

Meat / poultry / game

Barbecues The local wine: Australian (sparkling SHIRAZ?), S African, Chilean, Argentina are right in spirit. Reds need tannin and vigour. Or the freshness of cru Beaujolais.

Beef (*see also* Steak), boiled Red: B'x (eg. Fronsac), Roussillon, Gevrey-Chambertin or Côte-Rôtie. Medium-ranking white burgundy is gd, eg. Auxey-Duresses. In Austria you may be offered skin-fermented TRAMINER. Mustard softens tannic reds, horseradish kills your taste; can be worth the sacrifice.

   roast An ideal partner for fine red of any kind. Even Amarone. *See above for* mustard. The silkier the texture of the beef (wagyu, Galician, eg.), the silkier the wine. Wagyu, remember, is about texture; has v. delicate flavour.

stew, daube Sturdy red: Pomerol or St-Émilion, Hermitage, Cornas, BARBERA, SHIRAZ, Napa CAB SAUV, Ribera del Duero, or Douro red.

stroganoff Dramatic red: Barolo, Valpolicella Amarone, Priorat, Hermitage, late-harvest ZIN. Georgian SAPERAVI or Moldovan Negru de Purkar.

Boudin blanc Loire CHENIN BL, esp when served with apples: dry Vouvray, Saumur, Savennières; mature red Côte de Beaune if without.

Boudin noir / morcilla Local SAUV BL or CHENIN BL (esp in Loire). Or Beaujolais cru, esp Morgon. Or light TEMPRANILLO. Or Fino.

Brazilian dishes Pungent flavours that blend several culinary traditions. Rhônish grapes work for red, or white with weight: VERDICCHIO, Californian CHARD. Or a Caipirinha (better not have two).

Cajun food Gutsy reds, preferably New World: ZIN, CARMENÈRE, SHIRAZ. Fish or white meat: off-dry RIES, MARSANNE, ROUSSANNE. Or, of course, cold beer.

Cassoulet Red from Southwest France (Gaillac, Minervois, Corbières, St-Chinian or Fitou) or SHIRAZ. But best of all Fronton, Beaujolais cru or young TEMPRANILLO.

Chicken / turkey / guinea fowl, roast Virtually any wine, incl v. best bottles of dry to medium white and finest old reds (esp burgundy). Sauces can make it match almost any fine wine (eg. coq au vin; the burgundy can be red or white, or *vin jaune* for that matter).

fried Sparkling works well.

chicken Kiev Alsace RIES, Collio, CHARD, Bergerac rouge.

Chilli con carne Young red: Beaujolais, TEMPRANILLO, ZIN, Argentine MALBEC, Chilean CARMENÈRE. Or beer.

Chinese dishes To the purist there's no such thing as Chinese food: food in China is regional – like Italian, only more confusing. It's easiest to have both white and red; no one wine goes with all. Peking duck is pretty forgiving. Champagne becomes a thirst quencher. Beer too.

Cantonese Big, slightly sweet flavours work with slightly oaky CHARD, PINOT N, off-dry RIES. GEWURZ is often suggested but rarely works; GRÜNER V is a better bet. You need wine with acidity. Dry sparkling (esp Cava) works with textures.

Shanghai Richer and oilier than Cantonese, not one of wine's natural partners. Shanghai tends to be low on chilli but high on vinegar of various sorts. German and Alsace whites can be a bit sweeter than for Cantonese. For reds, try MERLOT – goes with the salt. Or mature Pinot N, but a bit of a waste.

Szechuan VERDICCHIO, Alsace PINOT BL, or v. cold beer. Mature Pinot N can also work; but *see* above. The Creator intended tea.

Taiwanese LAMBRUSCO works with traditional Taiwan dishes if you're tired of beer.

Choucroute garni Alsace PINOT BL, PINOT GR, RIES, or lager.

Cold roast meat Generally better with full-flavoured white than red. Mosel Spätlese or Hochheimer and Côte Chalonnaise are v.gd, as is Beaujolais. Leftover Champagne too.

Confit d'oie / de canard Young, tannic red B'x, California CAB SAUV and MERLOT, Priorat cuts richness. Alsace PINOT GR or GEWURZ match it.

Coq au vin Red burgundy. Ideal: one bottle of Chambertin in the dish, two on the table. *See also* CHICKEN.

Dirty (Creole) rice Rich, supple red: NZ PINOT N, GARNACHA, Bairrada, MALBEC.

Duck / goose PINOT N is tops. Or rich white, esp for goose: Pfalz Spätlese or off-dry Grand Cru Alsace. With oranges or peaches, Sauternais propose Sauternes, others Monbazillac or RIES Auslese. Mature, weighty vintage Champagne handles accompanying red cabbage surprisingly well. So does decent Chianti.

Peking *See* CHINESE FOOD.

wild duck Worth opening good Pinot N. Austrian or Tuscan red (easy on the oak) is also gd.

with olives Top-notch Chianti or other Tuscans.

roast breast & confit leg with Puy lentils Madiran (best), St-Émilion, Fronsac.

Foie gras Sweet white: Sauternes, Tokaji Aszú 5 Puttonyos, late-harvest PINOT GR or RIES, Vouvray, Montlouis, Jurançon *moelleux*, GEWURZ. Old dry Amontillado can be sublime.

hot Mature vintage Champagne. But never CHARD, SAUV BL, or (shudder) red.

Game birds, young, roast The best red wine you can afford, but not too heavy. Partridge is more delicate than pheasant, which is more delicate than grouse. Up the weight of wine accordingly, starting with youngish PINOT N, BLAÜFRANKISCH, SYRAH, GARNACHA, and moving up.

older birds in casseroles Gevrey-Chambertin, Pommard, Châteauneuf, Dão, or Grand Cru Classé St-Émilion, Rhône.

well-hung game Vega Sicilia, great red Rhône, Château Musar.

cold game Best German RIES or mature vintage Champagne.

Game pie, hot Red: Oregon PINOT N, St-Émilion Grand Cru Classé. **Cold** Gd-quality white burgundy or German Erstes Gewächs, cru Beaujolais, Champagne.

Goat (hopefully kid). As for lamb. **Jamaican curry goat** *See* INDIAN DISHES.

Goulash Needs a flavoursome young red: Hungarian Kékoportó, ZIN, Uruguayan TANNAT, Douro red, MENCÍA, young Australian SHIRAZ, SAPERAVI. Dry Szamarodni from Tokaj.

Haggis Fruity red, eg. young claret, young Portuguese red, New World CAB SAUV or MALBEC or Châteauneuf. Or, of course, malt whisky.

Ham, cooked Softer red burgundies: Volnay, Savigny, Beaune; Chinon or Bourgueil; sweetish German white (RIES Spätlese); lightish CAB SAUV (eg. Chilean), or New World PINOT N. And don't forget the heaven-made match of ham and Sherry.

Hare Jugged hare calls for flavourful red: not-too-old burgundy or B'x, Rhône (eg. Gigondas), Bandol, Barbaresco, Ribera del Duero, Rioja Res. The same for saddle or for hare sauce with pappardelle.

Indian dishes Various options: dry Sherry is brilliant. Choose a fairly weighty Fino with fish, and Palo Cortado, Amontillado, or Oloroso with meat, according to weight of dish; heat's not a problem. The texture works too. Otherwise, medium-sweet white, v. cold, no oak: Orvieto *abboccato*, S African CHENIN BL, Alsace PINOT BL, TORRONTÉS, Indian sparkling, Cava or NV Champagne. Rosé is gd all-rounder. For tannic impact Barolo or Barbaresco, or deep-flavoured reds, ie. Châteauneuf, Cornas, Australian GRENACHE or MOURVÈDRE, or Valpolicella Amarone – will emphasize the heat. Hot-and-sour flavours need acidity.

Sri Lankan More extreme flavours, coconut. Sherry, rich red, rosé, or a mild white.

Japanese dishes A different set of senses come into play. Texture and balance are key; flavours are subtle. Gd mature fizz works well, as does mature dry RIES; you need acidity, a bit of body, and complexity. Dry FURMINT can work well. Umami-filled meat dishes favour light, supple, bright reds: Beaujolais perhaps, or mature PINOT N. Full-flavoured *yakitori* needs lively, fruity, younger versions of the same reds. KOSHU with raw fish – but why not sake? Orange Koshu with wagyu beef. *See also* SASHIMI, SUSHI, TERIYAKI.

Korean dishes Fruit-forward wines seem to work best with strong, pungent Korean flavours. PINOT N, Beaujolais, Valpolicella can all work: acidity is needed. Non-aromatic whites: GRÜNER V, SILVANER, VERNACCIA. But I drink beer.

Lamb, roast One of the traditional and best partners for v.gd red B'x, or its CAB SAUV equivalents from the New World. In Spain, finest old Rioja and Ribera del Duero Res or Priorat, in Italy ditto SANGIOVESE. Lay off mint sauce.

slow-cooked roast Flatters top reds, but needs less tannin than pink lamb. *See also* TAGINES.

milk-fed Delicate and deserves top, delicate B'x, burgundy, top (delicate) Spanish.

Liver Young red: Beaujolais-Villages, St-Joseph, Médoc, Italian MERLOT, Breganze CAB SAUV, ZIN, Priorat, Bairrada.

calf's Red Rioja Crianza, Fleurie. Or a big Pfalz RIES Spätlese.

Mexican food Californians favour RIES: Calavera restaurant in Oakland lists 33 RIES, mostly German. Beer for me.

Moussaka Red or rosé: Naoussa, SANGIOVESE, Corbières, Côtes de Provence, Ajaccio, young ZIN, TEMPRANILLO.

Mutton A stronger flavour than lamb, and not usually served pink. Needs a strong sauce. Robust red; top-notch, mature CAB SAUV, SYRAH. Sweetness of fruit (eg. Barossa) suits it.

'Nduja Calabria's spicy, fiery spreadable salumi needs a big, juicy red: young Rioja, Valpol, CAB FR, AGLIANICO, CARIGNAN, NERELLO MASCALESE.

Osso bucco Low-tannin, supple red such as DOLCETTO d'Alba or PINOT N. Or dry Italian white such as Soave.

Ox cheek, braised Superbly tender and flavoursome, this flatters the best reds: Vega Sicilia, St-Émilion. Best with substantial wines.

Oxtail Rather rich red: St-Émilion, Pomerol, Pommard, Nuits-St-Georges, Barolo, or Rioja Res, Priorat or Ribera del Duero, California or Coonawarra CAB SAUV, Châteauncuf, mid-weight SHIRAZ, Amarone.

Paella Young Spanish: red, dry white, or rosé: Penedès, Somontano, Navarra, or Rioja.

Pastrami Alsace RIES, young SANGIOVESE, or St-Émilion.

Pâté Chicken liver calls for pungent white (Alsace PINOT GR or MARSANNE), a smooth red eg. light Pomerol, Volnay, or NZ PINOT N. More strongly flavoured (duck, etc.) needs Gigondas, Moulin-à-Vent, Chianti Classico, or gd white Graves. Amontillado can be marvellous match.

Pigeon or squab PINOT N perfect; young Rhône, Argentine MALBEC, young SANGIOVESE. Try Franken SILVANER Spätlese. With luxurious squab, top quite tannic red.

Pastilla Depends on sweetness of dish. As above, or if authentically sweet, try RIES Spätlese, Alsace PINOT GR with some sweetness.

Pork A perfect rich background to a fairly light red or rich white.

roast Deserves ★★★ treatment; Médoc is fine. Portugal's suckling pig is eaten with Bairrada; S America's with CARIGNAN; Chinese is gd with PINOT N.

with prunes or apricots Something sweeter, eg. Vouvray.

pork belly Slow-cooked and meltingly tender, needs red with some tannin or acidity. Italian would be gd: Barolo, DOLCETTO, or BARBERA. Or Loire red, or lightish Argentine MALBEC. With Chinese spices, VIOGNIER, CHENIN BL.

pulled pork Comes with spicy sauce: juicy New World reds.

Pot au feu, bollito misto, cocido Rustic red wines from region of origin; SANGIOVESE di Romagna, Chusclan, Lirac, Rasteau, Portuguese Alentejo or Spain's Yecla, Jumilla.

Quail Succulent little chick deserves tasty red or white. Rioja Res, mature claret, PINOT N. Or mellow white: Vouvray, St-Péray.

Quiche Egg and bacon are not great wine matches, but one must drink something. Alsace RIES or PINOT GR, even GEWURZ, is a classical match. Beaujolais could be gd too.

Rabbit Lively, medium-bodied young Italian red, eg. AGLIANICO del Vulture, REFOSCO; Chiroubles, Chinon, Saumur-Champigny or Rhône rosé.

with prunes Bigger, richer, fruitier red.

with mustard Cahors.

as ragu Medium-bodied red with acidity: Aglianico, NEBBIOLO.

Satay McLaren Vale SHIRAZ, Alsace or NZ GEWURZ. Peanut sauce: problem for wine.

Sauerkraut (German) German RIES, lager or Pils. (But *see also* CHOUCROUTE GARNI.)

Singaporean dishes Part Indian, part Malay and part Chinese, Singaporean food has big and bold flavours that don't match easily with wine – not that that bothers the country's many wine-lovers. Off-dry RIES is as gd as anything. With meat dishes, ripe, supple reds: Valpolicella, PINOT N, DORNFELDER, unoaked MERLOT, or CARMENÈRE.

Steak Rare steak needs brisker, more tannic reds; well-done needs juicy, fruity reds, eg. young Argentine MALBEC. Fattier cuts need acidity, tannin.

au poivre A fairly young Rhône red or CAB SAUV. Nothing too sweetly fruity.

fillet Silky red: Pomerol or PINOT N.

ribeye, tomahawk, tournedos Big, pungent red: Barolo, Cahors, SHIRAZ, Rioja.

sirloin Suits most gd reds. B'x blends, Tuscans.

Fiorentina (bistecca) Chianti Classico Riserva or BRUNELLO.

Korean *yuk whe* (world's best steak tartare) Sake.

tartare Vodka or light young red: Beaujolais, Bergerac, Valpolicella. Aussies drink GAMAY with kangaroo tartare, charred plums, Szechuan pepper.

T-bone Reds of similar bone structure: Barolo, Hermitage, Australian CAB SAUV or Shiraz, Chilean SYRAH, Douro.

Wagyu Delicate, silky red, or orange wine.

from older cattle Has deep, rich savouriness. Top Italian, Spanish red.

Steak-&-kidney pie or pudding Red Rioja Res or mature B'x. Pudding (with suet) wants vigorous young wine. Madiran with its tannin is gd.

Stews & casseroles Burgundy such as Nuits-St-Georges or Pommard if fairly simple; otherwise lusty, full-flavoured red, eg. young Côtes du Rhône, BLAUFRÄNKISCH, Corbières, BARBERA, SHIRAZ, ZIN, etc.

Sweetbreads A rich dish, so grand white wine: Rheingau RIES or Franken SILVANER Spätlese, Grand Cru Alsace PINOT GR, or Condrieu, depending on sauce.

Tagines Depends on what's under the lid, but fruity young reds are a gd bet: Beaujolais, TEMPRANILLO, SANGIOVESE, MERLOT, SHIRAZ. Amontillado is great.

chicken with preserved lemon, olives VIOGNIER.

Tandoori chicken RIES or SAUV BL, young red B'x or light n Italian red served cool. Also Cava and NV Champagne, or, of course, Palo Cortado or Amontillado Sherry.

Thai dishes Ginger and lemon grass call for pungent SAUV BL (Loire, Australia, NZ, S Africa) or RIES (Spätlese or Australian). Most curries suit aromatic whites with a touch of sweetness: GEWURZ also gd.

Tongue Gd for any red or white of abundant character. Alsace PINOT GR, GEWURZ, gd GRÜNER V. Beaujolais, Loire reds, BLAUFRÄNKISCH, TEMPRANILLO and full, dry rosés.

Veal A friend of fine wine. Rioja Res, CAB blends, PINOT N, NEBBIOLO, German or Austrian RIES, Vouvray, Alsace PINOT GR, Italian GRECO di Tufo.

Venison Big-scale reds, incl MOURVÈDRE, solo as in Bandol or in blends. Rhône, Languedoc, B'x, NZ Gimblett Gravels or California CAB SAUV of a mature vintage; or rather rich white (Pfalz Spätlese or Alsace PINOT GR).

with sweet & sharp berry sauce Try a German Grosses Gewächs RIES, or Chilean CARMENÈRE, or SYRAH.

Vietnamese food Slanted Door (famous San Fran Vietnamese restaurant) favours RIES, dry or up to Spätlese, German, Austrian, NZ; also GRÜNER V, SÉM. For reds, PINOT N, CAB FR, BLAUFRÄNKISCH.

Vitello tonnato Full-bodied whites: CHARD. Light reds (eg. Valpolicella) served cool. Or a southern rosé.

Wild boar Serious red: top Tuscan or Priorat. NZ SYRAH. I've even drunk Port.

Vegetable dishes

(*See also* FIRST COURSES)

Agrodolce Italian sweet-and-sour, with pine kernels, sultanas, capers, vinegar and perhaps anchovies. Points to fresh white: VERDICCHIO, unoaked CHARD. Or orange.

Artichokes Not great for wine. Incisive dry white: NZ SAUV BL; Côtes de Gascogne or Greek (precisely, 4-yr-old MALAGOUSIA, but easy on the vinaigrette); VERMENTINO. Orange wine, yes; red no, unless you absolutely have to, in which case go for acidity: LAGREIN, DOLCETTO.

Asparagus is lightly bitter, and needs acidity. RIES is classic; SAUV BL (or English BACCHUS) echoes the flavour. Unoaked young SÉM or CHARD, esp Australian; Chard gd with melted butter, hollandaise. Dry MUSCAT, or Jurançon Sec. Argument for trying a really sweet wine too, maybe not Yquem.

Aubergine Comes in a multitude of guises, usually pungent. Sturdy reds with acidity are a gd bet: SHIRAZ, Greek, Lebanese, Turkish. Or structured white, eg, VERDICCHIO.

Avocado Not a wine natural. Dry to slightly sweet RIES Kabinett will suit the dressing. Otherwise, light and fresh: ALIGOTÉ, TREBBIANO, PINOT GRIGIO.

Beetroot Mimics a flavour found in red burgundy. You could return the compliment. New-wave (ie. light) GRENACHE/GARNACHA is gd, as well.

Cauliflower, roast, etc. Go by the other (usually bold) flavours. Try Austrian GRÜNER V, Valpolicella, NZ PINOT N.

cauliflower cheese Crisp, aromatic white: Sancerre, RIES Spatlese, MUSCAT, ALBARIÑO, GODELLO. CHARD too, and Beaujolais-Villages.

with caviar – yes, really. Vintage Champagne.

Celeriac, slow-roast or puréed Won't interfere with rest of dish. Acidity works well, so classic CAB blends, Beaujolais, PINOT N, according to dish.

remoulade with smoked ham Needs bright red: DOLCETTO, simple Valpolicella. Or white GRÜNER V.

Chestnuts In a slow-cooked daube or as purée, look for earthy, rich red: Tuscan or S Rhône.

Chickpeas Look at other flavours. Casserole works with TEMPRANILLO, s French reds.

hummus Any simple red, pink or white, or, of course, Fino.

Chilli Some like it hot, but not with your best bottles. Tannic wines become more tannic, if you like that, go for it. Light, fruity reds and whites are refreshing: TEMPRANILLO, Chilean MERLOT, NZ SAUV BL. Same for **harissa**. *See also* CHILLI CON CARNE, CHINESE DISHES, INDIAN DISHES.

Couscous with vegetables Young red with a bite: SHIRAZ, Corbières, Minervois; rosé; orange wine; Italian REFOSCO or SCHIOPPETTINO.

Dhal Comes with many variations, but all share aromatic earthiness. Simple, warm-climate reds work best: CARMENÈRE, Dão, s Italian.

Fennel-based dishes SAUV BL: Pouilly-Fumé or NZ; SYLVANER or English SEYVAL BL; or young TEMPRANILLO.

Fermented foods *See also* SAUERKRAUT, CHOUCROUTE, KOREAN. Kimchi and miso are being worked into many dishes. Fruit and acidity are generally needed. If in sweetish veg dishes, try Alsace.

Grilled Mediterranean vegetables Italian whites, or for reds Brouilly, BARBERA, TEMPRANILLO or SHIRAZ.

Lentil dishes Sturdy reds such as Corbières, ZIN or SHIRAZ. *See also* DHAL.

Macaroni cheese As for CAULIFLOWER CHEESE.

Mushrooms (in most contexts) A boon to most reds and some whites. Context matters as much as species. Pomerol, California MERLOT, Rioja Res, top burgundy or Vega Sicilia.

on toast Best claret, even Port.

button or Paris with cream Fine whites, even vintage Champagne.

ceps / porcini Ribera del Duero, Barolo, Chianti Rúfina, Pauillac or St-Estèphe, NZ Gimblett Gravels.

Onion / leek tart / flamiche Fruity, off-dry or dry white: Alsace PINOT GR or GEWURZ is classic; Canadian, Australian or NZ RIES; Jurançon. Or Loire CAB FR.

Peppers, cooked Mid-weight Rhône grapes, CARMENÈRE, Rioja; or ripe SAUV BL (esp with green Hungarian wax peppers).

stuffed Full-flavoured red, white, or pink: Languedoc, Greek, Spanish. Or orange.

Pickled foods & vinegar Vinegar and wine don't go, it's true, but pickled foods are everywhere. Try Alsace, German RIES with CHOUCROUTE/SAUERKRAUT. With pickled veg as part of a dish, just downgrade the wine a bit (no point in opening best bottles) and make sure it has some acidity. (Or have beer.) In dressings, experiment with vinegars: Sherry vinegar can work with Amontillado, etc., big reds; Austrian apricot vinegar is delicate; balsamic can work with rich Italian reds. Wine just has to work harder than it used to.

Pumpkin / squash ravioli or risotto Full-bodied, fruity dry or off-dry white: VIOGNIER or MARSANNE, demi-sec Vouvray, Gavi or S African CHENIN. Red: MERLOT, ZIN.

Radicchio, in a salad Eg. with game, points towards brisk, acid reds: NEBBIOLO, LAGREIN, or white VERMENTINO.

roast This is easier. Valpolicella, SANGIOVESE, BLAÜFRANKISCH.

Ratatouille (or Piperade) Vigorous young red: Chianti, NZ CAB SAUV, MERLOT, MALBEC, TEMPRANILLO, Languedoc. Or gd rosé.

Roasted root veg Sweet potatoes, carrots, etc., often mixed with eg. beetroot, cabbage, garlic, onions and others have plenty of sweetness. Rosé, esp with some weight, or orange wine. Pesto will tilt it towards white.

Saffron Found in sweet and savoury dishes, and wine-friendly. Rich white: ROUSSANNE, VIOGNIER, PINOT GR. Orange wines can be gd too. With desserts, Sauternes or Tokaji. *See also* TAGINES.

Salsa verde Whatever it's with, it points to more acidity, less lushness in the wine.

Seaweed (Nori) Depends on the context. *See also* SUSHI. Iodine notes go well with Austrian GRÜNER V, RIES.

Sweetcorn fritters Often served with a hot, spicy sauce. Rosé, orange or neutral white all safe.

Tahini Doesn't really affect wine choice. Go by rest of dish.

Tapenade Manzanilla or Fino Sherry, or any sharpish dry white or rosé. Definitely not Champagne.

Truffles Black truffles are a match for finest Right Bank B'x, but even better with mature white Hermitage or Châteauneuf. White truffles call for best Barolo or Barbaresco of their native Piedmont. With buttery pasta, Lugana. Or at breakfast, on fried eggs, BARBERA.

Watercress, raw Makes every wine on earth taste revolting.

Wild garlic leaves, wilted Tricky: a fairly neutral white with acidity will cope best.

Desserts

Apple pie, strudel, or tarts Sweet German, Austrian or Loire white, Tokaji Aszú, or Canadian Icewine.

Apples Cox's Orange Pippins with Cheddar cheese Vintage Port.

Russets with Caerphilly Old Tawny, or Amontillado.

Bread-&-butter pudding Fine 10-yr-old Barsac, Tokaji Aszú, Australian botrytized SEM.

Cakes *See also* CHOCOLATE, COFFEE, RUM. BUAL or MALMSEY Madeira, Oloroso or Cream Sherry. Asti, sweet Prosecco.

Cheesecake Sweet white: Vouvray, Anjou, or Vin Santo – nothing too special.

Chocolate A talking point. Generally only powerful flavours can compete. Texture matters. BUAL, California Orange MUSCAT, Tokaji Aszú, Australian Liqueur Muscat, 10-yr-old Tawny or even young Vintage Port; Asti for light, fluffy mousses. Or take a risk for a weightier partnership: VDN Banyuls, Maury or Rivesaltes. Médoc can match bitter black chocolate, though it's a bit of a waste of wine. Armagnac, or a tot of gd rum.

Christmas pudding, mince pies Tawny Port, Cream Sherry or that liquid Christmas pudding itself, PEDRO XIMÉNEZ Sherry. Tokaji Aszú. Asti, or Banyuls.

Coffee desserts Sweet MUSCAT, Australia Liqueur Muscats, or Tokaji Aszú.

Creams, custards, fools, syllabubs *See also* CHOCOLATE, COFFEE, RUM. Sauternes, Loupiac, Ste-Croix-du-Mont or Monbazillac.

Crème brûlée Sauternes or Rhine Beerenauslese, best Madeira or Tokaji Aszú.

Ice cream and sorbets Give wine a break.

Lemon flavours For dishes like tarte au citron, try sweet RIES from Germany or Austria or Tokaji Aszú; v. sweet if lemon is v. tart.

Meringues (eg. Eton Mess) Recioto di Soave, Asti, mature vintage Champagne.

Nuts (incl praliné) Finest Oloroso Sherry, Madeira, Vintage or Tawny Port (nature's match for walnuts), Tokaji Aszú, Vin Santo or Setúbal MOSCATEL. Cashews and Champagne. Pistachios with Fino.

    salted nut parfait Tokaji Aszú, Vin Santo.

Orange flavours Experiment with old Sauternes, Tokaji Aszú, or California Orange MUSCAT.

Panettone Vinsanto. Jurançon *moelleux*, late-harvest RIES, Barsac, Tokaji Aszú.

Pears in red wine Rivesaltes, Banyuls, or RIES Beerenauslese.

Pecan pie Orange MUSCAT or Liqueur Muscat.

Raspberries (no cream, little sugar) Excellent with fine reds which themselves taste of raspberries: young Juliénas, Regnié.

Rum flavours (baba, mousses, ice cream) MUSCAT – from Asti to Australian Liqueur, according to weight of dish.

Strawberries, wild (no cream) Serve with red B'x (most exquisitely Margaux) poured over.

    with cream Sauternes or similar sweet B'x, Vouvray *moelleux*, or Vendange Tardive Jurançon.

Summer pudding Fairly young Sauternes of a gd vintage.

Sweet soufflés Sauternes or Vouvray *moelleux*. Sweet (or rich) Champagne.

Tiramisú Vin Santo, young Tawny Port, MUSCAT de Beaumes-de-Venise, Sauternes, or Australian Liqueur Muscat. Better idea: skip the wine.

Trifle Should be sufficiently vibrant with its internal Sherry (Oloroso for choice).

Zabaglione Light-gold Marsala or Australian botrytized SEM, or Asti.

Wine & cheese

Counter-intuitively, white is a safer option than red. Fine red wines are slaughtered by strong cheeses. Principles to remember (despite exceptions): first, the harder the cheese, the more tannin the wine can have; second, the creamier the cheese, the more acidity is needed in the wine – and don't be shy of sweetness. Cheese is classified by its texture and the nature of its rind, so its appearance is a guide to the type of wine to match it.

Bloomy-rind soft cheeses: Brie, Camembert, Chaource Full, dry white burgundy or Rhône.

Blue cheeses The extreme saltiness of Roquefort or most blue cheeses needs sweetness: Sauternes, Tokaji, youngish Vintage or Tawny Port, esp with Stilton. Intensely flavoured old Oloroso, Amontillado, Madeira, Marsala and other fortifieds go with most blues. Dry red does not. Trust me.

Cooked cheese dishes

Frico Traditional in Friuli. Cheese baked or fried with potatoes or onions; high-acid local REFOSCO (r), or RIBOLLA GIALLA (w).

Mont d'Or Delicious baked, with potatoes. Neutral fresh white: GRÜNER V, Savoie.

fondue Trendy again. Light, fresh white as above.

macaroni or cauliflower cheese *See* VEGETABLE DISHES/CAULIFLOWER.

Fresh cream cheese, crème fraîche, mozzarella Light crisp white: Chablis, Bergerac, Entre-Deux-Mers; juicy rosé can work too.

Hard cheeses – Gruyère, Manchego, Parmesan, Cantal, Comté, old Gouda, Cheddar Hard to generalize, relatively easy to match. Gouda, Gruyère, some Spanish, and a few English cheeses complement fine claret or CAB SAUV and great SHIRAZ/SYRAH. But strong cheeses need less refined wines, preferably local ones. Granular old Dutch red Mimolette, Comté or Beaufort gd for finest mature B'x. Also for Tokaji Aszú. But try tasty whites too.

Natural rind (mostly goats cheese) – St-Marcellin Sancerre, light SAUV BL, Jurançon, Savoie, Soave, Italian CHARD; or young Vintage Port.

Semi-soft cheeses – Livarot, Pont l'Evêque, Reblochon, Tomme de Savoie, St-Nectaire Powerful white B'x, even Sauternes, CHARD, Alsace PINOT GR, dryish RIES, s Italian and Sicilian whites, aged white Rioja, dry Oloroso Sherry. The strongest of these cheeses kill almost any wines. Try marc or Calvados.

Washed-rind soft cheeses – Langres, mature Époisses, Maroilles, Carré de l'Est, Milleens, Münster Local reds, esp for Burgundian cheeses; vigorous Languedoc, Cahors, Côtes du Frontonnais, Corsican, s Italian, Sicilian, Bairrada. Also powerful whites, esp Alsace GEWURZ, MUSCAT. Gewurz with Munster, always.

Food & your finest wines

With v. special bottles, the wine guides the choice of food rather than vice versa. The following is based largely on gastronomic conventions, some bold experiments and much diligent and ongoing research.

Red wines

Amarone Classically, in Verona, risotto all'Amarone or pastissada. But if your butcher doesn't run to horse, then shin of beef, slow-cooked in more Amarone.

Barolo, Barbaresco Risotto with white truffles; pasta with game sauce (eg. pappardelle alla lepre); porcini mushrooms; Parmesan.

Great Syrahs: Hermitage, Côte-Rôtie, Grange; Vega Sicilia Beef, venison, well-hung game; bone marrow on toast; English cheese (Lincolnshire Poacher) but also hard goats' milk and ewes' milk cheeses such as England's Lord of the Hundreds. I treat Côte-Rôtie like top red burgundy.

Great Vintage Port or Madeira Walnuts or pecans. A Cox's Orange Pippin and a digestive biscuit is a classic English accompaniment.

Red Bordeaux v. old, light, delicate wines, (eg. pre-59) Leg or rack of young lamb, roast with a hint of herbs (not garlic); entrecôte; simply roasted (and not too well-hung) partridge; roast chicken never fails.

fully mature great vintages (eg. 59 61 82 85) Shoulder or saddle of lamb, roast with a touch of garlic; roast ribs or grilled rump of beef.

mature but still vigorous (eg. 89 90) Shoulder or saddle of lamb (incl kidneys) with rich sauce. Fillet of beef marchand de vin (with wine and bone marrow). Grouse. Avoid beef Wellington: pastry dulls the palate.

Merlot-based Beef (fillet is richest) or well-hung venison. In St-Émilion, lampreys.

Red burgundy Consider the weight and texture, which grow lighter/more velvety with age. Also the character of the wine: Nuits is earthy, Musigny flowery, great Romanées can be exotic, Pommard is renowned for its four-squareness.

Roast chicken or (better) capon is a safe standard with red burgundy; guinea fowl for slightly stronger wines, then partridge, grouse or woodcock for those progressively more rich, pungent. Hare and venison (chevreuil) are alternatives.

great old burgundy The Burgundian formula is cheese: Époisses (unfermented); a fine cheese but a terrible waste of fine old wines. *See* above.

vigorous younger burgundy Duck or goose roasted to minimize fat. Or faisinjan (pheasant cooked in pomegranate juice). Coq au vin, or lightly smoked gammon.

Rioja Gran Reserva, Top Duero reds Richly flavoured roasts: wild boar, mutton, saddle of hare, whole suckling pig.

White wines

Beerenauslese / Trockenbeerenauslese Biscuits, peaches, greengages. But TBAs don't need or want food.

Condrieu, Château-Grillet, Hermitage Blanc V. light pasta scented with herbs, tiny peas or broad beans. Or v. mild tender ham. Old white Hermitage loves truffles.

Grand Cru Alsace Ries Truite au bleu, smoked salmon, or choucroute garni.

Pinot Gr Roast or grilled veal. Or truffle sandwich (slice a whole truffle, make a sandwich with salted butter and gd country bread – not sourdough or rye – wrap and refrigerate overnight. Then toast it in the oven. Thanks, Dom Weinbach).

Gewurztraminer Cheese soufflé (Münster cheese).

Vendange Tardive Foie gras or tarte tatin.

Old vintage Champagne (not Blanc de Blancs) As an apéritif, or with cold partridge, grouse, woodcock. The evolved flavours of old Champagne make it far easier to match with food than the tightness of young wine. Hot foie gras can be sensational. Don't be afraid of garlic or even Indian spices, but omit the chilli.

late-disgorged old wines have extra freshness plus tertiary flavours. Try with truffles, lobster, scallops, crab, sweetbreads, pork belly, roast veal, chicken.

Sauternes Simple crisp buttery biscuits (eg. langues de chat), white peaches, nectarines, strawberries (without cream). Not tropical fruit. Pan-seared foie gras. Lobster or chicken with Sauternes sauce. Ch d'Yquem recommends oysters (and indeed lobster). Experiment with blue cheeses. Rocquefort is classic, but needs one of the big Sauternes.

Tokaji Aszú (5–6 puttonyos) Foie gras recommended. Fruit desserts, cream desserts, even chocolate can be wonderful. Roquefort. It even works with some Chinese, though not with chilli – the spice has to be adjusted to meet the sweetness. Szechuan pepper in gd. Havana cigars are splendid. So is the naked sip.

Top Chablis White fish simply grilled or *meunière*. Dover sole, turbot, halibut are best; brill, drenched in butter, can be excellent. (Sea bass is too delicate; salmon passes but does little for the finest wine.)

Top white burgundy, top Graves, top aged Riesling Roast veal, farm chicken stuffed with truffles or herbs under the skin, or sweetbreads; richly sauced white fish (turbot for choice) or scallops, white fish as above. Lobster, wild salmon.

Vouvray moelleux, etc. Buttery biscuits, apples, apple tart.

Fail-safe face-savers

Some wines are more useful than others – more versatile, more forgiving. If you're choosing restaurant wine to please several people, or just stocking the cellar with basics, these are the wines: **Red** Alentejo, BARBERA d'Asti/d'Alba, BLAUFRÄNKISCH, Beaujolais, Chianti, GRENACHE/ GARNACHA if not overextracted/overoaked, young MALBEC (easy on the oak), PINOT N, Valpolicella. **White** Alsace PINOT BL, Entre-Deux-Mers, ASSYRTIKO, unoaked or v. lightly oaked CHARD, Fino Sherry, GRÜNER V, RIES from Alsace, Germany (dry, fruity), Sancerre, gd Soave, VERDICCHIO.

France

More heavily shaded areas are
the wine-growing regions.

Abbreviations used in the text:

Al	Alsace
Beauj	Beaujolais
Burg	Burgundy
B'x	Bordeaux
Cas	Castillon-Côtes de Bordeaux
Chab	Chablis
Champ	Champagne
Cors	Corsica
C d'O	Côte d'Or
Fron	Fronsac
L'doc	Languedoc
Lo	Loire
Mass C	Massif Central
Prov	Provence
N/S Rh	Northern/Southern Rhône
Rouss	Roussillon
Sav	Savoie
SW	Southwest
AC	appellation contrôlée
ch, chx	château(x)
dom, doms	domaine(s)

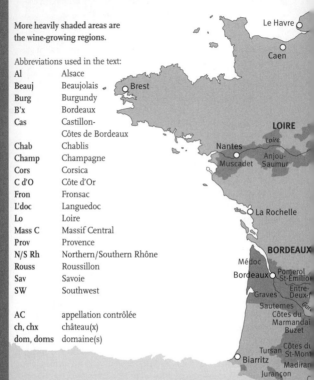

France has its work cut out. Does it make the best wines in the world? Sometimes, but so do other countries. Does it have the most adventurous, experimental winemakers in the world? No – but some of the world's most innovative winemakers just happen to be French.

France's reputation for making great wine, full stop, is almost an albatross round its neck: the country and its wine evolve faster than people's perception of them. A high proportion of the best-known wines (one might almost say a majority) sell purely on the value of the appellation: Chablis, Bordeaux, Burgundy, Champagne. People in search of bargains buy the cheapest bottles from these regions and wonder why what they get doesn't seem very special. If the reputation of France were like that of Spain or Italy they might be more adventurous and more prepared to do some research before they buy – and would get far nicer wines as a result.

In any famous region there will be some winemakers who coast along, not bothering much. France has a lot of famous regions. If you want great Chablis you will have to fork out for it. Burgundy prices have rocketed. So where can you find delicious wine for everyday, or at least

France entries also cross-reference to **Châteaux of Bordeaux**

Calais
Lille
Reims
CHAMPAGNE
Paris
Strasbourg
Marne
ALSACE
Orléans
Loire
Chablis
Pouilly-
Fumé
Dijon
Saône
Sancerre
raine
Côte d'Or
Cher
BURGUNDY
JURA
Côte Chalonnaise
Dordogne
Mâconnais
Geneva
Bugey
Beaujolais
Lyon
SAVOIE
N RHÔNE
Côte-Rôtie
Rhône
St-Joseph
Condrieu
Grenoble
Cornas
Hermitage
Crozes-Hermitage
S RHÔNE
Côtes du
Gigondas
Rhône-Villages
Beaumes-de-Venise
Châteauneuf-du-Pape
Cahors
Tarn
UTHWEST
Gaillac
Côtes du Frontonnais
LANGUEDOC
Nice
Minervois
St-Chinian
PROVENCE
ulouse
Montpellier
Corbières
Bandol
Marseille
JSSILLON
Fitou
Rivesaltes
Perpignan
Banyuls
erac
Dordogne

Bastia
CORSICA
Ajaccio

for Sundays? Every region has its own secrets. In Burgundy, the simple Bourgogne offered by top growers. In Bordeaux, from modest but proud châteaux. Champagne has bargains; Alsace has plenty of them. The South is full of gems – and this book aims to be your guide to finding them. Some might be hard to track down, so refer to www.winesearcher. com to find stockists. And enjoy the hunt: there is no wine as enjoyable as a treasure you had never heard of before.

Recent vintages of the French classics

Red Bordeaux

Médoc / Red Graves For many wines, bottle-age is optional: for these it is indispensable. Minor chx from light vintages may need only 1 or 2 yrs these days, but even modest wines of gd yrs can improve for 10 or so, and the great chx of these yrs can profit from double that time.

2019 Dark, concentrated Cab Sauv. Plentiful tannins. High alc but low pH, so balance. Potential to age. Low yields due to difficult flowering.

2018 Pure, aromatically intense Cab Sauv. Rich, powerful (alcs high) but balance there. Long-term potential, but yields uneven due to mildew, hail.

2017 Attractive wines: gd balance, fairly early-drinking. Volumes often small.

2016 Cab Sauv with colour, depth, structure. Vintage to look forward to.

2015 Excellent Cab Sauv yr, but not structure of 05 10. Some variation. Keep.

2014 Cab Sauv bright, resonant. Gd to v.gd; classic style, beginning to open.

2013 Worst since 92. Patchy at Classed Growth level. For early-drinking.

2012 Difficulties ripening Cab Sauv but early-drinking charm. Don't reject.

2011 Mixed quality, better than its reputation. Classic freshness, moderate alc. After an awkward phase, opening up.

2010 Outstanding. Magnificent Cab Sauv, deep-coloured, concentrated, firm. Keep for yrs.

2009 Outstanding yr. Structured wines, exuberant fruit. Accessible or keep.

Fine vintages: 08 06 05 00 98 96 95 90 89 88 86 85 82 75 70 66 62 61 59 55 53 49 48 47 45 29 28.

St-Émilion / Pomerol

2019 Drought hit. Merlot excellent on limestone, clay soils. Suffered on sandier reaches.

2018 Powerful but pure. Best from limestone, clay soils. Mildew affected yields.

2017 Gd balance, classic fruit-cake flavours. Will be quite early-drinking.

2016 Conditions as Méd. Some young vines suffered drought, overall excellent.

2015 Great yr for Merlot. Perfect conditions. Colour, concentration, balance.

2014 More rain than the Méd so Merlot variable. V.gd Cab Fr. Drinking now.

2013 Difficult flowering (so tiny crop), rot in Merlot. Modest yr, early-drinking.

2012 Conditions as Méd. Merlot marginally more successful. Drinking now.

2011 Complicated, as Méd. Gd Cab Fr. Pomerol best overall? Don't shun it.

2010 Outstanding. Powerful wines, high alc. Concentrated. Time in hand.

2009 Again, outstanding. Powerful wines (high alc) but seemingly balanced. Hail in St-Ém cut production at certain estates.

Fine vintages: 05 01 00 98 95 90 89 88 85 82 71 70 67 66 64 61 59 53 52 49 47 45.

Red Burgundy

Côte d'Or Côte de Beaune reds generally mature sooner than grander wines of Côte de Nuits. Earliest drinking dates are for lighter commune wines, eg. Volnay, Beaune; latest for GCs, eg. Chambertin, Musigny. Even the best burgundies are more attractive young than equivalent red B'x. It can seem magical when they really blossom yrs later.

2019 Another hot dry summer but with much smaller yields. Producers divided in preferring 18 or 19. Best will age endlessly.

2018 Set to be a legend but watch out for some overripe or flawed wines; those apart, many brilliant sumptuous reds, should drink well young or old.

2017 Survived frosts! Big crop, attractive wines, mostly ripe enough, stylish lighter wines, slight preference for Côte de Nuits.

2016 Short, frosted crop but some spectacular reds with great energy, fresh acidity. Keep them locked away, though.

2015 Dense, concentrated wines, as 05, but with additional juiciness of 10. Earning stellar reputation; don't miss. Lesser appellations attractive now.

2014 Beaune, Volnay, Pommard hailed once again. Attractive fresh reds, medium density, lovely fragrance. Take a look soon.

2013 Destined to be overlooked after the succession of fine vintages to follow. Côte de Beaune tricky after hail but some stars in Cote de Nuits, especially Grands Crus. Lesser wines ready.

2012 Côte de Beaune gd where not hail-damaged. Fine quality if small crop in Côte de Nuits, exuberant yet classy. Time to take a look except at top end.

2011 Some parallels with 07. Early harvest, lighter wines, now accessible. Start bringing them out of the cellar.

2010 Turning into great classic: pure, fine-boned yet also with impressive density. Village wines and some Premiers Crus open for business.

2009 Powerful hot vintage, can be awkward now in adolescence, some are flawed. But best will keep for another decade or two and should be left to mature.

Fine vintages: 05 03 02 99 96 (drink or keep) 95 93 90 88 85 78 71 69 66 64 62 61 59 (mature).

White burgundy

Côte d'Or White wines are now rarely made for ageing as long as they were in the past, but top wines should still improve for 10 yrs or more. Most Mâconnais and Chalonnais (St-Véran, Mâcon-Villages, Montagny) usually best drunk early (2-3 yrs).

2019 Small crop, may be high-octane but super-concentrated, could be legendary.

2018 Huge crop ripened well, wines more potential than 1st thought. Should be delicious from start yet with power to age.

2017 Turning out really well, decent crop of ripe, balanced, consistent wines with enough acidity. Best since 14.

2016 Small, frosted crops, inconsistent results. Undamaged v'yds did well, others lacking balance for long term.

2015 Rich, concentrated; warm, dry summer. Most picked early have done well, later wines may be too heavy. Similar to 09 but more successes.

2014 Finest, most consistent vintage for generation. White fruit flavours, elegant, balanced. Start drinking, but make sure to keep best for later.

2013 Tricky, often flabby. Drink soon, be careful.

2012 Not much made, but lively, concentrated. Worth a look now if you can find them, but no hurry.

2011 Fine potential for conscientious producers, some flesh, gd balance. Attractive early, start drinking up.

2010 Exciting; now developing exotic aromatics. Safest to drink soon.

Fine vintages (all ready): 05 02 99 96 93 92 85 79 73 59.

Chablis

GC Chablis of vintages with both strength and acidity really need at least 5 yrs, can age superbly for 15 or more; PCs proportionately less, but give them 3 yrs at least. Then, for the full effect, serve them at cellar temperature, not iced, and decant them. Yes, really.

2019 V. small crop, will be concentrated, too early to say how typical of Chab.

2018 Growers say vintage of century: wines are gd plus huge volumes. But actually 17 is even better.

2017 Despite frost again, classical style, much more exciting than 16. Gaining in stature, keep if you can.

2016 Hail, frost, more hail, hardly any wine, often unbalanced. Move along.

2015 Hot, dry summer; rich, ripe wines with concentrated core. Keep GCs.

2014 Excellent early crop, ideal balance, saline notes. Drink Chab, consider PCs.

Fine vintages: 12 10 08 02 00.

Beaujolais

19 Hot dry summer after early frost, should be gd. 18 Hot summer, another one of the big beasts, prefer to 15. 17 Large crop but hideous hail, esp Fleurie, Moulin-à-Vent. 16 Large crop of juicy wines, unless hailed (Fleurie). 15 Massive wines, best outstanding, others alc monsters. 14 Fine crop, enjoyable now. 13 Late vintage with mixed results.

Southwest France

2019 Spring storms brought havoc, esp Bergerac. Heatwave summer rescued what was left. Small crop, fine quality.

2018 Big yr. Dry white, rosés mostly ready 2020. Most reds (esp oaked) and sweet whites 2021 on. Heftier styles will need longer.

2017 Subtler yr than most. Excellent drinking 2021. Madiran, Cahors will keep.

2016 Satisfactory rather than brilliant; won't last much longer except big reds.

2015 Marvellous full, fruity yr. Sweet whites, big reds will keep for many yrs.

2014 Well-balanced sweet white; rest should be drunk now. Oaked styles will last a bit longer.

The Midi

2019 Hot dry summer; some sunburn. Severe hail in Pic St-Loup. Gd to v.gd, but small vintage.

2018 Mildew, but best are vibrant (r w). Pic St-Loup: powerful, rich, v.gd reds.

2017 Small, v.gd: Rouss possibly better than 15. Cabardes (r) balanced, will age, (w) acidity, fruit.

2016 Generally excellent. Reds balanced, will age; whites fresh, fruity.

2015 Reds: ripe fruit, elegant tannins. Some whites v.gd, some low acidity.

Northern Rhône

Depending on site and style, can be as long-lived as burgundies. White Hermitage can age as red. Hard to generalize, but don't be in a hurry (r or w).

2019 Tremendous reds; pleasing, tannins well absorbed, like 16 with more stuffing, acidity slightly better than 18. Whites concentrated, sunswept.

2018 V.gd; scaled up reds. Yr for top terroirs, which beat the climate. V. rich Hermitage (r), similar Côte-Rôtie. Crozes variable. Whites v. successful: depth, surprising freshness. NB Condrieu, Hermitage, St-Péray.

2017 V.gd, esp Côte-Rôtie. Full reds, deeper than 16, show sunshine, packed-in, firm tannins, so need time. Whites for hearty food, Condrieu variable.

2016 Gd–v.gd, reds harmonious, classic at Côte-Rôtie. NB Cornas, Crozes-Hermitage reds. Marvellous Hermitage whites, other whites gd, clean.

2015 Excellent, v. deep reds everywhere, lots of tannin, wonderful, v. long-lived Hermitage, Côte-Rôtie. Full whites, can be heady.

2014 Juicy reds, gained depth over time. Excellent whites: content, style, fresh.

2013 V.gd reds, with tight body, crisp tannin, freshness. Best still closed in 2020. 20–25 yrs life. Exceptional whites (Hermitage, St-Joseph, St-Péray).

2012 V.gd Hermitage, open-book Côte-Rôtie. Fresh reds come together well, will last 15 yrs+. Whites have style, freshness.

2011 Gd Hermitage, Cornas. Côte-Rôtie more body recently. Whites satisfactory.

2010 Wonderful. Reds: marvellous depth, balance, freshness. Long-lived. Côte-Rôtie as gd as 78. V.gd Condrieu, rich whites elsewhere.

Southern Rhône

2019 Reds marked by powerful Grenache (high alc). Ripe, thick tannins: some will find too much. Top names will demand patience. Whites full-bodied.

2018 Mixed, can be gd: cherry-pick. Shocking crop loss in Châteauneuf, esp organic. Note Valréas, Visan, Vinsobres: higher, later v'yds. V. full whites.

2017 V.gd, but can be variable. Rich, bold reds, tannins can be chewy from drought. Most strict top doms did best. Gd Rasteau. Full s whites.

2016 Wonder yr, excellent for all (Châteauneuf, old-vines Grenache triumph). Sensuous reds: fruit bonanza. Sun-filled whites; keep some to mature.

2015 V.gd: rich, dark, lots of body, firm tannins, often enticing flair. Quality high across board (Gigondas). V.gd, full whites.

2014 Gd in places, aromatic finesse returns to Châteauneuf. Stick to best names. NB Gigondas, Rasteau, Cairanne. Fresh whites.

2013 Tiny crop. Sparky, vibrant, slow-burn reds, v. low Grenache yields: atypical wines. Châteauneuf best from old vines. V.gd, long-lived whites.

2012 Dashing reds, lively, gd tannins. Open-book vintage. Food-friendly whites.

2011 Sunny, supple, can be fat, drink quite soon. Alc issue. Decent whites.

2010 Outstanding. Full-bodied, balanced, reds Tiptop Châteauneuf. Clear-fruited, well-packed tannins, long life ahead. Whites deep, long.

Champagne

2019 Spring frost reduced volume by c.20% (not as bad as Burg). Extreme heat spike (43°C/109°F) in July; showers in August. Blue chips happy.

2018 Record hot summer, looks great for Pinot N, esp n Montagne. Exceptional for Meunier too. Chard mixed, some lack acidity, definition.

2017 Difficult for Pinots N, M. Chards show surprising class, esp Charles Heidsieck, Krug Clos du Mesnil, bijou Lilbert.

2016 Purity of fruit, elegance: best for Pinot N. Underrated, classy, shrewd buy.

2015 Great Pinot N, refined M, some overripe Chard, others sumptuous.

2014 Chard best; ripe, classic as in Burg. Some fine vintage Blanc de Blancs.

2013 Potentially brilliant Côte des Blancs Chard. Pinot N patchy: glorious in Äy.

2012 Exquisite Pinot N, best since 52, fine Pinot M, Chard stolid, some lack verve. Stalwart NV base yr.

Fine vintages: 08 06 04 02' 00 98 96' 95 92 90 89 88' 82 76.

The Loire

2019 An unprecedented 6th v.gd yr in a row. But volume down, esp Muscadet and Anjou.

2018 Exceptional quality, esp reds. Gd sweet, less enthusiasm for dry whites. Some comparing with 47.

2017 Gd quality, April frosts again. Bourgueil, Menetou-Salon, Sancerre spared.

2016 Low quantity (April frosts), but quality gd for those who still had a crop. Sancerre spared.

2015 Gd–v.gd across range. Fine sweet wines in Layon and L'Aubance.

Alsace

2019 Another hot summer. E-facing hillsides, esp GCs, may be v.gd, ripe, dry (Ries, Pinot Gr, Gewurz). Wines from plains could be a problem.

2018 Warm, but fresh wines. Gewurz, Pinot Gr, Ries tops in high-altitude GCs.

2017 One of best since World War Two, but small crop. Recalls 71 08.

2016 Poised, finely balanced vintage and plenty of it, unlike 13 14 15.

2015 Rich vintage, one of driest ever. Great Pinot Gr; little volume but ripe, still fresh Sylvaner.

2014 Gewurz, Pinot Gr attacked by Suzuki fruit fly. Ries and Sylvaner are the winners.

2013 Classic vintage, opening up. Complex Ries.

> **AOP and IGP: what's happening in France**
> The Europe-wide introduction of AOP (Appellation d'Origine Protegée)
> and IGP (Indication Géographique Protegée) means that these terms
> may now appear on labels. AC/AOC will continue to be used, but for
> simplicity and brevity this book now uses IGP for all former VDP.

Abymes Sav w ★→★★ Hilly limestone zone s Chambéry, next to APREMONT. DYA Vin
de SAV AC (1973) cru (261 ha). Jacquère grape Try: 13 Lunes, A et M Quenard, des
Anges, Labbé, Perrier, Ravier.

AC or AOC (Appellation Contrôlée) / AOP Government control of origin and
production (but not quality) of most top French wines; around 45% of total. Now
being converted to AOP (Appellation d'Origine Protegée – which is much nearer
the truth than Contrôlée).

Ackerman Lo r p w (dr) (sw) sp ★→★★★ First SAUMUR sparkling (1811). Négociant/
estates: Celliers du Prieuré, Donatien-Bahuaud and Drouet Frères (Pays
Nantais), Hardières (Layon), Monmousseau, Perruche, Rémy-Pannier, Varière,
CH de SANCERRE. Group now called Orchidées.

Agenais SW Fr r p w ★ DYA IGP of Lot-et-Garonne. Beautiful country but wines
mostly dull. A few DOMS (Boiron, Campet, Lou Gaillot) put local co-ops to shame.

Agrapart Champ ★★★★ Pascal A makes fine-drawn CHAMP from scrupulously
tended v'yds. Rigorous focus on terroir, purity.

Alain Chabanon, Dom L'doc ★★★ MONTPEYROUX producer stays on top with
Campredon, Esprit de Font Caude, MERLOT-based Merle aux Alouettes. Delicious
whites Le Petit Trélans, pure VERMENTINO, age-worthy Trélans Vermentino/CHENIN.

Allemand, Thierry N Rh r ★★★ 90' 91' 95' 99' 01' 05' 06' 07' 08' 09' 10' 12' 13' 15' 16'
17' 18' 19' Quirky owner, compelling CORNAS from organic 5-ha DOM, low sulphur.
Two v. deep, rocky wines. Top is Reynard (profound, complex; 20 yrs+), Chaillot
(bursting fruit, floral) drinks earlier.

Alliet, Philippe Lo r w ★★★ 09' 10' 11 14' 15' 16 17' 18' Excellent CHINON with wife
Claude and son Pierre. Tradition, VIEILLES VIGNES and two steep s-facing v'yds e
of Chinon: L'Huisserie and Coteau de Noiré. Wines need time. Promising 19,
small crop.

Aloxe-Corton Burg r w ★★→★★★ 05' 09' 10' 12' 15' 17 18' 19' N end of CÔTE DE BEAUNE,
famous for GC CORTON, CORTON-CHARLEMAGNE, but less interesting at village or PC
level. Reds attractive if not overextracted. Recent warm vintages have helped.
Best DOMS: Follin-Arbelet, Rapet, Senard, TOLLOT-BEAUT.

Alquier, Jean-Michel L'doc r w ★★★ Stellar FAUGÈRES producer. MARSANNE/ROUSSANNE/
GRENACHE Des VIGNES au Puits; SAUV BL Pierres Blanches. Reds: SYRAH benefitting
from 340m (1115ft) altitude and schist for finesse; Les Bastides (old higher-
altitude Syrah) and Syrah-based Les Grandes Bastides d'Alquier longer in new
oak. Les Premières (younger vines); Maison Jaune, only in gd Grenache yrs.

Alsace (r) w (sw) (sp) ★★→★★★★ Sheltered e slope of Vosges and 1800 sun hrs
make France's Rhine wines: aromatic, fruity, full-strength, drier styles back in
vogue. Most sold by variety but rich diversity of terroirs (incl granite, sandstone,
limestone) is emerging, esp in wines for ageing, eg. RIES up to 20 yrs for GC.
PINOT BL *reaching high levels* (esp **16** 18' 19). *See* VENDANGE TARDIVE, SÉLECTION DES
GRAINS NOBLES.

Alsace Grand Cru Al ★★★→★★★★ 06 07 08' 10 12 13 14 (esp RIES) 15' 17' 18' 19 AC.
Restricted to 51 of best-named v'yds (approx 1600 ha, 800 in production) and
four noble grapes (PINOT GR, Ries, GEWURZ, MUSCAT). GC rules require greater min
ripeness. Local management specifies extra rules for each cru. Plans afoot for
PC in gd sites. Great houses (eg. LÉON BEYER, TRIMBACH) now embrace GC status.

Amiel, Mas Rouss r w sw ★★★ Leading MAURY, CÔTES DU ROUSS, IGP. Look for Vol de

Nuit from v. old CARIGNAN/GRENACHE Vers le Nord, Origine, Altaïr (w), others. Plus young *grenat* (Maury version of RIMAGE), venerable RANCIO VDN 20- to 40-yr-old Maury aged in 60-litre glass demijohns, intensely sweet and savoury.

Amirault, Yannick Lo ★★★→★★★★ 10' 11 12 14' 15' 16' 17' 18' 19' Organic. Top BOURGUEIL/ ST-NICOLAS-DE-BOURGUEIL. Best: La Mine (St-Nicolas), La Petite CAVE, Le Grand Clos, Les Quartiers (Bourgueil). 19 high quality.

André Jacquart Champ ★★★ Marie Doyard has 24 ha incl 18 ha in GC Le MESNIL. Flagship is Mesnil Experience NV; Vintage 13 15 18 19'; v.gd BLANC DE BLANCS NV.

Angerville, Marquis d' C d'O r w ★★★★ Bio superstar in VOLNAY, not just classy but classical too, esp legendary CLOS des Ducs (MONOPOLE). Enjoy Champans, Taillepieds as well. *See also* DOM DU PÉLICAN for Jura interests.

Anglès, Ch d' L'doc ★★★ LA CLAPE estate renovated by Eric Fabre, ex-technical director of CH LAFITE, captivated by MOURVÈDRE. Unoaked Classique (r p w), age-worthy oaked *grand vin* (r w).

Anjou Lo r p w (dr) (sw) (sp) ★→★★★★ Region and AC: ANJOU, SAUMUR. CHENIN BL dry whites from quaffers to complex: big move to dry whites in Layon, even in QUARTS DE CHAUME; juicy reds, incl GAMAY; fruity CAB FR-based Anjou Rouge; robust tannic ANJOU-VILLAGES, incl CAB SAUV. Mainly dry SAVENNIÈRES; lightly sweet to rich COTEAUX DU LAYON Chenin Bl; rosé (dr s/sw), esp CABERNET D'ANJOU, sparkling mainly CRÉMANT. Natural wines often VDF. April 19 badly frosted.

Anjou-Coteaux de la Loire Lo w sw s/sw ★★→★★★ 16 17 18' Tiny (15 producers, 30 ha) AOP for sweet CHENIN BL (w) of Angers; racier than COTEAUX DU LAYON. Esp CH de Putille, Delaunay, Fresche, Musset-Roullier (v.gd).

Anjou-Villages Lo r ★→★★★ 10' 14' 15' 16' 17' 18' (19) Structured red AC (CAB FR/ CAB SAUV, a few pure Cab Sauv). Can be tannic; needs bottle-age. Top wines gd value, esp Bergerie, Branchereau, Brizé, CADY, CH PIERRE-BISE, CLOS de Coulaine, Delesvaux, Ogereau, Sauveroy, Soucherie. Sub-AC Anjou-Villages-Brissac same zone as COTEAUX DE L'AUBANCE; look for Bablut, CH de Varière (part of ACKERMAN), Haute Perche, Montigilet, Princé, Richou, Rochelles. 19 badly frosted.

Aphillanthes, Dom Les S Rh r ★★→★★★ 16' 17 18' 19' Organic, bio DOM in PLAN DE DIEU. Two CUVÉES: des Galets, VIEILLES VIGNES (50S GRENACHE). RASTEAU 1921 (r); CÔTES DU RH-VILLAGES 3 CÉPAGES (r); CÔTES DU RH (r).

Apremont Sav w ★★ Largest cru of SAV (378 ha, 20.6% volume) just s of Chambéry. Fresh whites of Jacquère on limestone. Keep up to 5 yrs. Try: 13 Lunes, Dacquin, Giachino, Masson, Perrier, Ravier, Richel, Viallet.

Arbin Sav r ★★ SAV cru 39 ha. Dark, spicy MONDEUSE for aprés-skl. Drink to 8 yrs+. Try: l'Idylle, Magnin, Mérande, P Grisard, Quenard, Tosset.

Arbois Jura r p w (sp) ★★→★★★ 10' 12 14' 15' 16 18' 19 AC of n Jura, great spot for wine, cheese, walking and Louis Pasteur museum. CHARD, and/or SAVAGNIN whites, VIN JAUNE, reds from Poulsard, Trousseau or PINOT N. Try terroir-true Stephane TISSOT, fresh *ouillé* styles from DOM DU PÉLICAN, oxidative *(typés)* whites from Overnoy/ Houillon, plus all-rounders AVIET, Pinte, Renardières, Rolet.

Ardèche S Rh r p w ★→★★★ 17' 18 19' IGP Rocky hills w of Rh, dwindling selection; growers preferring VDF status. Quality up, often gd value. Fresh, clear reds, some oaked (unnecessary); VIOGNIER (eg. CHAPOUTIER, Mas de Libian), MARSANNE. Best from SYRAH, also GAMAY (often old vines), CAB SAUV (Serret). Restrained, burgundy-style Ardèche CHARD by LOUIS LATOUR (Grand Ardèche too oaky). CH de la Selve; DOMS de Vigier, du Grangeon, Flacher, JF Jacouton; Mas d'Intras (organic).

Ariège SW Fr r p w ★ 18 (19) Locally popular IGP s of Toulouse. Try Swiss-owned DOM Beau Regard. Gd SYRAH from Doms des Coteaux d'Engravies, Ribonnet.

Arjolle, Dom de l' L'doc ★★★ Large CÔTES DE THONGUE family-run estate. Range incl Equilibre, Equinoxe, Paradoxe, varietals and blends, and two original VDF: Z for ZIN, K for CARMENÈRE.

Arlaud C d'O r ★★★→★★★★ Leading MOREY-ST-DENIS estate energized by Cyprien A and siblings. Beautifully poised, modern wines with depth and class from exceptional BOURGOGNE Roncevie up to GCS. Fine range of Morey PCS, esp Ruchots.

Arlay, Ch d' Jura r p w sw ★★★ Historic producer with imposing CH, 25 ha v'yds, best wine VIN JAUNE.

Arlot, Dom de l' C d'O r w ★★→★★★ AXA-owned estate, stylish fragrant Nuits in both colours, esp red CLOS des Forêts St-Georges. CÔTE de Nuits Villages for value, ROMANÉE-ST-VIVANT for glory.

Armand, Comte C d'O r ★★★★ Graceful Pommard? Try MONOPOLE CLOS des Epeneaux for ageless wine. Gd value from AUXEY, VOLNAY too.

Arnoux-Lachaux C d'O r ★★★★ New star in VOSNE-ROMANÉE. Charles Lachaux has changed style from dark and oaky to ethereal. Utterly brilliant from 18, will soon be priced to match.

Aube Champ S v'yds of CHAMP, aka Côte des Bar, known for v.gd PINOT N used by great Reims houses eg. VEUVE CLICQUOT, KRUG, LANSON.

Aupilhac, Dom d' L'doc ★★★ MONTPEYROUX. Sylvain Fadat cultivates s-facing old-vine MOURVÈDRE, CARIGNAN: old-vine Le Carignan (his 1st wine, 1989), as well as n-facing, higher-altitude SYRAH and whites in Les Cocalières.

Auxey-Duresses C d'O r w ★★→★★★ (r) 15′ 16 17 18′ 19′ (w) 14′ 15′ 17′ 18 19′ CÔTE DE BEAUNE village in valley behind MEURSAULT. Similar *whites offer value*, reds now ripen in most yrs. Best: (r) COCHE-DURY, COMTE ARMAND, d'Auvenay (Les Boutonniers), Gras, Paquet, Prunier; (w) Lafouge, LEROUX, Paquet, VINCENT.

Aveyron SW Fr r p w ★ IGP DYA. Est AOPS ENTRAYGUES, Estaing, MARCILLAC for crisp light reds. Also experimental ★★ Nicolas Carmrans Rols, DOM Lafon, L'DOC's Olivier Jullien. ★ Doms Bertau, Bias (PINOT N), Prat.

Aviet, Lucien Jura ★★ Fine ARBOIS grower, who is nicknamed Bacchus. Gd value, eg. attractive light Poulsard and tangy SAVAGNIN.

Avize Champ ★★★★ Côte des Blancs GC CHARD, home to finest growers AGRAPART, Bonville, SELOSSE; Huge co-op Union CHAMP provides base wines to biggest houses. Its *Pierre Vaudon* brand is worth seeking out.

Aÿ Champ Revered PINOT N village, home of BOLLINGER, DEUTZ. Mix of merchants and growers' wines, some in barrel (eg. Claude Giraud, master of Argonne oak), some in tanks. *Gosset-Brabant* Noirs d'Aÿ excels. Aÿ Rouge (AC Coteaux Champenois) now excellent in ripe yrs (esp 15′ 18′). Wines typically have weight and power.

Ayala Champ Reborn Aÿ house, owned by BOLLINGER. Fine BRUT Zéro, BLANC DE BLANCS ace 13; Perlé d'Ayala 08′ 12′ 15′ 16 17′. Precision, purity. Energy under Caroline Latrive, chef de CAVE.

Bachelet Burg r w ★★ ·★★★★ Widespread family name in S C D'O. Look out for: B-Monnot (esp PULIGNY and BÂTARD-MONTRACHET), Bernard B (Maranges), Jean-Claude B (CHASSAGNE and ST-AUBIN, etc.). And no relation to Denis B (great GEVREY-CHAMBERTIN).

Bandol Prov r p (w) ★★★ Prov's top reds, for ageing. MOURVÈDRE, plus GRENACHE, CINSAULT. Real personality; different styles from limestone or clay soil. Rosé now main production; can be v. pale, Prov-style, or gutsier from Mourvèdre. Some white: CLAIRETTE, UGNI BL, occasionally SAUV BL. Tops: DOMS de la Bégude, du Gros'Noré, La Bastide Blanche, Lafran Veyrolles, La Suffrène, Mas de la Rouvière, *Pibarnon*, Pradeaux, *Tempier*, Terrebrune, Vannières.

Banyuls Rouss r p w br sw ★★→★★★ Undervalued (magic word) VDN, oxidized or not, from GRENACHE of all colours. Young, fresh RIMAGE, but stars are RANCIOS, long-aged, pungent, intense. Try with chocolate. Best: ★★★ DOMS du Mas Blanc, ★★★ la Rectorie, ★★★ la Tour Vieille, Les Clos de Paulilles, ★★ Coume del Mas, Madeloc, Vial Magnères. *See also* MAURY.

Baronne, Ch La L'doc ★★★ Bio CORRIÈRES, min intervention, amphorae, concrete eggs; presses all the buttons. Alaric, Les Chemins, Les Lanes and (CARIGNAN planted 1892) Pièce de Roche. IGP Hauterive. Sulphur-free Les Chemins de Traverse, and VDF VERMENTINO/Grenache Gris (w).

Barral, Dom Léon Rouss r w FAUGÈRES bio producer. Concentrated MOURVÈDRE, SYRAH, CARIGNAN. Valinière and Jadis top blends. Terret Blanc main white.

Barrique B'X (and Cognac) term for oak barrel holding 225 litres. Used globally, but global mania for excessive new oak now thankfully fading. Average price €750/barrel.

Barsac Saut w sw ★★→★★★★ 01' 05 09' 11' 15' 16' 18 Neighbour of SAUT with v. similar botrytized wines from lower-lying limestone soil; fresher, less powerful. 17 frost; 19 tricky. Top: CAILLOU, *Climens*, COUTET, DOISY-DAËNE, *Doisy-Védrines*, NAIRAC.

Barthod, Ghislaine C d'O r ★★★→★★★★ A reason to fall in love with CHAMBOLLE-MUSIGNY, if you haven't already Wines of perfume, delicacy yet depth, concentration. Unbeatable range of 11 different PCS, incl Charmes, Cras, Fuées, Les Baudes. No reason to expect change now that son Clément is taking over.

Bâtard-Montrachet C d'O w ★★★★ 04' 07' 08' 09' 10 12 14' 15 17' 18 19' 12-ha GC downslope from LE MONTRACHET. Grand, hefty wines that should need time; more power than neighbours BIENVENUES-B-M and CRIOTS B-M. Seek out: BACHELET-Monnot, BOILLOT (both H and JM), CARILLON, FAIVELEY, GAGNARD, LATOUR, LEFLAIVE, LEROUX, MOREY, OLIVIER LEFLAIVE, Pernot, Ramonet, SAUZET, VOUGERAIE.

Alsace soils: Ries likes granite, schist, volcanic; Gewurz limestone; Pinot Gr clay.

Baudry, Dom Bernard Lo r p w ★★→★★★ 14' 15' 16' 17' 18' 19' Mixed soils: sand, gravel, limestone. Fine CHINONS, from CHENIN BL whites to CAB FR (r p); drink Les Granges early; CLOS Guillot, Croix Boissée, Les Grézeaux. Organic. Top wines age 20 yrs+.

Baudry-Dutour Lo r p w (sp) ★★→★★★ 16' 17' 18' 19' CHINON's largest producer. Run by J-M Dutour and Christophe Baudry. Big, reliable range: light, early-drinking to age-worthy (r w). CLIX de St Louans (r w), La Grille, La Perrière, La Roncée, l'ainsi fait (no sulphur), 3 Coteaux (w) IGP SAUV BL.

Baumard, Dom des Lo r p w sw sp ★★→★★★ 15' 16' 17' 18' 19' 40 ha ANJOU dom, esp CHENIN BL whites, incl SAVENNIÈRES (CLOS St Yves, Clos du Papillon), Clos Ste Catherine. As cryoextraction now banned in QUARTS DE CHAUME will less sweet be made?

Baux-en-Provence, Les Prov r p w ★★→★★★ Tourist-trap village, super-pretty; some wines coasting. AC stipulates organic. White: CLAIRETTE, GRENACHE BL, Rolle, ROUSSANNE. Red: CAB SAUV, SYRAH, GRENACHE. TRÉVALLON still best. Others: Dalmeran, d'Estoublon, DOM Hauvette, Gourgonnier, Lauzieres, Mas de Carita, Mas de la Dame, Mas Ste Berthe, Romanin, Terres Blanches, Valdition, atypical Milan.

Béarn SW Fr r p w AOP (r) ★→★★ 17 18 (19) (p w) DYA. Reds from ★ DOMS de la Callabère, Lapeyre/Guilhémas. Whites from local Ruffiat de Moncade grape. Pinks from ★ Bellocq co-op all the rage in summer.

Beaucastel, Ch de S Rh r w ★★★★ 90' 95' 99' 05' 06' 07' 09' 10' 12' 13' 15' 16' 17' 18 19' Large, organic CHÂTEAUNEUF estate: old MOURVÈDRE, 100-yr-old ROUSSANNE. Dark-fruited, recently more polished wines, drink at 2 yrs or from 7–8. Intense, brilliant, top-quality 60% Mourvèdre Hommage à Jacques Perrin (r). **Wonderful, complex old-vine Roussanne**: enjoy over 5–25 yrs. Genuine, deep own-vines CÔTES DU RH Coudoulet de Beaucastel (r) has flair, lives to yrs+. Famille Perrin GIGONDAS (v.gd), RASTEAU, VINSOBRES (best) all gd, authentic. Note organic Perrin Nature Côtes du Rh (r w). Growing N Rh merchant venture, Maison Les Alexandrins (elegant). (*See also* Tablas Creek, California.)

Beaujolais r (p) (w) ★ DYA. Basic appellation of huge Beauj region. A love-or-hate wine: love for its simple fresh fruit, best from growers in hills. Dislike overcropped industrial examples. Can now be sold as COTEAUX BOURGUIGNONS.

Beaujolais Primeur / Nouveau Beauj More of an event than a drink. The BEAUJ of the new vintage, hurriedly made for release at midnight on the 3rd Wednesday in Nov. Enjoy juicy fruit but don't let it put you off real thing.

Beaujolais-Villages Beauj r ★★ 15' 17 18' 19' Next best v'yds after the ten named crus, eg. MOULIN-À-VENT. May specify best village, eg. Lantigné. Try CH de Basty, Ch des Vergers, F Berne, F Forest, JM BURGAUD, N Chemarinn.

Beaumes-de-Venise S Rh r (p) (w) br ★★ (r) 09' 10' 12' 13' 15' 16' 17' 18 19' (MUSCAT) DYA. Village nr GIGONDAS, mix high and plain v'yds, popular for VDN Muscat apéritif/dessert. Serve v. cold: grapey, musky, honeyed, eg. DOMS Beaumalric, Bernardins (complex, traditional), Coyeux, Durban (rich, long life), Fenouillet (brisk), JABOULET, Perséphone (stylish), Pigeade (fresh, v.gd), VIDAL-FLEURY, co-op Rhonéa. Also robust, grainy reds. CH Redortier, DOMS Cassan, Durban, de Fenouillet, la Ferme St-Martin (organic), les Baies Gouts, Mathiflo, St-Amant (gd w). Leave for 2–3 yrs. Simple whites (some dry Muscat, VIOGNIER).

Beaumont des Crayères Champ Co-op making model PINOT M-based Grande Rés NV. Vintage Fleur de Prestige top value 12' 13 14 15' 18'. Great CHARD-led CUVÉE Nostalgie 04' 08 10 13. New Fleur de Meunier BRUT NATURE 12' 15 18' ★★★. More Chard planned.

Beaune C d'O r (w) ★★★ 05' 09' 10' 11 12 14 15' 16 17 18' 19' Centre of Burgundy wine trade, classic merchants: BOUCHARD, Champy, CHANSON, DROUHIN, JADOT, LATOUR, Remoissenet and young pretenders Lemoine, LEROUX, Roche de Bellène; top DOMS Croix, DE MONTILLE, Dominique LAFON, LAFARGE, plus iconic HOSPICES DE BEAUNE. Graceful, perfumed PC reds offer value, eg. Bressandes, Cras, VIGNES Franches; more power from Grèves. Try Aigrots, CLOS St-Landry, and esp *Clos des Mouches (Drouhin)* for whites.

Beauregard, Ch de Burg (r) w ★★→★★★ Top POUILLY-FUISSÉ producer run by Frédéric Burrier. Wide range of profound, age-worthy single-v'yd wines, esp Vers Cras. Fine BEAUJ crus too.

Becker, Caves J Al r w ★★→★★★ Bio estate. Stylish wines, incl poised, taut GC Froehn in Zellenberg, prime RIES country. Exceptional 17' contrast with Ries GC Mandelberg **16** 17'. Excellent ripe *Sylvaner* 16 18'.

Belargus, Dom Lo ★★★ Ambitious venture by Parisian financier and CHENIN-lover Ivan Massonnat, bought Pithon-Paillé in QUARTS DE CHAUME. Mainly dry Chenin here, plus SAVENNIÈRES. Hit by April 19 frost.

Bellet Prov r p w ★★ Tiny (70 ha) AC within city of Nice. Gd Rolle white, age-worthy. Red from Folle Noire (DYA), Braquet for rosé. CH de Bellet best; also Ch de Cremat, Collet de Bovis, DOMS de la Source, de Toasc, Via Julia Augusta. CLOS St-Vincent v. highly regarded.

Bellivière, Dom de Lo r w sw (sp) ★★→★★★ 16' 17' 18' 19' Bio; v.gd JASNIÈRES, COTEAUX DU LOIR, peppery red Pineau d'Aunis. Les Arches de Bellivière (NÉGOCIANT).

Bergerac SW Fr r p w dr sw ★→★★★ 15' 16 17 18 (19) AOP Poor man's B'x. Huge variations in styles and quality. Best from ★★★ CLOS des Verdots, ★★★ *Tour des Gendres*, DOM de l'Ancienne Cure, ★★ Doms du Cantonnet, Fleur de *Thénac*, Jonc-Blanc, Julien Auroux. *See* sub-AOPs MONBAZILLAC, MONTRAVEL, PÉCHARMANT, ROSETTE, SAUSSIGNAC for some of best.

Berthet-Bondet, Jean Jura r w ★★ Biggest producer (still small) of CH-CHALON VIN JAUNE but reliably covers all bases for Jura red, white and CRÉMANT. CÔTES DU JURA tradition gd value.

Bertrand, Gérard L'doc r p w ★★ Ubiquitous grower-NÉGOCIANT, lots of bio. Flagship: CH l'Hospitalet (LA CLAPE) with top Hospitalis. Aspirational, expensive CLOS d'Ora

(MINERVOIS-La Livinière). Also zero-sulphur Prima Nature, Vallernajou (CORBIÈRES cru Boutenac), l'Aigle (LIMOUX), v.gd Sauvageonne rosé Terrasses du Larzac, Cabrières (new ultra-premium CLOS du Temple rosé), recently bought Ch de la Soujeole (Malepère).

Besserat de Bellefon Champ ★★ Épernay house specializing in gently sparkling CHAMP (old CRÉMANT style). Part of LANSON-BCC group. Respectable rising quality, always gd value, esp **13 14 15**, tiny but excellent 17'.

Beyer, Léon Al r w ★★→★★★ Top family AL house, arch-traditionalists delivering intense, dry gastronomic wines listed by many Michelin-starred restaurants. Best, and calling card, is lovely RIES Comtes d'Eguisheim 14 17'. With climate change, Beyer PINOT N is startlingly gd in poised in burgundian style, esp 16 18 19.

Bichot, Maison Albert Burg r w ★★ →★★★★ Major BEAUNE merchant/grower with DOMS as far afield as NUITS (CLOS Frantin), BEAUJOLAIS (Rochegres) and CHABLIS (LONG-DEPAQUIT). Most now BIO, all offer quality across range.

Bienvenues-Bâtard-Montrachet C d'O ★★★→★★★★ 04 07 08 09 10 12 14' 15 17' 18 19' Fractionally lighter version of BÂTARD, more initial grace, super-succulence. Best: BACHELET, CARILLON, FAIVELEY, LEFLAIVE, Ramonet, VOUGERAIE.

Billaud Chab w ★★★→★★★★ Difficult CHAB choice between DOM Billaud-Simon back on form under FAIVELEY ownership and Samuel B's sensational wine under his own label. Both brilliant.

Billecart-Salmon Champ ★★★★ New generation of family house. Quality irreproachable as ever, young chef de CAVE really on form in 200th Anniversary CUVÉE for long cellarage, to 2035. Cuvée Louis BLANC DE BLANCS is luscious in 06; classic, more complex, age-worthy in 07'. Superb CLOS St-Hilaire 98, ace 02'. NF Billecart perhaps greatest 02. Elisabeth Salmon Rosé 02 06'. Entry level 08 great value.

Bize, Simon C d'O r w ★★★ Chisa Bize has taken on mantle of late husband Patrick. Whole-bunch-style reds sensational, esp SAVIGNY Grands Liards and Guettes. Whites sound too.

Blagny C d'O r w ★★→★★★ 05' 09' 10' 12 14 15' 16' 17 18' 19' Hamlet on hillside above MEURSAULT and PULIGNY. Own AC for austere yet fragrant reds, diminishing volumes. Whites sold as Meursault-Blagny PC. Best v'yds: La Jeunelotte, Pièce Sous le Bois, Sous le Dos d'Ane. Best growers: (r) LEROUX, Matrot; (w) de Cherisey, JOBARD, LATOUR, Matrot.

Blanc de Blancs Any white wine made from white grapes only, esp CHAMP. Description of style, not quality.

Blanc de Noirs White (or slightly pink or "blush", or "gris") wine from red grapes, esp CHAMP: can be solid in style, but many now more refined; better PINOT N and new techniques.

Blanck, Paul & Fils Al r w ★★→★★★★ Old family of growers at Kientzheim. Now among best in Haut-Rhin. Finest from 6-ha GC Furstentum (RIES, GEWURZ, PINOT GR), sumptuous GC SCHLOSSBERG (great Ries 14 16 17'). Excellent Classique generics: tiptop, great value. Lovely MUSCAT 16 terroir-driven, chemical-free.

Blanquette de Limoux L'doc w sp ★★ Crisp, appley fizz, pretty, fun. 90% Mauzac plus CHARD, CHENIN BL. AC CRÉMANT de Limoux, more classic with Chard, Chenin Bl, PINOT N, and less Mauzac. Gd rosé too. Sieur d'Arques co-op is biggest, gd enough. Also Antech, Delmas, Laurens, RIVES-BLANQUES, Robert and several newcomers: DOMS Jo Riu, La Coume-Lumet, Les Hautes Terres, Monsieur S.

Blaye B'x r ★→★★ 11 12 14 15 16 17 (19) Designation for better reds (lower yields, higher v'yd density, longer maturation) from AC BLAYE-CÔTES DE B'X.

Blaye-Côtes de Bordeaux B'x r w ★→★★ 12 14 15 16' 17 (19) Mainly MERLOT-led red AC on right bank of Gironde. A little dry white (mainly SAUV BL). Best CHX: Bel Air la Royère, Bourdieu, Cantinot, des Tourtes, Gigault (CUVÉE Viva), Haut-Bertinerie,

Haut-Grelot, Jonqueyres, Monconseil-Gazin, Mondésir-Gazin, Montfollet, Peybonhomme Les Tours, Roland la Garde, Ste-Luce Bellevue. Also CAVE des Hauts de Gironde (Tutiac) CO-OP for whites.

Boeckel, Dom Al ★★★ VIGNERONS since 1600s; DOM started 1853, 23 ha, organic. RIES Wibbelsberg, CLOS Eugenie is rich, rounded 15 17' 18. Zotzenberg unique GC for top *Sylvaner*, Ries 15 16 19'. V.gd CRÉMANT.

Boillot C d'O r w Leading Burgundy family. Look for ★★★ Jean-Marc (POMMARD), esp fine, long-lived whites; ★★→★★★★ Henri (MEURSAULT), potent, stylish, both colours; ★★★ Louis (CHAMBOLLE) for great reds from both Côtes, and his brother Pierre (DOM Lucien B) ★★→★★★ (GEVREY). There may be more lurking...

Boisset, Jean-Claude Burg Ultra-successful merchant/grower group created over last 50 yrs. Boisset label from amazing new winery in NUITS and esp own v'yds *Dom de la Vougeraie* excellent. Recent additions to empire are (Burg) brands VINCENT GIRARDIN and Alex Gambal. Also projects in California (Gallo connection), Canada, Chile, Uruguay.

Boizel Champ ★★★ Exceptional value, rigorous quality, family-run. Well-aged BLANC DE BLANCS NV is a steal. CUVÉE Sous Bois shows expression without woodiness. Prestige Cuvée Joyau de France, esp Rosé 12 will drink sublimely in 2020.

Bollinger Champ ★★★★ Great classic house, ever-better quality, much fresher now. BRUT Special NV singing since 2012; RD 04; underrated Grande Année 07 08'. ★★★★ PINOT N-led, innovative Vintage Rosé 06, powerful, with high 30% of Côte aux Enfants rouge (14). VIEILLES VIGNES Françaises 2010. *See also* LANGLOIS-CH.

Bonneau du Martray, Dom C d'O (r) ★★★ (w) ★★★★ Reference producer for CORTON-CHARLEMAGNE, bought 2016 by Stanley Kroenke, owner of Screaming Eagle (California) and Arsenal FC (UK). Intense wines designed for long (c.10 yrs) ageing, glorious mix of intense fruit, underlying minerals. Small amount of fine red CORTON.

Bonnes-Mares C d'O r ★★★★ 90' 93 96' 99' 02' 05' 09' 10' 12' 15' 16' 18' 19' GC between CHAMBOLLE-MUSIGNY and MOREY-ST-DENIS with some of latter's wilder character. Sturdy, long-lived wines, less fragrant than MUSIGNY. Best: ARLAUD, Bart, Bernstein, BRUNO CLAIR, DE VOGÜÉ, Drouhin-Laroze, DUJAC, Groffier, JADOT, MORTET, ROUMIER, VOUGERAIE.

Bonnezeaux Lo w sw ★★★ →★★★★ 14 15' 16 17 18' 80 ha; 42 producers. Long-lived Lo sweet CHENIN BL from three sw-facing slopes (schist) in COTEAUX DU LAYON. Esp: CHX de Fesles, La Varière (Les Melleresses), DOMS de Mihoudy, Petit Val, Petite Croix, Les Grandes VIGNES. V.gd 18. Sadly difficult to sell.

Bordeaux r (p) w ★→★★ 18 (19) Catch-all AC for generic B'x (represents nearly half region's production). Most brands (*Dourthe*, Michel Lynch, MOUTON CADET, *Sichel*) are in this category. Co-ops a big source. *See* CHX Bauduc, Bonhoste, BONNET, Lamothe-Vincent, La Rame, Reignac, Tour de Mirambeau.

Bordeaux Supérieur B'x r ★→★★ 14 15 16 17 18 (19) Higher min alc, lower yield, longer ageing than the last. Mainly bottled at property. Consistent CHX: Camarsac, Fleur Haut Gaussens, Grand Village, Grée-Laroque, Jean Faux, Landereau, *Parenchère* (CUVÉE Raphaël), Penin (les Cailloux), *Pey la Tour* (Rés), Reignac, *Thieuley*, Turcaud (Cuvée Majeure).

Borie-Manoux B'x Admirable B'x shipper, CH-owner: BATAILLEY, BEAU-SITE, DOM DE L'EGLISE, LYNCH-MOUSSAS, TROTTEVIEILLE. Also owns NÉGOCIANT Mähler-Besse.

Bouchard Père & Fils Burg r w ★★→★★★★ Top BEAUNE merchant, quality v. sound, all-round, robust style. Whites best in MEURSAULT and GC, esp CHEVALIER-MONTRACHET. Flagship reds: Beaune VIGNE de L'Enfant Jésus, CORTON, *Volnay Caillerets Ancienne Cuvée Carnot*. Same group as WILLIAM FÈVRE (CHAB), CH de Poncié (BEAUJ).

Bouches-du-Rhône Prov r p w ★ IGP from Marseille environs. Simple, hopefully fruity, reds from s varieties, plus CAB SAUV, SYRAH, MERLOT.

Bourgeois, Henri Lo r p w ★★→★★★ 15′ 16′ 17′ 18′ 19′ Dynamic SANCERRE grower/merchant in Chavignol. Now rebranded as Famille Bourgeois with increased emphasis on wine tourism. Covers all Central Lo AOPs and Petit Bourgeois (IGP). Best: Etienne Henri, Jadis, La Bourgeoise (r w), MD de Bourgeois, Sancerre d'Antan. Top wines v. age-worthy. Also CLOS HENRI in Marlborough, NZ.

Bourgogne Burg r (p) w ★→★★ (r) 15′ 17 18′ 19′ (w) 14′ 17′ 18 19′ Ground-floor AC for Burg, ranging from mass-produced to bargain beauties. Sometimes comes with subregion attached, eg. CÔTE CHALONNAISE, HAUTES-CÔTES or latest addition, C D'O. Whites from CHARD unless B ALIGOTÉ. Reds from PINOT N unless declassified BEAUJ crus (sold as Bourgogne GAMAY) or B Passetoutgrains (Pinot/Gamay mix, must have 30%+ of former).

Bourgueil Lo r (p) ★★→★★★ 14′ 15′ 16 17′ 18′ 19′ Full-bodied TOURAINE reds and rosés mainly CAB FR. Gd vintages age 50 yrs+. AMIRAULT, Ansodelles, Audebert, Chevalerie, Courant, de la Butte, Gambier, Lamé Delisle Boucard, Ménard, Minière, Nau Frères, Omasson, Revillot, Rochouard. Excellent 18, promising 19 (small crop: frost/drought).

Bouscassé, Dom SW Fr r w ★★★ 14 15′ (17) (18) MADIRAN Palace. BRUMONT's Napa-Valley-style home. Reds a shade quicker to mature than oaky flagship Montus. ★★★ Petit Courbu-based dry PACHERENC best of its kind.

Bouvet-Ladubay Lo (r) p w sp ★★→★★★ SAUMUR sparkling CRÉMANT DE LO Run by Patrice Monmousseau (president), daughter Juliette (CEO). Big range incl Saphir, Trésor (p w), Taille Princesse Blanc de Gérard Depardieu SAUMUR-CHAMPIGNY Les Nonpareils best.

Bouzereau C d'O r w ★★→★★★ The B family infest MEURSAULT, in a gd way. DOM Michel B is leader but try also Jean-Marie B, Philippe B (CH de Citeaux), Vincent B or B-Gruère & Filles for gd-value whites.

Bouzeron Burg (r) w ★★ 17′ 18 19 CÔTE CHALONNAISE village with unique AC for ALIGOTÉ, esp golden version. Stricter rules and greater potential than straight BOURGOGNE Aligoté. BOUCHARD PÈRE, Briday, FAIVELEY, Jacqueson gd; *A & P de Villaine* outstanding. Red potential too.

Bouzy Rouge Champ r ★★★ 09 12 15′ 18′ Still red of famous PINOT N village. Formerly like v. light burg, now with more intensity (climate change, better viticulture), also refinement. VEUVE CLICQUOT, Colin and Paul Bara best producers.

Boxler, Albert Al ★★★★ Compact DOM, 13.5 ha, classic AL, as complex as great burgundy. Artisan precepts, brilliant RIES GC Sommerberg 14 and PINOT GR Res 10. Exceptional quality/price ratio PINOT BL 18.

Brocard, J-M Chab w ★★→★★★ Successful quality grower/merchant. Julien B has added bio methods to father Jean-Marc's flair. Sustainable mix of volume/value CHAB lines and high-class individual bottlings. Try GC Les Preuses.

Brouilly Beauj r ★★ 17 18′ 19′ Largest of ten BEAUJ crus: solid, rounded wines with some depth of fruit, approachable early but can age 3–5 yrs. Top growers: CHX de la Chaize (new owners), Thivin, DOMS Chermette, JC Lapalu, L&R Dufaitre, Piron.

Brumont, Alain SW Fr r w ★★★ MADIRAN's star still makes blockbusting BOUSCASSÉ, Le Tyre, MONTUS, but also easier-drinking wines eg. ★ Torus and range of quaffable IGPS. ★★★ PACHERENCS (dr sw) outstanding.

Brut Champ Term for dry classic wines of CHAMP. Most houses/growers have reduced dosage (adjustment of sweetness) in recent yrs. But great Champ can still be made at 8–9g residual sugar.

Brut Ultra / Zéro Term for bone-dry wines (no dosage) in CHAMP (also known as Brut NATURE); fashionable esp with sommeliers, quality generally improving. Needs ripe yr, old vines, max care, eg. *Pol Roger Pure*, ROEDERER Brut Nature Philippe Starck 12′, Veuve Fourny Nature.

Bugey Sav r p w sp ★→★★ AC 490 ha, 80 producers: light, fresh sparkling (56%), *pétillant*, still. Three distinct sectors, n–s: Belley, Cerdon, Montagnieu. Eight Bugey ACs. Whites (50% incl sp) mainly CHARD also incl ALIGOTÉ, Jacquère, Roussette. Rosé (33%). Reds (17%): GAMAY, MONDEUSE, PINOT N. Try: Angelot, Bugiste, Lingot-Martin, Monin, Peillot, Pellerin, Trichon.

Burgaud Beauj r ★★★ Man from MORGON, top bottlings of Côte du Py, Grands Cras, Charmes, etc., with fine ageing potential, but juicily attractive from day one.

Buxy, Caves de Burg r (p) w ★→★★ Leading CÔTE CHALONNAISE co-op for decent CHARD, PINOT N, source of many merchants' own-label ranges. Easily largest supplier of AC MONTAGNY.

Buzet SW Fr r (p) (w) AOP ★★ 15′ 17 18 Plummy cousin of B'X. Empire-building co-op (★★ CH de Guèze) cannot suppress ★★★ DOM du Pech (natural wines) way ahead of field, also ★★ Ch du Frandat, bio Dom Salisquet.

Cabernet d'Anjou Lo p s/sw ★→★★ DYA. Anjou's largest AOP. Lightly sweet rosé: CAB FR/CAB SAUV. Try: Bablut, Bergerie, CADY, CH PIERRE-BISE, Chauvin, Clau de Nell, de Sauveroy, Grandes VIGNES, Montgilet, Ogereau, Varière.

Cadillac-Côtes de Bordeaux B'x r ★→★★ 14 15 16 18 (19) Long, narrow, hilly zone on right bank of Garonne. Mainly MERLOT with CABS SAUV, FR. Medium-bodied, fresh reds; quality v. varied. Best: Alios de Ste-Marie, Biac, *Carsin*, CH Carignan, CLOS Chaumont, Clos Ste-Anne, de Ricaud, Grand-Mouëys, Lamothe de Haux, Laroche, Le Doyenné, Mont-Pérat, Plaisance, Réaut (Carat), *Reynon*, Suau.

Cady, Dom Lo r p w sw sp ★★→★★★ 15′ 16′ 17′ 18′ 19′ V.gd organic family estate in ANJOU with accent on CHENIN, incl COTEAUX DU LAYON, esp Chaume, Les Varennes. 19 frost.

Cahors SW Fr r ★★★ 12 15′ 16 17 (18′) (19) AOP on River Lot now wrestling back its MALBEC birthright. All red (some IGP w). Something for everyone here. Easy-drinking ★★ CHX Paillas, CLOS Coutale, Combel la-Serre. Traditional ★★★ *Clos de Gamot*. Modern ★★★ CH DU CÈDRE or trendy ★★★★ DOM Cosse-Maisonneuve. All styles from ★★★ Ch Chambert, *Clos Triguedina*, Clos Troteligotte, de la Bérengeraie, Haut-Monplaisir, Lo Domeni, ★★ Chx Armandière, Gaudou, La Coustarelle, Vincens. Otherwise try ★★★ Clos d'Un Jour, Dom du Prince, Lamartine, Les Croisille, Mas La Périé. For oak-lovers ★★★ La Reyne, ★★ Eugénie, La Caminade. Influential Vigouroux stable tending to Argentinian influence incl ★★ Chx Hautes-Serres, Léret-Monpézat, Mercuès.

Cailloux, Les S Rh r w ★★★ 78′ 81′ 90′ 98′ 03′ 05′ 09′ 10′ 16′ 18′ 19′ 21-ha CHÂTEAUNEUF DOM; elegant, profound, handmade reds, top value. Special wine Centenaire, oldest GRENACHE 1889 noble, dear, max elegant 16′. Also DOM André Brunel (esp gd-value CÔTES DU RH red Est-Ouest).

Cairanne S Rh r p w ★★→★★★ 10′ 15′ 16′ 17′ 18′ 19′ Great choice from *garrigue* soils, wines of character, dark fruits, mixed herbs, some refinement, esp CLOS Romane, DOMS Alary (stylish), *Amadieu* (pure, bio), Boisson (punchy), Brusset (deep), Cros de Romet, Escaravailles (flair), Grands Bois (organic), Grosset, Hautes Cances (true, traditional until 19), Jubain, *Oratoire St Martin* (detail, classy), Présidente, Rabasse-Charavin (punchy), *Richaud* (scale, great fruit). Food-friendly, 3D *whites*.

Cal Demoura, Dom L'doc r p w ★★★ Pure, meticulous TERRASSES DU LARZAC. Paroles de Pierre, L'Etincelle with six different grapes (w); L'DOC blends (r) Combariolles, predominantly GRENACHE Feu Sacré, Terres de Jonquières and Fragments, mainly old SYRAH.

Canard-Duchêne Champ House owned by ALAIN THIÉNOT. Now run by next generation. Improved Authentique Cuvée (organic) 09 12′ 13 15 17′ 18′. Single-v'yd Avize Gamin 12 13 16′ 17.

Canon-Fronsac B'x r ★★→★★★ 10′ 14 15 16 18 (19) Small enclave in FRON, otherwise same wines. 47 growers. Best: rich, full, finely structured. Try CHX Barrabaque,

Canon Pécresse, Cassagne Haut-Canon la Truffière, GABY, Grand-Renouil, La Fleur Cailleau, MOULIN PEY-LABRIE, Pavillon, Vrai Canon Bouché.

Carillon C d'O w ★★★ Contrasting PULIGNY brothers: Jacques unchangingly classical; try PC Referts. François for exciting modern approach. Try Combettes, Folatières. Village Puligny great from both.

Cassis Prov (r) (p) w ★★ DYA. From hills round port/resort e of Marseille; pricey because, well, it's Prov. Whites develop richness with age; savoury, saline. Based on BOURBOULENC, CLAIRETTE, MARSANNE, UGNI BLANC. Look for: *Clos Ste Magdeleine*, DOM de la Ferme Blanche, Fontcreuse, Paternel.

Castelnau, De Champ ★★★ Rising co-op. Cellarmaster Elisabeth Sarcelet insists on longer lees-ageing in excellent BLANC DE BLANCS and vintage 02. Innovative Prestige Collection Hors d'Age blended from best wines, different each yr. current release CCF2067 led by fine MEUNIER.

Castillon-Côtes de Bordeaux B'x r ★★→★★★ 10' 14 15 16 18 (19) Appealing e neighbour of ST-ÉM with improved quality; similar wines, usually less plump. 25% of growers organic or bio. Top: Alcée, Ampélia, Cap de FAUGÈRES, CLOS Les Lunelles, Clos Louie, *Clos Puy Arnaud*, Côte Montpezat, *d'Aiguilhe, de l'A*, de Pitray, Joanin Bécot, *La Clarière-Laithwaite*, l'Aurage, Le Rey, *l'Hêtre*, Montlandrie, Poupille, Roquevieille, Veyry.

Cathiard, Dom Sylvain C d'O r ★★★★ Sébastien C makes wines of astonishing quality from VOSNE-ROMANÉE, esp Malconsorts, Orveaux, Reignots, plus NUITS-ST-GEORGES Aux Thorey, Murgers. Seductive young, will mature to splendour.

Cave Cellar, or any wine establishment.

Cave coopérative Growers' co-op winery; over half of all French production. Wines often well-priced, probably not most exciting. Many co-ops closing down.

Cazes, Dom Rouss r p w sw ★★→★★★ Big, bio, part of huge Advini, gd value. Now favours Midi vines. Le Canon du Maréchal GRENACHE/SYRAH; Crédo CÔTES DU ROUSS-VILLAGES with Ego, Alter. Aimé Cazes is sensational aged RIVESALTES. CLOS de Paulilles BANYULS, COLLIOURE. Sulphur-free Hommage, powerful SYRAH; MAURY SEC. Ambre (sw w) made with GRENACHE BL.

Cédre, Ch du SW Fr r w ★★→★★★ 12 14 15' 16 (17) (18) (19) Verhaeghe brothers make best-known modern-style CAHORS. ★★★ Extra Libre easier than ★★ more serious top growths. Delicious ★★ VIOGNIER IGP.

Cépage Grape variety. *See* pp.14–24 for all.

Cérons B'x w sw ★★ 11 13 14 15' 16 18 Tiny 33-ha (2018) sweet AC next to SAUT. Less intense wines, eg, CHX de Cérons, CLOS Bourgelat, du Seuil, Grand Enclos.

Chablis w ★★→★★★ 14' 15 17' 18' 19' Both wine and region full of energy. Transcends CHARD grape with marine mineral infusion flavours.

Chablis Grand Cru Chab w ★★★→★★★★ 07' 08' 10' 12' 14' 15 17' 18' 19' Contiguous s-facing block overlooking River Serein, most concentrated CHAB, needs 5–15 yrs to show detail. Seven v'yds: Blanchots (floral), Bougros (incl Côte Bouguerots), CLOS (usually best), Grenouilles (spicy), Preuses (cashmere), Valmur (structure), Vaudésir (plus brand La Moutonne). Many gd growers.

Chablisienne, La Chab r w ★★→★★★ Exemplary co-op responsible for huge slice of CHAB production, esp supermarket own-labels. Gd individual CUVÉES too, eg. GC Grenouilles.

Chablis Premier Cru Chab w ★★★ 09 10' 12' 14' 15 17' 18' 19' Well worth premium over straight CHAB: better sites on rolling hillsides. Mineral favourites: Montmains, Vaillons, Vaucoupin, softer style Côte de Léchet, Fourchaume; greater opulence from Mont de Milieu, *Montée de Tonnerre*, Vaulorent.

Chambertin C d'O r ★★★★ 90' 93 96' 99' 02' 05' 09' 10' 12' 14 15' 16 17 18' 19' Called the King of Burg, but in touch with its feminine side. Imperious wine; amazingly dense, sumptuous, long-lived, expensive. Producers who match

---

Chablis
There is no better expression of the all-conquering CHARD than the full but tense, limpid but stony wines it makes on the heavy limestone soils of CHAB. Best makers use little or no new oak to mask the precise definition of variety and terroir. Investigate: Barat, B Defaix, Beru, ★ Bessin, ★ BILLAUD (Samuel), ★ Billaud-Simon, ★ Boudin, BROCARD, ★ C MOREAU, Collet, ★ Dampt family, ★ Droin, ★ DROUHIN, Duplessis, E&E Vocoret, Grossot, J Dauvissat, LAROCHE, Lavantureux, LONG-DEPAQUIT, Malandes, ★ Michel, MOREAU-Naudet, N Fèvre, ★ Picq, ★ Pinson, Piuze, Pommier, ★ RAVENEAU, Temps Perdu, Tribut, ★ V DAUVISSAT, ★ W FÈVRE. Simple, unqualified "Chablis" may be thin, though not from names above, and PETIT CHAB thinner; it's well worth premiums for PC or GC, and essential to mature 3–15 yrs. Co-op LA CHABLISIENNE has high standards (esp ★ Grenouille) and many different labels. (★ = outstanding.)

potential incl Bernstein, BOUCHARD PÈRE & FILS, Charlopin, Damoy, DOM LEROY, DUGAT-Py, DROUHIN, MORTET, ROSSIGNOL-TRAPET, ROUSSEAU, TRAPET.

Chambertin-Clos de Bèze C d'O r ★★★★ 90' 93 96' 99' 02' 03 05' 09' 10' 12' 14 15' 16 17 18' 19' Splendid neighbour to CHAMBERTIN, slightly more accessible in youth, velvet texture, deeply graceful. Best: Bart, B CLAIR, Damoy, D Laurent, DROUHIN, Drouhin-Laroze, Duroché, FAIVELEY (incl super-CUVÉE Les Ouvrées Rodin), Groffier, JADOT, Prieuré-Roch, ROUSSEAU.

Chambolle-Musigny C d'O r ★★★→★★★★ 93 99' 02' 05' 09' 10' 12' 15' 16 17 19' Silky, velvety wines from CÔTE DE NUITS: Charmes, Combe d'Orveau for substance, more chiselled from Cras, Fuées, seduction from Amoureuses, plus GCS BONNES MARES, MUSIGNY. Superstars: BARTHOD, MUGNIER, ROUMIER, VOGÜÉ. Try Amiot-Servelle, DROUHIN, Felettig, Groffier, HUDELOT-Baillet, Pousse d'Or, RION and Sigaut.

Champagne Sparkling wines of PINOTS N, M and CHARD: 33,805 ha, heartland c.145 km (90 miles) e of Paris. Sales 300 million bottles+/yr. Some PINOT BL in AUBE adds freshness. Other sparkling wines, however ace, cannot be called Champ.

Champagne le Mesnil Champ ★★★ Top-flight CO-OP in greatest GC CHARD village. Exceptional CUVÉE Sublime 08' 09 13 15 17' 19 from finest sites. Majestic Cuvée Prestige 05 triumphs against odds. Real value.

Chandon de Briailles, Dom C d'O r w ★★★ DOM defined by bio farming, min sulphur, whole bunches, no new oak. Brilliantly pure perfumed reds, esp CORTON-Bressandes, Île de Vergelesses, PERNAND-VERGELESSES.

★ *Chanson Père & Fils* Burg r w ★→★★★ Resurgent BEAUNE merchant, quality whites (CLOS DES MOUCHES, CORTON-Vergennes) and stylish idiosyncratic reds (whole-cluster aromatics) esp CLOS des Fèves.

Chapelle-Chambertin C d'O r ★★★ 99' 02' 05' 09' 10' 12' 14' 15' 16 18' 19' Lighter neighbour of CHAMBERTIN; thin soil does better in cooler, damper yrs. Fine-boned wine, less meaty. Top: Damoy, DROUHIN-Laroze, JADOT, PONSOT, ROSSIGNOL-TRAPET, TRAPET, Tremblay.

Chapoutier N Rh ★★ →★★★★ Vocal bio grower/merchant at HERMITAGE. Broadly stylish reds via low-yield, plot-specific, expensive CUVÉES. Dense GRENACHE CHÂTEAUNEUF – Barbe Rac, Croix de Bois (r); CÔTE-RÔTIE La Mordorée; Hermitage – L'Ermite (outstanding r w), Le Pavillon (deep r), Cuvée de l'Orée (w), Le Méal (w). Also ST-JOSEPH Les Granits (r w). *Hermitage whites* outstanding, 100% old-vine MARSANNE. Gd-value *Meysonniers Crozes*. V'yds in COTEAUX D'AIX-EN-PROV, CÔTES DU ROUSS-VILLAGES (gd DOM Bila-Haut), RIVESALTES. Also own Ferraton at Hermitage, BEAUJ house Trenel, CH des Ferrages (Prov), has AL v'yds and Australian joint ventures, esp Doms Tournon and Terlato & Chapoutier (fragrant); also Portuguese Lisboa project, Douro too, hotel, wine bar in Tain, list gets longer every yr.

Charbonnière, Dom de la S Rh r w ★★★ 05′ 09′ 10′ 16′ 17 18 17-ha CHÂTEAUNEUF estate run by sisters. Sound Tradition (r), deep, special wines: VIEILLES VIGNES (vigour, best), authentic Mourre des Perdrix, also Hautes Brusquières. V. elegant, pure white. Also peppery, small-quantity VACQUEYRAS red.

Chardonnay As well as a white wine grape, also the name of a MÂCON-VILLAGES commune, hence Mâcon-Chardonnay.

Charlemagne C d'O w ★★★ 13 14′ 15′ 17′ 18 19′ Almost extinct sister appellation to CORTON-C, revived by DOM DE LA VOUGERAIE from 2013. Same rules as sibling.

Charlopin C d'O r (w) ★★→★★★ Philippe C makes impressive range of reds from GEVREY base. BOURGOGNE C D'O, MARSANNAY for value, gd range GC for top of line. Son Yann C, DOM C-Tissier also exciting.

Charmes-Chambertin C d'O r ★★★★ 99′ 02′ 03 05′ 09′ 10′ 12′ 15′ 16 17 18′ 19′ GEVREY GC, 31 ha, incl neighbour MAZOYÈRES-CHAMBERTIN. Raspberries and cream plus dark-cherry fruit, sumptuous texture, fragrant finish. So many gd names: ARLAUD, BACHELET, Castagnier, Coquard-Loison-Fleurot, DUGAT, DUJAC, Duroché, LEROY, MORTET, Perrot-Minot, Roty, ROUSSEAU, Taupenot-Merme, VOUGERAIE.

Chartogne-Taillet Champ A disciple of SELOSSE, Alexandre Chartogne is new star. Ideal BRUT Ste Anne NV and single-v'yds Le Chemin de Reims and Les Barres; striking energy. High hopes for 19′.

Charvin, Dom S Rh r w ★★★ 98′ 99′ 01′ 06′ 07′ 09′ 10′ 12′ 15′ 16′ 17′ 18′ Terroir truth at 8-ha CHÂTEAUNEUF estate, 85% GRENACHE, no oak, only one handmade CUVÉE. Spiced, mineral, high-energy red, vintage accuracy. Recent gd white. Top value, genuine, long-lived CÔTES DU RH (r).

Chassagne-Montrachet C d'O r w ★★→★★★★ (w) 02′ 04 05′ 08′ 09′ 12′ 14′ 15 17′ 18′ 19′ Large village at s end of CÔTE DE BEAUNE. Great white v'yds Blanchot, Cailleret, Montrachet GCS. Try: Coffinet, COLIN, GAGNARD, MOREY, Pillot families plus DOMS Heitz-Lochardet, MOREAU, Niellon, Ramonet. Time for a red revival? Best reds: Blain-Gagnard, JM Gagnard, Morgeot, Pillot (esp CLOS St-Jean), Ramonet.

Château (Ch) Means an estate, big or small, gd or indifferent, particularly in B'X (*see* pp.100–121). Means, literally, castle or great house. In Burg, DOM is usual term.

Château-Chalon Jura w ★★★★ 96 99′ 00 05′ 09 10′ 12 Not a CH but AC and village, the summit of VIN JAUNE style from SAVAGNIN grape. An unfortified, winey version of Sherry. Min 6 yrs barrel-age, not cheap but gd value. Ready to drink (or cook a chicken in) when bottled, but gains with further age. Fervent admirers search out BERTHET-BONDET, MACLE, TISSOT or Bourdy for old vintages.

Château-Grillet N Rh w ★★★★ 01′ 04′ 07′ 09′ 10′ 12′ 14′ 15′ 16′ 17′ 18 19′ France's smallest AC. 3.7-ha amphitheatre nr CONDRIEU, sandy-granite terraces, improved v'yd care. Bought by F Pinault of CH LATOUR in 2011, prices way up, wine *en finesse*, less rich than before. Smooth, precise VIOGNIER: drink at cellar temperature, decanted, with refined dishes.

Châteaumeillant Lo r p ★→★★ Dynamic but isolated 82-ha AC s Bourges. GAMAY, PINOT N (only 40% permitted) for light reds (75% production), VIN GRIS (25%) with 10% PINOT GR allowed. BOURGEOIS, Chaillot, Gabrielle, Goyer, Joffre, Joseph MELLOT, Lecomte, Nairaud-Suberville, Roux, Rouzé, Siret-Courtaud.

Châteauneuf-du-Pape S Rh r (w) ★★★→★★★★★ 78′ 81′ 90′ 95′ 01′ 07′ 09′ 10′ 16′ 17 18′ 19′ Nr Avignon, about 50 gd DOMS (remaining 85 fair, uneven to poor). Up to 13 grapes (r w), headed by GRENACHE, plus SYRAH, MOURVÈDRE (increasing), Counoise.

A trio of Champagne growers to watch

Bauchet, Bisseuil: Origine a nice mix of tradition, innovation, at once precise, mineral yet gourmand. **Collard-Picard,** REUILLY nr Châtillon: CUVÉE Prestige a mix of PINOTS M/N, LE MESNIL CHARD. Mastery of oak. **J-B Geoffroy,** Cumières and AŸ: leading grower/winemaker. Try PC Volupté.

Warm, spiced, musky, textured, long-lived; should be fine, pure, magical, but till mid-2010s too many heavy, sip-only Parker-esque wines. Small, traditional names can be gd value. Prestige old-vine wines (v.gd Grenache 16'). To avoid: late-harvest, new oak, 16% alc, too pricey. Whites fresh, fruity, or sturdy, best can age 15 yrs+. For top names, *see* box, below.

Chave, Dom Jean-Louis N Rh r w ★★★★ 99' 00 01' 03' 04 05' 07' 09' 10' 11' 12' 13' 15' 16' 17' 18' 19' Excellent family DOM at heart of HERMITAGE. Classy, silken, long-lived reds (more polished recently), incl expensive, v. occasional Cathelin. V.gd, stylish, intricate white (mainly MARSANNE); occasional VIN DE PAILLE. ST-JOSEPH reds deep, great fruit, plot-specific CLOS Florentin (since 2015) v. stylish; also lively J-L Chave brand St-Joseph Offerus, jolly CÔTES DU RH Mon Coeur, sound value merchant Hermitage Farconnet (r), Blanche (w).

Chavignol Lo r p w SANCERRE village, v. steep v'yds Cul de Beaujeu/ Les Monts Damnés dominate picturesque village. Clay-limestone soil gives full-bodied, mineral whites and reds that age 15 yrs+. V. fine young producers: Matthieu Delaporte (Vincent D), Pierre Martin. Also: ALPHONSE MELLOT, Boulay (v.gd), BOURGEOIS, Cotat, DAGUENEAU, Paul Thomas, Thomas Laballe. Whites brilliant match with Crottin de Chavignol cheese.

Chénas Beauj r ★★★ 15' 16 18' 19' Smallest BEAUJ cru, between MOULIN-À-VENT and JULIÉNAS, gd value, meaty, age-worthy, merits more interest. Thillardon is reference DOM, but try also Janodet, LAPIERRE, Pacalet, Piron, Trichard, co-op.

Chêne Bleu S Rh ★★ Glossy, well-funded, part-bio DOM in hills e of SÉGURET. Extravagant, flashy, expensive wines, mainly VENTOUX. Stylish part-oaked rosé.

Chevalier-Montrachet C d'O w ★★★★ 04 08 09' 10 12 14' 15 17' 18' 19' Just above MONTRACHET on hill, just below in quality; brilliant crystalline wines, dancing white fruit and flowers. Long-lived but can be accessible early. Top grower is LEFLAIVE, special CUVÉES Les Demoiselles from JADOT, LOUIS LATOUR and La Cabotte from BOUCHARD; try also: Chartron, COLIN (P), Dancer, DE MONTILLE, Niellon, VOUGERAIE.

Cheverny Lo r p w ★→★★ 17 18' 19' AC s Blois. White from SAUV BL (majority)/ CHARD blend. Light reds mainly GAMAY, PINOT N (also CAB FR, CÔT). *Cour-Cheverny* (48 ha): traditional Romorantin – ages well. Esp Cazin, CLOS Tue-Boeuf, DOMS de la Desoucheriem, de Montcy, du Moulin, de Veilloux, Gendrier, Huards, H Villemade, Tessier. Much low-lying; frost prone.

Chevillon, R C d'O r ★★★ With GOUGES, the reference DOM for NUITS-ST-GEORGES PCS: Bousselots, Chaignots, Pruliers more accessible; Cailles, Les St-Georges, Vaucrains for the long term.

Chevrot C d'O r w ★★ Pablo and Vincent C great source for juicy MARANGES, esp Sur le Chênes, PC Croix Moines. Decent whites, CRÉMANT too. Lively wines, fair prices.

Chidaine, François Lo (r) w dr sw sp ★★★ 15' 16' 17' 18' 19' V.gd MONTLOUIS single-v'yds, bio from 1999. VOUVRAY but has to be labelled VDF, TOURAINE. Historic

---

Châteauneuf: kings of the castle

Top names in this enormous and confusing appellation: CHX DE BEAUCASTEL, Fortia, Gardine (also w), Mont-Redon, Nalys, RAYAS, Sixtine, Vaudieu; DOMS Barroche, Beaurenard (bio), Bois de Boursan, Bosquet des Papes, Chante Cigale, Chante Perdrix, CHARBONNIÈRE, CHARVIN, CLOS du Caillou, Clos du Mont-Olivet, CLOS DES PAPES, Clos St-Jean, Cristia, de la Biscarelle, de la Janasse, de la Vieille Julienne (bio), du Banneret, Fontavin, Font-de-Michelle, Grand Tinel, Grand Veneur, Henri Bonneau, LES CAILLOUX, Marcoux, Mas du Boislauzon, Pegaü, Pierre André (bio), Porte Rouge, P Usseglio, R Usseglio (bio), Roger Sabon, Sénéchaux, St-Préfert, Vieux Donjon, VIEUX TÉLÉGRAPHE.

CLOS Baudoin (Vouvray). Accent on SEC and DEMI-SEC. Also Le Chenin d'Ailleurs (LIMOUX) to compensate for LO frosts, and red/white from Spain.

Chignin Sav (r) w ★→★★ AC: Jacquère, MONDEUSE, GAMAY, PINOT N grapes. Chignin-Bergeron is ROUSSANNE. Try: Berthollier, Partage.

Chinon Lo r p (w) ★★→★★★ 14′ 15′ 16′ 17′ 18′ 19′ Light to rich TOURAINE CAB FR from sand, gravel, limestone. Best age 30 yrs+. A little dry CHENIN BL (some wood). Best: ALLIET, BAUDRY, BAUDRY-DUTOUR, Coulaine, Couly-Dutheil, Dozon, Grosbois, JM Raffault, Jourdan-Pichard, Landry, L'R, Moulin à Tan, Noblaie, Pain, Petit Thouars, P&B Couly; now extensive frost protection. 18 excellent, age-worthy; 19 gd, vines on sand hit by drought.

Chiroubles Beauj r ★★ 15′ 18′ 19′ BEAUJ cru in hills above FLEURIE: fresh, fruity, savoury wines. Growers: Berne, CH de Javernand, Cheysson, Lafarge-Vial, Métrat, Passot, Rousset, or merchants DUBOEUF, Trenel.

Chorey-lès-Beaune C d'O r (w) ★★ 09′ 10′ 12 15′ 17 18 19′ Village just s of BEAUNE. TOLLOT-BEAUT remains reference for this affordable, uncomplicated AC, DOM Gay promising. Try also Arnoux, DROUHIN, Guyon, JADOT, Rapet, ROUGET.

Chusclan S Rh r p w ★→★★★ 16′ 17′ 18′ 19′ CÔTES DU RH-VILLAGES with gd-quality Laudun-Chusclan co-op, incl gd fresh whites, often MARSANNE, ROUSSANNE, VIOGNIER. Soft reds, cool rosés. Best co-op labels (r) Chusclan DOM de l'Olivette, CÔTES DU RH Femme de Gicon (r), Enfant Terrible (w), Excellence, LIRAC Dom St-Nicolas. Also full CH Signac (best Chusclan, can age), Dom La Romance (fresh, organic), special CUVÉES from *André Roux*.

Clair, Bruno C d'O r p w ★★★→★★★★ Top-class CÔTE DE NUITS estate for supple, subtle, savoury wines. Gd-value MARSANNAY, old-vine SAVIGNY La Dominode, GEVREY-CHAMBERTIN (CLOS ST-JACQUES, Cazetiers) and standout CHAMBERTIN-CLOS DE BÈZE. Best whites from MOREY-ST-DENIS, CORTON-CHARLEMAGNE.

Clairet B'x between rosé/red. B'x Clairet is AC. Try CHX Fontenille, Penin, Turcaud.

Clairette de Die N Rh w dr s/sw sp ★★ NV Rh/low Alpine bubbly: flinty or (better) MUSCAT sparkling (s/sw). Underrated, muskily fruited, gd value, low degree; or dry CLAIRETTE, can age 3–4 yrs. NB: Achard-Vincent, Carod, Jaillance (value), J-C Raspail (organic IGP SYRAH), Poulet et Fils (terroir, Chatillon-en-Diois r). Must try.

Clape, Dom Pierre, Olivier N Rh r (w) ★★★→★★★★ 89′ 90′ 91′ 95′ 99′ 01′ 03′ 05′ 06′ 07′ 09′ 10′ 12′ 14′ 15′ 16′ 17′ 18′ 19′ The kings of CORNAS. Great location SYRAH v'yds, many old vines, gd soil work. Profound, complex reds, much vintage accuracy, need 6 yrs+, live 25↑. Clear fruit in youngish-vines label Renaissance. Superior CÔTES DU RH, VDF (r), *St-Péray* (gd style, improved). Auguste C RIP 2018.

Clape, La L'doc r p w ★★→★★★ Limestone massif on Med nr Narbonne. MOURVÈDRE hotspot, plus salty, herbal whites from BOURBOULENC, will age. CHX ANGLÈS, Camplazens, La Négly, *l'Hospitalet*, Mire l'Etang, Pech-Céleyran, *Pech-Redon*, Ricardelle, Rouquette-sur-Mer and Mas du Soleila, Sarrat de Goundy.

Climat Burg Refers to individual named v'yd at any level, esp in C D'O, eg. MEURSAULT Tesson, MAZOYÈRES-CHAMBERTIN. UNESCO World Heritage status.

Clos A term carrying some prestige, reserved for distinct (walled) v'yds, often in one ownership (esp AL, Burg, CHAMP).

Clos des Fées Rouss r w ★★ Organic philosophy, wines with character from range of terroirs. Barrel-fermented SYRAH/GRENACHE/MOURVÈDRE/CARIGNAN. De Battre mon Coeur s'est Arrêté is high-altitude Syrah. Gd IGP COTES DE CATALANES (w) old-vine GRENACHE BL.

Clos de Gamot SW Fr r ★★★ 12 14 15′ 16 (17) (18) CAHORS as it was before outsiders started messing about with it. ★★★★ Low-yield CUVÉE VIGNES Centenaires (best yrs only) sadly facing retirement. Drink while you can. ★★★ Mainstream red almost as gd; resisting deer, wild boar and changing fashions.

Clos des Lambrays C d'O r ★★★ 99′ 02 05′ 09′ 10′ 12′ 15′ 16′ 17 18′ 19′ All-but-

MONOPOLE GC v'yd at MOREY-ST-DENIS, now belongs to LVMH. Early-picked, spicy, stemmy style will evolve with new winemaker from 2019.

Clos du Mesnil Champ ★★★★ KRUG's famous walled v'yd in GC Le Mesnil. Long-lived, pure CHARD vintage, great mature yrs like 95 *à point* till 2023+; **02** and remarkable **03** 08' 13' will be classics, as will 17' 19'.

Clos des Mouches C d'O r w ★★★ (w) **02** 05' **09' 10'** 15 17' 18 19' PC v'yd in several Burg ACs. Mouches = honeybees; *see* label of DROUHIN's iconic BEAUNE bottling. Also BICHOT, CHANSON (Beaune), plus CLAIR, MOREAU, Muzard (SANTENAY), Germain (MEURSAULT).

Clos des Papes S Rh r w ★★★★ 99' 01' 03' 04' 05' 07' 09' 10' **14'** 15' 16' 17' 18' 19' Always top-class CHÂTEAUNEUF DOM of Avril family, v. small yields, Burg meets S Rh. Stylish red (mainly GRENACHE, MOURVÈDRE, drink 2–3 yrs or from 8+); *great white* (six varieties, complex, allow time, merits rich cuisine: 2–3 yrs, or 10–20).

Clos de la Roche C d'O r ★★★★ 90' 93' 96' 99' **02'** 05' 08 09' **10'** 15' 16' 17' 18' 19' Maybe finest GC of MOREY-ST-DENIS, as much grace as power, more savoury than sumptuous, blueberries. Needs time. DUJAC, H LIGNIER, PONSOT references but try Amiot, ARLAUD, Bernstein, Castagnier, Coquard, LEROY, LIGNIER-Michelot, Pousse d'Or, Remy, ROUSSEAU.

Clos du Roi C d'O r ★★→★★★ Frequent Burg v'yd name. The king usually chose well. Best v'yd in GRAND CRU CORTON (DE MONTILLE, Pousse d'Or, VOUGERAIE); top PC v'yd in MERCUREY, future PC (?) in MARSANNAY. Less classy in BEAUNE.

Clos Rougeard Lo r w (sw) ★★★★ 09 10 12 14 (15) (16) Iconic small DOM, reputation made by late DRUET and now departed Nady Foucault. Owned by M and O Bouygues (CH MONTROSE; *see* B'X). Jacques Toublanc in charge. Finesse, long-lived: COTEAUX DE SAUMUR, SAUMUR Blanc, SAUMUR-CHAMPIGNY.

Clos de Vougeot had one owner until the Revolution; now has 88.

Clos St-Denis C d'O r ★★★ 90' 93' 96' 99' **02'** 05' 09' 10' 12' 15' 16' 17 18' 19' GC at MOREY-ST-DENIS. Sumptuous in youth, silky with age. Try Amiot-Servelle, ARLAUD, Bertagna, Castagnier, Coquard-Loison-Fleurot, Heresztyn-Mazzini, DUJAC, JADOT, Jouan, LEROUX, PONSOT (Laurent P from 2016).

Clos St-Jacques C d'O r ★★★★ 90' 93 96' 99' **02'** 05' 09' 10' 12' 15' 16 17 18' 19' Hillside PC in GEVREY-CHAMBERTIN with perfect se exposure. Shared by five excellent producers: CLAIR, ESMONIN, FOURRIER, JADOT, ROUSSEAU; powerful, age-worthy, velvety reds ranked (and often priced) above many GCS.

Clos St-Landelin Al r w ★★→★★★★ Great name, esp fine, full-bodied GC **Vorbourg Ries** and PINOT GR **13 14**. **Pinot N Cuvée "V"** ripe, wine rich, is region's best, exceptional in 15'. Only surpassed by 18'.

Clos Ste-Hune Al w ★★★★ Legendary TRIMBACH single site from GC ROSACKER. Greatest RIES in AL? Super 08 10 13 16 17. Austere, needs a decade ageing; complex wine for gastronomy. GC SCHLOSSBERG more luxuriant, floral.

Clos de Tart C d'O r ★★★★ **02'** 05' 08' **10'** 13' 14 15' 16' 17 18' 19' Expensive MOREY-ST-DENIS GC. Now part of Pinault/Artemis empire (CH LATOUR, CH GRILLET, etc.) with new winemaker. Picking earlier than before, but will style lighten and freshen?

Clos de Vougeot C d'O r ★★★→★★★★ 90' 99' **02'** 05' **09' 10'** 12' 13' 15' 16 17 18' 19' CÔTE DE NUITS GC with many owners. Occasionally sublime, needs 10 yrs+ to show real class. Style, quality depend on producer's philosophy, technique, position. Top: ARNOUX-LACHAUX, BOUCHARD, Castagnier, CH de la Tour (stems), Coquard-Loison-Fleurot, DROUHIN, EUGÉNIE (intensity), *Faiveley*, Forey, GRIVOT, *Gros*, HUDELOT-NOËLLAT, JADOT, LEROY, LIGER-BELAIR (both), MÉO-CAMUZET, MONTILLE, MORTET, MUGNERET-Gibourg, PONSOT, *Vougeraie*, Y Clerget.

Clusel-Roch, Dom N Rh r w ★★★ 95' 99' 01 05' 09' 10' 11 12' 13' 14 15' 16' 17' 18' 19' Rare organic CÔTE-RÔTIE DOM, gd range v'yds, mostly Serine (local variation of

SYRAH). Patience rewarded. Les Schistes gd entry point, high-quality plot wines La Viallière, Les Grandes Places. Son Guillaume C makes v. drinkable CÔTEAUX DU LYONNAIS (r w).

Coche-Dury C d'O r w ★★★★ Top MEURSAULT DOM led by Raphaël C in succession to legend Jean-François. Exceptional whites from ALIGOTÉ to CORTON-CHARLEMAGNE; v. pretty reds too. Stratospheric prices. Gd-value cousin Coche-Bizouard (eg. Meursault Goutte d'Or) is sound, but not same style.

Colin C d'O (r) w ★★★→★★★★ Leading CHASSAGNE and ST-AUBIN family; current generation turning heads with brilliant whites, esp Pierre-Yves C-MOREY, DOM Marc C, Joseph C (from 2017) and their cousins Philippe C and Bruno C.

Colin-Morey Burg ★★★ Pierre-Yves C-M has made his name with vibrant tingling whites, esp from *St-Aubin* and CHASSAGNE PC, with their characteristic gun-flint bouquets. Great wines in youth and for ageing.

Collines Rhodaniennes N Rh r w ★★ N Rh IGP has character, quality, incl v.gd Seyssuel (schist), sparky granite hillside, plateau reds v.gd value, often from top estates. ("Rhodanienne" = "of the Rhône".) Mostly SYRAH (best), plus MERLOT, GAMAY, mini-CONDRIEU VIOGNIER (best). Reds: A Paret, A PERRET, Bonnefond, CLOS de la Bonnette (organic), E Barou, Hameau Touche Boeuf, *Jamet*, Jasmin, J-M Gérin, L Chèze, Monier-Pérreol (bio), N Champagneux, *S Ogier*, S Pichat, ROSTAING. Whites: Alexandrins, A Perret (v.gd), Barou, F Merlin, *G Vernay*, P Marthouret, X Gérard, Y Cuilleron.

Collioure Rouss r p w ★★ Same v'yds as BANYULS. Terraces overlooking sea. Table wines only, mainly GRENACHE of all colours. Rosé can be serious. Top: DOMS Bila-Haut, de la Rectorie, du Mas Blanc, La Tour Vieille, Madeloc, Vial-Magnères; Coume del Mas, Les CLOS de Paulilles. Co-ops Cellier des Templiers, l'Étoile.

Comté Tolosan SW Fr r p w ★ Usually DYA. Mopping-up IGP covering most of SW and multitude of sins. ★★ DOM de Ribonnet (all colours, huge variety of grapes) stands out among a host of mostly moderate wines. *See* PYRÉNÉES-ATLANTIQUES.

Condrieu N Rh w ★★★→★★★★ 16' 18' 19' Home of VIOGNIER; soft floral airs, pear, apricot flavours from sandy granite slopes. Best: pure, precise (16 over 15 17); but beware excess oak, sweetness, alc. Growers starting to adapt to v. hot yrs. Rare white *ami* for asparagus. 80 growers, not all gd. Best: A Paret, A PERRET (all three wines gd), Boissonnet, CHAPOUTIER, CLOS de la Bonnette (organic), C Pichon, DELAS, Faury (esp La Berne), F Merlin, F Villard (lighter recently), Gangloff (great style), GUIGAL (big), *G Vernay* (fine, incl top-class Coteau de Vernon), Monteillet, Niéro, ROSTAING, St Cosme, X Gérard (value), Y Cuilleron.

Corbières L'doc r (p) (w) ★→★★★ Characterful red, plenty of CARIGNAN, some v.gd white; styles reflect contrasts of terroir from coastal lagoons to dry foothills of Pyrénées. Try: CHX Aiguilloux, *Aussières*, Borde-Rouge, LA BARONNE, Lastours, la Voulte Gasparets, Les CLOS Perdus, Les Palais, *Ollieux Romanis*, Pech-Latt; DOMS de Fontsainte, de Villemajou, du Grand Crès, du Vieux Parc, Trillol, Villerouge; Clos de l'Anhel, Grand Arc, SERRES-MAZARD. Castelmaure an outstanding co-op.

Cornas N Rh r ★★★ 99' 01' 05' 09' 10' 12' 15' 16' 17' 18' 19' Top-quality N Rhô SYRAH, *très à la mode*. Dark, strongly fruited, always mineral-lined. Some made for lusty early fruit, really need 5 yrs+. Stunning 10 15. Top: ALLEMAND (top two), Balthazar (traditional), M Bourg, *Clape* (benchmark), Colombo (oak), Courbis (modern), *Delas*, *Dom du Tunnel*, Dumien Serrette, G Gilles, J&E Durand (racy fruit), Lemenicier, Lionnet (organic), M Barret (bio), P&V Jaboulet, Tardieu-Laurent (full, oak), Voge (swish, oak), V Paris.

Corsica / Corse r p w ★→★★★ Should be wonderful, with that scenery; sometimes is. Plenty of variety; altitude, sea winds give freshness. ACS Ajaccio, PATRIMONIO; plus crus Calvi, Coteaux du Cap Corse, Sartène. IGP: Île de Beauté and Mediterranée. Reds elegant, spicy from SCIACARELLO, structured from rarer NIELLUCCIO. Gd rosés;

tangy, herbal Vermentino whites. Also sweet MUSCATS. Top: Abbatucci, Alzipratu, Canarelli, CLOS Capitoro, Clos d'Alzeto, Clos Poggiale, Fiumicicoli, *Peraldi*, Pieretti, *Nicrosi*, Saperale, *Torraccia*, Vaccelli. Hard to find elsewhere.

Corton C d'O r (w) ★★★→★★★★ 90' 99' 02' 03' 05' 09' 10' 12' 15' 17 18' 19' Largely overpromoted GC, but can be underrated from best v'yds CLOS DU ROI, Bressandes, Renardes, Rognet. Wines can be fine, elegant, not all blockbusters. References: BONNEAU DU MARTRAY, BOUCHARD, CHANDON DE BRIAILLES, DRC, Dubreuil-Fontaine, FAIVELEY (CLOS des Cortons), Follin-Arbelet, MÉO-CAMUZET, Senard, TOLLOT-BEAUT. Under the radar and can be gd value: Bichot, Camille Giroud, Clavelier, DOM des Croix, H&G Buisson, Mallard, Pousse d'Or, Rapet, Terregelesses. Occasional whites, eg. HOSPICES DE BEAUNE, CHANSON, CH de MEURSAULT from Vergennes v'yd.

Corton-Charlemagne C d'O w ★★★ →★★★★ 04 05' 09' 10' 14' 15' 17' 18 19' Potentially scintillating GC, invites mineral descriptors, should age well. Sw- and w-facing limestone slopes, plus band round top of hill. Top: BIZE, *Bonneau du Martray*, BOUCHARD, CLAIR, *Coche-Dury*, FAIVELEY, HOSPICES DE BEAUNE, JADOT, LATOUR, Mallard, MONTILLE, P Javillier, Rapet, Rollin. *See also* CHARLEMAGNE.

Costières de Nîmes S Rh r p w ★→★★ r p w ★→★★ N of Rhône delta, sw of CHÂTEAUNEUF, similar v. stony soils, Mistral-blown. Gd quality, value. Red (GRENACHE, SYRAH) robust, spiced, can age. Best: CHX de Grande Cassagne, de Valcombe, d'Or et des Gueules (full), L'Ermitage, Mas Carlot (gd fruit), Mas des Bressades (top fruit), Mas Neuf, Montfrin (organic), Mourgues-du-Grès (organic), Nages, Roubaud, Tour de Béraud, Vessière (w); DOMS de la Patience (organic), du Vieux Relais, Galus, M Gassier, M KREYDENWEISS, Petit Romain, Terres des Chardons (bio). Gd, bright rosés; some stylish whites (ROUSSANNE).

Côte de Beaune C d'O r w ★★→★★★★ S half of C D.O. Also a little-seen AC in its own right applying to top of hill above BEAUNE itself. Try from DROUHIN: largely declassified Beaune PC. Also VOUGERAIE.

Côte de Beaune-Villages C d'O r ★★ 10' 12 15' 17 18' 19' Reds from lesser villages of s half of C D'O. Nowadays usually NÉGOCIANT blends.

Côte de Brouilly Beauj r ★★ 15' 17 18' 19' Variety of styles as soils vary on different flanks of Mont Brouilly. Merits a premium over straight BROUILLY. Reference is CH THIVIN, but try also Blain, Brun, Dufaitre, LAFARGE-Vial, Pacalet.

Côte Chalonnaise Burg r w sp ★★ Region immediately s of C D'O; always threatening to be rediscovered. Lighter wines, lower prices. BOUZERON for ALIGOTÉ, *Rully* for accessible, juicy wines in both colours; *Mercurey* and GIVRY have more structure and can age; MONTAGNY for leaner CHARD.

Côte de Nuits C d'O r (w) ★★→★★★★ N half of C D.O. Nearly all red, from CHAMBOLLE-MUSIGNY, MARSANNAY, FIXIN, GEVREY-CHAMBERTIN, MOREY-ST DENIS, NUITS-ST GEORGES, VOSNE-ROMANÉE, VOUGEOT.

Côte de Nuits-Villages C d'O r (w) ★★ 05' 09' 10' 12' 15' 16 17 18' 19' Junior AC for extreme n/s ends of CÔTE DE NUITS; can be bargains. Chopin, Gachot-Monot, Jourdan specialists. Top single-v'yd wines CLOS du Chapeau (Arlot), Croix Violette (FOURNIER), Faulques (Millot), Leurey (JJ Confuron), Meix Fringuet (TRAPET), Montagne (many), Robignotte (Jourdan), Vaucrains (JADOT).

Côte d'Or Burg Département name applied to central and principal Burg v'yd slopes: CÔTE DE BEAUNE and CÔTE DE NUITS. Not used on labels except for BOURGOGNE C D'O AC, finally introduced for 17 vintage.

Côte Roannaise Lo r p ★★ 17' 18' 19' V.gd AC, lower slopes of granite hills w of Roanne. GAMAY. Try: Bonneton, Désormière, Fontenay, Giraudon, Paroisse, Plasse, Pothiers, Sérol, Vial. V. interesting white IGP Urfé: ALIGOTÉ, CHARD, ROUSSANNE, VIOGNIER. 19 continues run of gd vintages.

Côte-Rôtie N Rh r ★★★→★★★★ 99' 01' 05' 09' 10' 12' 15' 16' 17' 18' 19' Most refined Rh red, mainly SYRAH, some VIOGNIER, style links to Burg. Violet airs, pure (esp

16 19), complex, v. fine with age (5–10 yrs+). Exceptional, v. long-lived 10 15. Top: *Barge* (traditional), B Chambeyron, Billon, Bonnefond (oak), Bonserine (esp La Garde), Burgaud, CHAPOUTIER, Clusel-Roch (organic), DELAS, DOM de Rosiers, Duclaux, Gaillard (oak), Garon, X Gérard, GUIGAL (long oaking), *Jamet*, Jasmin, Jean-Luc Jamet, J-M Gérin, J-M Stéphan (organic), Lafoy, Levet (traditional), *Rostaing* (fine), S Ogier (racy, oak), Semaska, VIDAL-FLEURY (La Chatillonne).

Coteaux d'Aix-en-Provence Prov r p w ★★ Lots of styles from big AC centred on Aix: CAB SAUV in cooler n: Pigoudet, Revelette, Vignelaure. Med grapes often more interesting in warmer spots. DOM du CH Bas, Chx Beaupré, Calissanne, La Realtière, Les Bastides, Les Béates. Eole (on Alpilles). *See* LES BAUX-EN-PROV, PALETTE.

Coteaux d'Ancenis Lo r p w (sw) ★→★★ 17 18' 19' AOP, both sides of Loire e of Nantes. Dry, DEMI-SEC, sweet CHENIN BL whites, age-worthy sweet *Malvoisie* (PINOT GR); light red, rosé mainly GAMAY, plus CABS SAUV, FR, esp Guindon, Landron-Chartier, Paonnerie (natural wine), Pléiade, Quarteron.

Coteaux de l'Aubance Lo w sw ★★→★★★ 15' 16 17 18' Small AC, age-worthy sweet CHENIN BL. Nervier, less rich than COTEAUX DU LAYON except SÉLECTION DES GRAINS NOBLES. S of LO nr Angers, less steep than Layon. Esp Bablut, CH Princé, Haute-Perche, Montgilet, Richou, Rochelles, Varière. V.gd 18; 19 small quantity.

Coteaux des Baronnies S Rh r p w ★→★★ DYA. Rh IGP in high hills e of VINSOBRES, nr Nyons. SYRAH (best), CAB SAUV, MERLOT, CHARD (gd value), plus GRENACHE, CINSAULT, etc. Genuine, fresh country wines: impressive easy reds, also clear VIOGNIER. NB: DOMS du Rieu-Frais, Le Mas Sylvia, Rosière (bio).

Coteaux Bourguignons Burg ★ DYA. Mostly reds, GAMAY, PINOT N. New AC since 2011 to replace BOURGOGNE Grand Ordinaire and to sex up basic BEAUJ. Market accepting the change. Rare whites ALIGOTÉ, CHARD, MELON, PINOTS BL/GR.

Coteaux de Chalosse SW Fr r p w ★ DYA. Local IGP from Les Landes. Rare local grapes incl Arriloba, Baroque, Egiodola. Active co-op at Tursan. Few independents incl ★ DOMS de Labaigt, Tastet. Rare outside region.

Coteaux Champenois Champ r (p) w ★★★ (w) DYA. AC for still wines of CHAMP, eg. BOUZY. Vintages as for Champ. Better reds with climate change (12'). Impressive range of Coteaux Champenois Grands Blancs based on 17' by CHARLES HEIDSIECK, as gd as fine white burg.

Coteaux du Giennois Lo r p w ★→★★ 17 18' 19' Small AC, scattered v'yds along n Lo, Cosne to Gien. Bright, lemony SAUV BL like lighter SANCERRE; can be v.gd and gd value. Light reds blend GAMAY/PINOT N. Best: Berthier, BOURGEOIS, Langlois, Charrier, Émile Balland, Paulat, Treuillet, Villargeau. 19 gd, decent volume.

Coteaux de Glanes SW Fr r p ★★ DYA IGP from Upper Dordogne. Nine-man co-op can't make enough wine to satisfy thirst of locals so rarely found far from base. Ségalin grape distinguishes it from other mainstream blends.

Coteaux du Layon Lo w sw ★★→★★★★ 10' 11' 14' 15' 16 17 18' Heart of ANJOU: top long-lived sweet CHENIN BL. Seven villages can add name to AC. Chaume now Layon PC. Top ACs: BONNEZEAUX, QUARTS DE CHAUME. Growers: Baudouin, BAUMARD, Breuil, *Ch Pierre-Bise*, Chauvin, Delesvaux, Forges, Guegniard, Juchepie, Ogereau, *Pithon-Paillé*, Soucherie. Great wine, difficult to sell, producers moving to dry Chenin.

Coteaux du Loir Lo r p w dr sw ★→★★★ 16 17 18' 19' Le Loir is a n tributary of La Loire: exciting region with Coteaux du Loir, *Jasnières*. Steely, fine, precise, long-lived CHENIN BL, reds: GAMAY, peppery Pineau d'Aunis – also fizz, plus Grolleau (p), CAB, CÔT. Best: Ange Vin, Breton, *Dom de Bellivière* (v.gd), Fresneau, Gigou, Janvier, Le Briseau, Les Maisons Rouges, Roche Bleue. 19 v. severe frost.

Coteaux du Lyonnais r p (w) ★ DYA Jnr BEAUJ. Best en PRIMEUR. NB: Guillaume Clusel.

Coteaux du Quercy SW Fr r p ★ 15' 16 17 (18) AOP s of CAHORS. Hearty country wines based on CAB FR keep well. Co-op challenged by independents: ★★ DOMS du Guillau, Merchien (IGP); ★ DOMS d'Ariès, Mazuc, Mystère d'Éléna (Dom de Revel).

Coteaux de Saumur Lo w sw ★★→★★★ 15' 16 17 18' Age-worthy late-harvest CHENIN BL – delicate, citrus, honeyed esp CH de Bréze, Champs Fleuris, GRATIEN & MEYER, Nerleux, Robert et Marcel, DOM de St-Just, Targé, Vatan.

Coteaux et Terrasses de Montauban SW Fr r p w ★→★★ DYA. Romain family invented and still dominate this local IGP with DOM de Montels' range of easy, gd-value wines often found in local markets. Also ★ Dom Biarnès, Mas des Anges.

Coteaux Varois-en-Provence Prov r p w ★→★★ Higher "Provence Vert" is cooler. Home of "Hollywood corner": Miraval by Brad Pitt and Angelina Jolie, and neighbouring Margui by George Lucas. Otherwise, SYRAH and VIOGNIER in cooler n, interesting IGP Coteaux du Verdon. Try CHX Duvivier, Trians, DOMS des Aspras, du Deffends, du Loou, Les Terres Promises, Routas, St Mitre. Correns is 1st all-organic commune in France.

Coteaux du Vendômois Lo r p w ★→★★ 17 18' 19' AC in Le Loir valley. VIN GRIS from Pineau d'Aunis and peppery reds, also blends of CAB FR, GAMAY, PINOT N. Whites: CHENIN BL, CHARD. Best: Brazilier, Four à Chaux, J Martellière, Montrieux, Patrice Colin; CAVE du Vendômois.

Côtes d'Auvergne Mass C r p (w) ★→★★ AC 18' 19' (410 ha). AC – n and s of Clermont-Ferrand. GAMAY, PINOT N, CHARD. Best reds age 3–4yrs. Villages: Boudes, Chanturgue, Châteaugay, Corent (p), Madargues (r). Producers: Bernard, CAVE St-Verny (v.gd), Maupertuis, Montel, Pradier, Sauvat, Tourlonias.

Côtes de Bordeaux B'x ★ AC launched in 2008 for reds. Embraces and permits cross-blending between CAS, FRANCS, BLAYE, CADILLAC and Ste-Foy. Growers who want to maintain *the identity of a single terroir* have stiffer controls (NB) but can put Cas, Cadillac, etc. before Côtes de B'x. BLAYE-CÔTES DE B'X, FRANCS-CÔTES DE B'X and Ste-Foy-Côtes de B'x also produce a little dry white. Around 1000 growers in group. Represents 10% of B'x production. Try CHX Dudon, Lamothe de Haux, Malagar.

Côtes de Bourg B'x r w ★→★★ 10' 14 15 16 18 (19) Solid, savoury reds, a little white from e bank of Gironde. Mainly MERLOT but 10% MALBEC (80% pre-phylloxera). Top CHX: Brûlesécaille, Bujan, Civrac, *Falfas*, Fougas-Maldoror, Grand-Maison, Grave (Nectar VIEILLES VIGNES), Haut-Guiraud, Haut-Macô, Haut-Mondésir, Macay, Mercier, Nodoz, *Roc de Cambes*, Rousset, Sociondo.

Côtes du Brulhois SW Fr r p (w) ★→★★ 16 17 18 (19) Lively AOP nr Agen; borrows its "black wine" tag from CAHORS. TANNAT (obligatory) and Prunelard notable features. Local co-op unusually supportive of a few independents: ★ CH la Bastide, DOMS Bois de Simon, des Thermes, du Pountet.

Côtes Catalanes Rouss r p w ★★ Wines much better than IGP status suggests: some of ROUSS's best. Innovative growers with v. old bush vines, often at altitude, esp GRENACHE (r w, gris), CARIGNAN (r w, gris). Best wines age well: DOMS Casenove, *Dom of the Bee*, GÉRARD GAUBY, *Jones*, L'Enfants, L'Horizon, La Préceptorie Centernach, Le Soula, Matassa, Olivier Pithon, Padié, Roc des Anges, Soulanes, TRELOAR, Vaquer.

Village life: the best of Côtes du Rhône-Villages

For best of this gd-value appellation, try: Gadagne, MASSIF D'UCHAUX, PUYMÉRAS, SIGNARGUES. CHX Fontségune, Signac; Doms Aure, Bastide St Dominique, *Biscarelle*, Bois St Jean, Cabotte (bio), Coulange, Coste Chaude, Crève Coeur (bio), Echevin (gd w), Grand Veneur, Grands Bois (organic), Gravennes, *Janasse*, Jérome, LES APHILLANTHES, Mas de Libian (bio), Montbayon, Montmartel, *Mourchon*, Pasquiers, Pique-Basse (gd w), Rabasse-Charavin, Réméjeanne, Renjarde, Romarins, Saladin (organic), St-Siffrein, Ste-Anne, Valériane, Viret; CAVE de RASTEAU, Les VIGNERONS d'Estézargues.

Côtes de Duras SW Fr r p w ★→★★★ 15′ 16 17′ 18′ (19) Affordable AOP s of BERGERAC. Nest of passionate organic growers pushing old-stagers aside. ★★★ DOMS Mont Ramé, Mouthes-les-Bihan, Nadine Lusseau, Petit Malromé (value); ★★ La Fon Longue, La Tuilerie la Brille, Les Cours, Les Hauts de Riquet, Mauro Guicheney. CH Condom's ★★★ sweet still outstanding. Traditional growers: ★★ Doms Chater, de Laulan, Grand Mayne.

Côtes du Forez Lo r p (sp) ★→★★ 18′ 19′ Most s Lo AC (147 ha), as far s as CÔTE-RÔTIE. V.gd GAMAY (r p). Bonnefoy, CLOS de Chozieux, Guillot, Mondon & Demeure, Real, Verdier/Logel. AC to explore, exciting IGP: CHARD, CHENIN BL, PINOT GR, RIES, ROUSSANNE, VIOGNIER. 19 continues run of gd yrs.

Côtes de Gascogne SW Fr (r) (p) w ★→★★ DYA IGP. ★★ Giant 900-ha+ DOM Tariquet has most market-share for these theatre-bar wines, light, often aromatic, party-style; boiled-sweet and nail-polish-remover flavours seem popular. ★★ Doms Chiroulet, de Cassagne, d'Arton, d'Espérance, de l'Herré, Horgelus, Ménard, Millet, Miselle, Pellehaut and CII des Cassagnoles. Or ★ Chx de Lauroux, de Magnaut, Papolle, Dom St-Lannes. NB Tariquet's fine Armagnacs: another story.

Côtes du Jura Jura r p w (sp) ★★→★★★ 05′ 09 10 12 14 15′ 16 18′ 19′ Revitalized region, big on natural wines; trendy with sommeliers, so pricey. Light perfumey reds from PINOT N, Poulsard, Trousseau. Try whites from fresh, fruity CHARD to deliberately oxidative SAVAGNIN or blends. Great food wines. *See:* AVIET, BERTHET-BONDET, *Ch d'Arlay*, GANEVAT, LABET, MACLE, MAIRE, *Pélican*, *Tissot*. Try also: *Bourdy*, J-M Petit, Pignier, Pinte. *See also* ARBOIS, CH CHALON, L'ÉTOILE ACS.

Côtes de Millau SW Fr r p w IGP ★ DYA. from nearby Gorges du Tarn. Foster's Millau viaduct celebrated in wines from popular co-op. ★ DOMS du Vieux Noyer, La Tour-St-Martin best of independents. *See* AVEYRON.

Côtes de Montravel SW Fr w sw ★★ 15 16′ 17 18′ (19) Sub-AOP of BERGERAC; attractive, unfashionable off-dry wines usually SÉM-based. Better with foie gras than most stickies; the local apéritif.

Côtes de Provence r p w ★→★★★ (p w) DYA. Prov rosé paradigm has conquered the world: pale, paler, palest, maybe at expense of flavour, but it's the colour of money. Whites (increasingly 100% Rolle) and reds (SYRAH, GRENACHE with MOURVÈDRE nearer coast) can be more interesting. Subzones: FRÉJUS, La Londe, PIERREFEU, STE-VICTOIRE, NOTRE DAME DES ANGES. Leaders: CLOS Cibonne (primarily Tibouren), Gavoty (superb); CHX d'Esclans (part-owned by LVMH), de Selle (Ott), Estandon VIGNERONS, Mascaronne, Gasqui (bio). *See* BANDOL, COTEAUX D'AIX, COTEAUX VAROIS.

Côtes du Rhône S Rh r p w ★→★★ 16′ 18′ 19′ The broad base of S Rh, 170 communes. Incl gd SYRAH of Brézème, St-Julien-en-St-Alban (N Rh). Ranges between enjoyable, handmade, high quality (esp CHÂTEAUNEUF estates, numbers rising) and dull, mass-produced. Lively fruit more emphasized. Mainly GRENACHE, also SYRAH, CARIGNAN. Usually best drunk young. Vaucluse best, then GARD (Syrah). Whites improving.

Côtes du Rhône-Villages S Rh r p w ★→★★★ 16′ 17′ 18 19′ Robust, spiced reds from 7700 ha, incl 21 named S Rh villages (St-Andéol new in 2018), numbers up and up. Best are generous, lively, gd value. Red heart is GRENACHE, plus SYRAH, MOURVÈDRE. Improving *whites*, often incl VIOGNIER, ROUSSANNE added to rich base CLAIRETTE, GRENACHE BL – *gd with food. See* CHUSCLAN, LAUDUN, PLAN DE DIEU, ST-GERVAIS, SABLET, SÉGURET (quality), VALRÉAS, VISAN (improving). New villages from 16 vintage: Ste-Cécile (gd range DOMS), Suze-la-Rousse, Vaison la Romaine. *See box, for best growers.*

Côtes du Roussillon-Villages Rouss r ★→★★★ Varied styles, often v.gd: lots of Carignan, also old-vine GRENACHE BL and Gris for whites. 32 villages: Caramany, Latour de France, Les Aspres, Lesquerde, Tautavel singled out on label.

Independent estates better than co-ops: try Boucabeille, CAZES, Charles Perez, CLOS DES FÉES, Clot de l'Oum, des Chênes, GAUBY, Les VIGNES de Bila-Haut, Mas Becha, Mas Crémat, Modat, Piquemal, Rancy, Roc des Anges, Thunevin-Calvet. Plain Côtes du Rouss to s makes simple warming reds. *See also* CÔTES CATALANES.

Côtes du Tarn SW Fr r p w ★ DYA. IGP. For GAILLAC nonconformists and growers just outside that AOP. Moelleux SAUV BL from ★★ DOM d'en Segur and Lou Bio range from Dom Vignes de Garbasses outstanding.

Côtes de Thongue L'doc r p w ★★ (p w) IGP. DYA, but best reds age. DOMS DE L' ARJOLLE, Condamine l'Evèque, l'Horte, la Croix Belle. Dom de Bassac one of 1st organic producers in the Midi.

Côtes de Toul Al r p w ★ DYA. V. light wines from Lorraine; mainly VIN GRIS.

Côtes du Vivarais S Rh r p w ★ 16' 18 Mostly DYA. Across hilly ARDÈCHE country w of Montélimar. Much improved: cool fruit, easy-drinking, based on GRENACHE, SYRAH; some more deep, oak-aged reds. NB: Gallety (best, full, lives well), Mas de Bagnols, *Vignerons de Ruoms* (v.gd value).

Coulée de Serrant Lo w dr sw ★★★ 10' 11 12 13 14 15 16 18' Historic, steep CHENIN BL AC monopole 7-ha site overlooking Lo in heart of AC SAVENNIÈRES. Nicolas Joly bio-pope, daughter Virginie in charge. Oxidative style, best decanted; marked bottle and vintage variation – unacceptably high?

Courcel, Dom de C d'O r ★★★ Leading POMMARD estate, fine floral wines using ripe grapes and whole bunches. Top, age-worthy PCS Rugiens and Épenots, plus interesting Croix Noires.

Crémant AC for quality classic-method sparkling from AL, B'X, BOURGOGNE, Die, Jura, LIMOUX, Lo, Luxembourg, SAV. Many gd examples.

Crépy Sav w ★★ AC 35 ha. Light w, s shore of Lake Geneva. 100% CHASSELAS. Much sold locally. Fichard, La Tour de Marignan, Mercier, Ripalle.

Criots-Bâtard-Montrachet C d'O ★★★ 09 10 12 14' 15 17' 18 19' Tiny MONTRACHET satellite. 1.57 ha. Less concentrated than BÂTARD-M. Try Belland, Blain-GAGNARD, Caroline MOREY, Fontaine G, LAMY, or d'Auvenay if you're v. rich.

Cros Parantoux Burg ★★★★ Cult PC in VOSNE-ROMANÉE made famous by the late Henri Jayer. Now made to great acclaim and greater price by DOMS ROUGET and MÉO-CAMUZET. But Brûlées better in cool vintages?

Crozes-Hermitage N Rh r w ★★→★★★ 10' 15' 16' 17' 18 19' SYRAH from mainly flat v'yds nr River Isère, hot summer challenges: dark-berry, liquorice, tar; mostly early-drinking (2–5 yrs). Stylish, complex, cooler from granite hills nr HERMITAGE: fine, red-fruited, can age. Best (simple CUVÉES) ideal for grills, parties. Some oaked, older-vine wines cost more, can age. Top: *A Graillot*, Aléofane (r w), Belle (organic), *Chapoutier*, Dard & Ribo (organic), *Delas* (Le CLOS v.gd, DOM des Grands Chemins), E Darnaud, G Robin; Doms Combier (organic), de *Thalabert* of JABOULET, des Entrefaux (oak), des Hauts-Châssis, des Lises (fine), *du Colombier*, Dumaine (organic), Fayolle Fils & Fille (v. stylish), Habrard (organic), Les Bruyères (bio, big fruit), Machon, Martinelles, Melody, Michelas St Jemms, Mucyn (fine), Remizières (oak), Rousset, Ville Rouge, Vins de Vienne, Y Chave. Drink *white* (MARSANNE) early, v.gd vintages recently. Value.

Cuve close Quicker method of making sparkling wine in a tank. Bubbles die away in glass much quicker than with *méthode traditionnelle*.

Cuvée Usually used to indicate a particular blend. In CHAMP, means 1st and best wines off the press.

Dagueneau, Didier Lo w ★★★→★★★★ 12' 13 14' 15' 16' 17 18' (19) Best POUILLY-FUMÉ DOM; from 2017 VDF due to legal dispute; precise, age-worthy SAUV BL. Run by Louis-Benjamin and Charlotte Dagueneau. CUVÉES: Buisson Renard, Pur Sang, Silex. SANCERRE (Le Mont Damné, CHAVIGNOL), Les Jardins de Babylone (JURANÇON).

Dauvissat, Vincent Chab w ★★★★ Imperturbable bio producer of CHAB using old barrels and local 132-litre *feuillettes*. Grand, age-worthy wines similar to RAVENEAU cousins. Best: La Forest, Les CLOS, Preuses, Séchet. Try DOM Jean D (no relation).

Deiss, Dom Marcel Al r w ★★★ Bio grower at Bergheim. Favours blends from v'yds with different varieties co-planted; mixed success, variable. Best wine RIES Schoenenbourg 13 14 17'.

Delamotte Champ Fine, small, CHARD-dominated house. Managed with SALON by LAURENT-PERRIER. BRUT, BLANC DE BLANCS, brilliant 07 08 CUVÉE Nicholas Delamotte. Newly appointed L-P cellarmaster Dominique Demarville (2020), ex-VEUVE CLICQUOT, should bring some v. rigorous thinking to SALON in particular.

Delas Frères N Rh r p w ★★★ N Rh v'yd owner/merchant, with v'yds in CONDRIEU, CROZES-HERMITAGE, CÔTE-RÔTIE, HERMITAGE. Quality rising, new cellars. Best: *Côte-Rôtie Landonne*, Hermitage DOM des Tourettes (r w), Les Bessards (r, terroir, v. fine, smoky, long life), ST-JOSEPH Ste-Épine (r, tight, interesting). S Rh: esp CÔTES DU RH St-Esprit (r), Grignan-les-Adhémar (r, value). Whites lighter recently. Owned by ROEDERER.

Delaunay, Edouard C d'O r w ★★→★★★ Old Burg name revived in NUITS and l'Étang-Vergy by Laurent D. Gd range NÉGOCIANT CUVÉES all price points. Off to gd (re-)start.

Demi-sec Half-dry: but in practice more like half-sweet (eg. CHAMP typically 45g/l dosage).

Derenoncourt, Stéphane B'x Leading international consultant; self-taught, focused on terroir, fruit, balance. Own property, *Dom de l'A* in CAS.

Deutz Champ One of top small CHAMP houses, ROEDERER family-owned; gets better and better. Improved BRUT Classic NV; Brut 06 08 12 lovely, harmonious. Top-flight CHARD CUVÉE Amour de Deutz 06, Amour de Deutz Rosé 06. *Superb Cuvée William Deutz* 95. New Parcelles d'AŸ 10 12'. BLANC DE BLANCS 13 brilliant value.

Dirler-Cadé, Dom Al Marriage (2000) of Jean Dirler/Ludivine Cadé meant 23 ha in warm sandstone soils. Substance and finesse. Some certified bio. Marvellous old-vines MUSCAT GC Saering 14 16 17' 18. Excellent Saering RIES, rich yet finely sketched 10 14' 16 17' 18 19.

Domaine (Dom) Property. *See* under name, eg. TEMPIER, DOM.

Dom Pérignon Champ Vincent Chaperon, now chef de CAVE, brings own style to this luxury CUVÉE of MOËT & CHANDON. Ultra-*consistent quality in incredible quantities*; maintained with creamy allure, esp with 10–15 yrs bottle-age. Plénitude releases have long bottle-age, recent disgorgement, huge price, at 7, 16, 30 yrs+ (P1, P2, P3); superb P2 98; still vibrant P3 70. More PINOT N focus in DP since 2000, esp exquisite 06 in both Blanc and Rosé.

Dopff au Moulin Al w ★★★ →★★★★ Blue-chip family producer, esp GEWURZ GCS Brand, Sporen 09 12 15' 16 17' 18' 19; lovely RIES SCHOENENBOURG 13 exceptional come 2020; *Sylvaner de Riquewihr* 14. Pioneer of AL CRÉMANT; v.gd CUVÉES: Bartholdi, Julien, Bio. Specialist in classic dry wines.

Doué, Didier Champ ★★ Refreshing CHAMP: finesse, no wood; from the AUBE.

Dourthe B'x Sizeable merchant/grower; wide range, quality emphasis. CHX: BELGRAVE, LA GARDE, LE BOSCQ, Grand Barrail Lamarzelle FIGEAC, PEY LA TOUR, RAHOUL, REYSSON. *Dourthe No 1* (esp w) well-made generic B'X.

Drappier, Michel Champ Great family-run AUBE house, characterful CHAMP. DOM of fine PINOT N, 60 ha+, certified bio; *Pinot-led NV*, BRUT ZÉRO, Brut *sans souffre*, Millésime d'Exception 12; ace Prestige CUVÉE Grande Sendrée 15 17 18'. Cuvée Quatuor (four CÉPAGES). Superb 95' 82 (magnums). Constant research into early C17 vines resistant to climate change.

DRC (Dom de la Romanée-Conti) C d'O r w ★★★★ Grandest estate in Burg (or world). This emperor is definitely wearing clothes. MONOPOLES ROMANÉE-CONTI and LA TÂCHE, major parts of ÉCHÉZEAUX, GRANDS-ÉCHÉZEAUX, RICHEBOURG, ROMANÉE-

ST-VIVANT and a tiny part of MONTRACHET. Also CORTON from 2009, CORTON-CHARLEMAGNE from 2019. Long-term patron Aubert de Vilaine is retiring. Crown-jewel prices. Keep top vintages for decades.

Drouhin, Joseph & Cie Burg r w ★★★→★★★★ Fine family-owned grower/NÉGOCIANT in BEAUNE; v'yds (all bio) incl (w) Beaune *Clos des Mouches*, MONTRACHET (Marquis de LAGUICHE) and large CHAB holdings. Stylish, fragrant reds from pretty CHOREY-LÈS-BEAUNE through great ranges in BEAUNE, CHAMBOLLE, VOSNE (NB Petits Monts) and now GEVREY. Also Domaine Drouhin Oregon, *see* US.

Duboeuf, Georges Beauj r w ★★→★★★ From hero – saviour of BEAUJ – to less so (too much BEAUJ NOUVEAU), always major player and still v. sound source for multiple Beauj crus and MÂCON bottlings. Georges D RIP 2020; son Franck has been in charge for some yrs.

Dugat C d'O r ★★★ Cousins Claude and Bernard (Dugat-Py) made excellent, deep-coloured GEVREY-CHAMBERTIN, respective labels. Both flourishing with new generation. Tiny volumes, esp GCS, huge prices, esp Dugat-Py. Collector territory. But try village Gevrey from either for a (just) affordable thrill.

Dujac, Dom C d'O r w ★★★→★★★★ MOREY-ST-DENIS grower originally noted for sensual, smoky reds, from unbeatable village Morey to outstanding GCS, esp CLOS DE LA ROCHE, CLOS ST-DENIS, ÉCHÉZEAUX. Slightly more mainstream these days. Gd *whites* from Morey and PULIGNY. Lighter merchant wines as D Fils & Père and DOM Triennes in COTEAUX VAROIS.

Dureuil-Janthial Burg r w ★★ Top DOM in RULLY in capable hands of Vincent D-J, with *fresh, punchy whites* and cheerful, juicy reds. Try Maizières (r w) or PC Meix Cadot (w).

Durup, Jean Chab w ★★ Volume CHAB producer as DOM de l'Eglantière and CH de Maligny, allied by marriage to Dom Colinot in IRANCY.

Duval-Leroy Champ Family-owned house, 200 ha of mainly fine CHARD crus. New generation now in charge, focusing on DOM wines. V.gd Fleur de CHAMP NV. Top Blanc de Prestige *Femme 13* ★★★★ is one of Champ's top prestige CUVÉES. Also great in 18' 19.

Échézeaux C d'O r ★★★ 99' 02' 05' 09' 10' 12' 15' 16 17 18' 19' GC next to CLOS DE VOUGEOT, but totally different style: lacy, ethereal, scintillating. Can vary depending on exact location. Best: ARNOUX-LACHAUX, Berthaut-Gerbet, Bizot, Coquard-Loison-Fleurot, DRC, DUJAC, EUGÉNIE, G NOËLLAT, GRIVOT, GROS, Guyon, Lamarche, LIGER-BELAIR, MÉO-CAMUZET, Millot, MUGNERET, Mugneret-Gibourg, Naudin-Ferrand, ROUGET, Tardy, Tremblay.

Ecu, Dom de l' Lo (r) w dr (sp) ★★★ 18' 19' Bio. Big range of varied wines. MUSCADET SÈVRE ET MAINE, GROS PLANT, VDF. Run by dynamic Fred Niger. Gd CAB FR, PINOT N. Big winery under construction 2019. Many amphorae. 18 quality, volume; frost again 19.

Edelzwicker Al w ★ DYA. Blended light white. CH d'Ittenwiller, HUGEL Gentil gd.

Eguisheim, Cave Vinicole d' Al r w ★★ Model co-op. Excellent value: fine GCS Hatschbourg, HENGST, Ollwiller, Spiegel. Owns Willm. Top label: WOLFBERGER. Best: Grande Rés 10, Sigillé, Armorié. Gd CRÉMANT, PINOT N, esp 15' 18.

Emmanuel Brochet Champ ★★★ Bijou producer, exceptional CHAMP from steep Mont Bernard. Extra BRUT pure, exhilarating, winemaking slow and patient, in barrel 9 mths. Certified organic, no fining or filtration. A true artisan.

Entraygues et du Fel and Estaing SW Fr r p w ★→★★ DYA. Two tiny AOP neighbours in almost vertical terraces above Lot Valley. Bone-dry CHENIN BL for whites, esp ★★ DOMS Laurent Mousset (gd reds, esp La Pauca, excellent rosé), Méjanassère ★★ Nicolas Carmarans making wines in and out of AOP.

Entre-Deux-Mers B'x w ★→★★ DYA. Often gd-value dry white B'x (drink *entre d* huitres) from between the Rivers Garonne and Dordogne. Best: CHX Beaure

Ducourt, BONNET, Chantelouve, Fontenille, Haut-Rian, Landereau, *Le Coin* (for Sauvignon Gris), Les Arromans, Lestrille, Marjosse, Martinon, Nardique-la-Gravière, Sainte-Marie, *Tour de Mirambeau*, Turcaud.

Esmonin, Dom Sylvie C d'O r ★★★ Rich, dark wines from fully ripe grapes, whole-bunch vinification and new oak. Best: GEVREY-CHAMBERTIN VIEILLES VIGNES, CLOS ST-JACQUES. Cousin Frédéric has gd Estournelles St-Jacques.

Etoile, L' Jura w ★★ AC of Jura best known for elegant CHARD grown on limestone and marl. VIN JAUNE and VIN DE PAILLE also allowed but not reds. DOM de Montbourgeau is reference, esp En Banode. Also try: Cartaux-Bougaud, Philippe Vandelle.

Eugénie, Dom C d'O r (w) ★★★→★★★★ Former DOM Engel in VOSNE, now owned by Artemis Estates. Powerful, dark-coloured wines. CLOS VOUGEOT, GRANDS-ÉCHÉZEAUX best, but try village CLOS d'Eugenie.

Faiveley, Dom Burg r w ★★ →★★★ More grower than merchant, revitalized by Erwan F since 2005, incl major overhaul of facilities. Now making high-class reds and sound whites. Leading light in CÔTE CHALONNAISE, but save up for top wines from CHAMBERTIN-CLOS DE BÈZE, CHAMBOLLE-MUSIGNY, CORTON *Clos des Cortons*, NUITS. Ambitious recent acquisitions throughout C D'O and DOM Billaud-Simon (CHAB).

Faller, Théo / Weinbach, Dom Al w ★★★ →★★★★ Probably finest v'yds in AL, wines of great *character and elegance*. Drier gastronomic style, supremely expressed in GC SCHLOSSBERG 10. Also RIES L'inedit 13 mineral complexity. SELECTION DES GRAINS NOBLES GEWURZ as gd as it gets.

Faugères L'doc r (p) (w) ★→★★★ Rare instance of single-soil AC. schist on s-facing foothills of Cevennes. Lots of organic v'yds; fresh, spicy reds, will age. Elegant whites from MARSANNE, ROUSSANNE, VERMENTINO, GRENACHE BL. Drink DOMS Cébène, Chaberts, Chenaie, des Trinités, JEAN-MICHEL ALQUIER, LÉON BARRAL, Mas d'Alezon, Ollier-Taillefer, St Antonin, Sarabande, many others.

Fevre, William Chab w ★★★ →★★★★ Biggest owner of CHAB GCS; Bougros Côte Bougerots and Les CLOS outstanding. Small yields, no expense spared, priced accordingly, top source for rich, age-worthy wines and some more humble. Cousins N&G Fèvre also sound.

Fiefs Vendéens Lo r p w ★→★★★ 16 17 18' 19' Mainly DYA AC. From the Vendée nr Sables d'Olonne, some serious age-worthy wines. CHARD, CHENIN BL, MELON, SAUV BL (w), CAB FR, CAB SAUV, GAMAY, Grolleau Gris, Négrette, PINOT N (r p). Esp: Coirier, DOM St-Nicolas (bio v.gd), Mourat (organic), Pheuré-la-Chaume (bio). V.gd 18, small gd 19.

Fitou L'doc r ★★ Mostly CARIGNAN, so expect spice, richness, chewiness. Character too. Two parts: schist on inland hills s of Narbonne; and chalk, limestone nr coast. Seek out CH de Nouvelles, Champs des Soeurs, DOM Bergé-Bertrand, de Rolland, Jones, Lérys.

Fixin C d'O r (w) ★★★ 05' 09' 10' 12' 14 15' 16 17 18' 19' Worthy, undervalued n neighbour of GEVREY-CHAMBERTIN. Sturdy, sometimes splendid reds, can be rustic but enjoying warmer vintages. Best v'yds: Arvelets, CLOS de la Perrière, Clos du Chapitre, Clos Napoléon. Top locals: Berthaut-Gerbet, Gelin, Joliet, Naddef plus v.gd Bart, Bichot, CLAIR, FAIVELEY, MORTET.

Fleurie Beauj r ★★★ 14 15' 18' 19' Top BEAUJ cru for perfumed, strawberry fruit, silky texture. Racier from La Madone hillside, richer below. Classic names: CHX Beauregard, Chatelard, de Poncié; DOMS Brun, Chignard, CLOS de la Roilette, Depardon, DUBOEUF, Métrat; co-op. Naturalists: Balagny, Dutraive, Métras, Pacalet, Sunier. Newcomers: Chapel, Clos de Mez, Dom de Fa, Hoppenot, Lafarge-Vial.

Fourrier, Domaine C d'O r ★★★★ Jean-Marie F has taken a sound GEVREY-CHAMBERTIN DOM to new levels, with cult prices to match. Sensual vibrant reds at all levels. Best CLOS ST-JACQUES, Combe aux Moines, GRIOTTE-CHAMBERTIN.

Francs-Côtes de Bordeaux B'x r w ★★ 10' 14 15 16 18 (19) Tiny B'X AC next to CAS. Fief of Thienpont (PAVIE-MACQUIN) family. Mainly red from MERLOT, but some gd white (eg. Charmes-Godard). 50% of growers organic or bio. Top CHX: Cru Godard, Franc Cardinal, Francs, La Prade, Marsau, Puyfromage, *Puygueraud*.

Fréjus Prov r p ★★ Volcanic soils in e subzone of CÔTES DE PROV give tight wines that age. Must incl high proportion of Tibourenc grape, which restricts use of the denomination. Try CH de Rouet. Ch d'Esclans is here, but doesn't use Fréjus label.

Fronsac B'x r ★★→★★★ 10' 14 15 16 18 (19) Underrated hilly AC w of ST-ÉM; great-value MERLOT-dominated red, some ageing potential. Top CH: Arnauton, DALEM, Fontenil, *la Dauphine*, la Rivière, la Rousselle, LA VIEILLE CURE, LES TROIS CROIX, Haut-Carles, Mayne-Vieil (CUVÉE Alienor), *Moulin-Haut-Laroque*, Puy Guilhem, Tour du Moulin, Villars. *See also* CANON-FRON.

Fronton SW Fr r p ★★ 15 16 18 AOP n of Toulouse. Must be based on rare (sometimes unblended) Négrette grape (flavours of violets, cherries, liquorice). ★★★ CHX Baudare, Caze, du Roc, Laurou, Plaisance (esp sulphur-free Alabets). Best-known ★★ Chx *Bellevue-la-Forêt*, *Bouissel*, but try Boujac, La Colombière, DOMS des Pradelles, Viguerie de Belaygues. No AOP for whites as yet.

Fuissé, Ch Burg w ★★→★★★ Smart operation in POUILLY-FUISSÉ with long track record. Concentrated oaky whites. Top terroirs Le CLOS, Combettes. Also BEAUJ crus, eg. JULIÉNAS.

Gagnard C d'O (r) w ★★★→★★★★ Clan in CHASSAGNE. Long-lasting wines, esp Caillerets, BÂTARD from Jean-Noël G; while Blain-G, Fontaine-G have full range incl rare CRIOTS-BÂTARD, MONTRACHET itself. Gd value offered by all Gagnards. Decent Chassagne reds all round.

Chambertin Clos de Beze was defined in yr 630; same boundaries today.

Gaillac SW Fr r p w dr sw sp ★→★★ 15' 16 18' (19') Rare grapes Braucol, Duras, Len de l'El, Mauzac, Prunelard distinguish this historic AOP w of Albi. Young bio producers stealing march on establishment. Look for ★★★ DOMS Brin, Causse-Marines, La Ramaye, La Vignereuse, Le Champ d'Orphée, Peyres-Roses, PLAGEOLES, Rotier. ★★ L'Enclos des Braves, L'Enclos des Roses, La Ferme du Vert; d'Escausses, du Moulin, Mayragues, More traditional from CH Larroque, Doms Labarthe, La Chanade, Mas Pignou. ★★ Ch Bourguet (w sw).

Ganevat Jura r w ★★★→★★★★ CÔTES DU JURA superstar. Single-v'yd CHARD (eg. Chalasses, Grand Teppes), pricey but fabulous. Also innovative reds.

Gauby, Dom Gérard Rouss r w ★★★ Bio, increasingly natural; lots of innovation; other producers followed Gauby to village of Calce. Son taking over. High-altitude v'yds up to 550m (1804ft), chalk for fresh acidity. Try Muntada; Les Calcinaires VIEILLES VIGNES. Dessert wine Le Pain du Sucre. Associated with DOM Le Soula.

Gers SW Fr r p w ★ DYA IGP usually sold as CÔTES DE GASCOGNE; comes to same thing.

Gevrey-Chambertin C d'O r ★★★ 99' 02' 05' 09' 10' 12' 15' 16 17 18' 19' Major AC for fine savoury reds at all levels up to great CHAMBERTIN and GC cousins. Top PCS Cazetiers, Combe aux Moines, Combottes, CLOS ST-JACQUES. Value from single-v'yd village wines (En Champs, La Justice) and VIEILLES VIGNES bottlings. Top: BACHELET, BOILLOT, Burguet, Damoy, DROUHIN, Drouhin-Laroze, DUGAT, Dugat-Py, Duroché, ESMONIN, FAIVELEY, FOURRIER, Géantet-Pansiot, Harmand-Geoffroy, Heresztyn-Mazzini, JADOT, LEROY, Magnien, Marchand-Grillot, MORTET, ROSSIGNOL-TRAPET, Roty, ROUSSEAU, Roy, SÉRAFIN, TRAPET.

Gigondas S Rh r p ★★→★★★ 01' 05' 06' 09' 10' 12' 13' 15' 16' 17' 18 19' Top S Rh red. Striking v'yds on stony clay-sand plain rise to alpine limestone hills e of Avignon; GRENACHE, plus SYRAH, MOURVÈDRE. Robust, clear, menthol-fresh wines; best offer fine dark-red fruit. Top 10 15 16. More oak recently, higher prices, but genuine local feel in many. Try: Boissan, Bosquets (gd modern), Bouïssière (punchy), Brusset, Cayron, CH de Montmirail, CH DE ST COSME (swish), **Clos des Cazaux** (value), CLOS du Joncuas (organic, traditional), DOM **Famille Perrin**, Goubert, Gour de Chaulé (fine), Grapillon d'Or, Longue Toque, Moulin de la Gardette (stylish), Notre Dame des Pallières, **Les Pallières**, P Amadieu (v. consistent), Pesquier (terroir), Pourra (robust, character), **Raspail-Ay**, Roubine, ST GAYAN (long-lived), Santa Duc (now stylish), Semelles de Vent, Teyssonières. Heady rosés.

Gimonnet, Pierre Champ ★★→★★★ 28 ha of GCS and PCS on n Côte des Blancs making beautifully consistent CHARD. Ace CUVÉE Gastronome for seafood 13 15 17'. Not a fan of single-v'yd CHAMP. Cuvées Fleuron, Club unmatched complex expression of great Chard 13 17' for long ageing due to intricate assembly of diverse v'yds.

Ginglinger, Dom Paul Al 13th-generation Michel makes pure, terroir-expressive RIES, PINOT BL. Excellent CRÉMANT.

Girardin, Vincent C d'O r w ★★→★★★ White-specialist MEURSAULT-based grower/ NÉGOCIANT, now under BOISSET ownership. Original coming back to Meursault as DOM Pierre Vincent: delicate, detailed wines.

Givry Burg r (w) ★★ 15' 17 18' 19' Top tip in CÔTE CHALONNAISE for tasty reds that can age. Better value than MERCUREY. Rare whites nutty in style. Best (r): CLOS Salomon, *Faiveley*, F Lumpp, JOBLOT, Masse, Thénard.

Goisot Burg r w ★★★ Guilhem & J-H G, outstanding bio producers of single-v'yd bottlings of ST-BRIS (SAUV BL) and Côtes d'Auxerre for CHARD, PINOT N. Nobody else comes close.

Gonon, Dom N Rh r w ★★★ 09' 10' 11 12 13' 14 15' 16' 17 18' 19' Top estate at ST-JOSEPH; bros Pierre and Jean work organically, hand-graft cuttings on 10 ha prime old v'yds. Mainly whole-bunch, old 600-litre casks, aromatic, peppered, iron-fused red, most savoury, textured Les Oliviers (w), both live 20 yrs.

Gosset Champ Oldest house, based in AŸ, owned by Cointreau and blossoming. Chef de CAVE Odilon de Varine is passionate about terroir. Grand Blanc de Meunier is a 1st from a Grande Marque, mainly 07, elegant, aged on CHARD lees. Prestige Celebris Bris Extra BRUT one of best. Classic in 04; sublime double-aged Gosset Célébresienne 95 ★★★★.

Gosset-Brabant Champ AŸ grower, great PINOT N v'yds. Nuits d'Aÿ 12 15 18, cathedral of top Pinot flavours.

Gouges, Henri C d'O r w ★★★ Reference point over several generations for meaty, long-lasting NUITS-ST-GEORGES with Grégory G now at helm. Great range of PC v'yds, eg CLOS des Porrets, Vaucrains and esp Les St Georges. Also rare, excellent *white Nuits*, from PINOT BL.

Graillot, Dom Alain N Rh r w ★★★ 13' 15' 16' 17 18' 19' Alain woke up CROZES-HERMITAGE in 1985. 20 ha organic v'yds, whole-bunch ferments. Racy CROZES (r): La Guiraude special selection. Crozes (w), ST-JOSEPH (r). Son Maxime: Crozes DOM des Lises, gd merchant range Equis.

Gramenon, Dom S Rh r (w) ★★→★★★ 18' 19' Organic for decades, bio since 2007, in lower Drôme. Fruit purity, v. low sulphur. CÔTES DU RH: Poignée des Raisins (GRENACHE), La Sagesse, Sierra du Sud (SYRAH).

Grand Cru (GC) Official term meaning different things in different areas. One of top Burg v'yds with its own AC. In AL, one of 51 top v'yds, each now with its own rules. In ST-ÉM, 60% of production is St-Ém GC, often run-of-the-mill. In MÉD there are five tiers of GC CLASSÉS. In CHAMP top 17 villages are GCs. Since 2011 in Lo for QUARTS DE CHAUME, and emerging system in L'DOC. Take with pinch of salt in Prov

Grande Rue, La C d'O r ★★★ 02' 03 05' 06 09' 10' 12' 15' 16 17 18' 19' MONOPOLE of DOM Lamarche, GC between LA TÂCHE, ROMANÉE-CONTI. Quality, consistency improved under Nicole L. Fascinating blood-orange hallmark across vintages.

Grands-Échézeaux C d'O r ★★★★ 90' 93 96' 99' 02' 03 05' 09' 10' 12' 15' 17 18' 19' Superlative GC next to CLOS DE VOUGEOT, but with a MUSIGNY silkiness. More weight than most ÉCHÉZEAUX. Top: BICHOT (CLOS Frantin), Coquard-Loison-Fleurot, DRC, DROUHIN, EUGÉNIE, G NOËLLAT, Lamarche, Millot.

Grange des Pères, Dom de la L'doc r w ★★★ Next door to MAS DE DAUMAS GASSAC, and IGP. Red from CAB SAUV, SYRAH, MOURVÈDRE; white 80% ROUSSANNE, plus MARSANNE, CHARD. Stylish wines with ageing potential.

Gratien, Alfred and Gratien & Meyer Champ ★★★ BRUT 93 12 13 15' 18' Small but beautiful CHAMP house, owned by Henkell Freixenet. Brut NV. CHARD-led Prestige CUVÉE Paradis Brut, Rosé (MV). Fine, v. dry, lasting, oak-fermented wines incl *The Wine Society's house Champagne*. Careful buyer of top crus, esp CHARD from favourite growers. Also Gratien & Meyer (r sp sw) in SAUMUR.

Graves B'x r w ★ →★★ 14 15 16 18 (19) Gravel soils provide the name; best of region is PESSAC-LÉOGNAN. Appetizing grainy reds from MERLOT, CAB SAUV, fresh SAUV/SÉM dry whites. Some of best values in B'x today. Top CHX: ARCHAMBEAU, Brondelle, CHANTEGRIVE, CLOS Bourgelat, *Clos Floridène*, CRABITEY, de Cérons, Ferrande, Fougères, Grand Enclos du Ch de Cérons, Haura, Magneau, Pont de Brion, RAHOUL, *Respide-Médeville*, Roquetaillade La Grange, St-Robert CUVÉE Poncet Deville, *Vieux Ch Gaubert*, Villa Bel Air.

Graves de Vayres B'x r w ★ DYA. Tiny AC within E-2-M zone. Red, white, *moelleux*.

Paris has urban winery, 58m (190ft) up Eiffel Tower. 1st release 2020.

Grignan-les-Adhémar S Rh r (p) w ★ →★★ 16' 17' 18 19' Mid-Rh AC; best reds hearty, herbal. Leaders: Baron d'Escalin, DELAS (value), La Suzienne (value), CHX Bizard, La Décelle (incl w Côtes du Rh), Mas Théo (bio), DOMS de Bonetto-Fabrol, de Montine (stylish r, gd p w, also r CÔTES DU RH), Grangeneuve best (esp VIEILLES VIGNES), St-Luc.

Griotte-Chambertin C d'O r ★★★★ 90' 96' 99' 02' 03 05' 09' 10' 12' 15' 16 17 18' 19' Small GC next to CHAMBERTIN; nobody has much volume. Brisk red fruit, with depth and ageing potential, from DROUHIN, DUGAT, Duroché, FOURRIER, *Ponsot (Laurent)*.

Gripa, Dom N Rh r w ★★★ 10' 13' 15' 16' 17' 18' 19' Top ST-JOSEPH, ST-PÉRAY DOM, whites a speciality. St-Joseph Le Berceau (w) 100% 60-yr+ MARSANNE; *St-Péray Les Figuiers*, mainly ROUSSANNE. Both St-Joseph reds gd, top Le Berceau tracks vintage: deep 15, pure-fruit 16, dense 17, bold 18.

Grivot, Jean C d'O r w ★★★→★★★★ VOSNE-ROMANÉE DOM that keeps on improving, likely to continue as Mathilde G starts to take over. Superb range of PCS (NB Beaux Monts and NUITS Boudots) topped by GCS CLOS DE VOUGEOT, ÉCHÉZEAUX, RICHEBOURG. Higher prices these days.

Gros, Doms C d'O r w ★★★→★★★★ Family of VIGNERONS in VOSNE-ROMANÉE with stylish wines from Anne (sumptuous RICHEBOURG), succulent reds from Michel (CLOS de Réas), Anne-Françoise (now in BEAUNE) and Gros Frère & Soeur (CLOS VOUGEOT En Musigni). Not just GCS; try value HAUTES-CÔTES DE NUITS. Also Anne's DOM Gros-Tollot in MINERVOIS.

Gros Plant du Pays Nantais Lo w (sp) ★ →★★ DYA. Much-improved AC for GROS PLANT (FOLLE BLANCHE); best racy, saline, ideal oyster partner. Try: Basse Ville, ECU, Haut-Bourg, Luneau-Papin, Poiron-Dabin, Preuille. Sparkling: pure or blended. Gd 19 but small crop.

Guigal, Ets E N Rh r w ★★→★★★★ Famous, always-expanding grower-merchant: CÔTE-RÔTIE mainly, plus CONDRIEU, CROZES-HERMITAGE, HERMITAGE, ST-JOSEPH, 52-ha CHÂTEAUNEUF CH de Nalys, plus two lots of 7-ha and 18-ha v'yds there.

Merchant: Condrieu, Côte-Rôtie, Crozes-Hermitage, Hermitage, S Rh. Owns DOM de Bonserine (sturdy Côte-Rôtie), VIDAL-FLEURY (fruit, quality on up). Top, v. expensive Côte-Rôties La Mouline, La Landonne, La Turque (ultra rich, new oak for 42 mths, so atypical), also v.gd Hermitage, St-Joseph VIGNES de l'Hospice; all reds dense. Standard wines: gd, esp **top-value Côtes du Rh** (r p w). Best whites: Condrieu, Condrieu La Doriane (oaky), Hermitage.

Hautes-Côtes de Beaune / Nuits C d'O r w ★★ ★★ (r) 15' 17 18' 19' (w) 17' 18 19' Generic BOURGOGNE ACS for villages in hills behind CÔTE DE BEAUNE/NUITS. Attractive lighter reds, whites for early-drinking. Best whites: Devevey, Duband, Féry, GROS, Jacob, Jouan, LIGER-BELAIR, Magnien, Naudin-Ferrand, Parigot, Verdet. Useful large co-op nr BEAUNE.

Haut-Médoc B'x r ★★ ★★★★ 10' 14 15 16' 18 (19) Prime source of dry, digestible CAB/MERLOT reds. Usually gd value. Plenty of CRUS BOURGEOIS. Wines usually sturdier in n; finer in s. Five Classed Growths (BELGRAVE, CAMENSAC, *Cantemerle*, *La Lagune*, LA TOUR-CARNET). Other top CHX: Arnauld, BELLE-VUE, CAMBON LA PELOUSE, Charmail, CISSAC, CITRAN, CLOS du Jaugueyron, COUFRAN, D'AGASSAC, *de Lamarque*, Gironville, Lamothe-Bergeron, LANESSAN, Larose Perganson, *Malescasse*, SÉNÉJAC, *Sociando-Mallet*.

Haut-Montravel SW Fr w sw ★★ 15 16' 17 18' Sweetest of three MONTRAVEL white AOPs, best (★★★ CH Puy-Servain-Terrement, DOMS Moulin Caresse ★★, bargain Libarde) should not be overlooked for better-known MONBAZILLAC, SAUSSIGNAC.

Haut-Poitou Lo r p w sp ★ ★★★ 18' 19' Best age 5–6 yrs+. AC for CAB SAUV, CAB Fr, GAMAY, CHARD, PINOT N, SAUV BL. Dynamic *Ampelidae* dominates with IGP. Also La Tour Beaumont, Morgeau La Tour, Sauvion.

Heidsieck, Charles Champ ★★★★ Iconic house, smaller than before, wines more brilliant than ever. *NV Brut* all purity and subtle ripe complexity. Peerless Blanc des Millénaires 04'. Older BRUT Vintage in prime mature form, esp 83 81; CHAMP Charlie Prestige CUVÉE likely to be reintroduced from 2023. New BLANC DE BLANCS NV, nicely priced, delicious.

Heidsieck Monopole Champ Once-great CHAMP house. Fair-quality, gd-price Gold Top 09 12 15. Part of VRANKEN group.

Hengst Al GC for powerful wines: top GEWURZ from JOSMEYER, ZIND-HUMBRECHT; also AUXERROIS, CHASSELAS, PINOT N.

Henri Abelé Champ ★★★ New name for Abel Lepître, oldest house, now focusing on exports, Best CUVÉE Sourire de Reims 12 15, new Sourire Rosé 15 voluptuous, from Les Riceys (AUBE), expansive burgundian style. Gd value.

Henriot Champ Fine family CHAMP house. BRUT Souverain NV much improved; ace BLANC DE BLANCS de CHARD NV; Brut 98' 02' 08; Brut Rosé 06 09. New long-aged *Cuve 38*, a solera of GC Chard since 1990. Exceptional new prestige CUVÉE Hemera 05. Still stocks of long-lived prestige cuvée Les Enchanteleurs 90. Also owns BOUCHARD PÈRE & FILS, FÈVRE.

Hermitage N Rh r w ★★★ ★★★★★ 99' 01' 05' 06' 07' 09' 10' 11' 12' 13' 15' 16' 17' 18' 19' (10 15 both brilliant). Part granite hill on e Rhône bank with grandest, deepest, most stylish SYRAH and complex, nutty/white-fruited, fascinating, v.-long-lived white (MARSANNE, some ROUSSANNE) best left for 6–7 yrs+. Best: Alexandrins, Belle (organic), *Chapoutier* (bio), Colombier, DELAS, Faurie (pure), GUIGAL, Habrard (w), *J-L Chave* (rich, elegant), M Sorrel (mighty Le Gréal r, retired 2018), *Paul Jaboulet Aîné*, Philippe & Vincent Jaboulet (r w), Tardieu-Laurent (oak). TAIN co-op gd (esp Gambert de Loche r, VIN DE PAILLE w).

Hortus, Dom de l' L'doc r w ★★★ Dynamic, family-run PIC ST-LOUP estate. Fine SYRAH-based reds: elegant Bergerie and oak-aged Grande CUVÉE (r). Intriguing Bergerie IGP Val de Montferrand (w) with seven grapes. Also CLOS du Prieur (r) in cooler TERRASSES DU LARZAC.

Hospices de Beaune C d'O Spectacular medieval foundation with grand charity auction of CUVÉES from its 61 ha for Beaune's hospital, 3rd Sunday in Nov, run since 2005 by Christie's. Individuals can buy as well as trade. Winemaker Ludivine Griveau making fine consistent wines. Try BEAUNE cuvées, VOLNAYS or expensive GCS, (r) CORTON, ÉCHÉZEAUX, MAZIS-CHAMBERTIN, (w) BÂTARD-MONTRACHET. Quality high; bargains unlikely; charity is the point.

Hudelot C d'O r w ★★★ VIGNERON family in CÔTE DE NUITS. New life breathed into H-NOËLLAT (VOUGEOT), one of new superstars, esp GCS ROMANÉE-ST-VIVANT, RICHEBOURG, while H-Baillet (CHAMBOLLE) is challenging with punchy reds.

Huet Lo w ★★★★ 10' 14 15' 16' 17 18' 19' Famous VOUVRAY estate, bio since 1990; CHENIN BL benchmark. Anthony Hwang also owns Királyudvar in Tokaji (*see* Hungary). Single v'yds: CLOS du Bourg and Le Mont on Première Côte, Le Haut Lieu nearby. V. age-worthy, esp sweet: 1919 21 24 47 59 89 90 96 97 03 05. Also *pétillant*. Long-serving winemaker/v'yd chief JB Berthomé retired Dec 2019.

Hugel & Fils Al r w sw ★★→★★★★ Revered AL house at Riquewihr, led by 12th-generation Jean-Philippe Hugel, no longer opposed to GC designation. Famed for late-harvest, esp RIES, GEWURZ VENDANGE TARDIVE, SÉLECTION DE GRAINS NOBLES. Superb Ries Schoelhammer 07 10 13 17 from GC site.

IGP (Indication Géographique Protegée) Potentially most dynamic category in France (with over 150 regions), allowing scope for experimentation. Successor to VDP, from 2009 vintage, but new terminology still not accepted by every area. Zonal names most individual: eg. CÔTES DE GASCOGNE, CÔTES DE THONGUE, Pays des Cévennes, Haute Vallée de l'Orb, among others. Enormous variety in taste and quality, but never ceases to surprise.

Irancy Burg r (p) ★★ 05' 15' 16 18' 19' Light though structured red made nr CHAB from PINOT N and more rustic local César. Elbows-on-table stuff. Best v'yds: Palotte, Mazelots. Best: Cantin, *Colinot*, *Dauvissat*, Goisot, Renaud, Richoux.

Irouléguy SW Fr r p (w) ★→★★★ 15' 16 18' (19) The only French Basque wines. Like better-mannered MADIRANS. Based on TANNAT softened by CAB FR. Best from ★★★ Ameztia, Arretxea, Bordaxuria, Brana, Mourguy, Ilarria. Fruity whites based on Petit Courbu and the two MANSENGS; ★★★ Xuri d'Ansa (from gd co-op), DOM Xubialdea. Rosés make perfect summer drinking.

Jaboulet Aîné, Paul N Rh r w Grower-merchant at Tain. V'yds organic, well worked, across HERMITAGE, CONDRIEU, CORNAS, CROZES-HERMITAGE, CÔTE-RÔTIE, ST-JOSEPH. Wines polished, modern, would love more soul. Once-leading producer of HERMITAGE, esp ★★★★ La Chapelle (fabulous 61 78 90), quality varied since 90s, some revival since 2010 on reds. Also CORNAS St-Pierre, *Crozes Thalabert* (can be stylish), Roure (sound). Merchant of other Rh, notably CÔTES DU RH *Parallèle 45*, VACQUEYRAS, VENTOUX (r, quality/value). Whites lack true Rh body, drink most young, range incl new v. expensive La Chapelle (w, not made every yr).

Jacquart Champ Simplified range from co-op-turned-brand, concentrating on what it does best: PC Côte des Blancs CHARD from member growers. Fine range of Vintage BLANC DE BLANCS 08 13' for restaurants. V.gd Vintage Rosé 06.

Jacquesson Champ ★★★★ Bijou Dizy house for precise, v. dry wines. Outstanding single-v'yd Avize CHAMP Caïn 08 12'. Corne Bautray, all CHARD. Dizy 10 13' excellent *numbered NV cuvées* 730' 731 732 733 734 735 738 739 740 (12 base: best yet) 741 742 743.

Jadot, Louis Burg r p w ★★→★★★★ Dynamic BEAUNE merchant making powerful whites (DIAM corks) and well-constructed age-worthy reds with significant v'yd holdings in BEAUJ, C D'O, MÂCON; esp POUILLY-FUISSÉ (DOM Ferret), MOULIN-À-VENT (CH des Jacques, *Clos du Grand Carquelin*).

Jamet, Dom N Rh r w ★★★★ 99' 05' 09' 10' 12 13' 14 15' 16' 17' 18' 19' Top CÔTE-RÔTIE, v.-long-lived, complex *vins de terroir* from many sites, mainly schist. Classic red intricate, dashing fruit, top Côte Brune (r) is mighty, mysterious. Also high-grade CÔTES DU RH (r w), COLLINES RHODANIENNES (r).

Jasnières Lo w dr (sw) ★★→★★★ 16 17' 18' 19' Long-lived CHENIN BL, lively, sharp, dry to sweet, AC, s-facing slopes Loir Valley. Try: Breton, DE BELLIVIÈRE, Gigou, Janvier, J-B Métais, L'Ange Vin (also VDF), Le Briseau, Les Maisons Rouges, Roche Bleue, Ryke. 19 frost.

Jobard C d'O r w ★★★ VIGNERON family in MEURSAULT. Top DOMS are Antoine J, esp long-lived Poruzots, Genevrières, CHARMES; and Rémi J for immediately classy Meursaults plus reds from MONTHÉLIE, VOLNAY. Former has added a POMMARD dom from 2019.

Joblot Burg r w ★★ Reference DOM for GIVRY, new CUVÉES Empreinte, Mademoiselle from 2016, plus classics La Servoisine and CLOS du Cellier Aux Moines.

Joseph Perrier Champ Fine family-run CHAMP house with v.gd PINOTS N, M v'yds, esp in own Cumières DOM. Ace Prestige CUVÉE Joséphine 02 08' 12'. Excellent BRUT Royale NV, as generous as ever but more precise with less dosage. Distinctive, tangy BLANC DE BLANCS 08 13' 15'. Owner Jean-Claude Fourmon, one of Champ's great characters, passing reins to son. Now drier, finer Cuvée Royale BRUT NV; distinctive tangy BLANC DE BLANCS 02 04 06 08 13'.

Josmeyer Al w ★★→★★★ Exceptional family AL house, pioneer of bio viticulture, centred on RIES GC Hengst 13 14 16 17. Intriguing single-v'yd PINOT AUXERROIS and entry-level Ries Dragon delicious, gd value. Untimely death of Jean Meyer and departure of his son-in-law, a fine winemaker, means question mark over future.

Juliénas Beauj r ★★★ 15' 16 17 18' 19' Deserves to be better known for deep-fruited BEAUJ, esp for CLIMATS Beauvernay, etc. DOM Perrachon leads the way, try also Audras (Clos de Haute Combe), Aufranc, Besson, Burrier, CH BEAUREGARD, CH FUISSÉ, Chignard, Dom Granit Doré, Trenel.

Jurançon SW Fr w dr sw ★→★★★ (sw) 15' 16' 18 (dr) 16 17' 18' (19) Separate AOPS for sweet and dry whites. Balance of richness, acidity the key to quality. Best sweets incl DAGUENEAU's tiny ★★★★ Jardins de Babylon, DOM de Souch, ★★★ Doms *Cauhapé*, Guirardel, Lapeyre, Larrédya, Larrouyat. ★★ CH Jolys, Doms Bellegarde, Bordenave, Castéra, du Ciinquau, CLOS Benguères, Nigri, Uroulat. ★ Gan co-op gd value.

Kaefferkopf Al w dr (sw) ★★★ The 51st GC of AL at Ammerschwihr. Permitted to make blends as well as varietal wines, possibly not top-drawer.

Kientzler, Andre Al w sw ★★→★★★★ 5th-generation family DOM. Lush sensual GEWURZ GC Kirchberg 16 18' 19. VENDANGE TARDIVE dessert wines. Exemplary care in v'yds.

Kreydenweiss, Marc Al w sw ★★→★★★★ Bio for decades. Rich diversity of soils: GC Moenchberg on limestone for PINOT GR and majestic RIES Kastelberg on black schist, ages for up to 20 yrs 07 10 13 17' 18 19 Also in COSTIÈRES DE NÎMES.

Krug Champ Supremely prestigious de luxe house. ★★★★ Grande CUVÉE, esp 160th Edition based on 04; 167th Edition (11 base) v. graceful despite questionable reputation of base yr. Vintage 98 02 04; Rosé; CLOS DU MESNIL 03' 02 CLOS D'AMBONNAY 95' 98' 00; Krug Collection 69 76' 81 85. Rich, nutty wines, oak-fermented; highest quality, ditto price. Vintage 03 a fine surprise.

Kuentz-Bas Al w sw ★★→★★★ Among highest AL v'yds, organic/bio, serious yet accessible wines: drier RIES 13 16 17'. Fine PINOT GR, GEWURZ VENDANGE TARDIVE 09 12 17.

Labet, Dom Jura ★★★ Key CÔTES DU JURA estate in s part of region (Rotalier). Best-known for range of single-v'yd CHARD whites, eg. En Billat, En Chalasse, La Bardette; gd PINOT N, VIN JAUNE.

Ladoix C d'O r w ★★ (r) 09' 10' 12 15' 17 18' 19' (w) 14' 15 17' 18 19' Explore here for fresh, exuberant whites, eg. Grechons, and juicy reds, esp Les Joyeuses. Key producers (w) Chevalier, FAIVELEY, Loichet (r) Mallard, CH de MEURSAULT (from 2018), Naudin-Ferrand, Ravaut. Those with more cash try Le Clou from Prieuré-Roch.

Ladoucette, de Lo (r) (p) w ★★★ 15 16 17 18' 19' Largest POUILLY-FUMÉ property at CH du Nozet. Premium Baron de L. Also SANCERRE Comte Lafond, CLOS de La Poussie (steep site above Bué: r p w) and Marc Brédif: CHINON, MUSCADET, TOURAINE, VOUVRAY etc.

Lafarge, Michel C d'O r (w) ★★★★ Classic VOLNAY bio estate run by Frédéric L, son of ever-present Michel, 112 vintages between them. Outstanding, long-lived PCS *Clos des Chênes*, Caillerets, CLOS du CH des Ducs. Also fine BEAUNE, esp Grèves (r) and Clos des Aigrots (w). Exciting new FLEURIE project, Lafarge-Vial.

Lafon, Dom des Comtes Burg r w ★★★★ Fabulous bio MEURSAULT DOM, with long-lasting red VOLNAY *Santenots* equally outstanding. Value from excellent Mâconnais wines under Héritiers L label, while Dominique L makes his own CÔTE de Beaune wines separately.

Laguiche, Marquis de C d'O w ★★★★ Largest owner of Le MONTRACHET and a fine PC CHASSAGNE, both excellently made by DROUHIN.

Lalande de Pomerol B'x r ★★ 09' 10' **14** 15 **16** 18 (19) Improving satellite neighbour of POM. Largely MERLOT. Varied terroir: clay, gravel, sand. Top CHX: Belles-Graves, Chambrun, Enclos de Viaud, Garraud, Grand Ormeau, Haut-Chaigneau, Jean de Gué, La Chenade, La Croix des Moines, LA FLEUR DE BOÜARD, La Sergue, Les Cruzelles, *Les Hauts Conseillants*, Pavillon Beauregard, Sabines, Samion, Siaurac, *Tournefeuille*.

La Livinière L'doc r ★★→★★★ AC from 2020; former cru of MINERVOIS. Stricter selection, lower yield, longer ageing. SYRAH with GRENACHE, CARIGNAN, MOURVÈDRE. V'yds up to 400m (1312ft): fresh acidity. Try CLOS Centeilles, Clos des Roques, Clos l'Ora, DOM Fauzon, CHX Faiteau, Maris, Mignan.

La Londe Prov r p ★★ Chanel has moved in here and bought DOM de l'Ille; LVMH owns Galoupet. Others: CLOS Mireille, Dom Perzinsky, Léoube, St Marguerite. Coastal schist subzone of CÔTES DE PROV, incl island of Porqueyrolle (three v'yds).

Lamy C d'O (r) w ★★★ DOM Hubert Lamy is go-to address for ST-AUBIN. Breathtakingly fresh and concentrated whites, often from higher-density plantings. Rapidly improving reds too. Also DOMS L-Caillat (intense w), more traditional L-Pillot in CHASSAGNE.

Landron, Doms Lo w dr sp ★★→★★★ 16 17 18' 19' V.gd bio MUSCADET SÈVRE ET MAINE: incl Amphibolite, Fief du Breil, La Louvetrie.

Langlois-Chateau Lo (r) (w) sp ★★→★★★ SAUMUR, SANCERRE (Fontaine-Audon/ Thauvenay). BOLLINGER-owned. Gd range CRÉMANT de Lo. Still wines incl fine Saumur Blanc VIEILLES VIGNES 15 16, SAUMUR-CHAMPIGNY, POUILLY-FUMÉ.

Languedoc r p w In the pyramid of MIDI ACS this is the bottom – not the worst, necessarily, but the least specific. Incl ACs of CORBIÈRES, MINERVOIS, ROUSS from Cabardès to Sommières. Hierarchy of superior crus is work in progress. Subregions incl Cabrières (historic rosé AC with high % CINSAULT), Grès de Montpellier, Pézenas, Quatourze, St-Saturnin: usual L'doc grapes. Clairette du L'doc tiny AC (five producers) for whites from CLAIRETTE. Not all ACs have much specific identity: go by grower.

Lanson Champ ★★★ Major house, owned by PAILLARD. Black Label NV; Rosé NV; Vintage BRUT on a roll, esp 02 **08** 12' 15. Ace prestige NV Noble CUVÉE BLANC DE BLANCS, rosé and vintage; Brut vintage single-vyd CLOS Lanson 08 **09** 12'. Extra Age multi-vintage, Blanc de Blancs esp gd. Experienced new winemaker Hervé Dantan (since 15) allows some malo for a rounder style.

La Peira L'doc r w English-owned DOM, AL winemaker, rich intense reds. La Peira

(SYRAH/GRENACHE), las Flors de la Peira (Syrah/Grenache/MOURVÈDRE), Obriers de la Peira (CARIGNAN/CINSAULT).

Lapierre, Marcel Beauj r ★★★ Mathieu and Camille L continue cult DOM making sulphur-free MORGON. Range of styles and CUVÉES on offer.

Laplace, Dom SW Fr r Oldest and at one time only producer of MADIRAN. Still one of best and most important. Top wine ★★★ CH d'Aydie, which needs lots of time. Odie d'Aydie less so. Aydie l'Origine harks back to family history. More commercial, lighter, all-TANNAT IGPS ★★ Les Deux Vaches, ★ Aramis. Excellent ★★★ PACHERENCS (dr sw). Sweet fortified Maydie (think BANYULS) gd with chocolate.

Laroche Chab w ★★ Major player in CHAB with quality St Martin blend, Vieille Voye special CUVÉE, exceptional GC *Res de l'Obediencerie* named for historic HQ (worth a visit). Winemaker Gregory Viennois involved in NEGOCIANT IRANCY project. Also Mas La Chevalière in L'DOC.

Latour, Louis Burg r w ★★→★★★★ Famous traditional family grower/merchant making full-bodied whites from c D'O v'yds (esp CORTON-CHARLEMAGNE), Mâconnais, Ardèche (all CHARD) while reds are looking classier – CORTON, ROMANÉE-ST-VIVANT. Also owns Henry Fessy in BEAUJ.

Latricières-Chambertin C d'O r ★★★★ 90' 93 96' 99' 03 05' 09' 10' 12' 15' 16 17 18' 19' GC next to CHAMBERTIN. Deep soil and cooler site gives rich earthy wines in warm dry yrs. Best: ARNOUX-LACHAUX, BIZE, Drouhin-Laroze, Duband, Duroché, FAIVELEY, LEROY, Remy, ROSSIGNOL-TRAPET, TRAPET.

Laudun S Rh r p w ★→★★ 16' 17' 18' 19' Sound CÔTES DU RH-VILLAGE, w bank. Clear, peppy whites. Red fruit/peppery reds (much SYRAH), lively rosés. Immediate flavours from Laudun-CHUSCLAN co-op. *Dom Pelaquié* best, esp elegant white. Also CHX Courac, de Bord, Juliette, St-Maurice; DOMS Duseigneur (bio), Carmélisa, Maravilhas (bio, r w), Olibrius.

Laurent-Perrier Champ Important house; family presence less obvious. BRUT NV (CHARD-led) perfect apéritif. V.gd skin-contact Rosé. Fine vintages: **02 08 09 12**. Grand Siècle CUVÉE multi-vintage on form, peerless Grand Siècle Alexandra Rosé 09. Arrival of eminent Dominique Demarville (ex-VEUVE CLICQUOT) as chef de CAVE a coup for L-P. Expect some rigorous rethinking.

Laurent Vervesin Champ Young grower rejoins family Oger DOM. Organic principles in Original BLANC DE BLANCS GC 13: 15% oak, clear, precise. Aubeline GC, fuller, liked in Scandinavia.

Lavantureux Chab w ★★ Specialist source for PETIT CHAB, CHAB, esp single-v'yd Vauprin. BOURGOGNE Epineuil reds to follow soon.

Leflaive, Dom Burg w ★★★★ Reference PULIGNY-MONTRACHET DOM restored to top quality and prices. Total revamp with new director and winemaker since 2016. Outstanding GC, incl LE MONTRACHET, CHEVALIER and *fabulous PCs*: Pucelles, Combettes, Folatières, etc. Value from developing s Burg range, eg. MÂCON Verzé.

Leflaive, Olivier C d'O r w ★★→★★★ White specialist NÉGOCIANT at PULIGNY-MONTRACHET.

Languedoc rising stars
A region where great terroirs are still being discovered, so a place to look for new names and attitudes. **Cabardès** Guilhem Barré; **Corbières** L'Espérou, Olivier Mavit; **Faugères** Grain Sauvage, Les Serrals, Mas Lou; **Fitou** Sarrat d'en Sol; **La Clape** La Combe de St Paul; **Limoux** Cathare, Monsieur S; **Pézenas** La Grange des Bouys, Ste Cécile du Parc; **Pic St-Loup** CH Fontanès, Mas Gourdou, Mas Peyrolle; **St Chinian** La Lauzeta, Lanye Barrac, Les Païssels, **Terrasses du Larzac** CLOS Aguilem, Clos Maïa, Clos Rouge, Les Chemins de Carabote, Mas Combarèla, Terre des Deux Source.

Smart wines of late, spot on with BOURGOGNE Les Setilles and own GC v'yds. Hoping to develop reds too. Also La Maison d'Olivier, hotel, restaurant, tasting room.

Leroux, Benjamin C d'O r w ★★★ Growing reputation as BEAUNE-based NÉGOCIANT equally at home in red or white. Poised, honest wines. C D'O only, strengths (w) in MEURSAULT with increasing DOM and (r) BLAGNY, GEVREY, VOLNAY.

Leroy, Dom C d'O r w ★★★★ Lalou Bize Leroy, bio pioneer, delivers extraordinary quality from tiny yields in VOSNE-ROMANÉE and from DOM d'Auvenay. Both fiendishly expensive even ex-dom. As is amazing treasure trove of mature wines from family NÉGOCIANT Maison L.

Liger-Belair, Comte C d'O r ★★★★ Comte Louis-Michel L-B makes brilliantly ethereal wines in VOSNE-ROMANÉE, an ever-increasing stable headed by LA ROMANÉE. Try also village La Colombière, PC Reignots and NUITS-ST-GEORGES crus. Also in Oregon, Chile.

Liger-Belair, Thibault C d'O (w) ★★★→★★★★ New winery in NUITS-ST-GEORGES, for succulent bio burg from generics to Les St-Georges and GC RICHEBOURG. More whites now as NÉGOCIANT. Also range of stellar old-vine single-v'yd MOULIN-À-VENT.

Lignier C d'O r w ★★★ Family in MOREY-ST-DENIS. Hubert L on form under son Laurent, whole range brilliant, esp CLOS DE LA ROCHE. V.gd PCS from Virgile L-Michelot, esp Faconnières, but DOM Georges L divides opinion.

Lilbert Champ Bijou DOM, highest standards. Young CHARDS hard as diamond, but age gracefully for 30 yrs. Entry-level Perlé GC NV a gd intro.

Limoux L'doc r w ★★ Still wine from BLANQUETTE, CRÉMANT de Limoux regions. Obligatory oak-ageing for white, from CHARD, CHENIN, Mauzac. Red based on MERLOT, plus SYRAH, GRENACHE, CABS. PINOT N in CRÉMANT and for IGP Haute Vallée de l'Aude. Growers: DOMS de Baronarques, de Fourn, Mouscaillo, RIVES-BLANQUES and Cathare, *Jean-Louis Denois*.

Lirac S Rh r p w ★★→★★★ 10' 15' 16' **17'** 18 19' Four villages nr TAVEL, stony, quality soils. Spiced red (can live 5 yrs+), impetus from CHÂTEAUNEUF owners via better fruit, more brio. Reds best, esp DOMS *de la Mordorée* (best, r w), Carabiniers (bio), Duseigneur (bio), Giraud, Joncier (bio, character), Lafond Roc-Epine, La Lôyane, La Rocalière (gd fruit), *Maby* (Fermade, gd w), Marcoux (stylish), Plateau des Chênes; CHX de Bouchassy (gd w), de Manissy (organic), de Montfaucon (v.gd w incl CÔTES DU RH), Mont-Redon, St-Roch; Mas Isabelle (handmade), P Usseglio, Rocca Maura, R Sabon. Whites convey freshness, body, go 5 yrs.

Listrac-Médoc H-Méd r ★★→★★★ 10' **14** 15 **16'** 18 (19) Much=improved AC for savoury red B'x; now more fruit, depth and MERLOT due to clay soils. Growing number of gd whites under AC B'x. Best CHX: Cap Léon Veyrin, CLARKE, Ducluzeau,

---

Jura jewels

What other region has so many styles? They seem to like confusing customers – sommeliers love the game. **Dry whites** from CHARD, many single-v'yd versions now. Also tangy blends with SAVAGNIN. Fresh or deliberately **oxidative** pure Savagnin exciting too; many made in natural style can be pretty gamey. *See* ARBOIS, CÔTES DU JURA, L'ETOILE. *Vin typé* on labels means heading towards Sherry. *Vin ouillé* means "ullaged": the barrel has been topped-up to avoid oxidation. **Light reds** from PINOT N, Poulsard, Trousseau, or blends. Some more rosé than red. Côtes du Jura, Arbois. Aged **sherrified whites** known as VIN JAUNE. *See* CH-CHALON. Intensely sweet **vin de paille** made fom red and white grapes. Fortified **Macvin**, local version of ratafia. Classic producers: Bourdy, MACLE, Overnoy, Puffeney. Avant garde: A&M TISSOT, GANEVAT, Pignier. Volume/value: co-ops (known here as CAVES Fruitières), Boilley, HENRI MAIRE, J Tissot, LABET.

FONRÉAUD, Fourcas-Borie, FOURCAS-DUPRÉ, FOURCAS-HOSTEN, l'Ermitage, LESTAGE, Mayne-Lalande, Reverdi, SARANSOT-DUPRÉ.

Long-Depaquit Chab w ★★★ Energetic CHAB DOM with famous flagship brand, GC La Moutonne. Part of BICHOT empire.

Lorentz, Gustave Al w ★★→★★★ Grower/merchant at Bergheim. RIES is strength in GCS Altenburg de Bergheim, Kanzlerberg, age-worthy 12 **13 14** 16 18 19. Young volume wines (esp *Gewurz*) well made. Fine new organic Evidence GEWURZ 16 **17**.

Lot SW Fr ★→★★ DYA IGP of Lot département increasingly useful to CAHORS growers for rosés and whites not allowed in AOP (eg. CLOS de Gamot, CH DU CÈDRE. Look beyond AOP for ★★ DOMS Belmont, Sully, Tour de Belfort.

Loupiac B'x w sw ★★ **13 14** 15 16 18 Minor SÉM-dominant *liquoreux*. Lighter, fresher than SAUT across River Garonne. Top CHX: CLOS Jean, Dauphiné-Rondillon, *de Ricaud*, Les Roques, *Loupiac-Gaudiet*, Noble.

Luberon S Rh r p w ★→★★ 16' 17' 18 19' Hilly adjunct to S Rh; terroir is arid, can be no more than okay. Too many technical wines. SYRAH has lead role. Whites improving. Bright star: CH de la Canorgue, Edem, Fontvert (bio, gd w), O Ravoire, Puy des Arts (w), St-Estève de Neri (improver), Tardieu-Laurent (rich, oak), DOMS de la Citadelle, Fontenille, La Cavale (swish), Le Novi (terroir), Marrenon, Maslauris (organic), Val-Joanis; gd-value *La Vieille Ferme* (can be VDF).

Lussac-St-Émilion B'x r ★★ 10' 15 **16** 18 (19) Lightest of ST-ÉM satellites; 2nd to MONTAGNE in size. Top CHX: Barbe Blanche, Bel-Air, Bellevue, Courlat, Croix de Rambeau, DE LUSSAC, La Rose-Perrière, Le Rival, LYONNAT, Mayne-Blanc.

Macération carbonique Traditional fermentation technique: whole bunches of unbroken grapes in a closed vat. Fermentation inside each grape eventually bursts it, giving vivid, fruity, mild wine, not for ageing. Esp in BEAUJ, though not for best wines; now much used in the MIDI and elsewhere, even CHÂTEAUNEUF.

Macle, Dom Jura ★★★ Legendary producer of CH-CHALON VIN JAUNE for long ageing. Also CÔTES DU JURA.

Mâcon Burg r (p) w DYA ★ Simple, juicy GAMAY reds and most basic rendition of Mâconnais whites from CHARD.

Mâcon-Villages Burg w ★★ **14'** 17' 18 19' Chief appellation for Mâconnais whites. Individual villages may also use their own names eg. Mâcon-Lugny. Co-ops at Lugny, Terres Secretes, Viré for quality-price ratio, plus *brilliant grower wines* from Guffens-Heynen, Guillot, Guillot-Broux, LAFON, LEFLAIVE, Maillet, Merlin. Also major NÉGOCIANTS, DROUHIN, LATOUR, etc.

Macvin From France, not Scotland. Grape juice is fortified by local marc to make a sweet apéritif between 16–22% alc. Most Jura producers make one. Usually white, can be red.

Madiran SW Fr r ★★→★★★ 00' 05' 10 12 15' 16 (18) (19) Gascon AOP. France's home of TANNAT grape. Lighter styles gaining ground over traditional dark heavyweights: ★★★ CHX BOUSCASSÉ and MONTUS (owner BRUMONT has 15% entire AOP), Laffitte-Teston, *Laplace*. Wide ranges from ★★★ DOMS Berthoumieu, Capmartin, CLOS Basté, Damiens, Dou Bernès, Labranche-Laffont, Laffont, Pichard; Chx Arricaud-Bordès and de Gayon. Doms ★★ Barréjat, ★★ Crampilh, Maouries not far behind.

Madura, Dom la L'doc r w ★★★ Ex-*régisseur* of B'X CH FIEUZAL created own estate in ST-CHINIAN. Stylish Classique, Grand Vin. Original blend of SAUV BL/PICPOUL.

Maillard, Nicolas Champ ★★★ Uses stainless steel, a little oak. PC Platine 12 15 18 19 ageing beautifully, esp in magnums.

Mailly Grand Cru Champ Top co-op, all GC grapes. Prestige CUVÉE des *Echansons* 08 **12'** for long ageing. Sumptuous Echansons Rosé 09; refined, classy L'Intemporelle 06. Sébastien Moncuit, cellarmaster since 14, a real talent.

Maire, Henri Jura r w sw ★ Former legend, creator of Vin Fou, still a huge producer, mostly from own v'yds, sometimes using DOM names eg. Sobief, Bregand, or supermarket brand Auguste Pirou. Part of BOISSET empire.

Maison Ventenac L'doc r (p) w ★★★ Cabardès; B'X and L'DOC varieties. Fresh acidity from Atlantic air gives elegance, freshness. Top wines in Dissident range, all VDF. Paul (CAB FR fermented in *jarres*), Le Paria (GRENACHE) has outstanding fruit and vibrancy, Idiote (MERLOT), Candide (oak-fermented CHENIN), Prejugé (oaked CHARD) all excellent.

Mann, Albert Al r w ★★→★★★★ Distinguished grower at Wettolsheim; irreproachable excellence. 20.5 ha, deft winemaking in stainless steel, larger casks and barriques. V.gd AUXERROIS, ace range of GCS: HENGST, SCHLOSSBERG (esp 14). Great red PINOT N Les Stes Claires in 15 18' 19. Immaculate bio v'yds.

Maranges C d'O r (w) ★★ 15' 17 18' 19' Name to watch. Robust well-priced reds from s end of CÔTE DE BEAUNE. Try PCS Boutière, Croix Moines, Fussière. Best: BACHELET-Monnot, Chevrot, Contat-Grangé, Giroud, MOREAU.

Marcillac SW Fr r p ★★ AVEYRON AOP based on Mansois (aka FER SERVADOU). Fruity, curranty/raspberry food wines you either love or hate. Best at 3 yrs. Try with strawberries as well as charcuterie or sausages. ★★ DOM du Cros largest independent grower (gd w IGPS too) also Doms des Boissières, Costes Rouges, Vieux Porche (Jean-Luc Matha). Excellent co-op. Local rosés as gd as reds.

Margaux H-Méd r ★★→★★★★ 00 09' **10' 15** 16' 18 (19) Most s MÉD communal AC. Famous for elegant, fragrant wines; reality is more diverse. Top CHX: BOYD-CANTENAC, BRANE-CANTENAC, DAUZAC, DU TERTRE, FERRIÈRE, GISCOURS, ISSAN, KIRWAN, LASCOMBES, MALESCOT-ST-EXUPÉRY, MARGAUX, PALMER, RAUZAN-SÉGLA. Gd-value Chx: ANGLUDET, Haut Breton Larigaudière, LABÉGORCE, LA TOUR DE MONS, Paveil de Luze, SIRAN.

Marionnet, Henry Lo r w ★★→★★★ 17 18' 19 TOURAINE DOM famous for ungrafted v'yds (sandy soil), rare vines; reputation made by Henry M, now run by son Jean-Sebastién. SAUV BL (top CUVÉE L'Origine), GAMAY, Provignage (v. old Romorantin), La Pucelle de Romorantin, Renaissance. Managing historic v'yd at CH de Chambord – 1st vintage 18 – incl Menu Pineau, Romorantin.

Marmande SW Fr r p (w) ★→★★ (r) **15'** 16 18 (19) Improving Gascon AOP. Abouriou grape almost an exclusivity. ★★★ Cult winemaker Elian da Ros features it at his eponymous DOM. ★★ Doms Beyssac, Bonnet, Cavenac and CH Lassolle blend it with usual B'x grapes. ★★ Ch de Beaulieu SYRAH-based. Co-ops (95% total production) still dull.

Marsannay C d'O r p (w) ★★→★★★ (r) 12' **15'** 17 18' 19' Most n AOC of CÔTE DE NUITS, hoping to get PCS (eg. CLOS du Roy, Longeroies, Champ Salomon). 1st step has gained new village-level v'yds from 2019, eg. Chapitre and Montre-cul. Accessible, crunchy, fruit-laden reds, from energetic producers: Audoin, Bart, Bouvier, Charlopin, CLAIR, Fournier, *Pataille*, TRAPET. V.gd *rosé* needs 1–2 yrs; whites less exciting.

Mas, Doms Paul L'doc r p w ★★ Ubiquitous, enormous: controls 1312 ha from Grès de Montpellier to ROUSS. Mainly IGP. Working on organics, bio and low sulphur. Innovative marketing with wine tourism, restaurant. Arrogant Frog range; also La Forge, Les Tannes, Les VIGNES de Nicole and DOMS Ferrandière, Crès Ricards in TERRASSES DU LARZAC, Martinolles in LIMOUX and CH Lauriga in ROUSS, Côté Mas brand from Pézenas.

Mas Bruguière L'doc ★★★ 7th-generation family estate in PIC ST-LOUP. L'Arbouse, La Grenadière and Le Septième.

Mas de Daumas Gassac L'doc r p w ★★★ MIDI star since 80s; now run by 2nd generation Samuel Guibert, who uses horses in v'yd. CAB-based age-worthy reds from apparently unique soil. Also perfumed white from CHENIN blend;

super-CUVÉE Émile Peynaud (r); rosé Frizant. V.gd sweet Vin de Laurence (MUSCAT/SERCIAL).

Mas Jullien L'doc ★★★ TERRASSES DU LARZAC star owned by Olivier Jullien: typical Larzac freshness. MOURVÈDRE, CARIGNAN red: Autour de Jonquières, Carlan, États d'Âme, Lous Rougeos from L'DOC varieties. Carignan Blanc and Gris, CHENIN BL.

Massif d'Uchaux S Rh r ★★ 16' 17' 18' **19'** Gd Rh village, brightly fruited, fresh, spiced reds, not easy to sell, but best true, stylish. NB: CH St-Estève (incl gd VIOGNIER), DOMS *Cros de la Mûre* (character, gd value), de la Guicharde, La Cabotte (bio, no form), Renjarde (sleek fruit).

Maury Rouss r w sw ★★→★★★ Sweet VDN from GRENACHES Noir, BL, Gris on island of schist. Ambré, tuilé and RANCIO styles, try old rancio with chocolate. Now characterful dry red, AC Maury SEC prompted by recent improvements, led by *Mas Amiel*. New estates incl *Dom of the Bee; Jones*. Sound co-op.

Mazel, Le S Rh r w ★★ 18' 19' Mother lode for Rh Vin Nature, no sulphur, S ARDÈCHE. Gérald Oustric 1st vinified in 1997, carbonic maceration, vats only (no wood), low alc GRENACHE, CARIGNAN, whites CHARD, VIOGNIER notably. All VDF.

Mazis (or Mazy-) Chambertin C d'O r ★★★★ 90' **93 96'** 99' 05' 09' 10' 12' 15' 16' 17 18' 19' GC of GEVREY-CHAMBERTIN, top class in upper part; intense, *heavenly wines*. Best: Bernstein, DUGAT-PY, FAIVELEY, HOSPICES DE BEAUNE, LEROY, Tawse (ex-Maume), MORTET, ROUSSEAU.

Mazoyères-Chambertin C d'O ★★★★ Can be sold as CHARMES-CHAMBERTIN, but more growers now labelling M as such. Style is different: less succulence, more stony structure. Try DUGAT-PY, LEROUX, Perrot-Minot, Taupenot-Merme, Tawse.

Médoc B'x r ★★ 10' 15 16 18 (19) AC for reds in low-lying n part of Méd peninsula (aka Bas-Méd). Often more guts than grace. Can be gd value but selective. Top CHX: Bournac, CLOS Manou, Fleur La Mothe, Fontis, *Goulée*, GREYSAC, *La Tour-de-By*, LES ORMES-SORBET, Lousteauneuf, *Patache d'Aux*, POITEVIN, *Potensac*, PREUILLAC, Ramafort, *Rollan-de-By* (HAUT-CONDISSAS), TOUR HAUT-CAUSSAN, TOUR ST-BONNET, Vieux Robin.

Meffre, Gabriel S Rh r w ★★→★★★ Consistent S Rh merchant, owns gd GIGONDAS DOM Longue Toque. Fruit quality up, reduced oak. Also CHÂTEAUNEUF (gd St-Théodoric, also small doms), VACQUEYRAS St-Barthélemy. Reliable to gd S/N Rh Laurus (new oak, gd 15s) range, esp CONDRIEU, HERMITAGE (W), ST-JOSEPH.

Mellot, Alphonse Lo r p w ★★→★★★★ 15' 16' 17' 18' 19' SANCERRE (r – a revelation, w), bio, incl La Moussière (r w), CUVÉE Edmond, Génération XIX (r w); gd single-vyds incl *Satellite* (from CHAVIGNOL): Demoiselle, En Champs; Les Pénitents (Côtes de La Charité IGP) CHARD, PINOT N. Run by Alphonse Jnr; Alphonse Snr now retired.

Menetou-Salon Lo r p w ★★→★★★ **17' 18'** 19' AOP nr SANCERRE; similar SAUV BL. Excellent 18, promising 19. Increasingly fine reds (PINOT N). Best: BOURGEOIS, *Clement* (Chatenoy), Gilbert (bio, gd r w), *Henry Pellé* (gd r w), Jacolin, Joseph Mellot, Jean-Max Roger, Teiller, Tour St-Martin.

Méo-Camuzet C d'O r w ★★★★ Noted DOM in VOSNE-ROMANÉE: icons Brûlées, CROS PARANTOUX, RICHEBOURG. Value from M-C Frère et Soeur (NÉGOCIANT branch) and plenty of choice in between. Sturdy, oaky wines that age well.

Merande, Ch de Sav r ★★ Owned by DOM Genoux (12 ha, bio). Gd value: MONDEUSE.

Mercurey Burg r (w) ★★→★★★ 15' 16 **17 18'** 19' Leading village of CÔTE CHALONNAISE, firmly muscled reds, aromatic whites. Try BICHOT, CH *de Chamirey*, CH de Santenay, de Suremain, FAIVELEY, *Juillot-Theulot*, Lorenzon, M Juillot, Raquillet.

Merlin Burg r w ★★→★★★ Stylish MÂCON La Roche Vineuse since 1987, Olivier and Corinne M added MOULIN-À-VENT and an increasing POUILLY-FUISSÉ range. Co-owners CH des Quarts with Dominique LAFON.

Mesnil-sur-Oger, Le Champ Top Côte des Blancs village, v. long-lived CHARD. Best:

André Jacquart, JL Vergnon (until 17), KRUG CLOS du Mesnil, Pierre Péters. Needs 10 yrs+ ageing.

Méthode Champenoise Champ Traditional method of putting bubbles into CHAMP by refermenting wine in its bottle. Outside Champ region, makers must use terms "classic method" or *méthode traditionnelle*.

Meursault C d'O (r) w ★★★ →★★★★ 09' 10' 11 12 14' 15 17' 18 19' Potentially great full-bodied whites from PCS: Charmes, Genevrières, Perrières, more nervy from hillside v'yds *Narvaux*, Tesson, *Tillets*. Producers: Ballot, Boisson-Vadot, BOUZEREAU, Boyer-Martenot, *Ch de Meursault*, COCHE-DURY, *de Montille*, Ente, Fichet, *Girardin*, *Javillier*, JOBARD, *Lafon*, Latour (V), LEROUX, Matrot, Mikulski, *P Morey*, Potinet-Ampeau, PRIEUR, *Roulot*. Try de Cherisey for M-BLAGNY.

Meursault, Ch de C d'O r w ★★★ Huge strides lately at this 61-ha estate of big-biz Halley family: decent red from BEAUNE, POMMARD, VOLNAY. Now world-class white, mostly MEURSAULT, also v.gd BOURGOGNE BLANC, PULIGNY PC. Going bio.

Minervois L'doc r (p) (w) ★★ Hilly AC region, one of L'DOC's best. Characterful, savoury reds, esp CHX Coupe-Roses, de Gourgazaud, La Grave, La Tour Boisée, Oupia, Ste Eulalie, St-Jacques d'Albas, Villerembert-Julien; DOMS CLOS Centeilles, Combe Blanche, l'Ostal Cazes; Abbaye de Tholomiès, Borie-de-Maurel, Laville-Bertrou, Maris. Gros and Tollot (from Burg) raising bar. Potential new crus Cazelles, Laure.

Miquel, Laurent L'doc ★★★ Specializes in aromatic whites, unusually for CORBIÈRES and ST-CHINIAN; often IGP. VIOGNIER, ALBARIÑO from PYRÉNÉES, impressive. Powerful, elegant CAB FR. Own v'yds plus NÉGOCIANT activity, with Solas, VENDANGES Nocturnes, Nord Sud.

Mis en bouteille au château / domaine Bottled at CH, property, or estate. NB: *dans nos* CAVES (in our cellars) or *dans la région de production* (in the area of production) often used but mean little.

Moët & Chandon Champ By far largest CHAMP house, impressive quality for such a giant. Fresher, drier BRUT Imperial NV. New rare prestige CUVÉE MCIII "solera" concept aimed at rich technophiles, addicts of exclusiveness. Better value in run of Grand Vintages Collection, long lees-aged; new sumptuous, elegant 09 08 a little severe. Ace 12' to come. Outposts across Europe and New World. *See also* DOM PÉRIGNON.

Monbazillac SW Fr w sw ★★→★★★ 15' 17 18 BERGERAC sub-AOP: ★★★★ *Tirecul-la-Gravière* up there with best SAUTERNES. ★★★ CLOS des Verdots, L'Ancienne Cure, Les Hauts de Caillavel, co-op's *Ch de Monbazillac*. ★★ CHX de Belingard-Chayne, Grande Maison, Kalian, Monestier la Tour, Pécoula.

Monopole A v'yd that is under single ownership.

Montagne-St-Émilion B'x r ★★ 10' 15 16 18 (19) Largest satellite of ST-ÉM. Solid reputation. Top CHX: Beauséjour, CLOS de Boüard, Corbin, Croix Beauséjour, Faizeau, La Couronne, Maison Blanche, Roudier, Teyssier, Tour Bayard, Vieux Bonneau, Vieux Ch Palon, *Vieux Ch St-André*.

Montagny Burg w ★★ 14' 15 16 17 18 19' CÔTE CHALONNAISE village with crisp whites, mostly in hands of CAVES DE BUXY but gd NÉGOCIANTS too, incl Lorenzon, LOUIS LATOUR, JM PILLOT, O LEFLAIVE. Reference local grower is *S Aladame* but try also Berthenet, Cognard, *Feuillat-Juillot*.

Montcalmès, Dom L'doc ★★★ TERRASSES DU LARZAC, so fresh, elegant. Talented brother/sister team makes white MARSANNE/ROUSSANNE, plus pure CHARD and intriguing blend VDF. Stylish SYRAH/GRENACHE/MOURVÈDRE, plus varietal Grenache and AC L'DOC Le Geai.

Monthélie C d'O r (w) ★★→★★★ 12 15' 16 17 18' 19' Pretty reds, grown uphill from VOLNAY, but a touch more rustic. Les Duresses best PC. Try BOUCHARD PÈRE & FILS, *Ch de Monthélie* (Suremain), *Coche-Dury*, Darviot-Perrin, Florent Garaudet, LAFON. Whites mostly neutral.

FRANCE

Montille, de C d'O r w ★★★ Dense, spicy, whole-bunch reds from BEAUNE, VOLNAY (esp Taillepieds), POMMARD (Rugiens), CÔTE DE NUITS (Malconsorts) and exceptional whites from MEURSAULT, plus outstanding PULIGNY-MONTRACHET Caillerets. Since 2017 CH de Puligny wines are incl under de Montille.

Montlouis sur Loire Lo w dr sw sp ★★ →★★★ 15' 16 17' 18 19' Impressive sister AC to VOUVRAY. Top CHENIN BL; sparkling incl Pétillant Originel. Top: Berger, Chanson, CHIDAINE, Delecheneau, Jousset, Merias, Moyer, Saumon, *Taille-aux-Loups* (Blot), Vallée Moray, Weisskopf.

Montpeyroux L'doc ★★→★★★ Lively village within TERRASSES DU LARZAC with growing number of talented growers. Aspiring to cru status. Try: Chabanon, DOM d'Aupilhac, Villa Dondona. Newcomers: Joncas, Mas d'Amile. Serious co-op.

Montrachet (or Le Montrachet) C d'O w ★★★★ 92' 02' 04 05' 08 09' 10 12 14' 15 17 18 19' GC V'yd lent name to both PULIGNY and CHASSAGNE. Should be greatest white burg for intensity, richness of fruit and perfumed persistence. Top: BOUCHARD, COLIN, DRC, LAFON, LAGUICHE (DROUHIN), LEFLAIVE, Ramonet.

Montravel SW Fr r p w dr ★★ (r) 12' 15' 18' (p w) DYA. Sub-AOP of BERGERAC. Modern-style reds must be oak-aged. ★★★ DOMS de Bloy, de Krevel. ★★ CHX Jonc Blanc, Masburel, Masmontet, Moulin-Caresse. ★★ Dry white from same and many other growers. *See* CÔTES DE MONTRAVEL, HAUT-MONTRAVEL for stickies.

Montus, Ch SW Fr r w ★★★ 00' 05 09 10 12' 14 15' 16 (18) (19) Alain BRUMONT's flagship property, famous for long-extracted oak-aged wines. Long-lived all-TANNAT reds, much prized by lovers of old-fashioned MADIRAN. Classy sweet and dry white barrel-raised PACHERENC-DU-VIC-BILH (drink at 4 yrs+) on same high level.

Vouvray sells 20 bottles of dry wine for every bottle of sweet. The joy of secs?

Mordorée, Dom de la S Rh r p w ★★★ 15' 16' 17' 18' 19' Top estate at TAVEL, rosés with flair; also LIRAC, La Reine des Bois (r w). Gd CHÂTEAUNUF-DU-PAPE La Reine des Bois (incl 1929 GRENACHE), La Dame Voyageuse (r).

Moreau Chab w ★★→★★★ Widespread family in CHAB, esp *Dom Christian M*, noted for PC Vaillons Cuvée Guy M and GC Les CLOS des Hospices. Louis M has more commercial range, while DOM M-Naudet makes concentrated wines for longer keeping.

Moreau C d'O r w ★★★→★★★★ At s end c n'e. Outstanding CHASSAGNE PCS from DOM Bernard M; fine La Cardeuse (r). Tidy range of SANTENAY, MARANGES from David M. Neither related to CHAB dynasty.

Morey, Doms C d'O (r) w ★★★ VIGNERON family in CHASSAGNE. Current generation incl Caroline M and husband Pierre-Yves COLIN-M, Sylvain, Thomas (v. fine pure whites), Vincent (plumper style), Thibault M-Coffinet (LA ROMANÉE). Also Pierre M in MEURSAULT for Perrières and BÂTARD.

Morey-St-Denis C d'O r (w) ★★★→★★★★ 99' 02' 05' 09' 10' 12' 15' 16' 17 18' 19' Terrific source of top-grade red burg, to rival neighbours GEVREY-CHAMBERTIN, CHAMBOLLE-MUSIGNY. GCS CLOS DE LA ROCHE, CLOS DE LAMBRAYS, CLOS DE TART, CLOS ST-DENIS. Many gd producers: Amiot, ARLAUD, Castagnier, Coquard-Loison-Fleurot, CLOS DE TART, *Clos des Lambrays*, *Dujac*, Jeanniard, H LIGNIER, LIGNIER-Michelot, Perrot-Minot, PONSOT, Remy, *Roumier*, Taupenot-Merme, Tremblay.

Morgon Beauj r ★★★ 14 15' 17 18' 19' Powerful BEAUJ cru, volcanic slate of Côte du Py makes meaty, age-worthy wine, clay of Les Charmes for earlier, smoother drinking. Grands Cras, Javernières of interest too. Try A Sunier, Burgaud, CH de Pizay, *Ch des Lumières* (JADOT), Chemarin, CLOS de Mez, Desvignes, Foillard, Gaget, Godard, Grange-Cochard, J Sunier, LAPIERRE, Piron.

Mortet, Denis C d'O r ★★★→★★★★ Arnaud M on song with powerful yet refined reds from BOURGOGNE Rouge to CHAMBERTIN. Key wines incl GEVREY-CHAMBERTIN

Mes Cinq Terroirs, PCS Lavaut St-Jacques and Champeaux. From 2016 separate Arnaud M label – equally brilliant, incl CHARMES- and MAZOYÈRES-CHAMBERTIN.

Moueix, J-P et Cie B'x Libourne-based NÉGOCIANT and proprietor named after legendary founder Jean-Pierre. Son Christian runs company with his son Edouard. CHX: BELAIR-MONANGE, HOSANNA, LA FLEUR-PÉTRUS, *La Grave à Pomerol*, LATOUR-À-POMEROL, *Trotanoy*. Distributes PETRUS. In California (*see* DOMINUS ESTATE).

Moulin-à-Vent Beauj r ★★★ 09' 11' 14 15' 18' 19' Grandest BEAUJ cru, transcending GAMAY grape. Weight, spiciness of Rh but matures towards rich, gamey PINOT flavours. Increasing interest in single-v'yd bottlings from eg. *Ch du Moulin-à-Vent*, DOM La Bruyère, JADOT's Ch *des Jacques*, Janin, Janodet, *Merlin* (La Rochelle), Rottiers. More and more interest from C D'O producers eg. BICHOT (Rochegres), L BOILLOT (Brussellions), T LIGER-BELAIR (Les Rouchaux).

Moulis H-Méd r ★★ →★★★ 09' 10' 15 16 18 (19) Tiny inland AC w of MARGAUX. Honest, gd-value wines. Best offer fruit and charm. Top CHX: Anthonic, Biston-Brillette, BRANAS GRAND POUJEAUX, BRILLETTE, *Chasse-Spleen*, Dutruch Grand Poujeaux, Garricq, *Gressier Grand Poujeaux*, MAUCAILLOU, *Mauvesin Barton*, *Poujeaux*.

Mourgues du Grès, Ch S Rh r p w ★★ →★★★ 16' 17 18' 19' Leading COSTIÈRES DE NÎMES address, organic, v.gd range for early-drinking (Dorés, Galets Rouges, Rosés). Firmer Capitelles: Terre d'Argence (mainly SYRAH), Terre de Feu (mainly GRENACHE).

Moutard Champ Original champion of local Arbanne grape. Also eaux de vie. Geatly improved quality, esp CHARD Persin 14 and CUVÉE des 6 CÉPAGES 11 15' 18 19'.

No more Pinot: Champenois now calling P Meunier grape plain "Meunier".

Mugneret C D'O r w ★★★→★★★★ VIGNERON family in VOSNE-ROMANÉE. Sublime, stylish wines from Georges M-Gibourg (esp ÉCHÉZEAUX), almost matched by steadily improving Gérard M. Try also Dominique M and DOM Mongeard-M.

Mugnier, J-F C D'O r w ★★★★ Outstanding grower of CHAMBOLLE-MUSIGNY *Les Amoureuses* and *Musigny*. Finesse, not muscle. Equally at home with MONOPOLE NUITS-ST-GEORGES CLOS de la Maréchale. Courageous decision to take young vintages of MUSIGNY off the market to avoid infanticide.

Mumm, GH & Cie Champ Huge house owned by Pernod Ricard. New chef de CAVE, talented Laurent Fresnet (ex-HENRIOT) 2019. Mumm de Verzenay BLANC DE NOIRS 08 12', ★★★ RSVR BLANC DE BLANCS 12'. Mumm de Cramant, renamed Blanc de Blancs, elegantly subtle. Cordon Rouge much improved. Also in Napa Valley.

Muscadet Lo w ★★ →★★★ 17 18' 19' Popular, bone-dry wine from nr Nantes. Ideal with fish, seafood. Often great value. Best SUR LIE. Choose zonal ACS: *see* following entries. Frost again 19. Must-try age-worthy MUSCADET CRUS COMMUNAUX. 10% CHARD allowed in generic Muscadet.

Muscadet-Coteaux de la Loire Lo w ★→★★ 17 18' 19' Small AC. Esp Carroget, Guindon, Landron-Chartier, Merceron-Martin, Pléiade, Quarteron, VIGNERONS de la Noëlle.

Muscadet Côtes de Grand Lieu Lo ★→★★★ 17 18' 19' MUSCADET zonal AOP by Atlantic. Best SUR LIE: Eric Chevalier, Herbauges (107 ha), Haut-Bourg, Malidain. 19 bad frost.

Muscadet Crus Communaux Lo ★★ →★★★ MUSCADET's new top level. To try. Long lees-aging from specified soil sites, startlingly gd, complex wines. Now seven crus: Clisson, Gorges, Goulaine, La Haye Fouassière, Le Pallet, Monnières-St Fiacre, Mouzillon-Tillières. Champtoceaux and Vallet in process and Côtes de Grandlieu starting.

Muscadet Sèvre et Maine Lo ★ →★★★ 17 18' 19' Largest and best MUSCADET zone. Increasingly gd and great value. Top: Bonnet-Huteau, CH Briacé, Caillé, *Chereau*

Carré, Cormerais, Delhommeau, DOM DE L'ECU, Dom de la Haute Fevrie, Douillard, *Gadais*, Gunther-Chereau, Huchet, Landron, Lieubeau, Luneau-Papin, Métaireau, Olivier and *Sauvion*. Can age a decade+. Try the seven CRUS COMMUNAUX.

Muscat de Frontignan L'doc sw ★★ NV MUSCAT VDN from small, flat, coastal AC; also unfortified but sweet IGP. Leader remains CH la Peyrade. Nearby Muscat de Lunel (DOM du CLOS de Bellevue) and Muscat de Mireval (Dom de la Rencontre) v. similar.

Muscat de Rivesaltes Rouss w sw ★★ Sweet grapey fortified MUSCAT VDN from large AC centred on town of Rivesaltes. Often old-vine Muscats d'Alexandrie and Petits Grains. Muscat SEC IGP increasing as demand for sweet VDN declines. Look for *Corneilla*, DOM CAZES, Treloar; Baixas co-op.

Muscat de St-Jean de Minervois L'doc w sw ★★ Tiny AC for fresh, honeyed VDN MUSCAT. Try DOM de Barroubio, CLOS du Gravillas, Clos Bagatelle. Village co-op prefers dry Muscat.

Musigny (Le Musigny) C d'O r (w) ★★★★ 90' 93 96' 99' 02' 05' 09' 10' 12' 15' 17 18 19' Most beautiful red burg. GC lent its name to CHAMBOLLE-MUSIGNY. Hauntingly fragrant but with sinuous power beneath. Best: DE VOGÜÉ, DROUHIN, FAIVELEY, JADOT, LEROY, *Mugnier*, PRIEUR, ROUMIER, VOUGERAIE.

Nature Unsweetened, esp CHAMP: no dosage. Fine if v. ripe grapes, raw otherwise.

Négociant-éleveur Merchant who "brings up" (ie. matures) the wine.

Noëllat C d'O r ★★★ Noted VOSNE-ROMANÉE family. Georges N in news since arrival of Maxime Cheurlin in 2010, has expanded fast but now getting on top of his brief. NUITS Boudots, Vosne Petits-Monts and GC ÉCHÉZEAUX outstanding, but also some gd-value lesser appellations. Cousins at Michel N offer sound range. *See also* v. stylish HUDELOT-N in VOUGEOT.

Notre Dames des Anges Prov r p ★★ Subzone of CÔTES de PROV. Newest zone, hot central valley. Try CH Rimauresq, DOM Communion.

Nuits-St-Georges C d'O r ★★→★★★★ 99' 02' **03 05'** 09' 10' 12' 15' 16 17 18' 19' Three parts to this major AC: Premeaux v'yds for elegance (various CLOS: de la Maréchale, des Corvées, des Forêts, St-Marc), centre for dense dark plummy wines (Cailles, Les St-Georges, Vaucrains) and n side for the headiest (Boudots, Cras, Murgers). Many fine growers: Ambroise, ARLOT, ARNOUX-LACHAUX, CAILLARD, Chauvenet, Chevillon, Confuron, *Faiveley*, Gavignet, GOUGES, GRIVOT, Lechéneaut, LEROY *Liger Belair*, Machard de Gramont, Michelot, Millot, *Mugnier*, *Rion*,

Ollier-Taillefer, Dom L'doc r p w ★★★ Steep terraced v'yds in FAUGÈRES; delicious Allegro (VERMENTINO/ROUSSANNE); Collines (r p); Grand Rés (r) from old vines, and oak-aged Castel Fossibus (r). CUVÉE (r) Le Rêve de Noé, SYRAH/MOURVÈDRE blend.

Oratoire St Martin, Dom de l' S Rh r w ★★★ 10' 12' **13' 15'** 16' 17' 18' 19' 28 ha choice v'yds at CAIRANNE. Top-class bio reds, Les Douyes (1905 GRENACHE, MOURVÈDRE), Haut Coustias (vines c.70 yrs). Gd table white (CLAIRETTE).

Orléans Lo r p w ★ DYA. Important in C19; less so now. Mostly CHARD, VIN GRIS, rosé, reds (PINOT N, esp PINOT M) Try: Chant d'Oiseaux, CLOS St Fiacre, Deneufbourg.

Orléans-Clery Lo r ★ DYA. Lo Tiny AOP; light CAB FR, 18 more structured. Low-lying, sandy soil. Try: Chante d'Oiseaux, CLOS St-Fiacre, Deneufbourg.

Ostertag, Dom Al r w ★★★ Celeb bio grower, more interested in terroir than varietal expression. Great RIES, esp Muenchberg 10 **14**. Barrique-fermented, intense Muenchberg PINOT GR 15. Lovely PINOT N Fronholtz **12** 15 18'. 19 small, but cd be gd surprise.

Pacherenc du Vic-Bilh SW Fr w dr sw ★★→★★★ White AOP for wines from MADIRAN area. Gros, Petit MANSENG, sometimes Petit Courbu and local Arruffiac produce dry and sweet styles. Made by most Madiran growers but note too ★★ CH de Mascaaras. Dry DYA, but sweet, esp if oaked, can be matured.

Paillard, Bruno Champ ★★★→★★★★ Recent grande marque has built empire. Top-quality BRUT Première CUVÉE NV, Rosé Première Cuvée; refined style, esp in long-aged BLANC DE BLANCS 04, NPU 04 02'. Brut NATURE, clever use of PINOT M. Bruno P heads LANSON-BCC group of mainly family houses; daughter Alice taking over reins at Paillard.

Palette Prov r p w ★★★ Characterful reds, AC nr Aix. MOURVÈDRE, GRENACHE; fragrant rosés, intriguing forest-scented whites; also oddities like FURMINT. Traditional, serious *Ch Simone*, Crémade.

Partagé, Dom Sav ★★ V.gd tiny bio DOM (4.5 ha) in CHIGNIN. Formerly called Gilles Berlioz. Altesse, JACQUÈRE, MONDEUSE, ROUSSANNE. Wines: El-hem, La Deuse, Le Jaja, Les Filles, Les Fripons.

Patrimonio Cors r p w ★★→★★★ AC. Some of island's finest, from dramatic limestone hills in n. Individual reds from NIELLUCCIO, intriguing whites, even *late-harvest Vermentino*. Top: Antoine Arena, CLOS de Bernardi, *Gentile*, Montemagni, Pastricciola, Yves Leccia at E Croce. Worth the journey.

Pauillac H-Méd r ★★★→★★★★ 90' 96' 00' **05' 09' 10'** 16' 18 (19) Communal AC in n MÉD with 18 Classed Growths, incl LAFITE, LATOUR, MOUTON. Famous for long-lived wines, the acme of CAB SAUV. Other top CHX: CLERC MILON, GRAND-PUY-LACOSTE, LYNCH-BAGES, PICHON-BARON, PICHON-LALANDE, PONTET-CANET. Gd-value CHX: FONBADET, La Fleur Peyrabon, PIBRAN.

Pays d'Oc, IGP L'doc r p w ★→★★★ Largest IGP, covering whole of L'DOC-ROUSS. Extremes of quality; best are innovative, exciting. Recent focus on varietal wines; 58 different grapes allowed. NIELLUCCIO latest addition (Summer of Love), ALBARIÑO (Laurent Miquel, Foncalieu) CARIGNAN, esp old vines, increasingly popular. Big players incl: Bruno Andreu, DOM PAUL MAS, GÉRARD BERTRAND, Jeanjean, CO-OP FONCALIEU. Try DOMS Calmel & Joseph, Gayda, Yeuse.

Pécharmant SW Fr r ★★ 12 15' 17 (18) BERGERAC inner AOP on edge of town. Iron and manganese in soil generate biggest, longest-living wines of area. Veteran ★★★ CH deTiregand, DOM du Haut-Pécharmant, Les Chemins d'Orient; ★★ Chx Beauportail, Champarel, Corbiac, de Biran, du Rooy, Hugon, Terre Vieille; Dom des Bertranoux; CLOS des Côtes, La Métairie.

Pélican, Dom du Jura ★★★ Retired legend Jacques Puffeney's v'yds now run by VOLNAY'S MARQUIS D'ANGERVILLE to make fine fresh styles. More to come.

Pernand-Vergelesses C d'O r w ★★★ (r) **05' 09' 10'** 12 **14** 15' 17 18' 19' (w) 14' 15' 17' 18' 19' Village next to ALOXE-CORTON, incl part of CORTON-CHARLEMAGNE, CORTON. Reds can be austere, but try chiselled, precise whites. Best PC v'yds: (r) Île des Vergelesses; (w) Sous Fretille. Local DOMS CHANDON DE BRIAILLES, Dubreuil-Fontaine, Rapet and Rollin lead the way but try also from CHANSON, JADOT, P-Y *Colin-Morey*.

Perret, André N Rh r w ★★★ 09' 10' 11' 12 15' 16' 17 18' 19' Impeccable CONDRIEU DOM, three wines, classic, stylish CLOS Chanson, rich, lingering Chéry. Also ST-JOSEPH (r w), bright-fruit classic red, serious, flowing old-vine Les Grisières (r). Gd COLLINES RHODANIENNES (r w) also.

Perrier-Jouët Champ 1st (in C19) to make dry CHAMP for UK market; strong in GC CHARD, best for gd vintage and de luxe Belle Epoque 95 **04'** 07 08 12' 15 18, Rosé 06, in painted bottle. BRUT NV; Blason de France NV; Blason de France Rosé NV; Brut 02 **04** 08. Fine new BLANC DE BLANCS. Owned by Pernod-Ricard.

Pessac-Léognan B'x r w ★★★→★★★★ **05' 09' 10' 15** 16 18 (19) AC created in 1987 for best part of n GRAV, incl all Crus Classés (1959): HAUT-BAILLY, HAUT-BRION, LA MISSION-HAUT-BRION, PAPE-CLÉMENT, etc. Aspiring unclassified: LES CARMES HAUT-BRION. Firm, full-bodied, earthy reds; B'X's finest dry whites. Value from Brown, DE ROCHEMORIN, Haut-Vigneau, Lafont-Menaut, Le Sartre.

Petit Chablis Chab w ★ DYA. Fresh and easy, would-be CHAB from outlying v'yds mostly not on kimmeridgian clay. Priced too close to full Chablis. Best wines from Billaud, BROCARD, DAUVISSAT, Defaix, Lavantureux, RAVENEAU and LA CHABLISIENNE co-op.

Pfersigberg Al GC in two parcels; v. aromatic wines. GEWURZ does v. well. RIES from BRUNO SORG, DOM PAUL GINGLINGER, LÉON BEYER (Comtes d'Eguisheim).

Philipponnat Champ ★★→★★★★ Small house, intense, esp in pure Mareuil-sur-Aÿ CUVÉE under careful oak since 18. Now owned by LANSON-BCC group. NV, Rosé NV, BRUT, Cuvée 1522 04. Famous for majestic single-v'yd *Clos des Goisses* 04, CHARD-led 08; exceptional late-disgorged vintage 09.

Picpoul de Pinet L'doc w ★→★★★ DYA. "MUSCADET OF MIDI". AC for PICPOUL around Pinet. S maritime slopes produce wines with salty tang, perfect *with oysters*, further inland more almond-*garrigue* notes. Best producers pursuing depth: reduced yields, harvest dates, ageing on lees. Oak a mistake. Best: Félines-Jourdan, La Croix Gratiot, St Martin de la Garrigue; co-ops Pinet, Pomérols.

Pic St-Loup L'doc r (p) ★★→★★★ Coolest, wettest part of L'DOC. Dramatic scenery; some high v'yds. AC relates to Rh with much SYRAH, plus GRENACHE, MOURVÈDRE. Reds for ageing; white potential considerable but still AC L'doc or IGP Val de Montferrand. Growers: Bergerie du Capucin, Cazeneuve, CLOS de la Matane, Clos Marie, de Lancyre, *Dom de l'Hortus*, Gourdou, Lascaux, *Mas Bruguière*, Mas Peyrolle, Valflaunès.

Pierre-Bise, Ch Lo r p w ★★→★★★★ 14' 15 16 17 18' 19' V.gd DOM in COTEAUX DU LAYON, incl Chaume, QUARTS DE CHAUME, SAVENNIÈRES (*Clos de Grand Beaupréau*, ROCHE-AUX-MOINES). ANJOU-GAMAY, ANJOU-VILLAGES (both CUVÉE Schist, Spilite), ANJOU Blanc Haut de la Garde. Claude Papin, architect of Quarts de Chaume GC, now semi-retired, son René in charge. 19 gd.

Pierrefeu Prov r p ★★ Subzone of CÔTES DE PROV: warmer maritime zone n of La Londe, between the Massifs de Maure and Ste Baume. Try CH la Gordonne, DOM Croix-Rousse.

Pierre Péters Champ ★★★★ Tiptop Côte des Blancs estate. Les Chétillons probably longest-lived in CHAMP: 05 07 13 14 17'. No wood; pristine. Linking with Prov's CH Mireval to make rosé Champ.

Pillot C d'O (r) w ★★★ Talented family in CHASSAGNE, known for whites. Look for F&L P, Jean-Marc P (CLOS St Marc), esp DOM Paul P for sublime PCE LA ROMANÉE, Grandes Ruchottes etc.

Pinon, François Lo w sw sp ★★★ 09 10' 11 14' 15' 16 17 18' 19' V.gd wines from Cousse Valley, Vernou, VOUVRAY. François and son Julien.

Piper-Heidsieck Champ ★★★ Historic house on surging wave of quality. Dynamic Brut Essentiel with more age, less sugar, floral yet vigorous; great with sushi, sashimi. Prestige Rare, now made as separate brand in-house, is a jewel, precise, pure, refined 98 02 08; 1st release of Rare Rosé 07.

Plageoles, Dom SW Fr r w sp Defenders and rebels guarding true GAILLAC style. Rare local grapes rediscovered incl Ondenc (base of ace sweet ★★★ Vin d'Autan), ★★ Prunelard (r, deep fruity), Verdanel (dr w, oak-aged) and countless subvarieties of Mauzac. More reds from Duras, Braucol (local name for FER SERVADOU). ★★★ Brilliant dry sparkler Mauzac Natur.

Plan de Dieu S Rh r ★→★★★ 15' 16' 17' 18 19' Rh village nr CAIRANNE with stony, wind-blown *garrigue* plain. Heady, robust, peppery, mainly GRENACHE wines; drink with game, casseroles. Gd choice. Best: CH la Courançonne, CLOS St Antonin, LE PLAISIR; DOMS APHILLANTHES (character), Arnesque, Durieu (full), Espiguette, Favards (organic), La Bastide St Vincent, Longue Toque, Martin (traditional), Pasquiers, St-Pierre.

Pol Roger Champ ★★★★ Family-owned Épernay house. BRUT Rés NV excels, dosage lowered since 2012; Brut 02' 04 06 08', lovely 09 12'; Rosé 09; BLANC DE

> **Vine archaeology in the southwest**
> France's far sw has greater diversity of vines than anywhere else in the country: travellers and pilgrims passed through here on their way to and from Spain, and local vines were never replaced with CHARD and CAB SAUV. PRODUCTEURS DE PLAIMONT has project of finding and identifying old and almost extinct vines by DNA, slotting them into the vine family tree, to discover what was being grown when. So far it can draw family trees back to the Middle Ages. Before that, which vines were grown is guesswork. We know that PINOT N is over 1000 yrs old, but vines were domesticated 8000 yrs ago. A lot happened in between that we will never know.

BLANCS 09. Fine *Pure* (no dosage). Sumptuous CUVÉE Sir Winston Churchill **88** 02 04 **08'** always a blue-chip choice for long ageing, best value of prestige cuvées.

Pomerol B'X r ★★★ →★★★★ 98' 00' 01' 05' 09' 10' **15** 16 18 Tiny, pricey AC bordering ST-ÉM; MERLOT-led, plummy to voluptuous styles, but long life. Top CHX on clay, gravel plateau: CLINET, HOSANNA, L'ÉGLISE-CLINET, L'ÉVANGILE, LA CONSEILLANTE, LAFLEUR, LA FLEUR-PÉTRUS, LE PIN, PETRUS, TROTANOY, *Vieux-Ch-Certan*. Occasional value (Bourgneuf, de Sales, La Pointe, Mazeyres).

Pommard C d'O r ★★★★↘) 90' **96'** 99' 03 05' 09' 10' 12 15' 16' 17 18' 19' Antithesis of neighbour VOLNAY; potent, tannic wines to age 10 yrs+. Doing well in recent hot vintages. Best v'yds: Rugiens for power, Epenots for grace. Growers: BICHOT (DOM du Pavillon), CH de Pommard, Clerget, COMTE ARMAND, COURCEL, DE MONTILLE, HOSPICES DE BEAUNE, Huber-Vedereau, J-M BOILLOT, L Boillot, Lejeune, Parent, Pothier-Rieusset, Rebourgeon, Violot-Guillemard.

Pommery Champ ★★ Historic house with spectacular cellars; brand now owned by VRANKEN. BRUT NV steady bet, no fireworks; Rosé NV; Brut **04** 08 09 12'. Once outstanding CUVÉE Louise 02 04 less striking recently. Planting in England.

Ponsot, Dom C d'O r w ★★→★★★★ Idiosyncratic, top-quality MOREY-ST-DENIS DOM. Rose-Marie P now in charge. Key wines: *Clos de la Roche*, PC Monts Luisants (ALIGOTÉ). No significant changes in style. Now set up as NÉGOCIANT; outstanding from 16.

Ponsot, Laurent C d'O r w ★★→★★★★ Man who made DOM PONSOT wines for 30 yrs left family business to create his own Haute Couture label nearby in 2016. Kept the sharecropping contracts incl amazing GRIOTTE-CHAMBERTIN and CLOS ST-DENIS and has added fine range in both colours. Buying vines too. Watch these spaces.

Pouilly-Fuissé Burg w ★★→★★★ 14' 15 17 18 19' Top AC of MÂCON; potent, rounded but intense whites from around Fuissé, more mineral style in Vergisson. Enjoy young or with age. We are STILL waiting for a PC classification to happen. Top: Barraud, Bouchacourt, Bret, CH de Beauregard, CH DE FUISSÉ, Ch des Quarts, Ch des Rontets, Cordier, Cornin, Drouin, Ferret, Forest, Merlin, Paquet, Robert-Denogent, Rollet, Saumaize, Saumaize-Michelin, VERGET.

Pouilly-Fumé Lo w ★★→★★★ 17' 18' 19' E-bank neighbour of SANCERRE. SAUV BL. V.gd 18/19. Shows best with 2–3 yrs in bottle. Growers: Bain, Belair, BOURGEOIS, Cailbourdin, Champeau, CH de Favray, Chatelain, DIDIER DAGUENEAU (VDF from 2017), Edmond and André Figeat, Jean Pabiot, Jonathan Pabiot (v.gd), Jolivet, Joseph Mellot, LADOUCETTE, Masson-Blondelet, Redde, Saget, Serge Dagueneau & Filles, Tabordet, Treuillet.

Pouilly-Loché Burg w ★★ 14' 15 17 18 19' Least known of Mâconnais' Pouilly family. Reference: CLOS des Rocs. Try also Bret Bros, Tripoz, local CAVE des GCS Blancs.

Pouilly-sur-Loire Lo w ★★ DYA. In C19 Pouilly supplied Paris with CHASSELAS table grapes. Same area as POUILLY-FUMÉ. Only 27 ha remain, but Gitton, Jonathan Pabiot, Landrat-Guyollot, Masson-Blondelet, Redde, Serge Dagueneau & Filles all brave souls upholding tradition.

Pouilly-Vinzelles Burg w ★★ 14′ 15 17 18 19′ Between POUILLY-LOCHÉ and POUILLY-FUISSÉ geographically and in quality. Outstanding v'yd: Les Quarts. Best: Bret Bros, DROUHIN, Valette. Volume from CAVE des GCS Blancs.

Premier Cru (PC) First Growth in B'X; 2nd rank of v'yds (after GC) in Burg; 2nd rank in LO: one so far, COTEAUX DU LAYON Chaume.

Premières Côtes de Bordeaux B'x w sw ★→★★ 15′ 16 18 Same zone as CADILLAC-CÔTES DE B'x but for sweet whites only. SÉM-dominated *moelleux*. Generally early-drinking. Best CHX: Crabitan-Bellevue, du Juge, Fayau, *Suau*.

Prieur, Dom Jacques C d'O ★★★ Major MEURSAULT estate with range of underplayed GCS from MONTRACHET to MUSIGNY. Style aims at weight from late-picking and oak more than finesse. Owners Famille Labruyère also have CHAMP and MOULIN-À-VENT projects plus CH ROUGET, POM.

Prieuré de St-Jean de Bébian L'doc ★★★ Pézenas estate with high-end restaurant, CHÂTEAUNEUF varieties. Owned by Russians, Aussie winemaker Karen Turner. Complex soils: gravel, volcanic, clay, limestone. Three levels: La Chapelle, La Croix, Prieuré (r w); old-vines red 1152.

Primeur "Early" wine for refreshment and uplift; esp from BEAUJ; VDP too. Wine sold en primeur is still in barrel, for delivery when bottled.

Producteurs Plaimont SW Fr France's most dynamic co-op, bestriding SAINT MONT, MADIRAN and CÔTES DE GASCOGNE like the colossus it is. Has abandoned B'x varieties for grapes traditional to sw, incl some pre-phylloxera discoveries. All colours, styles, mostly ★★, all tastes, purses.

Propriétaire récoltant Champ Owner-operator, literally owner-harvester.

Puisseguin St-Émilion B'x r ★★ 10′ 15 16 18 (19) Most e of four ST-ÉM satellites; MERLOT-led wines firm, solid. Top CHX: Beauséjour, Branda, Clarisse, DES LAURETS, de Môle, Durand-Laplagne, Fongaban, Guibot la Fourvieille, Haut-Bernat, La Mauriane, Le Bernat, Soleil.

Puligny-Montrachet C d'O (r) w ★★★→★★★★ 09′ 10′ 12 14′ 15 17′ 18 19′ Floral, fine-boned, tingling white burg. Decent at village level, Enseignières v'yd exceptional, outstanding PCS, esp: Caillerets, Champ Canet, Combettes, Folatières, Pucelles, plus amazing MONTRACHET GCS. Producers: *Bouchard Père & Fils*, CARILLON, Chartron, CH de Puligny, *Dom Leflaive*, *Drouhin*, Ente, JADOT, *J-M Boillot*, *O Leflaive*, Pernot, *Sauzet*, Thomas-Collardot.

Puyméras S Rh r w ★ 16′ 17′ 18 19′ Sound, secluded village, high v'yds, supple plum-fruited reds centred on GRENACHE, fair whites, decent co-op. Limited choice. Note CAVE la Comtadine, DOM du Faucon Doré (bio), Puy du Maupas.

Pyrénées-Atlantiques SW Fr Mostly DYA. IGP in far sw for wines outside local AOPG. ★★★ CH Cabidos in middle of nowhere (superb dr and sw w PETIT MANSENG varietals that will age), ★★ DOM Moncaut (a JURANÇON in all but name nr Pau), ★ BRUMONT's commercial range of non-AOP varietals and blends. Otherwise pot luck.

Quarts de Chaume Lo w sw ★★★→★★★★ 07′ 10′ 11′ 14′ 15′ 16 17 18′ Slopes close to Layon, CHENIN BL. Best richly textured. Best: Baudouin, Belargus, Bellerive, Branchereau, CH PIERRE-BISE, FL, Guegniard, Ogereau, Suronde (same ownership Minière, BOURGUEIL). 19, little made: dry Chenin more saleable.

Quincy Lo w ★→★★ 17′ 18′ 19′ Revived AOP (308 ha; 60 ha in 1990), SAUV BL on low-lying sand/gravel banks. Growers: Mardon, Portier, Rouzé, Siret-Courtaud, Tatin-Wilk – DOMS Ballandors, Tremblay, Villalin.

Rancio Rouss ★★★ Refers to flavour of these VDNS: pungent, tangy. Usually a fault, but here, a virtue. Reminiscent of Tawny Port, or old Oloroso Sherry. Found in BANYULS, MAURY, RASTEAU, RIVESALTES, wood aged and exposed to oxygen, heat. Can be a grand experience; don't miss.

Rangen Al Most s GC of AL at Thann. V. steep (average 90%) slopes, volcanic soils.

Top: majestic RIES ZIND-HUMBRECHT (CLOS St Urbain **05' 08' 10' 17'**), SCHOFFIT (St-Théobald 08' **10 15'** 17'). Fascinating contrast with Z-H St-Théo: supreme finesse, no oak.

Rasteau S Rh r (p) (w) br (dr) sw ★★ 10' 12' 15' 16' 17' **18** 19' Full-on reds from mainly clay soils, mostly GRENACHE. Best in hot yrs. NB: Beaurenard (serious, age well), *Cave Ortas* (gd), CH La Gardine, *Ch du Trignon*, Famille Perrin; DOMS Beau Mistral, Collière, Combe Julière, Coteaux des Travers (bio), Didier Charavin, Elodie Balme (soft), Escaravailles, Girasols, Gourt de Mautens (talented, IGP from 2010), Grand Nicolet (character), Grange Blanche, Rabasse-Charavin (full), M Boutin, Soumade (polished), ST GAYAN, Trapadis. Grenache dessert VDN: quality on the up (Doms Banquettes, Combe Julière, Coteaux des Travers, Escaravailles, Trapadis). Rasteau doms also gd source CÔTES DU RH (r).

Raveneau Chab w ★★★★ Along with DAUVISSAT cousins, greatest CHAB producers, using classic methods for *extraordinary long-lived wines*. A little more modern while still growing in stature of late. Excellent value (except in secondary market). Look for PC Butteaux, Chapelot, Vaillons and GC Blanchots, Le Clos.

Rayas, *Ch* S Rh r w ★★★★ 98' **99** 05' 06' 07' 09' 10' 11' 15' 16' 17' 19' Brilliant, lost-in-time 13-ha CHÂTEAUNEUF estate, tiny yields, sandy soils. Pale, subtle, aromatic, sensuous reds (100% GRENACHE) whisper quality, offer delight, age superbly. White Rayas (GRENACHE BL, CLAIRETTE) v.gd over 18 yrs+. Stylish second wine, *Pignan*, still empties your wallet. Supreme CH Fonsalette CÔTES DU RH, incl marvellous long-lived SYRAH. Decant them all; each is an occasion. No 18 (mildew). Also gd CH des Tours VACQUEYRAS (peppery), VDP.

Each vine in Burg needs attention c.30 times a yr. And 10,000 vines/ha.

Regnié Beauj r ★★ **15'** 17 18' **19** Most recent BEAUJ cru, lighter wines on sandy soil, meatier nr MORGON. Starting to get some gd growers now. Try A Sunier, Burgaud, Chemarin, de la Plaigne, Dupré, J Sunier, Rochette.

Reuilly Lo r p w ★ ·★★★ 16 17 18' 19' Revived AC (274 ha, from 30 ha in 1990) neighbour of QUINCY. SAUV BL, rosés and *Vin Gris* PINOT N and/or *Pinot Gr*. Some gd Pinot N reds. Best: Claude Lafond (run by daughter Natalie), *Jamain*, Mardon, Renaudat, Rouze, Sorbe.

Riceys, *Les* Champ p DYA. Key AC in AUBE for notable PINOT N rosé. Producers: *A Bonnet*, Jacques Defrance, Morize. Great 09; v. promising **14** after lean period 11–13; ace **15'**.

Richebourg C d'O r ★★★★ 90' 93' **96'** 99' 02' **03** 05' 09' 10' 12' 15' 16 17 18' 19' VOSNE-ROMANÉE GC. Supreme burg with great depth of flavour; vastly expensive. Growers: DRC, GRIVOT, GROS, HUDELOT-NOËLLAT, LEROY, LIGER-BELAIR and MÉO-CAMUZET.

Rimage Rouss A growing mode: vintage VDN, super-fruity for drinking young. Think gd Ruby Port. Grenat is MAURY version.

Rion C d'O r (w) ★★ ·★★★ Related DOMS in NUITS-ST-GEORGES and VOSNE-ROMANÉE. Patrice R for excellent Nuits CLOS St-Marc, Clos des Argillières and CHAMBOLLE-MUSIGNY. Daniel R for Nuits and Vosne PCS; Bernard R more Vosne-based. All fairly priced.

Rivesaltes Rouss r w br dr sw ★★ NV or solera, also vintage, young and old VDN from large area in n ROUSS. Grossly underappreciated; deserves revival. V.-long-lasting wines, esp RANCIOS. Look for: Boucabeille, des Chênes, des Schistes, DOM CAZES, Rancy, Roc des Anges, Sarda-Malet, Vaquer. You won't be disappointed.

Rives-Blanques, *Ch* L'doc w sp ★★★ LIMOUX. Irish-Dutch couple with son Jean Ailbe make BLANQUETTE and CRÉMANT. Limoux white (and rosé), incl unusual 100% MAUZAC, Occitania, blend Trilogie and age-worthy CHENIN BL Dédicace. Dessert Lagremas d'Aur.

Roche-aux-Moines, La Lo w sw ★★→★★★ 14′ 15′ 16 18′ 19 33-ha cru of SAVENNIÈRES. Strict rules, top CHENIN BL. Try: aux Moines, *Ch Pierre-Bise*, CLOS de la Bergerie (Joly), FL, Forges, Laureau.

Roederer, Louis Champ ★★★★ Peerless family-owned house/DOM. Enviable v'yds: 240 ha, much organic/bio. BRUT Premier NV all finesse, flavour; Brut 08 **12**, BLANC DE BLANCS 12 **13** 15, Brut Saignée Rosé 09. Magnificent *Cristal* bio (since 2012) 08. Superb Cristal Vinothèque Blanc 95 and Rosé 95. Brut NATURE Philippe Starck (all Cumières 09 12 15). Also owns DEUTZ, CH PICHON-LALANDE. *See also* California.

Rolland, Michel B'x Veteran French consultant winemaker and MERLOT specialist (B'x and worldwide). Owner of Fontenil in FRON. *See also* Argentina (Clos de los Siete).

Rolly Gassmann Al w sw ★★★ Revered DOM, esp Moenchreben v'yd. Off-dry, rich, sensuous GEWURZ CUVÉE Yves 08 09 12 **15** 17′ 18 19. Now into bio, more finesse. Mineral zesty RIES 13 16 17′ 18 19′. Fine PINOT N 15 intense, gentle tannins.

Romanée, La C d'O r ★★★★ 09′ 10′ 12′ 15′ 16′ 17 18′ 19′ Tiniest GC in VOSNE-ROMANÉE, MONOPOLE of COMTE LIGER-BELAIR. Exceptionally fine, perfumed, intense: now on peak form and understandably expensive.

Romanée-Conti, La C d'O r ★★★★ 85′ 89′ 90′ 93′ 96′ 99′ 00 02′ **03** 05′ 09′ 10′ 12′ 14′ 15′ 16′ 17 18′ 19′ GC in VOSNE-ROMANÉE, MONOPOLE of DRC. Most celebrated GC in Burg, gold dust. On fabulous form these days. Patience required, for 10–20 yrs. But beware geeks bringing fake gifts.

Romanée-St-Vivant C d'O r ★★★★ 90′ 99′ 02′ 05′ 09′ **10′** 12′ 15′ 16′ 17 18′ 19′ GC in VOSNE-ROMANÉE. Downslope from LA ROMANÉE-CONTI, haunting perfume, delicate but intense. Ready a little earlier than famous neighbours. Growers: if you can't afford DRC or LEROY, or indeed ARNOUX-LACHAUX, CATHIARD or HUDELOT-NOËLLAT now, try ARLOT, Follin-Arbelet, JJ Confuron, LATOUR, Poisot. Nobody letting side down.

Rosacker Al GC at Hunawihr. Limestone/clay makes some of longest-lived RIES in AL (CLOS STE-HUNE).

Rosé d'Anjou Lo p ★→★★ DYA. Rosé – off-dry to sw (mainly Grolleau). Big AOP. Unfashionable but popular, increasingly well made. Look for: Bougrier, Clau de Nell, Bergerie, Grandes VIGNES, Mark Angeli (VDF).

Rosé de Loire Lo p ★→★★ DYA. Dry rosé: six grapes Grolleau Gris/Noir, CAB FR, CAB SAUV, GAMAY; PINOT N. AC. Best: Bablut, Bois Brinçon, Branchereau, GADT, CAVE de SAUMUR, CH PIERRE-BISE, Ogereau, Passavant Richou, Soucherie.

Rosette SW Fr w s/sw ★★ Tiny AOP DYA. Birthplace of BERGERAC, now reviving traditional off-dry aperitif whites. Also gd with foie gras or mushrooms. Avoid oaked versions that deny the style. CLOS Romain, CHX Combrillac, de Peyrel, Monplaisir, Puypezat-Rosette, Spingulèbre; DOMS de Coutancie, de la Cardinolle, du Grand-Jaure.

Rossignol-Trapet C d'O r ★★★ Equally bio cousins of DOM TRAPET, with healthy holdings of GC v'yds, esp CHAMBERTIN. Gd value across range from GEVREY VIEILLES VIGNES up. Also some BEAUNE v'yds from Rossignol side.

Rostaing, Dom N Rh r w ★★★ 99′ 01′ 05′ 09′ 10′ 12′ 13′ 15′ 16′ 17′ 18′ 19′ High-quality CÔTE-RÔTIE DOM: five tightly bound wines, all v. fine, pure, clear, discreet oak, wait 6+ yrs, decant. Son Pierre took over 2015. Complex, enticing, top-class Côte Blonde (5% VIOGNIER), Côte Brune (iron), also La Landonne (dark fruits, 15–20 yrs). Tangy, firm *Condrieu*, also IGP COLLINES RHODANIENNES (r w), L'DOC Puech Noble (r w).

Rouget, Domaine C d'O r ★★★★ Renamed as DOM R rather than Emmanuel R with new generation refreshing dom famed for Henri Jayer connection and CROS PARANTOUX v'yd. Fine NUITS-ST-GEORGES, VOSNE-ROMANÉE, as well as GCS.

Roulot, Dom C d'O w ★★★ →★★★★ Jean-Marc R leads outstanding MEURSAULT DOM, now cult status so beware secondary-market prices. Great PCS, esp CLOS des

Bouchères, Perrières; value from top village sites Luchets, Meix Chavaux, esp Clos du Haut Tesson.

Roumier, Georges C d'O r ★★★★ Reference DOM for BONNES-MARES and other *brilliant Chambolle* wines (incl Amoureuses, Cras) from Christophe R. Long-lived wines but still attractive early. Cult status means hard to find now at sensible prices. Best value is MOREY CLOS de la Bussière.

Rousseau, Dom Armand C d'O r ★★★★ Unmatchable GEVREY-CHAMBERTIN DOM thrilling with balanced, fragrant, refined, age-worthy wines from village to GC, esp CLOS ST-JACQUES. No changes expected anytime soon.

Roussette de Savoie Sav w ★★ Regional AOC: same area as AOC SAVOIE and 10% of its wines. 100% ROUSSETTE. 150 ha. Can age. Try: Curtet, de la Mar, Grisard, Mérande, Ravier, Quénard.

Roussillon Often linked with L'DOC, and incl in AC L'doc. Combination of altitude and maritime influence moderates heat. High % of old vines. GRENACHE key variety. Original, traditional VDN (eg. BANYULS, MAURY, RIVESALTES). Younger vintage RIMAGE/Grenat now competing with aged RANCIO. Also serious age-worthy table wines (r w). *See also* COLLIOURE, CÔTES DU ROUSS-VILLAGES, MAURY SEC, IGP CÔTES CATALANES.

Ruchottes-Chambertin C d'O r ★★★★ 99' 02' 05' 09' 10' 12' 15' 16 17' 18' 19' Tiny GC neighbour of CHAMBERTIN. Less weighty but ethereal, intricate, lasting wine of great finesse. Top growers: MUGNERET-Gibourg, ROUMIER, ROUSSEAU. Try also CH de MARSANNAY, H Magnien, Marchand-Grillot, Pacalet.

Ruinart Champ ★★★★ Oldest house? (1729). High standards going higher still. Rich, elegant wines. "R" de Ruinart BRUT NV; Ruinart Rosé NV; "R" de Ruinart Brut 08. Prestige CUVÉE *Dom Ruinart* one of two best vintage BLANC DE BLANCS in CHAMP (viz 90' esp in magnum, 02 04' 07. DR Rosé also v. special 06 04'. NV Blanc de Blancs much improved. High hopes for 13', classic cool lateish vintage.

Rully Burg r w ★★ (r) 15' 16 17 18' 19' (w) 14' 16' 17' 18 19' CÔTE CHALONNAISE village. *Light, fresh, tasty, gd-value whites*. Reds all about the fruit, not structure. Try *C Jobard*, Devevey, DROUHIN, *Dureuil-Janthial*, FAIVELEY, Jacqueson, Jaeger-Defaix, Ninot, *Olivier Leflaive*, Rodet.

Sablet S Rh r (p) w ★★ 16' 17' 18' 19' CÔTES DU RH-VILLAGE on plain near GIGONDAS. Easy wines, some serious. Sandy soils, trim red-berry reds, esp CAVE co-op Gravillas, CH Cohola (organic), *du Trignon*; DOMS de Boisson (organic, full), Les Goubert (r w), Pasquiers (full), Piaugier (r w). *Gd full whites* for apéritifs, food, NB: Boissan, SAINT GAYAN.

St-Amour Beauj r ★★ 15' 17 18' 19' Most n BEAUJ cru: mixed soils, so variable character. Try: DOM de Fa, Cheveau, *Patissier*, Pirolette, Revillon.

St-Aubin C d'O r w ★★★ (w) 14' 15 17 18' 19' Fine source for *lively, refreshing whites*, challenging PULIGNY and CHASSAGNE, esp on price. Also pretty reds mostly for early-drinking. Best v'yds: Chatenière, *En Remilly*, Murgers Dents de Chien. Best growers: COLIN-MOREY, JC BACHELET, Joseph Colin, *Lamy*, Marc COLIN. Also Prudhon for value.

St-Bris Burg w ★ DYA. Unique AC for SAUV BL in n Burg. Fresh, lively, but also worth keeping from GOISOT or de Moor. Try also Bersan, Davenne, Simonnet-Febvre.

St-Chinian L'doc r p ★ ★★★ Large hilly area nr Béziers with schist in nw, clay and limestone in se. Sound reputation. Incl CRUS of Berlou (mostly CARIGNAN) Roquebrun (mostly SYRAH) on schist. Warm, spicy reds, based on Syrah, GRENACHE, Carignan, MOURVÈDRE. Whites from ROUSSANNE, MARSANNE, VERMENTINO, GRENACHE BL. Gd co-op Roquebrun; CH Viranella Madura, DOMS Borie la Vitarèle, des Jougla, la Dournie, la Madura, Navarre, Rimbert; CLOS Bagatelle, Mas Champart. Several new estates.

St Cosme, Ch de S Rh r ★★★ 98' 99' 06' 09' 10' 11' 12' 13' 14' 15' 16' 17' 18' 19' 15 ha bio estate at GIGONDAS, wines with flair, oak. Plot-specific Gigondas Le Poste, Valbelle, CÔTES DU RH Les Deux Albion (r). Owner CH de Rouanne, VINSOBRES (2018). Gd N Rh merchant range, esp CÔTE-RÔTIE, CONDRIEU.

St-Émilion B'x r ★★→★★★★ 98' 01' 05' 09' 10' 15' 16 18 (19) Big MERLOT-led district on B'x's Right Bank, currently on a roll. CAB FR also strong. UNESCO World Heritage site. ACS St-Ém and (lots of) St-Ém GC. Top designation St-Ém PREMIER GRAND CRU CLASSÉ. Warm, full, rounded style but much diversity due to terroir, winemaking and blend. Best firm, v. long-lived. Top CHX: ANGÉLUS, AUSONE, CANON, CHEVAL BLANC, FIGEAC, PAVIE. Many chx are value.

St-Estèphe H-Méd r ★★→★★★★ 95 00' 05' 09' 10' 15 16' 18 (19) Most n communal AC in the MÉD. Solid, structured wines for ageing; happy hunting ground for value. Five Classed Growths: CALON-SÉGUR, COS D'ESTOURNEL, COS-LABORY, LAFON-ROCHET, MONTROSE. Top unclassified estates: CAPBERN, HAUT-MARBUZET, LE BOSCQ, LE CROCK, LILIAN LADOUYS, MEYNEY, ORMES-DE-PEZ, PHÉLAN-SÉGUR.

St-Gall Champ Brand of Union-CHAMP, top co-op at AVIZE. BRUT NV; Extra Brut NV; Brut BLANC DE BLANCS NV; Brut Rosé NV; Brut Blanc de Blancs 08; CUVÉE Orpale Blanc de Blancs 02' 08' 17'. Fine-value PINOT-led *Pierre Vaudon NV*. Makes top *vins clairs* for some great houses.

Saint Gayan, Dom S Rh r (w) ★★★ 90' 95' 98' 99' 01' 05' 06' 07' 10' 15' 16' 17' 18' 19' Top 16 ha GIGONDAS estate; 80% GRENACHE Origine great value. V.gd RASTEAU Ilex (r), charming SABLET L'Oratory (w).

Each 4g sugar added at bottling = 1 bar of pressure in bottle of Champagne.

St-Georges d'Orques L'doc r p ★★→★★★ Most individual and historic part of sprawling Grès de Montpellier, aspiring to cru status. Low-lying v'yds on sandstone, GRENACHE, SYRAH, MOURVÈDRE. Try Belles Pierres, CH l'Engarran, DOMS Henry, La Marfée, La Prose.

St-Georges-St-Émilion B'x r ★★ 10' 15 16 18 (19) Tiny ST-ÉM satellite. Sturdy, structured. Best CHX: Calon, CLOS Albertus, Macquin-St-Georges, St-André Corbin, ST-GEORGES, TOUR DU PAS-ST-GEORGES.

St-Gervais S Rh r (p) (w) ★→★★ 16' 17' 18 19' W-bank Rh village; gd soils but v. limited range. Co-op low-key; best is long-lived (10 yrs+) DOM Ste-Anne red (direct, firm, MOURVÈDRE liquorice flavours), gd VIOGNIER. Also Dom Clavel (Regulus r).

St-Jacques d'Albus, **Ch** L'doc r p w ★★ Dynamic MINERVOIS estate. English owner, Oz winemaker, SYRAH, GRENACHE on clay and limestone. Fruit-forward wines: Le Petit St-Jacques, Le DOM, Le CH and Syrah-dominant La Chapelle (all r). Coteaux de Peyriac from VIOGNIER/VERMENTINO/ROUSSANNE.

St-Joseph N Rh r w ★★→★★★ 99' 05' 09' 10' 12' 15' 16' 17' 18' 19' 64 km (40 miles) of mainly granite v'yds, some high, along w bank of N Rh. SYRAH reds. Best, oldest v'yds nr Tournon: stylish, red-fruited wines; further n darker, peppery, younger oak. More complete than CROZES-HERMITAGE, esp CHAPOUTIER (Les Granits), DOM Gonon (top class), *Gripa*, GUIGAL (VIGNES de l'Hospice), *J-L Chave* (gd style); also Alexandrins, Amphores (bio), A PERRET (Grisières), Chèze, Courbis (modern), Coursodon (racy, modern), Cuilleron, E Darnaud, *Delas*, J&E Durand (fruit), Faury, Ferraton, F Villard, Gaillard, P Marthouret (traditional), Monier-Perréol (organic), P-J Villa, S Blachon, Vallet, Vins de Vienne. Gd food-friendly *white (mainly Marsanne)*, esp A PERRET, Barge, *Chapoutier* (Les Granits), Cuilleron, Curtat (stylish), Dom Faury, Gonon (fab), Gouye (traditional), *Gripa*, J Pilon, Vallet.

St-Julien H-Méd r ★★★→★★★★ 96' 00 05' 09' 10' 15 16' 18 (19) Super-stylish mid-MÉD communal AC. 11 classified (1855) estates own most of v'yd area. Incl three LÉOVILLES, BEYCHEVELLE, DUCRU-BEAUCAILLOU, GRUAUD-LAROSE, LAGRANGE, TALBOT. Epitome of harmonious, fragrant, savoury red.

Saint Mont SW Fr r p w ★★ (r) 15' 16 17 18 (p w) DYA AOP from Gascon heartlands. Huge PRODUCTEURS PLAIMONT would like to take over appellation as a brand, but saxophonist J-L Garoussia (★ DOM de Turet), CH de Bergalasse and ★★ Dom des Maouries are the resistance.

St Nicolas de Bourgueil Lo r p ★→★★★ 16' 17' 18' 19' Similar to BOURGUEIL: CAB FR. Largely sand/gravel, light wines; fuller-bodied from limestone slopes. Try: Amirault, David, Delanoue, Jamet, Frédéric Mabileau, Laurent Mabileau, Lorieux, Mabileau-Rezé, Ménard, Mortier, Taluau-Foltzenlogel, Vallée, *Yannick Amirault*. 19 v.gd but small (frost, drought).

St-Péray N Rh w sp ★★ 16' 17' 18' 19' On-the-up white (MARSANNE/ROUSSANNE) from hilly granite, some lime v'yds opposite Valence, lots of fast new planting. Once *famous for fizz*; classic-method bubbles well worth trying. (A Voge, J-L Thiers, R Nodin, TAIN co-op). Still white should have grip, be smoky, flinty. Best: CHAPOUTIER, *Clape* (pure), *Colombo* (stylish), Cuilleron, *du Tunnel* (v. elegant), Gripa (v.gd), J&E Durand, J-L Thiers, L&C Fayolle, R Nodin, TAIN co-op, Vins de Vienne, Voge (oak).

St-Pourçain Mass C r p w ★→★★ AOC 18' 19'. Mainly light red and rosé (GAMAY, PINOT N), pure Pinot N banned, white from local Tressalier and/or CHARD, SAUV BL. Growers: Bérioles (rising star), DOM de Bellevue, Grosbot-Barbara, Laurent Nebout, Pétillat, Ray; VIGNERONS de St-Pourçain (the majority of production).

St-Romain C d'O r w ★★ (w) 15 17' 18' 19' *Crisp whites* from side valley of CÔTE DE BEAUNE. Excellent value by Burg standards. Best v'yds Sous le CH, Sous la Roche, Combe Bazin. Specialists Alain Gras, de Chassorney, H&G Buisson, but most NÉGOCIANTS have a gd one. Some fresh reds too.

St-Véran Burg w ★★ 15 17 18' 19 S AOC either side of POUILLY-FUISSÉ. Try CH de Beauregard, Chagnoleau, Corsin, Deux Roches, Litaud, Merlin for single-v'yd CUVÉES. Gd-value DUBOEUF, Poncetys, TERRES SECRETES co-op.

Ste-Croix-du-Mont B'x w sw ★★ 10' 11' 13 15 16 18 AC making sweet, white *liquoreux*. Faces SAUTERNES across R. Garonne. Best: rich, creamy, can age. Top CHX: Crabitan-Bellevue, du Mont, La Caussade, La Rame, *Loubens*.

Ste-Victoire Prov r p ★★ Subzone of CÔTES DE PROV, s limestone slopes of Montagne Ste-Victoire: gets much-needed freshness in hotter vintages. Broad creamy acidity. DOMS de St Ser, Gassier both benefit from high altitude. Dom Richeaume, v.gd, is IGP.

Salon Champ ★★★★ Original BLANC DE BLANCS, from LE MESNIL in Côte des Blancs. Tiny quantities. Awesome reputation for long-lived luxury-priced wines: in truth, inconsistent. On song recently, viz 83' 90 97', but 99 disappoints. *See also* DELAMOTTE. Both owned by LAURENT-PERRIER, whose new cellarmaster is Dominique Demarville. Watch closely.

Sancerre Lo r (p) w ★→★★★★ 14' 15' 16' 17' 18' 19' Touchstone SAUV BL, increasingly gd PINOT N. Impressively dynamic top producers. Series of gd/v.gd vintages since 14. Best: ALPHONSE MELLOT, Boulay (v.gd), *Bourgeois*, Claude Riffault, Cotat (variable), Dezat, François Crochet, Fleuriet, Fouassier (bio), Jean-Max Roger, Jolivet, *Joseph Mellot*, L Crochet, Natter, Neveu, Paul Prieur, Pinard, Pierre Martin, Raimbault, *P & N Reverdy*, Roblin, Thomas, Thomas Laballe, *Vacheron*, Vatan, Vattan, V Delaporte.

Sang des Cailloux, Dom Le S Rh r w ★★★ 10' 12' 13' 15' 16' 17' 18' 19' Best VACQUEYRAS DOM, 17 ha, bio, *garrigue* drive. Classic red rotates name every 3 yrs, Doucinello (17), Azalaïs (18), Floureto (19). Top Lopy red ages well. Firm Un Sang Blanc.

Santenay C d'O r (w) ★★→★★★ 05' 09' 12 14 15' 16 17 18' 19' S end of CÔTE DE BEAUNE, potential for fine reds; don't overlook. Best v'yds: CLOS de Tavannes,

Clos Rousseau, Gravières (r **w**). Producers: Belland, Camille Giroud, Chevrot, J Girardin, JADOT (now incl dom Prieur-Brunet) LAMY, MOREAU, Muzard, Vincent. Some gd whites too, eg. CHARMES.

Saumur Lo r p w sp ★ →★★★★ 16 17 18' 19' Big AC. Whites, light to age-worthy; mainly easy reds except SAUMUR-CHAMPIGNY; Saumur Rosé (Cabs). Home of Lo fizz: CRÉMANT most important, Saumur Mousseux. Saumur-Le-Puy-Notre-Dame AOP for CAB FR (mainly). Best: Antoine Foucault, BOUVET-LADUBAY, CH de Brézé, CLOS Mélaric, CLOS ROUGEARD, Ditterie, Guiberteau, Nerleux, Paleine, Parnay, René-Hugues Gay, ROBERT ET MARCEL, Rocheville, St-Just, Targé, VILLENEUVE, Yvonne.

Saumur-Champigny Lo r ★★★★ 15' 16' 17 18' 19' Attractive CAB FR from nine-commune AC, gd vintages age 15–20 yrs+. Best: Antoine Sanzay, Bonnelière (value), Bruno Dubois, CH DE VILLENEUVE, Ch Yvonne, Champs Fleuris, CLOS Cristal, CLOS ROUGEARD (cult), Cune, Ditterie, *Filliatreau*, Hureau, Nerleux, Petit St-Vincent, Robert et Marcel (co-op), Roches Neuves, Rocheville, St Just, St-Vincent, Seigneurie, *Targé*, Vadé, Val Brun. 19 gd but reduced by frost, drought.

Saussignac SW Fr sw ★★ 15 16' 17 18' (19) BERGERAC sub-aop, adjoining MONBAZILLAC, producing similar sweet wines perhaps with shade more acidity. Best: ★★★ DOMS de Richard, La Maurigne, Les Miaudoux, Lestevénie; ★★ CHX Le Chabrier, Le Payral, Le Tap.

Sauternes B'x w sw ★★ →★★★★ 01' 05' 07' 09' 11' 15' 16 18 AC making France's best *liquoreux* from "noble rotted" grapes. Luscious, golden and age-worthy. 19 small volume. Classified (1855) CHX: CLOS HAUT-PEYRAGUEY, GUIRAUD, *Lafaurie-Peyraguey*, LA TOUR BLANCHE, RIEUSSEC, SIGALAS-RABAUD, SUDUIRAUT, D'YQUEM. Gd value from DOM de l'Alliance, *Fargues*, HAUT-BERGERON, Les Justices, *Raymond-Lafon*.

Sauzet, Etienne C d'O w ★★★★ Leading DOM in PULIGNY with superb range of PCS (Combettes, Champ Canet best) and GC BÂTARD-M. Concentrated, lively wines, certified bio, once again capable of ageing.

Savennières Lo w dr (sw) ★★ →★★★★ 12 14' 15' 16 18' 19' Small ANJOU AC, high reputation, variable style and quality; v. long-lived dry whites (CHENIN BL) with marked acidity – a few DEMI-SEC. Baudouin, *Baumard*, BELARGUS, Bergerie, Boudignon, *Ch d'Epiré*, *Ch Pierre-Bise*, CH Soucherie, Closel, DOM FL, Laureau, Mahé, Mathieu-Tijou, Morgat, Ogereau. Top sites: CLOS Picot, COULÉE DE SERRANT, ROCHE-AUX-MOINES. Fine 18; 17 19 frost.

Savigny-lès-Beaune C d'O r (w) ★★★ 05' 09' 10' 12 14 15' 18' 19' Important village next to BEAUNE; similar mid-weight wines, savoury touch (but can be rustic). Top v'yds. Dominode, Guettes, Lavières, Marconnets, Vergelesses. Growers: *Bize*, Camus, *Chandon de Briailles*, Chenu, CLAIR, DROUHIN, Girard, Guillemot (w), Guyon, LEROY, Pavelot, Rapet, *Tollot-Beaut*.

Savoie r w sp ★★ →★★★ Alpine wines. AC (2129 ha) three ACs, 20 crus, incl APRÉMONT, CHIGNIN, CRÉPY, Jongieux, Ripaille. Regional ACs: CRÉMANT de Sav, ROUSSETTE DE SAV (Altesse), SEYSSEL. Reds mainly GAMAY, MONDEUSE; whites: Altesse, CHASSELAS, Jacquère, Mondeuse Bl, ROUSSANNE.

Schlossberg Al GC at Kientzheim famed since C15. Glorious compelling RIES from FALLER 10 and new TRIMBACH; 15 is great Ries yr here.

Schlumberger, Doms Al w sw ★ →★★★ Vast, top-quality AL DOM owning approx 1% of all Al v'yds. Rich wines. GCS Kitterlé and racy Saering 13 15 16 exceptional 17' 18', Spiegel. Rare RIES, signature CUVÉE Ernest and now GC Kessler GEWURZ lovely in 16; great PINOT GR.

Schoenenberg Al V. rich, successful Riquewihr GC: PINOT GR, RIES, v. fine VENDANGE TARDIVE, SÉLECTION DES GRAINS NOBLES, esp DOPFF AU MOULIN. Also v.gd MUSCAT. HUGEL Schoelhammer is from here.

Schoffit, Dom Al w ★★★★ Exceptional Colmar grower, superb late-harvest GEWURZ, PINOT GR VENDANGE TARDIVE GC RANGEN CLOS St-Théobald 10' 15' on volcanic soil.

Contrast with RIES GC Sonnenberg 13 15 16 17' on limestone. Delicious Harth CHASSELAS. No oak – super-elegance, esp drier styles.

Sec Literally means dry, though CHAMP so called is medium-sweet (and can be welcome at breakfast, teatime, weddings).

Séguret S Rh r p w ★★ 16' 17' 18 19' Hillside village nr GIGONDAS in Rh-Villages top three. V'yds on both warm plain and cool heights. Mainly GRENACHE, peppery, deep reds; bright whites. Esp CH la Courançonne (gd w), DOMS Amandine, Crève Coeur (bio), de Cabasse (elegant), *de l'Amauve* (fine, gd w), Fontaine des Fées (organic), Garancière, J David (organic), Maison Plantevin (organic), Malmont (stylish), *Mourchon* (robust), Pourra (intense, time), Soleil Romain.

Sélection des Grains Nobles Al Term coined by HUGEL for AL equivalent to German Beerenauslese, subject to ever-stricter rules. "Grains nobles" are grapes with "noble rot" for v. sweet wines.

Selosse, Anselme Champ ★★★★ Leading grower, an icon for many. Vinous, oxidative style, oak-fermented. Son Guillaume adding finesse: Version Originale still vibrant after 7 yrs on lees. Top probably MESNIL Les Carelles: saline, complex, akin to MEURSAULT Perrières with bubbles. 99 v. stylish; 02 still a baby, but worth waiting for.

Sérafin, Dom C d'O r ★★★ Now niece Frédérique is in charge, continuing Christian S recipe: deep colour, intense flavours and new wood: Look for back vintages as wines need to age. Try GEVREY-CHAMBERTIN VIEILLES VIGNES, Cazetiers and CHARMES-CHAMBERTIN.

Serres Mazard L'doc r p w ★★ CORBIÈRES estate with local varieties incl MACABEU, Terret and old CARIGNAN. Oak for long-ageing red (Joseph Mazard, Annie), also acacia (Jules w).

Seyssel Sav W sp ★★ Small regional AC. Feather-light wines, 72 ha: 55 ha white, 17 ha sparkling. Grapes: Altesse, CHASSELAS (sp only), Molette. Try: de la Brune, Lambert de Seyssel (organic), Mollex.

Sichel & Co B'x r w Respected B'x merchant est in 1883 (Sirius a top brand). Family-run: 6th generation at helm. Interests in CHX ANGLUDET, Argadens, PALMER and in CORBIÈRES (Ch Trillol).

Signargues S Rh ★→★★ 16' 17' 18 19' Modest CÔTES DU RH village, parched soils between Avignon and Nîmes (w bank). Spicy, sizeable reds to drink within 4–5 yrs. NB: Bellevue, CAVE Estézargues (punchy, gd range), CH Terre Forte (bio), CLOS d'Alzan, Haut-Musiel, La Font du Vent (fruit); DOMS des Romarins (deep), Valériane.

Simone, Ch Prov r p w ★★★ Historic estate outside Aix, where Churchill painted Mont STE-VICTOIRE. Rougier family for nearly two centuries. Virtually synonymous with AC PALETTE. N-facing slopes on limestone with clay and gravel contribute freshness. Many vines over 100 yrs old. Age-worthy whites well worth seeking out; characterful rosé, elegant reds from GRENACHE and MOURVÈDRE, with rare grape varieties Castet, Manosquin (r).

Sipp, Louis Al w sw ★★→★★★ Trades in big volumes of young wines, but also two GCS: fine RIES GC Kirchberg 13 16. luscious GEWURZ GC Osterberg VT 09 15 16 18 19. Gd classic dry wines too.

Sipp-Mack Al w sw ★★→★★★ Fine trad DOM in Hunawihr, great ALS village for peerless dry mineral wines (CLOS STE HUNE). Similar quality here but cheaper. Also RIES GC ROSACKER 13 and expansive PINOT GR. Charming holiday lets.

Sorg, Bruno Al w ★★★ Small grower at Eguisheim, GCS Florimont (RIES 13 14 16' great 17') and PFERSIGBERG (MUSCAT) 18'. Immaculate eco-friendly v'yds.

Sur lie "On the lees". Most MUSCADET is bottled straight from the vat, for max zest, body and character.

Tâche, La C d'O r ★★★★ 90' 93' 96' 99' 02' 03 05' 09' 10' 12' 14 15' 16' 17 18' 19'

GC of VOSNE-ROMANÉE, MONOPOLE of DRC. Firm in its youth, but how glorious with age. Headily perfumed, luxurious.

Tuille-aux-Loups, Dom de la Lo r w sw sp ★★→★★★★ 15' 16 (r) 17 18' (r) 19' Jacky Blot, top producer, with son Jean-Philippe. Barrel-fermented MONTLOUIS, VDF (aka VOUVRAY) mainly dry (*Remus*), esp single-v'yds: CLOS Mosny, Michet (Montlouis); Venise (Vouvray); *Triple Zéro* Montlouis *pétillant* (p w); v.gd BOURGUEIL DOM de la Butte (15ha). 19 v. enthusiastic for dry white.

Tain, Cave de N Rh ★★→★★★ Top N Rh co-op, many mature v'yds, incl 25% of HERMITAGE. Sound to v.gd red Hermitage, esp Epsilon (oak), Gambert de Loche (best), bountiful white Hermitage Au Coeur des Siècles, offer value. Gd ST-JOSEPH (r w), ST-PÉRAY (two wines), interesting Bio (organic) range (St-Joseph), others modern, mainstream. Gd recent CROZES reds, eg. Saviaux. Distinguished VIN DE PAILLE.

Taittinger Champ ★★★★ Family-run Reims house, exquisite airy wines. BRUT NV, Rosé NV, Brut 06 08 09, Collection Brut 89 90 95'. Epitome of apéritif style, exquisite weightlessness. Ace luxury *Comtes de Champagne* 95' 99, lovely 06 08'; Comtes Rosé also shines in 06 12. Excellent single-v'yd La Marquetterie. New cellarmaster fills Loec Dupont's big shoes. New English bubbly project in Kent, DOM Evremond. (*See also* Dom Carneros, California.)

Tavel S Rh p ★★ DYA. Historic GRENACHE rosé, aided by white grapes for texture, should be bright red, full, for vivid Med dishes. Now many lighter, Prov style, often for apéritif; a pity. Top: DOM de l'Anglore (no sulphur), *Dom de la Mordorée* (full), Corne-Loup, GUIGAL (gd), Lafond Roc-Epine, Maby, Moulin-la-Viguerie (organic, traditional), Prieuré de Montézargues (fine), Rocalière (v. fine), Tardieu-Laurent (deep), VIDAL-FLEURY; CHX Aquéria, de Manissy (organic), La Genestière, Ségriès, *Trinquevedel* (fine).

Tempier, Dom Prov r p w ★★★★ Probably the best, and best-known, BANDOL – where Lucien Peyraud revived AC in the 30s. Still tops for elegance, concentration and longevity.

Terrasses du Larzac L'doc r ★★→★★★ Rugged, high AC with cold nights makes fresh, stylish reds. Attracts innovative growers who assemble small plots of vines; over half organic/bio. Look for: CAL DEMOURA, CLOS des Serres, Jonquières, LA PEIRA, Mas Conscience, Mas de l'Ecriture, Mas Jullien, MONTCALMÈS, Pas de l'Escalette. White, rosé AC L'DOC or IGP.

Thénard, Dom Burg r w ★★→★★★★ Historic producer with large holding of MONTRACHET, mostly sold on to NÉGOCIANTS. Should be better known for v.gd reds from home base in GIVRY.

Thévenet, Jean Burg r w sw ★★★ Top MÂCONNAIS purveyor of rich, some semi-botrytized CHARD, eg. CUVÉE Levroutée at *Dom de la Bongran*. Also owns DOMS de Roally and Emilian Gillet.

Thézac-Perricard SW Fr r p w ★★ 18' (19) IGP. Lighter version of adjoining CAHORS (reds from MALBEC, MERLOT). Sandrine Annibal's ★★ DOM de Lancement the one independent. Lively co-op nearly as gd.

Thiénot, Alain Champ Young house, new generation in charge. Ever-improving quality; fairly priced ★★ BRUT NV; Rosé NV Brut; vintage Stanislas 02 04 06 08' 09 12' 13 15. Voluminous VIGNE aux Gamins (single-v'yd AVIZE 02 04 06). CUVÉE Garance CHARD 07 sings, classic 08 for long haul. Also owns CANARD-DUCHÊNE, JOSEPH PERRIER, CH Ricaud in LOUPIAC.

Thivin, Ch Beauj r (w) ★★→★★★ Eight generations of Geoffray family make great Côte de Brouilly. Single-v'yd bottlings cover soil types. Sept VIGNES also a winner.

Thomas, André & fils Al w ★★★ Rijou DOM, 6 ha in Ammerschwihr. PINOT BL from 50-yr-old vines. Excellent RIES Kaefferkopf 10 13 16 17'. Superb GEWURZ VIEILLES VIGNES 05 09 15 17' 18. Organic precepts.

Tissot Jura Dominant family around ARBOIS. ★★ Jacques T offers volume, value. ★★★ Stephane T (also as André & Mireille Tissot), cult pioneer of single-v'yd CHARD, VIN JAUNE using bio/natural methods; top CRÉMANT du Jura, Indigène.

Tollot-Beaut C d'O r (w) ★★★ Consistent CÔTE DE BEAUNE grower with 20 ha in BEAUNE (Grèves, CLOS du Roi), CORTON (Bressandes), SAVIGNY (esp MONOPOLE PC Champ Chevrey) and at CHOREY-LÈS-BEAUNE base (NB: Pièce du Chapitre). Easy-to-love fruit-and-oak combo. Gd CORTON-CHARLEMAGNE too.

Touraine Lo r p w dr sw sp ★→★★★ 17 18' 19' Big region, many AOPs (eg. BOURGUEIL, CHINON, VOUVRAY) plus umbrella AC of variable quality: fruity reds (CAB FR, CÔT, GAMAY, PINOT N), whites (SAUV BL), rosés, fizz. Touraine Village ACs: Azay-le-Rideau, Chenonceaux, Mesland, Noble-Joué, Oisly. Try: Biet, Bois-Vaudons, Cellier de Beaujardin, Corbillières, Echardières, Garrelière, Gosseaume, Jacky Marteau, Joël Delaunay, La Chapinière, Lacour, Mandard, *Marionnet*, Morantin, *Presle*, Prieuré, Puzelat, Ricard, Roussely, Tue-Boeuf, Villebois.

Touraine-Amboise Lo r p (w) ★→★★★ Village AC. François 1er pop blend (GAMAY/ CÔT/CAB FR); but Côt is top red; CHENIN BL top white. Best: Bessons, Closerie de Chanteloup, Dutertre, Frissant, Gabillière, Grange Tiphaine, Mesliard, Plou, Truet. Amboise cru for Chenin, Côt on blocks.

Touraine-Azay-le-Rideau Lo p w (sw) ★→★★ TOURAINE sub-AC. Mostly rosé (Grolleau 60% min); white, dry and off-dry CHENIN BL. Best: Aulée, Bourse, de la Roche, Grosbois, Nicolas Paget.

Touraine-Mesland Lo r p w ★→★★★ 16 17 18' 19' TOURAINE villages AC (nine producers), mainly red, rosé. Try: Grandes Espérances, Rabelais.

Touraine-Noble Joué Lo p ★→★★ DYA. V.gd rosé: three PINOTS (N, M, GR). AOP. Best: Astraly, Blondeau, Cosson, Dupuy, Rousseau. Also MALVOISIE, VDF. Frost-prone.

Trapet C d'O r ★★★ Long-est GEVREY-CHAMBERTIN DOM making sensual bio wines from eye-catching Passetoutgrains to GC CHAMBERTIN plus AL whites by marriage. *See also* cousins ROSSIGNOL-TRAPET.

Treloar, Dom Rouss r p w ★★★ Anglo-NZ owners. Emphasis on old vines, trad varieties and organic. Terre Promise IGP COTES CATALANES wild ferment in barrel; MACABEU/GRENACHE Gris/Carignan Blanc (w) and One Block Grenache of special interest. New: skin-contact MUSCAT d'Alexandrie.

Trévallon, Dom de Prov r w ★★★ Famous DOM at LES BAUX, created by Eloi Dürrbach; joined by daughter Ostiane. No GRENACHE, so must be IGP Alpilles. Huge reputation fully justified: meticulous viticulture. Intense, age-worthy CAB SAUV/ SYRAH. Barrique-aged white MARSANNE/ROUSSANNE, drop of CHARD, now GRENACHE BL.

Bio along the Rhône

Bio in the Rh is growing yr by yr, from single-handed small DOMS to much larger ones. Quality is high. Look for – **Cairanne**: Doms DE L'ORATOIRE ST MARTIN, des Amadieu; **Châteauneuf-du-Pape**: Doms de Beaurenard, de Cristia, de la Vieille Julienne, de Marcoux, Pierre André, Raymond Usseglio & Fils; **Côtes du Rh**: Dom GRAMENON, Mas de Libian; **Crozes-Hermitage**: David Reynaud; **Gigondas**: CH DE SAINT COSMÉ, CLOS du Joncuas; **Grignan-les-Adhémar**: Mas Théo; **Hermitage**: Ferraton Père & Fils, CHAPOUTIER; **Lirac**: Doms des Carabiniers, des Maravilhas, du Joncier; **Luberon**: Ch Fontvert; **Massif d'Uchaux**: Dom La Cabotte; **Plan de Dieu**: Dom LES APHILLANTHES; **Rasteau**: Doms des Coteaux des Travers, du Trapadis; **St-Joseph**: Doms des Amphores, des Miquettes, Monier-Perréol, La Ferme des Sept Lunes; **Séguret**: Dom de Crève Coeur; **Ventoux**: CHÊNE BLEU; **Vinsobres**: Doms Chaume-Arnaud, de la Péquélette, Vallot; **Visan**: Doms Dieulefit, Roche-Audran; **IGP**: Doms des Accoles, Gourt de Mautens.

Trimbach, FE Al w ★★★★ Matchless grower of AL RIES on limestone soils at Ribeauvillé, esp austere CLOS STE-HUNE: 71 89 still great. 10' 13 16 17' classically cool; almost-as-gd (much cheaper) *Frédéric Emile* 10 12 13 14 16 17. Dry and elegant wines for great cuisine. Look out for 1st GC label: from v'yds of Couvent de Ribeauville.

Tursan SW Fr r p w ★★ Mostly DYA. AOP in Landes. Super-chef Michel Guérard makes ★★★ lovely wines in chapel-like cellar at CH de Bachen, but not traditional Tursan. Real thing from ★★ DOM de Perchade. Lovely dry white from ★★ Dom de Cazalet (two MANSENGS plus rare local Baroque). Worthy co-op rather outclassed.

Vacqueyras S Rh r (p) w ★★ 07' 09' 10' 15' **16' 17'** 18 19' Hearty, spicy, GRENACHE-filled neighbour of GIGONDAS, hot v'yds; for game, big flavours. Lives 10 yrs+. NB: JABOULET; CHX de Montmirail, *des Tours* (v. fine); *Clos des Cazaux* (gd value); DOMS Amouriers, Archimbaud-Vache, Charbonnière, CLOS de Caveau (organic), Couroulu (v.gd, traditional), Famille Perrin, Font de Papier (organic), Fourmone, Garrigue (traditional), Grapillon d'Or, Monardière (v.gd), Montirius (bio), Montvac (elegant), Roucas Toumba (organic), SANG DES CAILLOUX (v.gd esp Lopy), Semelles de Vent, Verde. *Full whites* (Ch des Roques, *Clos des Cazaux*, SANG DES CAILLOUX).

Val de Loire Lo r p w mainly DYA. One of France's four regional IGPS, formerly Jardin de la France.

Valençay Lo r (p) w ★→★★ 18' 19' AOP, TOURAINE; SAUV BL, (CHARD); reds CÔT, GAMAY, PINOT N. Try: Delorme, Jourdain, Lafond, Preys, Roy, Sinson, Vaillant, VIGNERONS de Valençay.

Valréas S Rh r (p) w ★★ 16' 17' 18' **19'** CÔTES DU RH-VILLAGE In breezy n Vaucluse truffle area, quality rising; large co-op. Spicy, sometimes heady, red-fruited mostly GRENACHE red, improving white. Esp CH la Décelle, CLOS Bellane, Mas de Ste-Croix, DOMS GRAMENON (bio, stylish), des Grands Devers, du Séminaire (organic), du Val des Rois (best, organic), Prévosse (organic).

VdF (Vin de France) Replaces VDT, but with mention of grape variety, vintage. Often blends of regions with brand name. Can be source of unexpected delights if talented winemaker uses this category to avoid bureaucractic hassle. Eg. DIDIER DAGUENEAU, Yves Cuilleron VIOGNIER (N Rh). S ARDÈCHE, ANJOU hotbeds of VdF.

VDN (Vin Doux Naturel) Rouss Sweet wine fortified with wine alc so sweetness natural, not strength. Speciality of ROUSS based on GRENACHE N, Bl or Gris, or MUSCAT, Top, esp aged RANCIOS, can finish a meal on a sublime note.

VDP (Vin de Pays) *See* IGP.

VdT (Vin de Table) Category of standard everyday table wine now VDF.

Vendange Harvest. **Vendange Tardive:** late-harvest; AL equivalent to German Auslese but usually higher alc.

Venoge, de Champ ★★★ Venerable house, precise, more elegant under LANSON-BCC ownership. Gd niche blends: Cordon Bleu Extra-BRUT, Vintage BLANC DE BLANCS 00 04 06 08 12 13 14 16 17. Excellent Vintage Rosé 09 CUVÉE 20 Ans, Prestige Cuvée Louis XV 10-yr-old BLANC DE NOIRS.

Ventoux S Rh r p w ★★ 16' 17' 18' **19'** Widespread AC curves around Mont Ventoux between Rh and Prov. A few front-running DOMS v.gd value. Juicy, tangy red (GRENACHE/SYRAH, café-style to deeper, peppery, rising quality), rosé, gd white (more oak). Best: CH Unang (gd w), Ch Valcombe, CLOS des Patris, La Ferme St Pierre (p w), Gonnet, *La Vieille Ferme* (r, can be VDF), St-Marc, Terra Ventoux, VIGNERONS Mont Ventoux; Doms Allois (organic), Anges, Berane, Brusset, Cascavel, Champ-Long, Croix de Pins (gd w), du Tix, *Fondrèche*, Grand Jacquet, Martinelle, Murmurium, Olivier D, PAUL JABOULET, *Pesquié*, Pigeade, St-Jean du Barroux (organic), Terres de Solence, Verrière, VIDAL-FLEURY, Vieux Lazaret, Vignobles Brunier; co-op Bédoin.

Vernay, Dom Georges N Rh r w ★★★★ 15' 16' 17' 18' 19' Top CONDRIEU name; three wines, cool, stylish; Terrasses de l'Empire *apéritif de luxe*; Chaillées d'Enfer, richness; Coteau de Vernon, mysterious, intricate, supreme style, lives 20 yrs+. CÔTE-RÔTIE, ST-JOSEPH (r) clear fruit, restrained. V.gd IGP COLLINES RHODANIENNES (r w).

Veuve Clicquot Champ ★★★★ Historic house. Singing Yellow Label NV, a soupçon of 5% oak. Best DEMI-SEC NV, new CUVÉE Extra BRUT extra age based on Res wines, 2010–1990. Vintage Res 04 **06** 08. 12' is a magical PINOT N-led wonder of perfect maturity, acidity: best yr for PINOT N since 1952? Luxe La Grande Dame (GD) 12' is 92% Pinot N yet so elegant. Older vintages of GD stay course effortlessly in **04** (no 02, deemed too butch for GD) gloriously in 89 and firmly in **71**. GD Rosé 06 delicious, ready.

Veuve Devaux Champ ★★ Premium brand of powerful Union Auboise co-op. Excellent aged Grande Rés NV, and Œil de Perdrix Rosé, Prestige CUVÉE D 08, BRUT Vintage 09 **11 12** 15' 17.

Vézelay Burg r w ★→★★ Age 1–2 yrs. Lovely location (with abbey) in nw Burg. Promoted to full AC for tasty whites from CHARD. Also try revived MELON (COTEAUX BOURGUIGNON) and light PINOT (generic BOURGOGNE). Best: DOM de la Cadette, des Faverelles, Elise Villiers, La Croix Montjoie.

Vidal-Fleury N Rh r w sw ★★→★★★ GUIGAL-owned Rh merchant/grower of CÔTE-RÔTIE. Top-quality, tight, v. stylish *La Chatillonne* (12% VIOGNIER; oak, wait min 7 yrs). Broad range, on the up. Gd CAIRANNE, CHÂTEAUNEUF (r), CÔTES DU RH (r p), MUSCAT DE BEAUMES-DE-VENISE, ST-JOSEPH (r w), TAVEL, VENTOUX.

Vieille Ferme, La S Rh r p w ★→★★ Reliable v.gd-value brand from Famille Perrin of CH DE BEAUCASTEL; much now labelled VDF, with VENTOUX (r), LUBÉRON (p w) in some countries (France, Japan). Back on form recently, incl rosé.

Vieilles Vignes Old vines, which should make the best wine. Eg. DE VOGÜÉ, MUSIGNY, Vieilles Vignes. But no rules about age and can be a tourist trap.

Vieux Télégraphe, Dom du S Rh r w ★★★ 01' 05' 07' **09'** 10' **12' 14'** 15' 16' 17' 18' 19' High-quality estate; classic big-stone soils, tight, slow burn red CHÂTEAUNEUF; top two wines La Crau (crunchy, packed), since 2011 Pied Long et Pignan (v. pure, elegant, no 18). Also rich whites *La Crau* (v.gd 15 16 18), CLOS La Roquète (stylish, great with food, note 15 16 18). Owns fine, slow-to-evolve, complex *Gigondas Dom Les Pallières* with US importer Kermit Lynch.

Vigne or vignoble Vineyard (v'yd), vineyards (v'yds).

Vigneron Vine-grower.

Villeneuve, Ch de Lo r w ★★★ 10' 14' 15' 16' 17' 18' **19'** Excellent estate (25 ha) modern winery attached to renovated old cellars. Great SAUMUR Blanc (age-worthy Les Cormiers), SAUMUR-CHAMPIGNY (esp VIEILLES VIGNES, Grand CLOS). Organic. 19 another gd vintage.

Vin de paille Wine from grapes dried on straw mats, so v. sweet, like Italian passito. Esp in the Jura. *See also* CHAVE, VIN PAILLÉ DE CORRÈZE.

Vin gris "Grey" wine is v. pale pink, made of red grapes pressed before fermentation begins – unlike rosé, which ferments briefly before pressing. Or from eg. PINOT GR, not-quite-white grapes. "Œil de Perdrix" means much the same; so does "blush".

Vin jaune Jura w ★★★ Speciality of Jura; inimitable yellow wine. SAVAGNIN, 6 yrs+ in barrel without topping up, develops flor, like Sherry but no added alc. Expensive to make. Separate AC for top spot, CH-CHALON. Sold in unique 62cl Clavelin bottles. S TISSOT specializes in single-v'yd bottlings.

Vin paillé de Corrèze SW Fr r w 25 small growers and a tiny co-op once more making a wine once recommended to breast-feeding mothers. Not-too-ripe grapes laid on straw to make pungent wine for brave. Try ★ Christian Tronche.

Vinsobres S Rh r (p) (w) ★★ 15' 16' 17' **18'** 19' Low-profile AC notable for quality SYRAH, mix hillside, high-plateau v'yds. Best reds give punch, to drink with red meats, age 10 yrs. Leaders: CAVE la Vinsobraise; CH Rouanne; DOMS Chaume-Arnaud (bio), Constant-Duquesnoy, Famille Perrin (*Hauts de Julien* top class, Cornuds value), Jaume (modern, consistent), Moulin (traditional, gd r w), Péquélette (bio), Peysson (organic), Vallot (bio).

Viré-Clessé Burg w ★★ 14' 15 17' 18 19' AC based around two of best white villages of MÂCON. Known for exuberant rich style, sometimes late-harvest. Try Bonhomme, Bret Bros, Chaland, DOM de la Verpaille, Gandines, Gondard-Perrin, Guillemot, J-P Michel, LAFON, *Thévenet*.

Visan S Rh r (p) (w) ★★ 16' 17' 18' 19' Progressive RH VILLAGE: peppery mainly GRENACHE reds, sound filling, clear fruit; some softer, plenty organic. Whites okay. Best: DOMS Coste Chaude (organic, gd fruit), Dieulefit (bio, low sulphur), Florane, Fourmente (bio esp Nature), Guintrandy (organic), Montmartel (organic), Philippe Plantevin, Roche-Audran (organic), VIGNOBLE Art Mas (organic).

Vogüé, Comte Georges de C d'O r w ★★★★ Aristo CHAMBOLLE estate with lion's share of LE MUSIGNY. Great from barrel, but takes many yrs in bottle to reveal glories. Unique white Musigny.

Volnay C d'O r ★★★→★★★★ 90' 99' 02' 05' 09' 10' 15' 16 17' 18 19 Top CÔTE DE BEAUNE reds, except when it hails or gets too hot. Can be structured, should be silky and astonishing with age. Best v'yds: Caillerets, Champans, CLOS des Chênes, Santenots (more clay), Taillepieds and MONOPOLES Clos des Ducs, Clos du Ch des Ducs, Clos de la Bousse-d'Or, Clos de la Chapelle. Best growers: Bitouzet-Prieur, Buffet, *Clerget*, D'ANGERVILLE, *de Montille*, H BOILLOT, HOSPICES DE BEAUNE, JM&T Bouley, *Lafarge*, Lafon, N Rossignol, P Bouley, Pousse d'Or.

Vosne-Romanée C d'O r ★★★→★★★★ 90' 93' 96' 99' 02' 05' 09' 10' 12 15' 16' 17 18' 19' Village with Burg's grandest crus (eg. ROMANÉE-CONTI, LA TÂCHE) and outstanding PCS Malconsorts, Beaumonts, Brûlées, etc. There are (or should be) no common wines in Vosne. Just question of price... Top names: ARNOUX-LACHAUX, Bizot, CATHIARD, Coquard-Loison-Fleurot, DRC, EUGÉNIE, GRIVOT, GROS, Lamarche, LEROY, LIGER-BELAIR, MÉO-CAMUZET, MUGNERET, NOËLLAT, ROUGET. I really like Clavelier, Forey, Guyon, Tardy too.

Vougeot C d'O r w ★★★ 99' 02' 05' 09' 10' 12 15 16 17 18 19 Mostly GC as CLOS DE VOUGEOT but also village and PC, Cras, Petits Vougeots, and outstanding white MONOPOLE *Clos Blanc de V.* Clerget, Fourrier, HUDELOT-NOËLLAT, LEROUX, *Vougeraie* best

Vougeraie, Dom de la C d'O r w ★★★→★★★★ Bio DOM uniting all BOISSET's v'yd holdings. Fine-boned, perfumed wines, most noted for sensual GCS, esp BONNES-MARES, CHARMES-CHAMBERTIN, MUSIGNY. Fine whites too, with unique *Clos Blanc de Vougeot* and four GCS incl unique CHARLEMAGNE.

Vouvray Lo w dr sw sp ★★★→★★★★ (dr) 15' 16 17 18' 19' (sw) 08 09' 10 11 15' 16 18' AC. Limestone bluffs on n bank of Loire produce top (though variable dr/sw) wines. DEMI-SEC is classic style, but in best yrs *moelleux*, richly sweet balanced by acidity, is almost immortal. Fizz variable (60% production): *pétillant*, local speciality. Best: Aubuisières, Autran, Bonneau, Brunet, Carême, *Champalou*, CLOS Baudoin (VDF), Florent Cosme, Fontainerie, Foreau, F PINON, Gaudrelle, *Huet*, Mathieu Cosme, Meslerie (Hahn), Perrault-Jadaud, Rouvre, *Taille-aux-Loups* (VdF), Vigneau-Chevreau. Old vintages a must-try, esp sweet. 19 promising, little sweet.

Zind Humbrecht, Dom Al w sw ★★★★ One of greats. V'yds incl GC Brand, HENGST and volcanic RANGEN at Thann. All expensive. For easier prices, MUSCAT GC Goldert from ancient v'yd, dry, floral, structured and great with asparagus.

Châteaux of Bordeaux

Abbreviations used in the text:

B'x	Bordeaux
Bar	Barsac
Cas	Castillon-Côtes de Bordeaux
E-2-M	Entre-Deux-Mers
Fron	Fronsac
Grav	Graves
H-Méd	Haut-Médoc
L de P	Lalande de Pomerol
List	Listrac
Mar	Margaux
Méd	Médoc
Mou	Moulis
Pau	Pauillac
Pe-Lé	Pessac-Léognan
Pom	Pomerol
Saut	Sauternes
St-Ém	St-Émilion
St-Est	St-Estèphe
St-Jul	St-Julien

AC	appellation contrôlée
ch(x)	château(x)
dom(s)	domaine(s)

How does one buy Bordeaux in an era of luxury goods, when château owners are raking in so much that commissioning world-famous architects to design swanky new cellars is commonplace? And why should we care about their new cellars, anyway? Two questions, two answers. Yes, Bordeaux at the top end is about perfection and high prices. Anyone looking for wines to drink, rather than for investment, can take advantage of the trickle-down of technical perfection by picking the second wines of top châteaux, the odd well-priced vintage, or properties that regularly make substantial quantities of delicious wines. Prices of Bordeaux, because of the way the market works, are more variable than

Châteaux of Bordeaux entries also cross-reference to France.

for any other wine: www.winesearcher.com is a great way of comparing prices. As for restaurants, they seem to see Bordeaux as a licence to pile on juicy margins. And the cellars? They matter, believe it or not. When a property installs a new cellar, the tanks will usually be smaller and more numerous, allowing smaller parcels to be fermented separately; this makes for greater precision in the wine. The most gentle treatment also gives more delicacy; so when we mention a new cellar, we're flagging a probable rise in quality (and often price, of course).

2019 has no early recommendations here because Covid-19 stopped all tastings. The growing season was tricky, with heat spikes and drought. Yields were down due to poor fruit set and tiny berries, but the harvest was generally healthy and devoid of rot. Alcohols were again high and there's some variation in quality but as a whole the wines have depth of fruit, colour, structure and the potential to age. The luscious 09s are tempting at whatever level (but don't open them too soon). The 08s have come into their own. The 10s are just opening (as are the "classic" 14s), although the Grands Crus need longer. For early-drinking try the often-charming 12s or often-underrated 11s; austere at first but now improving with bottle-age. Mature vintages to look for are 96 (best Médoc), 98 (particularly Right Bank), 00, 01 (Right Bank again), 02 (top-end Médoc) and 04. The 06s are beginning to open, as are some of the splendid 05s, although patience is still a virtue here. Dry white Bordeaux remains consistent in quality and value with another good vintage in 19 (fresher than 18). Remember that fine white Graves ages as well as white burgundy – sometimes better. And Sauternes continues to offer an array of remarkable years, 19 small in quantity and more variable in quality. The problem here is being spoilt for choice. Even moderate years like 02 and 06 offer approachability and a fresher touch while the great years (01 09 11 15) have the concentration and hedonistic charm that makes them indestructible.

A, Dom de L' Cas r ★★ 09' 10' 11 12 14 15 16 18 MERLOT-led (20% CAB FR) CAS property owned by STÉPHANE DERENONCOURT and wife. Consistent quality.

Agassac, D' H-Méd r ★★ 09' 10' 11 12 14 15 16' 18 Consistent CH in s H-MÉD. Modern, accessible. "Precision" is old-vine (60 yrs) CUVÉE.

Aiguilhe, D' Cas r ★★ 10' 11 12 14 15 16 18 Vast estate on high plateau. Sisters CANON LA GAFFELIÈRE, LA MONDOTTE. MERLOT-led, *power and finesse*. Also Le Blanc d'Aiguilhe.

Andron-Blanquet St-Est r ★★ 09' 10' 11 14 15 16' 18 Sister to COS-LABORY. Often value.

Angélus St-Ém r ★★★★ 00' 01 02 03 04 05 06 07 09 10' 15' 16' 18' PREMIER GRAND CRU CLASSÉ (A) since 2012 so prices high. Pioneer of modern ST-ÉM; dark, rich, sumptuous. Lots of CAB FR (min 40%). Organic conversion. Also owns hotel, restaurants. Second label: Carillon d'Angélus (new, high-tech cellar from 2019).

Angludet Marg r (w) ★★ 09' 10' 11 12 14 15' 16' 18' Owned/run by NÉGOCIANT SICHEL family since 1961. CAB SAUV, MERLOT, 13% PETIT VERDOT. Fragrant, stylish. Gd value.

Archambeau Grav r w dr (sw) ★★ (r) 10 11 15 16 18 (w) 16 17 18 Owned by Dubourdieu family; v'yd in single block on hill at Illats. Gd *fruity dry white*; fragrant MERLOT/CAB (50/50) reds. Also rosé.

Arche, D' Saut w sw ★★ 05' 07 09' 10' 11 14 15 16 17' 18 Gd-value Second Growth on edge of SAUT. Environmental certification.

Armailhac, D' Pau r ★★★ 05' 06 08 09' 10' 11 12 14 15' 16' 18 Substantial Fifth Growth. (MOUTON) ROTHSCHILD-owned (1934). On top form, fair value. CAB SAUV-led.

Aurelius St-Ém r ★★ 10 12 14 15 16 18 Top CUVÉE from the go-ahead ST-ÉM co-op. Grapes from different terroirs in AC. Modern, MERLOT-led, new oak, concentrated.

Ausone St-Ém r ★★★★ 00′ 01′ 02 03′ 04 05′ 06′ 07 08 09′ 10′ 11 12 13 14 15′ 16′ 17 18′ Tiny, illustrious PREMIER GRAND CRU CLASSÉ (A) named after Roman poet. Only c.1500 cases. V'yd s and se facing, sheltered from winds. Lots of CAB FR (55%). Long-lived wines with volume, texture, finesse. At a price. Second label: Chapelle d'Ausone (500 cases). *La Clotte*, FONBEL, MOULIN-ST-GEORGES, Simard sister estates.

Balestard la Tonnelle St-Ém r ★★ 10 11 12 14 15 16 17 18 Historic DOM on limestone plateau, owned by Capdemourlin family. Modern, MERLOT-led.

Barde-Haut St-Ém r ★★→★★★ 01 02 03 05′ 06 07 08 09 10 11 13 14 15′ 16 18 GRAND CRU CLASSÉ. Sister property of CLOS L'ÉGLISE, HAUT-BERGEY, Branon in Léognan.

Bulk wines represent 40% of wines purchased on B'x market.

Bastor-Lamontagne Saut w sw ★★ 07 09′ 10 11 13 14 15 16 17 18 Large SÉM-led Preignac estate. Owned by Grands Chais de France since 2018. Earlier-drinking style. Organic certification from 2016. Second label: Les Remparts de Bastor.

Batailley Pau r ★★★ 03 04 05′ 06 08 09′ 10′ 11 12 13 14 15 16 17 18 Consistent, gd-value Fifth Growth. Second label: Lions de Batailley.

Beaumont H-Méd r ★★ 01 02 04 05′ 08 09′ 10′ 12 14 15 16 18 Large n H-MÉD estate (c.42,000 cases). Sister to BEYCHEVELLE; early-maturing, *easily enjoyable wines*.

Beauregard Pom r ★★★ 01 02 04 05′ 08 09′ 10′ 12 14 15 16′ Organically cultivated, mid-weight POM. More depth since 2015. Owned by SMITH-HAUT-LAFITTE and Galeries Lafayette families. Second label: Benjamin de Beauregard. Also Pavillon Beauregard in L DE P.

Beau-Séjour-Bécot St-Ém r ★★★ 00′ 01 02 03 04 05 06 08 09′ 10′ 11 12 14 15 16 18 Distinguished, family-owned PREMIER GRAND CRU CLASSÉ (B) on limestone plateau. Old quarried cellars for bottle storage. Lighter touch these days but still gd ageing potential.

Beauséjour-Duffau St-Ém r ★★★ 01 02 03 04 05′ 06 08 09′ 10′ 11 12 14 15 16 17 18′ Tiny PREMIER GRAND CRU CLASSÉ (B) estate on côtes. Owned by Duffau-Lagarrosse family; managed by Nicolas Thienpont (PAVIE-MACQUIN). Rich, cellar-worthy.

Beau-Site St-Est r ★★ 03 04 05 06 08 09 10 11 12 15 16 18 Gd-value CRU BOURGEOIS property owned by BORIE-MANOUX. C18 cellar. Supple, fresh, accessible.

Bélair-Monange St-Ém r ★★★ 00′ 01′ 02 03 04 05′ 06 08 09′ 10′ 11 12 13 14 15 16′ 17 18 PREMIER GRAND CRU CLASSÉ (B) on limestone plateau and côtes. Owned by J-P MOUEIX since 2008. Investment in v'yd. Environmental certification. Fine, riper and more intense these days. Second label: Annonce de Bélair-Monange.

Belgrave H-Méd r ★★ 02′ 03 04′ 05′ 06 08 09′ 10′ 11 12 13 14 15 16 17 18 Consistent n H-MÉD Fifth Growth. Environmental certification. CAB SAUV dominant. Modern-classic in style. Second label: Diane de Belgrave.

Bellefont-Belcier St-Ém r ★★ 04 05′ 06 07 08′ 09′ 10′ 12 14 15′ 16′ 17 18 ST-ÉM GRAND CRU CLASSÉ owned by Hong Kong businessman Peter Kwok (VIGNOBLES K). Suave.

Belle-Vue H-Méd r ★★ 05 06 08 09 10 11 12 14 15′ 16 17 18 Consistent, gd-value CRU BOURGEOIS. Dark, dense but firm, fresh, aromatic. 15–25% PETIT VERDOT in blend.

Berliquet St-Ém r ★★ 01 02 04 05′ 06 08 09 10 11 14 15′ 16′ 17 18 Tiny GRAND CRU CLASSÉ on plateau and côtes. Same ownership as CANON (added investment). Fresh, elegant, ages well. Second label: Les Ailes de Berliquet.

Bernadotte H-Méd r ★★→★★★ 01 02 03 04 05′ 06 08 09′ 10′ 11 14 15′ 16′ 17 Environmentally friendly CRU BOURGEOIS. Owned by a Hong Kong-based group. Hubert de Boüard (ANGÉLUS) consults. Savoury style, lately with more finesse.

Beychevelle St-Jul r ★★★ 00′ 01 02 03 04 05′ 06 07 08 09′ 10′ 11 12 13 14 15′ 16′ 17 18 Fourth Growth owned by Castel and Suntory. Wines of consistent *elegance* rather than power. New glass-walled winery. Second label: Amiral de Beychevelle.

Biston-Brillette Mou r ★★ 05′ 06 08 09 10′ 11 12 14 15 16′ 18 Family-owned CRU BOURGEOIS. 50/50 MERLOT, CAB SAUV. Gd-value, attractive, early-drinking wines.

Bonalgue Pom r ★★ 05 06 08 08 09 10 11 12 14 15 16 18 Dark, meaty. 90% MERLOT, 10% CAB FR from sand, gravel, clay soils. Gd value for POM. Owned by Libourne NÉGOCIANT JB Audy. Sisters CLOS du Clocher; CH du Courlat in LUSSAC-ST-ÉM.

Bonnet B'x r w ★★ (r) 10 12 14 15 16 18 (w) DYA. RIP owner André Lurton; son Jacques now at helm. Big producer of some of best E-2-M and red (Rés) B'X. LA LOUVIÈRE, **Couhins-Lurton**, ROCHEMORIN and Cruzeau in PE-LÉ same stable.

Bon Pasteur, Le Pom r ★★★ 02 03 04 05′ 06 08 09′ 10′ 11 12 13 14 15′ 16 17 18 Tiny property on ST-ÉM border. Former owner, MICHEL ROLLAND, makes the wine. Vinification intégrale in barrel. Ripe, opulent, seductive wines guaranteed.

Boscq, Le St-Est r ★★ 04 05′ 06 08 09′ 10 11 12 14 15′ 16′ 17 18 Quality-driven CRU BOURGEOIS managed by DOURTHE since 1995. Consistently great value.

Bourgneuf Pom r ★★ 03 04 05′ 06 08 09 10 11 12 14 15′ 16′ 17 18 MERLOT-led (80%) DOM. V'yd to w of POM plateau. Subtle, savoury wine. As gd value as it gets.

Bouscaut Pe-Lé r w ★★★ (r) 01 04 05 06 08 09 10′ 11 12 14 15 16′ 18 (w) 06 07 08 09 10′ 11 12 13 14 15 16 17 GRAV Classed Growth. MERLOT-based reds with 10% MALBEC. Sappy, age-worthy SAUV BL/SÉM **whites**.

Boyd-Cantenac Marg r ★★★ 02 03 04 05′ 06 08 09′ 10′ 11 12 14 15 16 18 Tiny Cantenac-based Third Growth. Owned by the Guillemet family since 1932. CAB SAUV-dominated. Gd value. POUGET same stable. Second label: Jacques Boyd.

Branaire-Ducru St-Jul r ★★★ 01 02 03 04 05′ 06 08 09′ 10′ 11 12 13 14 15 16′ 17 18′ Consistent Fourth Growth; regularly gd value; ageing potential. Owned by the Maroteaux family since 1988. Second label: *Duluc*.

Branas Grand Poujeaux Mou r ★★ 05′ 06 08 09 10 11 12 14 15′ 16′ 18 Tiny neighbour of CHASSE-SPLEEN, POUJEAUX. Owned by Justin Onclin since 2002. Investment. Rich, modern style. Sister to Villemaurine in ST-ÉM. Second label: Les Eclats de Branas.

Brane-Cantenac Marg r ★★★→★★★★ 00′ 01 02 04 05′ 06 08 09′ 10′ 11 12 13 14 15′ 16′ 18 CAB SAUV-led Second Growth owned by Henri Lurton. Classic, fragrant MARG with structure to age. Second label: *Baron de Brane*, value and consistency.

Brillette Mou r ★★ 05 06 08 09 10 11 12 14 15 16′ 18 Reputable, family-owned CRU BOURGEOIS with v'yd on gravelly soils. Gd depth. Second label: Haut Brillette.

Cabanne, La Pom r ★★ 05 06′ 09 10 11 14 15′ 16 18 MERLOT-dominant (94%) POM w of plateau. Firm when young; needs bottle-age. Second label: DOM de Compostelle.

Caillou Saut w sw ★★ 01′ 02 03′ 05 06 07 09′ 10′ 11′ 13 15′ 16 (18) Family owned/-run Second Growth BAR for pure liquoreux. 90% SÉM. Second label: Les Erables.

Calon-Ségur St-Est r ★★★★ 00′ 01 02 03′ 04 05′ 06 07 08 09′ 10′ 11 12 13 14 15′ 16′ 17 18′ Third Growth with historic (1855) v'yd intact. 5 yrs renovation, now on flying form, more CAB SAUV. Firm but fine, complex. Second label: Le Marquis de Calon.

Cambon la Pelouse H-Méd r ★★ 05 08 09 10′ 11 12 14 15 16′ 18 Big, reliable and fruit-forward s H-MÉD CRU BOURGEOIS. Acquired by Aussie group TWE in 2019.

Camensac, De H-Méd r ★★ 05 06 08 09 10′ 11 12 14 15 16 18′ Fifth Growth in n H-MÉD. Owned by Merlaut family (GRUAUD-LAROSE, CHASSE-SPLEEN) since 2005. Steady improvement. Eric Boissenot consults. Second label: Second de Camensac.

Cab Sauv accounts for 22% plantings in B'x.

Canon St-Ém r ★★★→★★★★ 01 02 03 04 05′ 06 07 08′ 09′ 10′ 11 12 13 14 15′ 16′ 18′ Esteemed PREMIER GRAND CRU CLASSÉ (B) on limestone plateau. Wertheimer-owned, like BERLIQUET, RAUZAN-SÉGLA, ST-SUPÉRY and DOM de l'Ile in Provence (2019). Now flying; elegant, long-lived. Second label: Croix Canon (was CLOS Canon to 2011).

Canon-la-Gaffelière St-Ém r ★★★ 00′ 01 02 03 04 05′ 06 08 09′ 10′ 11 12 13 14 15′ 16 18 PREMIER GRAND CRU CLASSÉ (B) on s foot slope. Von Neipperg-owned; also owns CLOS DE L'ORATOIRE, D'AIGUILHE, LA MONDOTTE. Lots of CABS FR (40%) and SAUV (10%). Stylish, impressive. Organic certification.

Cantemerle H-Méd r ★★★ 04 05′ 06 08′ 09′ 10′ 11 12 13 14 15 16 18 Large Fifth

Growth in s H-MÉD with beautiful wooded park. Renovated and replanted over last 40 yrs. Environmental certification. On gd form and gd value too.

Cantenac-Brown Marg r ★★→★★★ 01 02 03 04 05' 06 08 09' 10' 11 12 14 15 16 18 Third Growth. Sustainable approach. 65% CAB SAUV. More voluptous, refined these days. Dry white AltO (90% SAUV BL). Second label: BriO de Cantenac-Brown. Sold 2019: new investment, new winery planned.

Capbern St-Est r ★★ 04 05 06 08' 09' 10' 11 12 13 14' 15 16' 18 Capbern-Gasqueton until 2013. Same ownership, management as CALON-SÉGUR. Great form. Gd value.

Cap de Mourlin St-Ém r ★★→★★★ 01 03 04 05 06 08 09 10 11 12 14 15 16 18 GRAND CRU CLASSÉ on n slopes. Named after owning Capdemourlin family. MERLOT-led (65%). MICHEL ROLLAND consults. More power, concentration than in past.

Carbonnieux Pe-Lé r w ★★★ 02 04 05' 06 08 09' 10 11 12 15' 16' 18 GRAV Classed Growth making sterling red and white; large volumes of both. *Whites*, 65% SAUV BL, eg. 16 17 18, have ageing potential. Red can age as well. Second label: La Croix de Carbonnieux. Also CH Tour Léognan.

Carles, De B'x r ★★★ 05' 06 07 08 09 10 11 12 14 15 16 18 FRON property. Haut-Carles is prestige CUVÉE partly vinified in 500l oak barrels. Opulent, modern style.

Carmes Haut-Brion, Les Pe-Lé r ★★★ 04 05' 06 07 08 09' 10' 11 12' 14 15 16' 18' Tiny walled-in v'yd in heart of B'x city. CAB FR (40%+) and SAUV-led wines: structured but suave. Philippe Starck-designed winery. DERENONCOURT consults. Second label: Le C des Carmes Haut-Brion from grapes grown in Martillac.

Caronne-Ste-Gemme H-Méd r ★★ 05 06 08 09' 10' 11 12 14 15 16 18 Sizeable n H-MÉD estate. CAB SAUV-led (60%) wines; fresh, structured; more depth recently.

Carruades de Lafite Pau ★★★ Second label of CH LAFITE. 20,000 cases/yr. Second Growth prices. Refined, smooth, savoury; more MERLOT. Accessible earlier. try 09.

Carteau Côtes-Daugay St-Ém r ★★ 05 08 09 10' 11 14 15 16' 18 Gd-value ST-ÉM GRAND CRU; full-flavoured, supple wines. Same ownership for five generations.

Certan-de-May Pom r ★★★ 00' 01' 04 05' 06 08 09' 10' 11 12' 14 15' 16' 17 18 Tiny v'yd on the POM plateau. New *cuvier* in 2018. Long-ageing.

Chantegrive, De Grav r w ★★ →★★★ 05' 08 09' 10' 11 12 14 15 16 18 Leading GRAV estate. Sizeable so volume. V.gd quality, value. Rich, finely oaked reds. CUVÉE Caroline is top, *fragrant white* 16 17 18.

Chasse-Spleen Mou r (w) ★★★ 04 05' 06 08 09' 10' 12 14 15 16' 17 18 Big (100 ha), well-known MOU estate. Often outstanding, long-maturing wine; classical structure, fragrance. Environmental certification. A little Blanc de C-S. Second label: L'Oratoire de Chasse-Spleen.

Chauvin St-Ém r ★★ 05 06 08 09 10' 11 12 14 15 16' 18' GRAND CRU CLASSÉ owned by Sylvie Cazes since 2014. MERLOT-led (85%); constant progression. New Cupid label.

Cheval Blanc St-Ém r ★★★★ 01' 02 03 04 05' 06 07' 08 09' 10' 11 12 13 14 15' 16' 17 18' PREMIER GRAND CRU CLASSÉ (A) superstar of ST-ÉM, easier to love than buy. High percentage of CAB FR (60%). Firm, fragrant wines verging on POM. Press wine never used. Delicious young; lasts a generation, or two. Second label: Le Petit Cheval (small production). New 100% SAUV BL Le Petit Cheval Blanc from 2014.

Chevalier, Dom de Pe-Lé r w ★★★★ 00 01' 02 03 04 05' 06 07 08 09' 10' 11 12 13 14 15' 16' 17 18' Tripled in size from when Bernard family bought in 1983. Pure, dense, finely textured red. Impressive, complex, long-ageing white has remarkable consistency; wait for rich flavours 08 14 15' 16' 17' 18. CLOS des Lunes, DOM de la Solitude, Lespault-Martillac same stable. Shareholder in GUIRAUD.

Cissac H-Méd r ★★ 00 05 08 09 10 11 12 14 15 16' 17 18 CRU BOURGEOIS. Classic CAB SAUV-led; used to be austere, now purer fruit. Second label: Reflets du CH Cissac.

Citran H-Méd r ★★ 00 04 05' 06 08 09 10' 14 15 16 18 Sizeable s H-MÉD estate. Medium-weight, ageing up to 10 yrs. Second label: Moulins de Citran.

Clarence de Haut-Brion, Le Pe-Lé r ★★★ 00 01 02 03 04 05 06 07 08 09' 10' 11 12

14 15 16' 17 18 Second label of CH HAUT-BRION, previously known as Bahans Haut-Brion. Blend changes considerably with each vintage (usually MERLOT-led), same suave texture, elegance as *grand vin*. Le Clarence restaurant, Paris same stable.

Clarke List r (p) (w) ★★→★★★ 05' 06 08 09' 10' 11 12 14 15 16 17 18 Leading LIST acquired by Edmund de Rothschild in 1973; now owned by son Benjamin. V.gd MERLOT-based (70%) red. Dark fruit, fine tannins. Also dry white: Le Merle Blanc du CH Clarke. Ch Malmaison in MOU same stable.

Clauzet St-Est r ★★ 06 08 09 10 11 12 14 15 16' 17 Gd-value CRU BOURGEOIS. CAB SAUV-led, consistent quality. Property sold in 2018; vines acquired by CH LILIAN LADOUYS, brand and buildings by CH La Haye.

Clerc Milon Pau r ★★★ 00 01 02 04 05' 06 07 08 09 10' 11 12 13 14 15 16' 17 18' V'yd tripled in size since (MOUTON) Rothschilds purchased in 1970. 1% CARMENÈRE in blend. Consistent quality but prices up. Second label: Pastourelle de Clerc Milon.

Climens Saut w sw ★★★★ 00 01' 02 03' 04 05 06 07 08 09' 10' 11' 12' 13' 14 15 16' BAR Classed Growth managed with aplomb by Bérénice Lurton and technical director Frédéric Nivelle (since 1988). Concentrated wines with vibrant acidity; ageing potential guaranteed. Certified bio. Second label: Les Cyprès (gd value).

Clinet Pom r ★★★ 01 02 03 05' 06 07 08 09' 10 11 12 14 15' 16 17 18 Family-owned/-run estate on POM plateau. 90% MERLOT, 10% CABS SAUV/FR. Sumptuous, modern style. Owner now president of Union des Grands Crus de Bordeaux (UGCB).

Clos de l'Oratoire St-Ém r ★★ 04 05' 06 08 09 10' 11 12 14 15 16 18 Supple GRAND CRU CLASSÉ, ne slopes of ST-ÉM. Part of von Neipperg stable (CANON-LA-GAFFELIÈRE, etc.).

Clos des Jacobins St-Ém r ★★★★ 01 02 03 04 05' 06 07 08 09 10' 11 12 14 15 16 18 Côtes GRAND CRU CLASSÉ at top of game. Renovated, modernized; great consistency; powerful, modern style. CHX de Candale and la Commanderie same stable.

Clos du Marquis St-Jul r ★★ 01 03 04 05' 06 07 08 09' 10' 11 12 13 14 15' 16' 17 18' More typically ST-JUL than lofty stablemate LÉOVILLE-LAS-CASES; CAB SAUV-led. Second label: La Petite Marquise.

Clos Floridène Grav r w ★★ (r) 09' 10' 11 12 14 15' 16 18 (w) 09 10 11' 12 13 14 15 16 17' 18 Creation of late Denis Dubourdieu, now run by his sons, Fabrice and Jean Jacques. SAUV BL/SÉM from limestone provides *fine medium white* GRAV; much-improved, CAB SAUV-led red. CHX DOISY-DAËNE, Haura, REYNON in same stable.

Clos Fourtet St-Ém r ★★★ 00 01 02 03 04 05' 06 07 08 09' 10' 11 12 14 15' 16 17 18 PREMIER GRAND CRU CLASSÉ (☒) on limestone plateau owned by Cuvelier family. 12 ha of quarried cellars below. Classic, stylish ST-ÉM. Consistently gd form. POUJEAUX in MOULIS same stable. Second label: La Closerie de Fourtet.

Clos Haut-Peyraguey Saut w sw ★★★ 00 01' 02 03' 04 05' 06 07 08 09' 10' 11' 12 13 14 15 16 17 18 First Growth SAUTERNES in the hands of magnate Bernard Magrez (FOMBRAUGE, PAPE-CLÉMENT same ownership) since 2012. Elegant, SÉM-led (95%), harmonious wines. Second label: Symphonie.

Clos l'Église Pom r ★★★ 00 01 02 03 04 05' 06 07 08 09' 10 11 12 13 14 15' 16 18 Top-flight, consistent POM on edge of plateau. Elegant wine that will age. Owned by the Garcin-Cathiard family since 1997. Second label: Esprit de l'Église.

Clos Puy Arnaud Cas r ★★ 04 05' 06 08 09' 10 11 12 13 14 15 16' 17 18 Leading CAS estate run with passion by Thierry Valette. Certified bio. Wines of depth, bright acidity. Earlier-drinking CUVÉE Bistrot.

Clos René Pom r ★★ 01 04 05' 06 08 09 10 11 12 14 15' 16 18 Family-owned for

---

Architectural wonders

A number of extravagant new cellar projects under construction these last few yrs and are now (ta-dah!) complete. Make sure you book to view latest in C21 winery architecture and technology at BÉLAIR-MONANGE, FIGEAC, HAUT-BAILLY, LE DÔME, LYNCH-BAGES, TROPLONG MONDOT. Few discounts.

generations. MERLOT-led with a little spicy MALBEC. Less sensuous, celebrated than top POM, but great value for AC.

Clotte, La St-Ém r ★★ 05 06 08 09' 10' 11 12 15' 16' 17 18 AUSONE ownership since 2014. Steady improvement. Second label: L de La Clotte.

Conseillante, La Pom r ★★★★ 01 02 03 04 05' 06' 07 08 09' 10' 11 12 13 14 15' 16' 17 18 Owned by Nicolas family since 1871; v'yd surface area same since this time. Some of noblest, most fragrant POM with structure to age. Organic leaning. Second label: Duo de Conseillante.

Corbin St-Ém r ★★ 04 05 07 08 09 10' 11 12 14 15' 16 18 Consistent, gd-value GRAND CRU CLASSÉ. Power, finesse. Second label: Divin de Corbin.

Cos d'Estournel St-Est r ★★★★ 00 01 02 03 04 05' 06 07 08 09' 10' 11 12 13 14 15' 16' 17 18 Big Second Growth. Refined, suave, high-scoring. Cutting-edge cellars. Too-pricey SAUV BL-dominated white; now more refined. Former CH Pomys now La Maison d'Estournel boutique hotel (2019). Second label: Les Pagodes de Cos.

Cos-Labory St-Est r ★★ 00 02 03 04 05 06 07 08 09' 10' 11 12 14 15 16' 17 18 Small Fifth-Growth neighbour of COS D'ESTOURNEL owned by Audoy family; gd value and consistent. Second label: Charme de Cos Labory.

Coufran H-Méd r ★★ 05 06 08 09' 10 11 12 14 15 16 17 18 Atypical n HAUT-MÉD estate with 85% MERLOT. Owned by Miailhe family since 1924. Supple wine, but can age. Second label: N°2 de Coufran.

Couhins-Lurton Pe-Lé r w ★★ ·★★★ (r) 05 06 08 09 10' 11 12 14 15 16 17 (w) 05 06 07 08' 09 10 11 12 13 14 15' 16' 17 18 *Fine*, tense, long-lived Classed Growth *white* from SAUV BL (100%). Polished, MERLOT-led (up to 85%) red.

Couspaude, La St-Ém r ★★★ 04 05 06 07 08 09' 10' 11 12 14 15 16 17 18 GRAND CRU CLASSÉ on limestone plateau. Rich, creamy, MERLOT-led (75%), lashings of spicy oak.

Coutet Saut w sw ★★★ 01' 02 03' 04 05 07 09' 10' 11' 12 13 14' 16 17 18 Large property owned and run by Baly family since 1977. Consistently v. fine. CUVÉE Madame: v. rich, old-vine selection 01 03 09. Second label: La Chartreuse de Coutet. V.gd dry white, Opalie.

Couvent des Jacobins St-Ém r ★★ 00' 01 03 04 05 06 08 09 10 11 12 14 15 16 18 GRAND CRU CLASSÉ vinified within walls of town. Family-owned; Xavier Jean now at helm. MERLOT-led with CAB FR and PETIT VERDOT. Lighter style but can age.

Crabitey Grav r w ★★ (r) 08 09 10 11 12 14 15 16 17 18 (w) 11 12 13 14 15 16 17 18 Portets estate with v'yd restructured, replanted in 80s. Owner Arnaud de Butler now making harmonious MERLOT/CAB SAUV, small volume of lively SAUV BL (70%)/SÉM.

Crock, Le St-Est r ★★ 04 05 06 08 09' 10 11 12 14 15 16' 18 Gd-value CRU BOURGEOIS. Same stable as LÉOVILLE-POYFERRÉ. Solid, fruit-packed, can age.

Croix, La Pom r ★★ 04 05 06 08 09 10 11 12 14 15 16 18 Owned by NÉGOCIANT Janoueix. Organically run. MERLOT-led (90%). Also La Croix St-Georges, HAUT-SARPE.

Croix-de-Gay, La Pom r ★★★ 01' 04 05 06 08 09' 10' 12 14 15 16 17 18 Tiny MERLOT-dominant (95%) v'yd on POM plateau. La Fleur-de-Gay from separate parcels.

Croix du Casse, La Pom r ★★ 05 06 08 09 10 11 12 14 15' 16 MERLOT-based (90%+) POM on sandy/gravel soils. Medium-body; gd value. Owned by BORIE-MANOUX (2005).

Croizet-Bages Pau r ★★·★★★ 03 04 05 06 08 09 10' 11 12 14 15 16 17 18 Striving,

---

Cru Bourgeois hierarchy – again

A three-tier classification for CRUS BOURGEOIS was reintroduced from 2020. It sees the return of three quality levels: Cru Bourgeois, Cru Bourgeois Supérieur and Cru Bourgeois Exceptionnel. Classification is based on a quality assessment conducted by an independent organization, with technical and environmental factors considered. It will be renewable every 5 yrs. Applicants also submitted five vintages for tasting selected from 2008–16.

CAB SAUV-led Fifth Growth; still work to be done. Lately, more consistency but fails to excite. Same owner as RAUZAN-GASSIES.

Cru Bourgeois Méd Certificate awarded annually. 226 CHX in 2017. Quality variable.

Cruzelles, Les L de P r ★★ 08 09 10 11 12 14 15' 16' 17 18 Part of Denis Durantou stable (L'ÉGLISE-CLINET, La Chenade, Saintayme). Consistent, gd-value wine. Ageing potential in top yrs.

Dalem Fron r ★★ 05' 06 08 09' 10' 11 12 14 15 16 17 18 MERLOT-dominated (90%) property. Smooth, ripe, fresh. Environmental certification.

Dassault St-Ém r ★★ 03 04 05' 06 07 08 09' 10 11 12 14 15 16 18 Consistent, modern GRAND CRU CLASSÉ on sand, clay soils. 70% MERLOT, 30% CABS FR/SAUV. CHX La Fleur, Trimoulet, Faurie de Souchard same stable. Second label: D de Dassault.

There are 5834 growers in B'x (2018), less than half number of 20 yrs ago.

Dauphine, De la Fron r ★★ →★★★ 03 04 05 06' 08 09' 10' 11 12 14 15 16 17 18 Substantial estate. Sweeping change from 2000. Renovation, additional land acquired, organic certification, bio practices. Now more substance, finesse. Best of Wine Tourism International Award (2020). Second label: Delphis. Rosé.

Dauzac Marg r ★★ →★★★ 00' 01 02 04 05 08' 09' 10' 11 12 14 15 16' 18 Fifth Growth at Labarde; now dense, rich, dark wines. New owner 2019 but same CEO (Laurent Fortin) since 2013. Second label: La Bastide Dauzac. Also fruity Aurore de Dauzac. D de Dauzac is vegan.

Desmirail Marg r ★★ →★★★ 03 04 05 06 07 09' 10' 11 15 16' 17 Third Growth. CAB SAUV-led. Fine, delicate style. Second label: Initial de Desmirail. Visitor-friendly.

Destieux St-Ém r ★★ 03' 04 05' 06 07 08 09' 11 12 14 15 16 18 GRAND CRU CLASSÉ in St-Hippolyte. Powerful, needs time in bottle. New visitor centre; cellars renovated.

Doisy-Daëne Bar (r) w dr sw ★★★ 01 02 03 04 05' 06 07 08 09 10' 11' 12 13' 14 15' 17' 18' Second Growth owned by Dubourdieu family (CLOS FLORIDÈNE). Produces *fine, sweet Barsac*. L'Extravagant 16 17' 18' intensely rich, expensive, 100% SAUV BL CUVÉE. Also dry white Doisy-Daëne SEC.

Doisy-Védrines Saut w sw ★★★ 01' 03' 04 05' 07 09 10' 11' 12 13 14 15' 16' 18' BAR estate with v'yd in one block (80% SÉM). Richer style than DOISY-DAËNE. *Long-term fave*; delicious, gd value. Second label: Petit Védrines.

Dôme, Le St-Ém r ★★★ 05 06 08 09 10' 11 12 13 14 15 16 17 Microvinos rich, modern, powerful. Two thirds old-vine CAB FR, nr ANGÉLUS. New Norman Foster-designed winery for 2021. CH Teyssier (value) same stable.

Dominique, La St-Ém r ★★★ 00' 01 04 05' 06 08 09' 10' 11 12 14 15 16' 17 18' GRAND CRU CLASSÉ; 50 yrs of Fayat family ownership in 2019. Rich, juicy, MERLOT-led (81%). Jean Nouvel-designed winery with rooftop restaurant (La Terrasse Rouge) and shop. Environmental certification. Second label: Relais de la Dominique.

Ducru-Beaucaillou St-Jul r ★★★★ 01 02 03 04 05' 06 07 08 09' 10' 11 12 13 14 15' 16 17 18' Outstanding Second Growth in astute hands of Bruno Borie. Majors in CAB SAUV (85%+). Excellent form; classic cedar-scented claret, suited to long ageing. Organic conversion underway. La Croix de Beaucaillou sister estate.

Duhart-Milon Rothschild Pau r ★★★ 00' 01 02 03 04 05' 06 07 08 09' 10' 11 12 13 14 15 16' 17 18' Fourth Growth owned by LAFITE ROTHSCHILDS since 1962. CAB SAUV-dominated (65–80%). V. fine quality, esp in last 10 yrs. Cellars in heart of PAU.

Durfort-Vivens Marg r ★★★ 00 02 03 04 05 06 08 09' 10' 11 12 13 14 15' 16' 17 18 Much-improved MARG Second Growth; now lively and consistent. CAB-SAUV-dominated (70%). Organic and bio certification. Amphorae for vinification. Second labels: Vivens and Relais de Durfort-Vivens.

Eglise, Dom de l' Pom r ★★ 00 01 02 03 04 05' 06 08 09 10' 11 12 14 15 16 17 18 Oldest v'yd in POM (1589). Owned by BORIE-MANOUX since 1973. Clay/gravel soils of plateau. Consistent, fleshy of late.

Église-Clinet, L' Pom r ★★★→★★★★ 00' 01' 02 03 04 05' 06 07 08 09' 10' 11' 12 13 14 15' 16' 17 18 Tiny but high-flying. Great consistency; full, concentrated, fleshy but expensive. Second label: La Petite Église. LES CRUZELLES same stable.

Evangile, L' Pom r ★★★★ 01' 02 03 04 05 06 07 08 09' 10' 11 12 13 15' 16' 17 18 Rothschild (LAFITE)-owned property since 1990. Early-ripening site. Lots of investment. MERLOT-dominated (80%) with CAB FR. Consistently rich and opulent. Second label: Blason de L'Evangile.

Fargues, De Saut w sw ★★★★ 01 02 03' 04 05' 06 07 08 09' 10' 11' 13 14 15' 16' 17 Unclassified but top-quality (and price) SAUT owned by Lur-Saluces. Classic Saut: rich, unctuous but refined. Badly hit by hail 2018.

Faugères St-Ém r ★★→★★★ 00' 03 04 05 06 07 08 09' 10' 11 12 14 15 16 18 Sizeable GRAND CRU CLASSÉ. Rich, modern wines. Sister CH Péby Faugères (100% MERLOT) also classified. Vincent Cruège (ex-André Lurton) new (2019) technical director.

Ferrand, De St-Ém r ★★→★★★ 00 01 04 05 06 08 09 10' 12 14 15 16 18 Big St-Hippolyte GRAND CRU CLASSÉ owned by Pauline Bich Chandon-Moët. MERLOT-led (75%) with CABS FR/SAUV. Fresh, firm, expressive. Visitor-friendly.

Ferrande Grav r (w) ★★ 08 09 10 11 12 14 15 16' 17 18 Substantial GRAV property owned by NÉGOCIANT Castel (1991). Much-improved; easy red; fresh white.

Ferrière Marg r ★★★ 00' 02 04 05 06 08 09 10' 12 14 16' 17 18 Confidential Third Growth in MARG village. Bio certification (2018). Dark, firm, perfumed wines; gaining in finesse.

Feytit-Clinet Pom r ★★ →★★★ 00 01 04 05' 06 08 09' 10' 11 12 13 14 15 16' 18 Tiny 6-ha property owned by the Chasseuil family since 2000. 90% MERLOT on clay-gravel soils. Top, consistent form. Rich, seductive. Relatively gd value for a POM.

Fieuzal Pe-Lé r (w) ★★★ (r) 01 06 07 08 09' 10' 11 12 14 15 16 18 (w) 07 08 09' 10' 11 12 13 14 15 16 18 Classified PE-LÉ estate owned by Irish Quinn family. Vats named after grandchildren. Rich, ageable white; generous red. Second label: L'Abeille de Fieuzal (r w).

Figeac St-Ém r ★★★★ 00' 01' 02 03 04 05' 06 07 08 09' 10' 14 15' 16' 17' 18' Large PREMIER GRAND CRU CLASSÉ (B) currently on roll (magnificent 18 17 16). Classical CH, gravelly v'yd with unusual 70% CABS FR/SAUV. Now richer but always elegant wines; need long ageing. Major new winery complex inaugurated with the 2020 vintage. Second label: Petit-Figeac.

Filhot Saut w dr sw ★★ 01' 02 03' 04 05 07 09' 10' 11' 12 13 14 15 16 17' 18 Second Growth created in 1709 by Romain de Filhot. 60% SÉM, 36% SAUV BL, 4% MUSCADELLE. Richer, purer style from 2009.

Fleur Cardinale St-Ém r ★★ 04 05' 06 07 08 09' 10' 11 12 14 15 16 18 GRAND CRU CLASSÉ at St-Étienne-de-Lisse. In overdrive for last 15 yrs. Ripe, unctuous, modern style. Cellars recently extended. Visitor-friendly.

Fleur de Boüard, La B'x r ★★ →★★★ 05 06 07 08 09 10 11 12 13 14 15 16' 17 18 Leading estate in L DE P. Owned by de Boüard family (ANGÉLUS). Dark, dense, modern. Special CUVÉE, Le Plus: 100% MERLOT; more extreme. Winery with inverted, truncated, cone-shaped vats. Second label: Le Lion.

Fleur-Pétrus, La Pom r ★★★★ 00' 01 02 03 04 05' 06 08 09' 10' 11 12 13 14 15 16

Acacia mode

Could acacia barrels be the next big thing for ageing B'X's whites? Cheaper than the oak equivalent (wood is sawn rather than split, so less wastage), they impart little tannin and are aromatically fairly neutral (light toasting is essential). Advantage is mid-palate texture without any accompanying woody notes. CH DE FIEUZAL already uses 15% acacia barrels and the idea is being trialled at CARBONNIEUX and LA LOUVIÈRE, so maybe it will catch on.

17' 18 Top-of-range J-P MOUEIX property on POM plateau. 91% MERLOT, 6% CAB FR, 3% PETIT VERDOT. Finer style than PETRUS or TROTANOY. Needs time.

Fombrauge St-Ém r ★★ →★★★ 00' 01 02 03 04 05 06' 08 09 10 11 14 15 16' 18 A Bernard Magrez (PAPE-CLÉMENT) estate. Largest of GRANDS CRUS CLASSÉS (58.6 ha). Rich, dark, creamy, opulent. Magrez-Fombrauge is special red CUVÉE; also name for dry white B'X. Second label: Prélude de Fombrauge.

Fonbadet Pau r ★★ 03 04 05' 06 08 09' 10' 12 14 15 16' 17 18 Small non-classified estate; average v'yd age: 50 yrs. CAB SAUV-led (60%). Less long-lived but reliable.

Approx 300 négociants sell 70% of B'x wine.

Fonbel, De St-Ém r ★★ 07 08 09 10 11 12 14 15 16 18 Consistent source of juicy, fresh, gd-value ST-ÉM. Same stable as AUSONE, MOULIN-ST-GEORGES (shares cellar).

Fonplégade St-Ém r ★★ 04 05 06 08 09' 10 12 14 15 16' 18' American-owned GRAND CRU CLASSÉ. Previously concentrated, modern; now more fruit, balance. Bio conversion. Visitor-friendly. Second label: Fleur de Fonplégade.

Fonréaud List r ★★ 00' 03 04 05 06 08 09' 10' 11 12 14 15 16' 17 18 One of bigger, better LIST for satisfying, savoury wines. Family-owned and run. Some ageing potential. Small volume v.gd dry white: Le Cygne.

Fonroque St-Ém r ★★★ 03 04 05 06 08 09' 10' 12 14 15 16 18 Côtes GRAND CRU CLASSÉ nw of ST-ÉM town. Organic, bio certification. Firm but fresh and juicy.

Fontenil Fron r ★★ 05 06 08 09' 10' 11 12 14 15' 16 17 18' Leading FRON, owned by MICHEL ROLLAND since 1986. Ripe, opulent, balanced.

Forts de Latour, Les Pau r ★★★★ 00' 01 02 03 04' 05 06 07 08 09 10 11 12 13 16' 17 18 Second label of CH LATOUR (c 40% production), authentic flavour in slightly lighter format; high price. No more en PRIMEUR sales; only released when deemed ready to drink (2013 in 2019) but another 10 yrs often pays off.

Fourcas-Dupré List r ★★ 01 02 03 04 05 06 08 09 10' 11 12 15' 16' 17 18 Well-run property, fairly consistent. New owner 2019. SAUV BL (67%)/SÉM (33%) white.

Fourcas-Hosten List r ★★ →★★★ 03 05 06 08 09 10' 11 12 14 15 16 17 18 Large LIST estate. Organic certification from 2021. Considerable investment since 2006; now more precision, finesse. Also SAUV BL-led dry white.

France, De Pe-Lé r w ★★ (r) 04 05 06 08 09 10 11 12 14 16' 18 (w) 07 08 09 10 11 12 13 14 15 16 17 18 Unclassified Léognan estate; environmental certification. Ripe, modern, consistent wines. White fresh and balanced.

Franc-Mayne St-Ém r ★★ 01 03 04 05 08 09 10' 11 12 14 15' 16 18 Tiny GRAND CRU CLASSÉ on côtes. New ownership from 2018. Fresh and structured. Small hotel.

Gaby Fron r ★★ 04 05 06 08 09 10 12 14 15 16 17 18 Splendid s-facing slopes in CANON-FRON. MERLOT-dominated, age well. Special CUVÉE Gaby as well. Organic.

Gaffelière, La St-Ém r ★★★ 00' 01 02 03 04 05' 06 07 08 09 10' 11 12 13 14 15' 16 18 Family-owned First Growth at foot of côtes. Investment, improvement; part of v'yd replanted. Elegant, long-ageing wines. 75% MERLOT, 25% CAB FR. Second label: CLOS la Gaffelière.

Garde, La Pe-Lé r w ★★ (r) 05 06 08 09' 10' 11 12 14 15 16' 18 (w) 08 09 10' 11 12 13 14 16 17 18 Owned by DOURTHE (since 1990); supple, CAB SAUV/MERLOT reds. Tiny production of SAUV BL (90%)/SÉM white. Second label: La Terrasse de La Garde.

Gay, Le Pom r ★★★ 03 04 05 06 07 08 09' 10' 11 12 14 15 16 17 18 Major investment, MICHEL ROLLAND consults. Racy and suave with ageing potential. CH Montviel, La Violette same stable. Second label: Manoir de Gay.

Gazin Pom r ★★★ 00' 01 02 03 04 14 15' 16' 17 18 Large MERLOT-led (90%) estate. On v.gd form; generous, long ageing. Second label: L'Hospitalet de Gazin.

Gilette Saut w sw ★★★ 86 88 89 90 96 97 Extraordinary small Preignac CH. Family-owned since C18. Vintages back to 1953. Stores its sumptuous wines in concrete vats for 16–20 yrs. Ch Les Justices (SAUT) is sister estate.

Giscours Marg r ★★★ 00' 01 02 03 04 05 06 08 09 10' 11 12 14 15 16' 17 18 Substantial Third Growth. Composed of three gravelly hillocks. Full-bodied, long-ageing MARG capable of greatness (eg. 1970). Dutch owned since 1995; start of renaissance. Second label: La Sirène de Giscours. Little B'x rosé. Tertre is sister CH.

Glana, Du St-Jul r ★★ 04 05 06 08 09 15 16' 17 18 Big, unclassified CAB SAUV-led (65%) estate. Undemanding; robust; value. Second label: Pavillon du Glana.

Gloria St-Jul r ★★ →★★★ 00 02 03 04 05 06 07 08 09' 10' 11 12 14 15 16 17 18 V'yd in three zones of ST-JUL, average age 40 yrs. CAB SAUV-dominant (65%). Unclassified but sells at Fourth-Growth prices. Same stable as ST-PIERRE. Superb form recently.

Grand Corbin-Despagne St-Ém r ★★ →★★★ 00' 01 03 04 05 06 08 09' 10' 11 12 13 14 15 16' 18 Gd-value, family-owned GRAND CRU CLASSÉ in n ST-ÉM. 7th generation at helm. Aromatic wines now with riper, fuller edge. Environmental certification. CH Ampélia (CAS) sister estate. Second label: Petit Corbin-Despagne.

Grand Cru Classé ST-ÉM 2012: 64 classified; reviewed every 10 yrs.

Grand-Mayne St-Ém r ★★★ 00' 01' 02 03 04 05 06 07 08 09' 10' 11 12 14 15 16' 17 18 Impressive GRAND CRU CLASSÉ. Louis Mitjavile (TERTRE-RÔTEBOEUF) consults. Consistent, full-bodied, structured wines. Second label: Filia de Grand Mayne.

Grand-Puy-Ducasse Pau r ★★★ 00 01 02 03 04 05' 06 07 08 09' 10' 11 12 14 15' 16' 17 18 Fifth Growth showing a steady rise in quality in recent yrs. 60% CAB SAUV, 40% MERLOT. Reasonable value. Plans for new cellars in PAU. Second label: Prélude à Grand-Puy-Ducasse.

Grand-Puy-Lacoste Pau r ★★★ 00' 01 02 03 04 05' 06 07 08 09' 10' 11 12 13 14 15' 16' 17 18' Fifth Growth famous for CAB SAUV-driven (75%+) PAU to lay down. Owned by François-Xavier Borie. V'yd in one block around CH. Second label: Lacoste-Borie.

Grave à Pomerol, La Pom r ★★★ 01 02 04 05 06 08 09' 10 11 12 14 15 16 17' 18 Small J-P MOUEIX property on w slope of POM plateau. Mainly gravel soils. MERLOT-dominant (85%). Gd value. Can age.

Greysac Méd r ★★ 05' 06 08 09 10' 11 12 14 15 16' 18 Owned by Jean Guyon since 2012. Clay-gravel soils. MERLOT-led (65%), fine, fresh, consistent quality.

Gruaud-Larose St-Jul r ★★★★ 00' 01 02 03 04 05' 06 07 08 09' 10' 11 12 14 15' 16' 17 18 One of biggest, best-loved Second Growths. Owned by Jean Merlaut. Eric Boissenot consults. Vigorous claret to age. Second label: *Sarget de Gruaud-Larose.*

Guadet St-Ém ★★ 04 05 06 08 09 10 11 12 14 15 16' 18 Tiny GRAND CRU CLASSÉ, 7th generation. Better form last 10 yrs. DERENONCOURT consults. Organic, bio.

Guiraud Saut (r) w (dr) sw ★★★ 02 03 04 05' 06 07 08 09 10' 11' 13 14 15' 16' 17' Substantial organic neighbour of YQUEM. Unusual 35% SAUV BL. Restaurant La Chapelle. Visitor-friendly. Dry white G de Guiraud. Second label: Petit Guiraud.

Gurgue, La Marg r ★★ 04 05' 06 08 09' 10 11 12 14 15 16' 18 Same ownership, management as FERRIÈRE, HAUT-BAGES-LIBÉRAL. Organic, bio certification. Gd value.

Hanteillan H-Méd r ★★ 04 05' 06 09' 10 12 14 15 16 18 Large CRU BOURGEOIS. Early-drinking. Progress since 2012. DERENONCOURT consults. Second label: CH Laborde.

Haut-Bages-Libéral Pau r ★★★ 00 01 02 03 04 05' 06 08 09' 10' 11 12 14 15' 16 17 Medium-bodied Fifth Growth (next to LATOUR). CAB SAUV-led (70%). Eric Boissenot consults. Reasonable value. Second label: Le Pauillac de Haut-Bages-Libéral.

Sustainably conscious
One way of showing environmental concern is to sign up for HVE-level 3 (Haute Valeur Environnementale) certification. It's a government-supported sustainability programme that demands biodiversity conservation, reduction in use of pesticides and other agrochemicals and management of fertilizers and water resources. CHX that have credentials: CALON-SÉGUR, FIGEAC, HAUT-BAILLY, HAUT-BRION, LAFITE, LYNCH-BAGES, YQUEM. Classed Growths leading by example.

Haut-Bailly Pe-Lé r ★★★★ 00' 01 02 03 04 05' 06 07 08' 09' 10' 11 12 14 15' 16' 17 18' Top-quality PE-LÉ Classed Growth. Refined, elegant, CAB SAUV-led red (parcel of v. old, 100-yr+ vines). New winery in 2020. Second label: Haut-Bailly II from 2018 (previously La Parde de H-B). CH Le Pape (Pe-Lé) in same ownership.

Haut-Batailley Pau r ★★★ 00 02 03 04 05 06 07 08 09' 10' 11 12 13 14 15 16' 17' 18' Fifth Growth owned by Cazes family of CH LYNCH-BAGES. A sure thing. Steady progression in last 10 yrs. Second label: Verso (from 2017).

Ch Guiraud in Sauternes also grows 482 varieties of tomatoes.

Haut-Beauséjour St-Est r ★★ 04 05 08 09 10 11 14 15 16 17 Property created, improved by CHAMPAGNE ROEDERER; sold in 2017. MERLOT-led (60%+). Juicy but structured.

Haut-Bergeron Saut w sw ★★ 02 03 04 05 06 07 09 10 11 13 14 15' 16' 17' 18 Consistent non-classified SAUT owned and run by Lamothe family. V.-old vines (average 50 yrs). Mainly SÉM (90%). Rich, opulent, gd value.

Haut-Bergey Pe-Lé r (w) ★★→★★★ (r) 04 05 06 07 09 10 11 12 16 18 (w) 14 15 16 18 Non-classified property with Classed Growth pretensions. Rich, bold red. Fresh, concentrated dry white. Organic, bio certification. CLOS L'ÉGLISE (POM) sister estate.

Haut-Brion Pe-Lé r ★★★★ 00' 01 02 03 04 05 06 07 08 09' 10' 11' 12 13 14 15' 16' 17' 18' Only non-MÉD First Growth in list of 1855, owned by American Dillon family since 1935. RIP Jean-Bernard Delmas, general manager 1961–2004. Documented evidence of wine from 1521. Deeply harmonious, wonderful texture, for many no.1 or 2 choice of all clarets. Constant renovation: next project the *cuvier*. A little *sumptuous dry white* (SAUV BL/SÉM) for tycoons: 15' 16' 17 18. Also new La Clarté white from both H-B and La Mission H-B. *See* LA MISSION HAUT-BRION, LE CLARENCE DE HAUT-BRION, QUINTUS.

Haut Condissas Méd r ★★★ 06 07 08 09' 10' 11 12 14 15 16 17 18 Top wine from Jean Guyon stable (*see also* GREYSAC). 5000 cases annually. Sister to CH Rollan-de-By. Rich, concentrated, consistent. MERLOT-led plus 20% PETIT VERDOT.

Haut-Marbuzet St-Est r ★★→★★★ 01 02 03 04 05' 06 08' 09 10' 11 12 16 18 Started in 1952 with 7 ha; now 70; owned by Duboscq family. Easy to love, but unclassified. 60%+ sold directly by CH. Scented, unctuous wines matured in new oak barriques. CAB SAUV-led with MERLOT, CAB FR, PETIT VERDOT. Second label: MacCarthy.

Haut-Sarpe St-Ém r ★★ 04 05 06 08 09 10 11 12 14 15 16 18 GRAND CRU CLASSÉ owned by Janoueix family (1930). Hubert de Boüard (ANGÉLUS) consults. Modern wine.

Hosanna Pom r ★★★★ 00 01 03 04 05' 06 07 08 09 10' 11 12 14 15 16' 17 18 Tiny v'yd in heart of POM plateau. PETRUS a neighbour. Created by J-P MOUEIX in 1999 from parcels of former CH Certan Guiraud. Power, purity, balance, needs time.

Issan, D' Marg r ★★★ 01 02 03 04' 05' 06 07 08 09' 10' 11 12 13 14 15' 16' 18' Third Growth with fine moated CH. Fragrant wines; at top of game. CAB SAUV-led (60%). Cruse family owners since 1945. Second label: Blason d'Issan.

Jean Faure St-Ém r ★★ 06 08 09 10 11 12 14 15 16 18 GRAND CRU CLASSÉ on clay, sand, gravel soils. C18 origins. Organic certification (2017). Horses for ploughing. 50% CAB FR gives fresh, elegant style. Also 10% MALBEC.

Kirwan Marg r ★★★ 00' 01 02 03 04 05' 06 07 08 09 10' 11 12 14 15' 16' 17 18 Third Growth. Modern Stolnikoff-designed winery. CAB SAUV-dominant (60%). Dense, fleshy in 90s; now more finesse. Second label: Charmes de Kirwan.

Labégorce Marg r ★★→★★★ 03 04 05' 07 08 09 10' 11 12 14 15 16' 18 Substantial unclassified MARG owned by Perrodo family. Considerable investment, progress. Fine. Ch Marquis-d'Alesme same stable. Second label: Zédé de Labégorce.

Lafaurie-Peyraguey Saut w sw ★★★ 01' 02 03' 04 05' 06 07 09' 10' 11 13 14 15' 16' 17' 18 Leading Classed Growth owned by Lalique crystal-owner Silvio Denz (*see* FAUGÈRES). Rich, harmonious, sweet 90% SÉM. Relais & Châteaux hotel/restaurant. Second label: La Chapelle de Lafaurie-Peyraguey. Also SÉM-led dry white B'X.

Lafite-Rothschild Pau r ★★★★ 00' 01' 02 05' 06 07 08' 09' 10' 11' 12 13 14' 15' 16' 17 18' Big (112 ha) First Growth of famously elusive perfume and style, never great weight, although more dense, sleek these days. Great vintages need keeping for decades. Joint ventures in MIDI, Argentina, California, Chile, China (DOM de Long Dai, 1st vintage 17), Italy, Portugal. Second label: CARRUADES DE LAFITE. Also owns CHX DUHART-MILON, L'EVANGILE, RIEUSSEC.

Bottle of Lafite 1868, yr Rothschilds bought ch, sold for $123,000 in 2019.

Lafleur Pom r ★★★★ 01' 02 03 04' 05' 06 07 08 09' 10' 11 12 13 14 15' 16' 17' 18' Superb but tiny family-owned/-managed property. Elegant, intense wine for maturing. Expensive. New cellar 2018. Second label: *Pensées de Lafleur*. Also Les Champs Libres 100% SAUV BL.

Lafleur-Gazin Pom r ★★ 04 05 06 08 09 10 11 12 14 15 16' 18' Small, gd-value J-P MOUEIX estate. 85% MERLOT, 15% CAB FR. Fine, fragrant, accessible.

Lafon-Rochet St-Est r ★★★ 00' 01 02 03' 04 05' 06 07 08 09' 10' 11 12 13 14 15' 16' 17 18 Fourth Growth neighbour of COS LABORY. On gd form. Former PETRUS winemaker consults. Eye-catching canary-yellow buildings and label. Owner Basile Tesseron new president of ST-EST winemakers association. Second label: Les Pélerins de Lafon-Rochet.

Lagrange St-Jul r ★★★ 01 02 03 04 05' 06' 08' 09' 10' 11 12 13 14 15' 16' 17 18 (118 ha) Substantial Third Growth owned since 1983 by Suntory. Classic MÉD style. Much investment in v'yd, cellars. Eric Boissenot consults. Dry white Les Arums de Lagrange. Second label: Les Fiefs de Lagrange (gd value).

Lagrange Pom r ★★ 00 01 04 05 06 09 10 14 15' 16' 17 18 Tiny POM v'yd; clay, gravel soils. Owned by J-P MOUEIX (1953). 95% MERLOT. Supple, accessible early.

Lagune, La H-Méd r ★★★00' 02 03 04 05' 07 08 09' 10' 11 12 14 15' 16' 17 Third Growth in v. s of MÉD. Dipped in 90s; now on form. Fine-edged with more structure, depth. Bio certification (2019). Second label: Moulin de La Lagune. Also CUVÉE Mademoiselle L from v'yd in Cussac-Fort-Méd.

Lamarque, De H-Méd r ★★ 00' 04 05' 06 08 09' 10' 11 12 14 15 16 17 18 Medium-sized H-MÉD estate with medieval fortress. Eric Boissenot consults. Competent, mid-term wines, charm, value. Second label: D de Lamarque.

Lanessan H-Méd r ★★ 04 05 08 09 10' 12 11 12 14 15' 16' 18 Gd-value classic claret. Property located just s of ST-JUL. Hubert de Boüard (ANGÉLUS) consults. Environmental certification.

Langoa-Barton St-Jul r ★★★ 00' 01 02 03 04' 05' 06 07' 08' 09' 10' 11 12 13 14 15' 16' 17 18 Small Third-Growth sister CH to LÉOVILLE-BARTON. CAB SAUV-led (57%); charm, elegance. 9th-generation Barton at helm; assisted by 10th.

Larcis Ducasse St-Ém r ★★★ 00 02 03 04 05' 06 07 08 09' 10 11 12 13 14 15' 16' 17 18 PREMIER GRAND CRU CLASSÉ owned by Gratiot family. S-facing terraced v'yd. 80% MERLOT, 20% CAB FR. On top form. Second label: Murmure de Larcis Ducasse.

Larmande St-Ém r ★★ 04 05 06 07 08 09' 10 12 14 15 16 18 GRAND CRU CLASSÉ with a history dating back to 1585. V'yd planted to MERLOT (65%), CAB FR (30%), CAB SAUV (5%). Sound but lighter weight.

Laroque St-Ém r ★★→★★★ 04 05 06 08 09' 10' 11 12 14 15 16 17 18 Large GRAND CRU CLASSÉ. Terroir-driven wines. Second label: Les Tours de Laroque.

Larose-Trintaudon H-Méd r ★★ 05 06 07 08 09' 10 11 12 14 15 16 18 Largest v'yd in MÉD (165 ha). 75,000 cases/yr. Sustainable viticulture. Mostly for early-drinking. Second label: Les Hauts de Trintaudon. Also CHX Arnauld, Larose Perganson.

Laroze St-Ém r ★★ 00 01 05 06 07 08 09' 10' 12 14 15 16' 18 GRAND CRU CLASSÉ on sandy soils. MERLOT and 35% CABS FR/SAUV. Environmental certification. Lighter-framed wines; more depth of late. Second label: La Fleur Laroze.

Larrivet-Haut-Brion Pe-Lé r w ★★★ (r) 03 04 05' 06 08 09 10' 11 12 14 16 18

Unclassified PE-LÉ property owned by Gervoson family; v'yd expanded from 17 to 75 ha. Bruno Lemoine (ex-MONTROSE) winemaker. Visitor-friendly. Rich, modern red. Voluptuous, aromatic *white* 16 17 18. Second label: Les Demoiselles (r w).

Lascombes Marg r (p) ★★★ 00 01 02 03 04 05' 06 07 08 09 10' 11 12 14 15' 16 17 18 Large (120-ha) Second Growth with chequered history. Renovated C18 chartreuse. Wines rich, dark, concentrated, modern with touch of MARG perfume. Lots of MERLOT (50%+). Second label: Chevalier de Lascombes.

Latour Pau r ★★★★ 00' 01 02 03' 04 05' 06 07 08 09' 10' 11 12 14 15' 16' 17 18' First Growth considered grandest statement of B'X. Profound, intense, almost immortal wines in great yrs; even weaker vintages have unique taste and run for many yrs. Organic certification; part of historical "Enclos" section is bio. Ceased en PRIMEUR sales in 2012; wines now only released when considered ready to drink (08 and 11 in 19: still too soon). New cellars for more storage. Owned by Pinault family; vines also in Burgundy, Rhône, Napa. Second label: LES FORTS DE LATOUR; *third label: Pauillac;* even this can age 20 yrs.

Latour-à-Pomerol Pom r ★★★ 00' 01 02 04 05' 06 07 08 09' 10' 11 12 14 15' 16' 17 18 Managed by J-P MOUEIX (1962). Extremely consistent, well-structured wines that age. Relatively gd value.

Latour-Martillac Pe-Lé r w ★★ (r) 00 01 02 03 04 05' 06 08 09' 10' 11 12 14 15' 16 17 18' GRAV Cru Classé owned by Kressmann family. New *cuvier* 2019. Regular quality; gd value at this level; (w) 16 17 18. Second label: Lagrave-Martillac (r w).

Laurets, Des SI-Ém r ★★ 08 09 10 12 14 15 16 18 Substantial property owned by Benjamin de Rothschild. Hearty, MERLOT-led wine. Also special CUVÉE Baron. CH CLARKE same stable.

Laville Saut w sw ★★ 06 07 09' 10 11' 13 14 15 16 18 Non-classified Preignac estate run by Jean-Christophe Barbe; also lectures at Bordeaux University. SÉM-dominated (85%) with a little SAUV BL, MUSCADELLE. Lush, gd-value, botrytized wine.

Léoville-Barton St-Jul r ★★★★ 00' 01 02 03' 04 05' 06 07 08 09' 10' 11 12 13 14' 15' 16' 17 18 Second Growth with longest-standing family ownership; in Anglo-Irish hands of Bartons since 1826. Lilian Barton Sartorius 9th generation to run property; assisted by her children, Mélanie and Damien. Harmonious, classic claret; CAB SAUV-dominant (74%). Second label: La Rés de Léoville Barton.

Léoville Las Cases St-Jul r ★★★★ 00' 01 02 03' 04' 05' 06 07 98 99' 10' 11' 12 13 14 15' 16 17 18' Largest Léoville and original "Super Second"; *grand vin* from Grand Enclos v'yd. CAB SAUV dominant, also CAB FR. Elegant, complex wines built for long ageing Second label: Le Petit Lion, CLOS DU MARQUIS separate wine in ST-JUL.

Léoville-Poyferré St-Jul r ★★★★ 01 02 03 04 05' 06 07 08 09' 10' 11' 12 13 14 15' 16' 18 Owned by Cuvelier family (1920). Sara Lecompte Cuvelier now at helm.

St-Émilion classification – current version

The latest classification (2012) incl a total of 82 CHX: 18 PREMIERS GRANDS CRUS CLASSÉS and 64 GRANDS CRUS CLASSÉS. The new classification, now legally considered an exam rather than a competition, was conducted by a commission of seven, nominated by INAO, none from B'X. CHX ANGÉLUS and PAVIE were upgraded to Premier Grand Cru Classé (A) while added to the rank of Premier Grand Cru Classé (B) were CANON LA GAFFELIÈRE, LA MONDOTTE, LARCIS DUCASSE and VALANDRAUD. New to status of Grand Cru Classé were Chx BARDE-HAUT, CLOS de Sarpe, EAClos la Madeleine, Côte de Baleau, DE FERRAND, DE PRESSAC, FAUGÈRES, FOMBRAUGE, JEAN FAURE, La Commanderie, La Fleur Morange, Le Chatelet, Péby Faugères, QUINAULT L'ENCLOS, Rochebelle and SANSONNET. Although a motivating force for producers, the classification (reviewed every 10 yrs, since 1955) is still unwieldy guide for consumers.

> **Space, the final frontier**
> In 2019 a dozen bottles of (unidentified) wine were rocketed to the
> International Space Station to test their capacity for ageing in space.
> On return to earth they will be compared to a control sample to see
> how radiation and micro-gravity have affected polyphenols, crystals
> and tannins. It's all in the name of science, but not 1st time wine has
> travelled in the cosmos. In 1985 a half-bottle of CH LYNCH-BAGES was
> launched into space on board a NASA Discovery shuttle and eventually
> made it safely back to earth. No tasting note has been found.

"Super Second" level; dark, rich, spicy, long-ageing wines. *Ch Moulin-Riche* is a
separate 21-ha parcel. Second label: Pavillon de Léoville-Poyferré.

Lestage List r ★★ 04 05 06 08 09 10 11 12 14 15 16' 18 MERLOT-led CRU BOURGEOIS.
Same stable as FONRÉAUD. Firm, slightly austere. Also dry white La Mouette.

Lilian Ladouys St-Est r ★★ 04 05 06 07 08 09' 10' 11 12 14 15 16 18 Sizeable (80
ha) CRU BOURGEOIS. Environmental certification. More finesse in recent vintages.
Same stable as PÉDESCLAUX. Second label: La Devise de Lilian.

Liversan H-Méd r ★★ 10 14 15 16 18 CRU BOURGEOIS in n H-MÉD. V'yd in single block.
Owned by Advini group. Round, savoury, early-drinking.

Loudenne Méd r ★★ 04 05 07 08 09 10 12 14 15 16 18 Large CRU BOURGEOIS, once
Gilbeys, where I met my wife. Then Lafragette family, now Chinese-owned and
labelled Loudenne Le CH. Landmark C17 pink-washed *chartreuse* by river. Visitor-
friendly. 50/50 MERLOT/CAB SAUV reds. SAUV BL-led white.

Louvière, La Pe-Lé r w ★★★ (r) 04 05' 06 07 08 09' 10' 11 12 14 15 16' 18 (w) 16 17 18
Vignobles André Lurton property. Excellent *white* (100% SAUV BL), red of Classed
Growth standard (60/40 CAB SAUV/MERLOT). *See also* BONNET, COUHINS-LURTON.

Lussac, De St-Ém r ★★ 05 06 08 09 10 11 14 15 16 18 Top estate in LUSSAC-ST-ÉM.
Plenty of investment. Supple red and rosé, Le Blanc dry white. B&B as well.

Lynch-Bages Pau r (w) ★★★★ 00' 01 02 03 04' 05' 06 07 08 09' 10' 11 12 13 14 15
16' 17 18 Always popular, now a star, far higher than its Fifth-Growth rank. Rich,
dense CAB SAUV-led wine. Second label: Echo de Lynch-Bages. Gd white, *Blanc de
Lynch-Bages*, now fresher style. New winery designed by architect Chien Chung
Pei inaugurated in 2020. HAUT-BATAILLEY, ORMES-DE-PEZ same stable.

Lynch-Moussas Pau r ★★ 01 02 03 04 05' 07 08 09 10' 11 12 14 15 16 17 18 Fifth
Growth owned by Castéja family. Lighter PAU (75% CAB SAUV), much improved.

Lyonnat St-Ém r ★★ 06 08 09 10 12 14 15 16 18 Leading LUSSAC-ST-ÉM owned by
Milhade family. MERLOT-led; more precision lately. Also special CUVÉE Emotion.

Malartic-Lagravière Pe-Lé r (w) ★★★ (r) 09 10 14 15' 16 18 (w) 16 17 18 GRAV Classed
Growth. Owned and run by Bonnie family. Loads of investment. Visitor-friendly.
Rich, modern, CAB SAUV-led red; a little lush *white* (majority SAUV BL). CH Gazin
Rocquencourt (PE-LÉ) same stable.

Malescasse H-Méd r ★★ 04 05 06 08 09 10 11 12 14 15' 16' 17 18 CRU BOURGEOIS nr
MOULIS. Recent investment, upgrade. MERLOT-led. Supple, value wines.

Malescot-St-Exupéry Marg r ★★★ 00' 01 02 03 04 05' 07 08 09' 10' 11 12 14 15' 16
17 18 MARG Third Growth owned by Zuger family (1955). Ripe, fragrant, finely
structured wines. Second label: Dame de Malescot.

Malle, De Saut r w dr sw ★★★ (w sw) 01' 02 03' 05 06 09 10' 11' 13 14 15' 16' 17' 18
Preignac Second Growth (1855) with classical mansion owned by de Bournazel
family. Fine, medium-bodied SAUT. Second label: Les Fleurs de Malle.

Margaux, Ch Marg r (w) ★★★★ 01' 02 03' 04' 05' 06' 07 08 09' 10' 11 12 13 14 15
16' 17 18' First Growth; most seductive, fabulously perfumed, consistent wines.
Owned by Mentzelopoulos family (1977). In-house cooperage produces a third of
barrels required. Recent addition: Norman Foster-designed cellars. Second label:

Pavillon Rouge 04' 09 16' 18'. Third label: Margaux du Ch Margaux from 2009. *Pavillon Blanc* (100% SAUV BL): best white of MÉD, recent vintages fresher 15' 16' 18'.

Marojallia Marg r ★★★ 02 03 04 05' 06 07 08 09' 10 11 12 15 16 17 18' Micro-CH looking for big prices for big, rich, un-MARG-like wines. CAB SAUV-led (70%). Second label: CLOS Margalaine.

Marquis-de-Terme Marg r ★★ →★★★ 00' 01 02 03 04 05' 06 08 09' 10' 11 12 14 15' 16 17 18 Fourth Growth with v'yd dispersed around MARG. Owned by Sénéclauze family. Previously solid wine, now more seductive. Second label: La Couronne.

Maucaillou Mou r ★★ 05 06 08 09 10 11 12 14 15' 16 18 Large, consistent MOU estate. Güntzian gravel soils. Clean, value wines. Second label: N°2 de Maucaillou.

Mayne Lalande List r ★★ 08 09 10 11 12 14 15 16 18 Leading LIST estate. Bio conversion. Full, finely textured. B&B too.

Mazeyres Pom r ★★ 04 05' 06 08 09 10 14 15 16' 18 Lighter but consistent POM. Earlier-drinking, MERLOT-led. Organic, bio certification. Second label: Le Seuil.

Meyney St-Est r ★★ →★★★ 00 01 02 03 04 05' 06 08 09' 10' 11' 12 14 15' 16' 18 Big river-slope v'yd, superb site next to MONTROSE. Structured, age-worthy. CAB SAUV-led (60%). Gd value at this level. Same stable as GRAND-PUY-DUCASSE. Hubert de Boüard (ANGÉLUS) consults. Second label: Prieur de Meyney.

Mission Haut-Brion, La Pe-Lé r ★★★★ 00' 01 02 03 04 05' 06 07 08 09' 10' 11' 12 13 14 15' 16' 17' 18' Owned by Dillon family of neighbouring HAUT-BRION. Same technical team. Consistently grand-scale, full-blooded, long-maturing wine. Second label: La Chapelle de la Mission. Magnificent SÉM-dominated white: previously Laville-Haut-Brion; renamed La Mission-Haut-Brion Blanc 16 17' 18. Second label: La Clarté (fruit from both H-Bs).

Monbousquet St-Ém r (w) ★★★ 03 04 05' 06 07 08 09' 10' 11 12 14 15 16' 18 GRAND CRU CLASSÉ on sand/gravel soils. Transformed by Gérard Perse (*see* PAVIE). Concentrated, oaky, voluptuous. Rare *v.gd Sauv Bl/Sauvignon Gris* (AC B'X). Second label: Angélique de Monbousquet.

Monbrison Marg r ★★ →★★★ 01 02 04 05' 06 08 09' 10' 11 12 14 15 16 17 18 Tiny (13.2 ha), family-owned property at Arsac. CAB SAUV-led (60%). Delicate, fragrant MARG.

Mondotte, La St-Ém r ★★★★ 00' 01 02 03 04 05' 06 07 08 09' 10' 11 12 13 14 15' 16 17 18 Tiny (4.5 ha) PREMIER GRAND CRU CLASSÉ on limestone-clay plateau. Intense, firm, virile wines. Organic certification. Same von Neipperg stable as AIGUILHE, CANON-LA-GAFFELIÈRE, CLOS DE L'ORATOIRE.

Montrose St-Est r ★★★★ 00' 01 02 03 04 05' 06 07 08 09' 10' 11 12 13 14 15' 16' 17 18' Second Growth with riverside v'yd. Famed for forceful, long-ageing claret. Vintages 1979–85 were lighter. Bouyges brothers owners. Massive environmental programme. Second label: *La Dame de Montrose*. Third label: Le St-Estèphe de Montrose. CLOS ROUGEARD (Loire) same stable.

Moulin du Cadet St-Ém r p ★★ 01 03 05 08 08 09 10' 11 12 14 15 16 17' 18 Tiny GRAND CRU CLASSÉ; 100% MERLOT; clay-limestone soils. Same owner as SANSONNET. Formerly robust, now more finesse.

Value of 1 ha generic B'x is €16,000; in Pom it's €1.5 million.

Moulinet Pom r ★★ 05 06 08 09 10 11 12 15 Large CH for POM: 25 ha. MERLOT-led. Lighter style. Chinese-owned (2016).

Moulin-Haut-Laroque Fron r ★★ 06 08 09' 10' 11 12 14 15' 16 17 18' Leading FRON property. MERLOT-led (65%) with 5% MALBEC. Consistent, structured, can age.

Moulin Pey-Labrie Fron r ★★ 04 05' 08 09' 10' 11 12 15 16 17 18 MERLOT-led CANON-FRON with 5% MALBEC. Organic. Sturdy, well structured wines that can age.

Moulin-St-Georges St-Ém r ★★★ 03 04 05' 06 08 09' 10' 11 12 13 14 15' 16' 17 18 Shares a cellar with FONBEL. 80% MERLOT, 20% CAB FR. Dense, polished wines. Gd value at this level.

> **Turn up the sound**
> Bio-certified Third Growth (1855) CH PALMER is experimenting with a
> sound box in the v'yd. The lyrical gizmo emits "carefully calibrated
> vibration waves" to test a theory that sound waves can help flowering
> and vegetative growth. After a downy mildew-induced 11 hl/ha in 2018
> it's worth giving ear to everything.

Mouton Rothschild Pau r (w) ★★★★ 00' 01' 02 03 04' 05' 06 07 08' 09' 10' 11
12 13 14 15' 16' 17' 18' Rothschild-owned (1853); current generation Camille,
Philippe, Julien. Most exotic, voluptuous of PAU First Growths; at top of game.
2017 label "Hallelujah" created by French artist Annette Messager. White Aile
d'Argent (SAUV BL/SÉM) now more graceful. Second label: *Le Petit Mouton. See also*
D'ARMAILHAC, CLERC MILON.

Nairac Saut w sw ★★ 01' 02 03' 04 05' 06 07 09 10 11 13 14 15 BAR Second Growth
run by brother and sister, Nicolas and Eloïse Heeter Tari. C17 origins. Rich but
fresh. Second label: Esquisse de Nairac.

Nénin Pom r ★★★ 00' 01 02 03 04 05 06 07 08 09' 10' 11 12 13 14 15' 16' 17 18 Large
POM estate. Same owner as LÉOVILLE-LAS-CASES. Plenty of investment; evidently paid
off as wines now restrained but generous, precise, built to age. Increase in CAB FR
(40%). Gd-value second label: Fugue de Nénin.

Olivier Pe-Lé r w ★★★ (r) 02 04 05' 06 08 09' 10' 11 12 13 14 16' 17 18 (w) 16 17
18 Beautiful GRAV Classed Growth owned by Bethmann family. Significant
investment last 15 yrs. Sructured red (55% CAB SAUV), juicy SAUV BL-led (75%) white.

Ormes-de-Pez St-Est r ★★ 02 03 04 05 06 07 08' 09' 10' 11 12 14 15' 16' 17 18 CAZES
family of LYNCH-BAGES own (1940). Consistent, full, age-worthy.

Ormes-Sorbet, Les Méd r ★★ 04 05 06 08 09' 10' 11 12 14 15 16' 17 18 Reliably
consistent CRU BOURGEOIS. 9th-generation family owners. CAB SAUV (65%), MERLOT
(30%), PETIT VERDOT (5%). Elegant, gently oaked wines.

Palmer Marg r ★★★★ 00 01' 02 03 04 05' 06' 07 08 09' 10' 11 12 13 14 15' 16' 17
18' Third Growth on par with "Super Seconds" (occasionally Firsts). Voluptuous
wine of power, complexity and much MERLOT (40%). Dutch (Mähler-Besse) and
British (SICHEL) owners. Philosophy and certification bio. Second label: *Alter Ego
de Palmer*. Original Vin Blanc de Palmer (Loset/MUSCADELLE/Sauvignon Gris).
Experiment with SYRAH.

Pape-Clément Pe-Lé r (w) ★★★★ (r) 00' 01 02 03 04 05 06 07 08 09' 10' 11 12 13 14
15' 16' 17 18 (w) 11 12 13 14 15 16 17' 18 Historic estate in B'x suburbs (B&B, wine
shop, tastings). Owned by magnate Bernard Magrez. Dense, long-ageing reds.
Tiny production of rich, oaky white. Second label (r w): Clémentin.

Patache d'Aux Méd r ★★ 05' 06 09 10 11 12 14 15 16' 18 Sizeable CRU BOURGEOIS.
Reliable, CAB SAUV-led (50%). DERENONCOURT consults. Part of Advini group.

Pavie St-Ém r ★★★★ 00' 01 02 04 05' 06 07 08 09' 10' 11 12 13 14 15' 16' 17 18
PREMIER GRAND CRU CLASSÉ (A) splendidly sited on plateau and s côtes. Considerable
investment since 1998. Intense, strong wines for ageing; recent vintages less
extreme. MERLOT-led but now more CABS FR/SAUV. Second label: Arômes de Pavie.

Pavie-Decesse St-Ém r ★★★ 01' 02 03 04 05 06 07 08 09' 10' 11 12 14 15' 16 17 18
Tiny (3.5 ha) Classed Growth. 90% MERLOT. As powerful, muscular as sister PAVIE.

Pavie-Macquin St-Ém r ★★★ 00' 01 02 03 04 05' 06 07 08 09' 12 15 16' 17 18'
PREMIER GRAND CRU CLASSÉ (B) with v'yd on limestone plateau. 80% MERLOT, 20%
CABS FR/SAUV. Winemakers Nicolas (since 1994) and son Cyrille Thienpont.
Sturdy, full-bodied wines need time. Second label: Les Chênes de Macquin.

Pédesclaux Pau r ★★ 03 04 05 06 09 10' 11 12 13 14' 15 16' 18 Underachieving Fifth
Growth revolutionized by owner Jacky Lorenzetti. Extensive investment since
2014; cellars, more v'yds. More fruit/flavour. Organic, bio disposition.

Petit-Village Pom r ★★★ 00' 01 03 04 05 06 07 08 09' 10' 11 12 13 14 15' 16' 17 18' Much-improved estate on POM plateau. New owner from 2020: CH BEAUREGARD (Pom). Winemaking team will stay. Suave and dense, with increasingly finer tannins.

Petrus Pom r ★★★★ 00' 01 02 03 04 05' 06 07 08 09' 10' 11' 12 13 14 15' 16' 17' 18' (Unofficial) First Growth of POM: MERLOT solo *in excelsis*. 11.5 ha v'yd on blue clay gives 2500 cases of massively rich, concentrated wine for long ageing. One of 50 most expensive wines in world. Olivier Berrouet winemaker since 2007 (succeeded father, Jean-Claude). No second label.

Pey La Tour B'x r ★★ 10 11 12 14 15 16 18 Large (176 ha) DOURTHE property. Quality-driven B'X SUPÉRIEUR. Three red CUVÉES: Rés du CH (MERLOT-led) top; rosé, dry white.

Peyrabon H-Méd r ★★ 05 06 09' 10 11 12 15 16 17 18 Savoury CRU BOURGEOIS owned by NÉGOCIANT Millésima. Also La Fleur-Peyrabon in PAU.

Pez, De St-Est r ★★★ 03 04 05' 06 07 08 09' 10' 11 12 13 14 15' 16' 17 18' Dense, reliable cru owned by ROEDERER. MERLOT/CAB SAUV with a little CAB FR, PETIT VERDOT. Aged 12–18 mths in oak barrel.

Phélan-Ségur St-Est r ★★★ 03 04 05' 06 07 08 09 10' 11 12 13 14 15' 16' 17 18' Reliable, top-notch, unclassified CH with Irish origins; long, supple style. CAB SAUV/MERLOT blend but experimental MALBEC, CARMENÈRE planted. Second label: Frank Phélan.

Pibran Pau r ★★ 03 04 05 06 08 09' 10' 11 12 13 14 15 16' 17 18 Earlier-drinking PAU allied to PICHON-BARON. MERLOT/CAB SAUV blend.

Pichon-Baron Pau r ★★★★ 00 01 02 03' 04 05' 06 07 08 09' 10' 11 12 13 14 15' 16' 17 18 Owned by AXA; formerly CH Pichon-Longueville (until 2012). Revitalized Second Growth; CAB SAUV-led (75%+). Powerful, long-ageing PAU at a price. Second labels: Les Tourelles de Longueville (approachable: more MERLOT); Les Griffons de Pichon Baron (Cab Sauv-dominant).

Pichon-Longueville Comtesse de Lalande (Pichon Lalande) Pau r ★★★★ 00 01 02 03' 04 05' 06 07 08 09' 10' 11 12 13 14 15' 16' 17 18' ROEDERER-owned Second Growth, neighbour of LATOUR. Always among top performers; long-lived wine of famous breed. MERLOT-marked in 80s, 90s; more CAB SAUV in recent yrs (71% in 2018). New high-tech, gravity-fed winery. Second label: *Rés de la Comtesse*.

Pin, Le Pom r ★★★★ 00 01' 02 04 05' 06' 07 08' 09' 10' 11 12 14 15 16' 17 18' The original B'X cult wine (inspiration for *garagistes* in 90s). Only 2.8 ha. Tiny modern winery in v'yd. 100% MERLOT; almost as rich as its drinkers; prices are out of sight. Ageing potential. L'If (ST-ÉM), L'Hêtre (CAS) are stablemates.

Plince Pom r ★★ 04 05 06 08 09 10 11 12 14 15 16' 18 Neighbour of NÉNIN and LA POINTE. Lighter style of POM on sandy soils.

Pointe, La Pom r ★★ 04 05 06 07 08 09' 10 11 12 14 15' 16 18 Large (for POM), well-managed estate; progress in last 10 yrs. Gd value. Hubert de Boüard consults.

Poitevin Méd r ★★ 08 09 10 11 12 14 15 16 17 18 Supple, elegant CRU BOURGEOIS. MERLOT/CAB SAUV with 5% PETIT VERDOT. Consistent quality.

Cost of a new French oak barrel nowadays: €750 and upwards.

Pontet-Canet Pau r ★★★★ 00' 01 02' 03 04' 05' 06' 07 08 09' 10' 11 12 13 14 15 16' 17 18 Large, fashionable, bio-certified, Tesseron-family-owned Fifth Growth. Radical improvement has seen prices soar. CAB SAUV-dominant (60%+). New *cuvier* with 32 amphora-shaped vats. Second label: Les Hauts de Pontet-Canet.

Potensac Méd r ★★ 02 03 04 05' 07 08 09' 10' 11 12 13 14 15 16' 17 18' Same stable as LÉOVILLE LAS CASES. Firm, long-ageing wines; gd value. MERLOT-led (45%) but lots of old-vine CAB FR. Second label: Chapelle de Potensac.

Pouget Marg r ★★ 02 03 04 05' 06 08' 09' 10' 12 14 15 16 18 Obscure Fourth Growth sister of BOYD-CANTENAC. Blend varies. Sturdy; needs time.

Poujeaux Mou r ★★ 01 03 04 05 08 09 10 11 12 14 15' 16' 17 18 Same stable as CLOS

> **Don't drop that barrel**
> A barrel made of Lalique crystal has been produced to celebrate 400th anniversary of SAUTERNES First Growth (1855) CH LAFAURIE-PEYRAGUEY. The 225-litre barrique contains the estate's 13 vintage and took Lalique craftsmen 2 yrs to design and create.

FOURTET. DERENONCOURT consults. Full, robust wines with ageing potential. Second label: La Salle de Poujeaux.

Premier Grand Cru Classé St-Ém 2012: 18 classified; ranked into A (4) and B (14).

Pressac, De St-Ém r ★★ 06 08 09 10 11 12 14 15' 16' 17 18 GRAND CRU CLASSÉ with part-terraced v'yd. MERLOT-LED but CABS FR/SAUV, MALBEC, CARMENÈRE. Value.

Preuillac Méd r ★★ 06 08 09 10 11 12 14 15 16 18 Savoury, structured CRU BOURGEOIS. DERENONCOURT consults. MERLOT-led (54%). Second label: Esprit de Preuillac.

Prieuré-Lichine Marg r ★★★ 00' 01 02 03 04 05 06 07 08 09' 10' 11 12 14 15' 16' 17 18' Fourth Growth owned by Ballande group; put on map in 60s by Alexis Lichine. Parcels in all five MARG communes. Fragrant Marg currently on gd form. DERENONCOURT consults. Gd SAUV BL/SÉM as well.

Puygueraud B'x r ★★ 03' 05' 06 08 09 10 11 12 14 15' 16' 17' 18 Leading CH of this tiny FRANCS-CÔTES DE B'X AC. Nicolas Thienpont and son Cyrille winemakers (*see* PAVIE-MACQUIN). MERLOT-led. Oak-aged wines of surprising class. Also MALBEC/CAB FR-based cuvée George. A little white (SAUV BL/Sauvignon Gris).

Quinault L'Enclos St-Ém r ★★→★★★ 09 10 11 12 14 15 16' 17 18' GRAND CRU CLASSÉ located in Libourne. Same team and owners as CHEVAL BLANC. MERLOT-led but 20% CAB SAUV; more freshness, finesse. Aged in 500l casks (50% new).

Quintus St-Ém r ★★★ 11 12 13 14 15 16' 17 18' Created by Dillons of HAUT-BRION from former Tertre Daugay and L'Arrosée v'yds. Gaining in stature; 18 16 best yet. Price has soared. Second label: Le Dragon de Quintus.

Rabaud-Promis Saut w sw ★★ →★★★ 01' 02 03' 04 05' 06 07 09' 10 11 12 13 14 15' 16 17' 18 Family-owned First Growth. Quality, gd value. CUVÉE Raymond-Louis from selected parcels.

Rahoul Grav r w ★★ (r) 08 09' 10 11 12 14 15 16' 17 18 Owned by DOURTHE; reliable, MERLOT-led red. SÉM-dominated white 16 17 18. Gd value.

Ramage-la-Batisse H-Méd r ★★ 05' 08 09 10 11 12 14 15 16 18 Reasonably consistent, widely distributed CRU BOURGEOIS. CAB SAUV-led with MERLOT, PETIT VERDOT, CAB FR.

Rauzan-Gassies Marg r ★★★ 02 03 04 05' 06 07 09 08 09' 10 11 12 15 16' 17 18 Second Growth owned by Quié family. Improvement over last 10 yrs but lags behind top MARGS. Second label: Gassies.

Rauzan-Ségla Marg r ★★★★ 00' 01 02 03 04' 05' 06 07 08 09' 10' 11 12 14 15' 16' 17 18' Leading MARG Second Growth long famous for its fragrance; owned by Wertheimers of Chanel. Managed by Nicolas Audebert. CAB SAUV-led (62%). Second label: Ségla (value).

Raymond-Lafon Saut w sw ★★★ 01' 02 03' 04 05' 06 07' 09' 10 11' 13 14 15' 17' 18 Unclassified SAUT owned by Meslier family producing First-Growth quality. Rich, complex wines that age; gd value.

Rayne Vigneau Saut w sw ★★★ 01' 03 05' 07 09' 10' 11' 13 14 15 16' 17 18' Substantial First Growth owned by Trésor du Patrimoine group. Wines suave, age-worthy. Also special CUVÉE Gold. Visitor-friendly. Second label: Madame de Rayne.

Respide Médeville Grav r w ★★ (r) 06 08 09 10 11 12 14 15 16 18 (w) 11 12 13 14 15 16 18 Top GRAV property; elegant red, complex *white*. Second label (r) Dame de Respide.

Reynon B'x r w ★★ Leading CADILLAC-CÔTES DE B'X estate. Owned by Dubourdieu family (*see* CLOS FLORIDÈNE). MERLOT-led red 15 16' 18'. B'x white SAUV BL (DYA).

Reysson H-Méd r ★★ 06 08 09' 10' 11 12 14 15 16 18 CRU BOURGEOIS owned by NÉGOCIANT DOURTHE. Mainly MERLOT (88%), CAB FR, PETIT VERDOT. Rich, modern style.

Rieussec Saut w sw ★★★★ 01′ 02 03′ 04 05′ 06 07 09′ 10′ 11′ 13 14 15′ 16 17 18′ First Growth with substantial v'yd in Fargues, owned by (LAFITE) Rothschilds since 1984. Regularly powerful, opulent, SÉM-dominant (90%), SAUV BL, MUSCADELLE; a bargain. Second label: Carmes de Rieussec.

Rivière, De la Fron r ★★ 03 04 05′ 06 08 09′ 10 12 14 15 16′ 18 Largest (65 ha), most impressive FRON property with C16 CH. MERLOT-led. Chinese-owned. Formerly big, tannic; now more refined. Claude Gros consults. Second label: Les Sources.

Roc de Cambes B'x r ★★★ 01 05 10 12 15′ 16 17 18 Undisputed leader in CÔTES DE BOURG; MERLOT (80%) and v. old (50 yrs) CAB SAUV. Savoury, opulent but pricey. Also DOM de Cambes.

Rochemorin, De Pe-Lé r w ★★→★★★ (r) 06 08 09′ 10′ 11 12 14 15 16 18 (w) 10 11 12 13 14 15 16 17 18 Large property at Martillac owned by VIGNOBLES André Lurton (see BONNET, COUHINS-LURTON, LA LOUVIÈRE). Fleshy, MERLOT-led (55%) red; aromatic white (100% SAUV BL). Fairly consistent quality.

Rol Valentin St-Ém r ★★★ 03 04 05′ 06 07 08 09′ 10′ 11 12 13 14 15 16′ 17 18 Once garage-sized; now bigger v'yd with clay-limestone soils. MERLOT-led (90%).

Rouget Pom r ★★ 03 04 05′ 06 07 08 09′ 10′ 11 12 14 15 16′ 17 18 Go-ahead estate on n edge of POM. Labruyère family owned (1992); Edouard at helm. MICHEL ROLLAND consults. Rich, unctuous wines. Second label: Le Carillon de Rouget.

St-Georges St-Ém r ★★ 03 04 05′ 06 08 09 10 11 14 15 16 18 V'yd represents 25% of ST-GEORGES AC. MERLOT-led (80%) with CABS FR/SAUV. Gd wine sold direct to public. Visitor-friendly. Second label: Puy St-Georges.

St-Pierre St-Jul r ★★★ 00′ 01′ 02 03 04 05′ 06 07 08 09′ 10′ 11 12 13 14 15′ 16′ 17 18′ Tiny Fourth Growth to follow, same stable as GLORIA. Stylish, consistent, classic. Second label: Esprit de St-Pierre.

Sales, De Pom r ★★ 04 05 06 08 09 10′ 12 15 16 18 Biggest v'yd of POM (10,000 cases). Younger generation at helm. Ex-PETRUS winemaker consults. Honest, drinkable; can age. Second label: CH Chantalouette (5000 cases).

Sansonnet St-Ém r ★★ 03 04 05′ 06 08 09′ 10′ 11 12 13 14 15 16′ 17 18 Ambitious GRAND CRU CLASSÉ. Modern but refreshing. Second label: Envol de Sansonnet. MOULIN DU CADET same stable.

Saransot-Dupré List r (w) ★★ 03 04 05 06 09′ 10′ 11 12 15 16 18 Small property owned by Yves Raymond. Fleshy, MERLOT-led with 24% CAB SAUV. Dry white B'x.

Sénéjac H-Méd r (w) ★★ 04 05′ 06 08 09 10′ 11 12 14 15 16′ 18 S H-MÉD (Pian) cru. Consistent, well-balanced wines. Drink young or age. Sister to TALBOT.

Serre, La St-Ém r ★★ 04 05 06 08 09′ 10 11 12 14 15 16′ 17 18′ Small GRAND CRU CLASSÉ on limestone plateau. New winery 2018. Fresh, stylish wines with fruit.

Sigulus-Rabaud Saut w sw ★★★ 01′ 02 03 04 05 07′ 09′ 10′ 11 12 13 14 15′ 16 17′ 18 Tiny First Growth owned by Laure de Lambert. *V. fragrant and lovely.* Second label: Le Lieutenant de Sigalas. Also La Sémillante dry white.

Siran Marg r ★★→★★★ 02 03 04 05 06 07 08 09′ 10′ 11 12 14 15′ 16′ 17 18 Surprisingly unclassified MARG estate in Labarde. 160 yrs in same family hands in 2019. Visitor-friendly. Wines have substance, fragrance. Environmental disposition. Second label: S de Siran.

Touriga? Why not?

The B'x and B'x SUPÉRIEUR wine-growers association has approved use of several new grape varieties on an experimental basis. They must not exceed 10% of blend, and incl (r) Arinarnoa, TOURIGA N, MARSELAN and Castets; (w) ALVARINHO, PETIT MANSENG and Liliorila. Global warming is reason for test. Institut National de l'Origine et de la Qualité (INAO) still has to give green light, but be prepared for innovative twist to your usual MERLOT/CABS SAUV/FR mix.

Smith Haut Lafitte Pe-Lé r (p) (w) ★★★★ (r) 00' 01 02 03 04 05' 06 07 08 09' 10' 11 12 13 14 15' 16' 17 18 (w) 16 17 18 Celebrated Classed Growth with spa hotel (Caudalie), regularly one of PE-LÉ stars. White is full, ripe, sappy; red precise/ generous. Second label: Les Hauts de Smith. Also CAB SAUV-based Le Petit Haut Lafitte. Organic certification. New spa hotel in Loire (2020).

Sociando-Mallet H-Méd r ★★★ 00' 01' 02 03 04 05 06' 07 08 09' 10' 11 12 14 15' 16' 18 Large H-MÉD estate in St-Seurin-de-Cadourne; RIP founder Jean Gautreau; 50th anniversary ownership 2018. 54% MERLOT, 46% CABS SAUV/FR. Classed-Growth quality. Wines for ageing. Second label: La Demoiselle de Sociando-Mallet.

Sours, De B'x r p w ★★ Valid reputation for popular B'x rosé (DYA). Gd white; acceptable B'x red. Owned by Jack Ma of Alibaba fame.

Soutard St-Ém r ★★★ 00' 01' 05 06 07 08 09 10 11 12 14 15 16' 17 18 *Potentially excellent* GRAND CRU CLASSÉ on limestone plateau. Massive investment, making strides but still room for improvement. MERLOT-led, CAB FR/SAUV and 2% MALBEC. Visitor-friendly. Second label: Petit Soutard.

Suduiraut Saut w sw ★★★★ 01' 02 03' 04 05' 06 07' 09' 10' 11' 13 14 15' 16' 17' 18 One of v. best SAUT. SÉM-dominant (90%+); Preignac-based. Greater consistency, luscious quality. Second labels: Castelnau de Suduiraut; Lions de Suduiraut (fruitier). Dry wines "S" and entry-level Le Blanc Sec.

Taillefer Pom r ★★ 03 04 05' 06 08 09' 10 11 12 14 15 16 18 Owned by branch of MOUEIX family. MERLOT-led (75%). Sandier soils. Lighter weight, but polished and refined.

Talbot St-Jul r (w) ★★★ 00' 01 02 03 04 05' 08 09' 10' 11 12 14 15 16 17 18 Substantial Fourth Growth in heart of AC ST-JUL. Wine rich, *consummately charming, reliable*. 100 yrs of Cordier family ownership in 2018 (special bottle). Second label: Connétable de Talbot. Approachable SAUV BL-based: Caillou Blanc.

Tertre, Du Marg r ★★★ 03 04 05' 06 08 09' 10' 11 12 14 15' 16' 17 18' Fifth Growth isolated s of MARG, sister-CH to GISCOURS. Fragrant (20% CAB FR) fresh, fruity but structured wines. Visitor-friendly. Second label: Les Hauts du Tertre. Also Tertre Blanc VDF dry white CHARD/VIOGNIER/GROS MANSENG/SAUV BL.

Tertre-Rôteboeuf St-Ém r ★★★★ 00' 01 02 03' 04 05' 06' 07 08 09' 10' 11 12 13 14' 15' 16' 17 18 Tiny, unclassified, family-owned, ST-ÉM côtes star; concentrated, exotic, MERLOT-based. V. consistent; can age. Frightening prices. V.gd ROC DE CAMBES.

Thieuley B'x r p w ★★ E-2-M supplier of consistent quality AC B'x (r w); oak-aged CUVÉE Francis Courselle (r w). Run by sisters Marie and Sylvie Courselle.

Tour-Blanche, La Saut (r) w sw ★★★ 01' 02 03' 04 05' 06 07' 08 09' 10' 11' 13 14 15 16 17' 18 Excellent First Growth SAUT; rich, bold, powerful wines on sweeter end of scale. SÉM-dominant (83%), some MUSCADELLE (7%) Environmental certification. Second label: Les Charmilles de La Tour-Blanche.

Tour-Carnet, La H-Méd r ★★★ 01 02 03 04 05' 06 08 09' 10' 11 12 14 15 16' 17 18 Substantial H-MÉD Classed Growth owned by Bernard Magrez (*see* FOMBRAUGE, PAPE-CLÉMENT). Concentrated, opulent wines. B&B as well. Second label: Les Pensées de La Tour Carnet. Also dry white B'x Blanc de La Tour Carnet.

China is now B'x's top export market in volume and value.

Tour-de-By, La Méd r ★★ 03 04 05' 06 08 09 10 11 12 14 15' 16 17 18 Extensive family-run estate in n MÉD. Popular, sturdy, reliable, CAB SAUV-led (60%) wines with 5% PETIT VERDOT. Also rosé and special CUVÉE Héritage Marc Pagès.

Tour de Mons, La Marg r ★★ 04 05' 06 08 09' 10 11 12 14 15' 16' 18 MARG CRU BOURGEOIS. Same stable (since 2019) as Marquis d'Alesme. MERLOT-led (55%). Steady improvement.

Tour-du-Haut-Moulin H-Méd r ★★ 03 04 05' 06 08 09 10 14 15' 16' 18 Family-owned property (since 1870) in n H-MÉD. Intense, structured wines to age.

Tour-du-Pas-St-Georges St-Ém r ★★ 04 05 06 08 09' 10 11 12 14 15 16 18 ST-GEORGES-ST-ÉM estate run by Delbeck family. Six permitted red varieties planted. Classic.

Tour Figeac, La St-Ém r ★★ 02 04 05' 06 07 08 09' 10' 11 12 14 15' 16' 18 GRAND CRU CLASSÉ in "graves" sector of ST-ÉM. Gd proportion of CAB FR (30%). Fine, floral.

Tour Haut-Caussan Méd r ★★ 06 08 09' 10' 11 12 14 15 16' 17 18 Consistent, Courrian family-run CRU BOURGEOIS. 50/50 CAB SAUV/MERLOT. Value.

Tour-St-Bonnet Méd r ★★ 06 08 09' 10' 11 12 14 15 16 18 MÉD cru at St-Christoly. MERLOT, CAB SAUV/FR, PETIT VERDOT, MALBEC. Reliable. Value.

7867 ha in B'x certified organic in 2018; plus 2950 in official conversion.

Tournefeuille L de P r ★★ 03 04 05' 06 07 08 09 10' 11 12 14 15 16 18 Reliable L DE P on clay and gravel soils. 70% MERLOT, 30% CAB FR. Round, fleshy wine.

Trois Croix, Les Fron r ★★ 07 08 09 10 11 12 13 14 15 16 17 18 Owned by Léon family (1995). Fine, balanced wines from consistent producer. Clay-limestone soils. 80% MERLOT. Gd value.

Tronquoy-Lalande St-Est r ★★★ 05 06 07 08 09' 10' 11 12 14 15' 16' 17 18 Same stable as MONTROSE; v'yd in one block. MERLOT-led wines; consistent, dark, satisfying. Organic leaning. Second label: Tronquoy de Ste-Anne. A little B'x white.

Troplong-Mondot St-Ém r ★★★ 00' 01' 02 03 04 05' 06 07 08 09 10 11 12 14 15' 16 17 18' PREMIER GRAND CRU CLASSÉ (B) on limestone plateau. Lots of investment since 2017; new winery in 2020. *Wines of power, depth* with increasing elegance (and price) these days (earlier picking, less new oak). Second label: Mondot.

Trotanoy Pom r ★★★★ 00' 01' 02 03 04' 05' 06 07 08 09' 10' 11 12 13 14 15' 16' 17 18' One of jewels in J-P MOUEIX crown. 90% MERLOT, 10% CAB FR on clay and gravel soils. Power, elegance, long ageing.

Trottevieille St-Ém r ★★★ 00' 01 03' 04 05' 06 07 08' 09' 10' 11 12 14 15 16' 17 18 PREMIER GRAND CRU CLASSÉ (B) on limestone plateau. Much improved since 2010; wines long, fresh, structured. Lots of CAB FR (50%+) incl some pre-phylloxera vines. Second label: La Vieille Dame de Trottevieille.

Valandraud St-Ém r ★★★★ 00' 01' 02 03 04 05' 06 07 08 09' 10 11 12 13 14 15 16' 17 18 PREMIER GRAND CRU CLASSÉ (B) in e ST-ÉM. Garage wonder turned First Growth. Formerly super-concentrated; now rich, dense but balanced. Virginie de Valandraud from unclassified land, Valandraud Blanc (SAUV BL/Sauvignon Gris).

Vieille Cure, La Fron r ★★ 05' 06 08 09' 10' 11 12 13 14 15 16 17 18 Leading FRON estate; appetizing wines. Value. Second label: La Sacristie de La Vieille Cure.

Vieux-Ch-Certan Pom r ★★★★ 00' 01' 02 04 05' 06 07 08 09' 10' 11' 12 13 14 15' 16' 17' 18' Different in style to neighbour PETRUS; *elegance, harmony, fragrance.* Plenty of old-vine CAB FR/SAUV (30%) one of reasons. Alexandre Thienpont and son Guillaume at helm. Second label: La Gravette de Certan.

Vieux Ch St-André St-Ém r ★★ 06 08 09' 10 11 12 14 15' 16' 18 Small MERLOT-based v'yd in MONTAGNE-ST-ÉM. Owned by former PETRUS winemaker and son. *Gd value.*

Villegeorge, De H-Méd r ★★ 05 06 08 09' 10 12 14 15 16 18 Tiny s H-MÉD owned by Marie-Laure Lurton. CAB SAUV-led (63%). Light but elegant wines.

Vray Croix de Gay Pom r ★★★ 04 05' 06 08 09' 10 11 12 14 15 16' 17 18' Tiny v'yd in best part of POM. More finesse of late. Environmental certification. LATOUR owner a shareholder. CHX Siaurac (L DE P), Le Prieuré (ST-ÉM) same stable.

Yquem Saut w sw (dr) ★★★★ 00 01' 02 03' 04 05' 06' 07 08 09' 10' 11' 13' 14 15' 16' 17' King of sweet, *liquoreux*, wines. Strong, intense, luscious; kept 3 yrs in barrel. Most vintages improve for 15 yrs+, some live 100 yrs+ in transcendent splendour. 100 ha in production (75% SÉM/25% SAUV BL). No Yquem made 51, 52, 64, 72, 74, 92, 2012. No second label (rejected wine sold to NÉGOCIANTS). Makes small amount (800 cases/yr) of off-dry, Sauv Bl (75%), Sém (25%) "Y" (pronounced "ygrec").

Italy

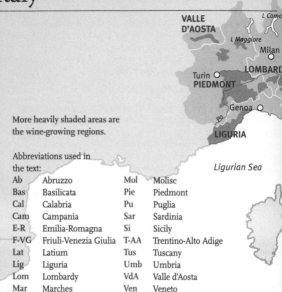

More heavily shaded areas are
the wine-growing regions.

Abbreviations used in
the text:

Ab	Abruzzo	Mol	Molise
Bas	Basilicata	Pie	Piedmont
Cal	Calabria	Pu	Puglia
Cam	Campania	Sar	Sardinia
E-R	Emilia-Romagna	Si	Sicily
F-VG	Friuli-Venezia Giulia	T-AA	Trentino-Alto Adige
Lat	Latium	Tus	Tuscany
Lig	Liguria	Umb	Umbria
Lom	Lombardy	VdA	Valle d'Aosta
Mar	Marches	Ven	Veneto

Start by asking yourself: "Is Italy really just one country?" Obviously yes; look at the map: the boot from the Alps almost to Africa could hardly have a clearer outline. Obviously yes, when you hear the accent and see the body language. And it all started so long ago; Etruscans, then colonizing Greeks, then centuries of Roman Empire. But then what happened? Geographical Italy was divided into a score of city-states, the Pope at its heart. Spain took chunks. France took chunks. Austria took more chunks. Centuries passed until, a mere 170 years ago, Italy the country was born. Was it likely to be uniform, homogeneous, organized? Precisely.

Then there's the topography; mountains down the middle, mountains at the top, volcanoes, the sea all round. Every sort of rock, soil, slope and exposition. Every sort of weather – in short, terroir. Add their inventive genius, their sense of style and the depth of disdain Italians feel for any products but their own ("Il vero"), and you can see that herding cats is an easier option. Once we thought of Italy as being limited to pretty variable red wine, with white limited to the northeast. Now many of the best come from as far south as you can go.

And then there are the grapes. One third of all the world's wine grape varieties are Italian. They are often blended. Need I say more? Then is there any unifying Italian style or character? Mysteriously, yes. It is its aptitude for, its real need for, food. Italian wines, almost all of them, have acidity, or astringency, or both; enough to make them appetizing. They were created to be the other vital half of the great Italian feast.

Recent vintages

Amarone, Veneto & Friuli

2019 Low quantity. Delayed harvest. Veneto: gd for Amarone, Soave. Friuli: fair quality.

2018 Optimal weather conditions. Gd for quantity and quality, esp fresh whites.

2017 V. difficult (non-stop rain); weedy Amarones a risk, green reds in general.

2016 V. hot summer; round but low-acid, big reds, chunky whites.

2015 Quantity gd, quality better; v.gd Friuli reds. Fresher than initially thought.

2014 Not memorable for Amarone, better Soave, Friuli whites, Valpolicellas v.gd.

2013 Gd whites. Reds, esp passito, suffered Oct/Nov rain/hail.

2012 Prolonged heat, drought made showy but tough tannic reds and blowsy whites.

Campania & Basilicata

2019 Classic, balanced vintage. Best for Aglianico, gd for whites.

2018 Rainy, but whites fresh, lively; sleek reds (Aglianico best).

2017 Low-volume yr of reds plagued by gritty tannins. Whites often flat; Greco best.

2016 Cold spring delayed flowering, hot summer allowed catch-up. Best for Fiano.

2015 Hot, dry early summer means ripe but at times tough reds, broad whites; drink up.

2014 Lots of rain; spotty quality. Avoid green Aglianicos, Piedirossos. Whites pretty gd.

2013 Classic perfumed fresh whites, thinner late-picked Aglianicos.

2012 Too hot; Sept rains saved whites. Indian summer made memorable Aglianicos.

Marches & Abruzzo

2019 Difficult yr. Rainy, cold spring. Low quantity, medium quality.

2018 Patchy spring. More balanced than 17. V. high volume, gd quality (r w).

2017 Best to forget: hot, droughty. Reds gritty, whites overripe. Low volume.

2016 Rain, cold, lack of sun = difficult yr. Lemony, figgy Pecorino probably best.

2015 Hot summer, whites fresher than expected (esp Trebbiano), reds ripe, not cooked.

2014 Marches: sleek reds, classic whites. Abruzzo: best for Pecorino.

2013 Whites classically mineral and age-worthy; reds refined, classic, not lean.

2012 Too hot: tough reds, blowsy whites.

Piedmont

2019 Classic vintage, more balanced than 18. Lower quantity but higher quality.

2018 Despite difficult spring, potentially classic Barolo/Barbaresco.

2017 Among earliest harvests in living memory. Can lack depth.

2016 Potentially top vintage; classic, perfumed, age-worthy Barolo/Barbaresco.

2015 Outstanding Barolo/Barbaresco. Should be long-lived. Barbera/Dolcetto gd; Grignolino less so.

2014 Rain: later-picked grapes thrived, early ones (eg. Dolcetto) didn't. Barbaresco (not Barolo) best.

2013 Bright, crisp (r w), improving with time; classic vintage, steely, deep.

2012 Overrated Barolo/Barbaresco, unfailingly green, gritty tannins. Beware hype.

Earlier fine vintages: 10 08 06 04 01 00 99 98 97 96 95 90 89 88.

Vintages to keep: 01 99 96. Vintages to drink up: 09 03 00 97 90 88.

Tuscany

2019 Maybe one of best since 2000. Classic, balanced. Gd quality, quantity.

2018 Gd reds of steely personality, age-worthiness.

2017 V. difficult hot vintage. Better in Chianti Classico than coastal Maremma.

2016 Hot summer, fresh autumn; success from Chianti to Montalcino to Maremma. Small crop.

2015 Rich flavourful reds of gd ripeness; some soft.

2014 Quantity up, quality patchy, buyer beware.

2013 Uneven ripening. Not great yr, but some peaks.

2012 Drought and protracted heat; overrated wines, mostly.

Earlier fine vintages: 08 07 06 04 01 99 97 95 90. Vintages to keep: 01 99.

Vintages to drink up: 03 00 97 95 90.

Abrigo, Orlando Pie ★★★ Some of most mineral, steely, refined BARBARESCOS. Little-known crus such as Montersino, Meruzzano diamonds in rough.

Accornero Pie ★★★ Some of Italy's best medium-bodied reds. GRIGNOLINO sings in Bricco del Bosco (steel-aged) and Bricco del Bosco Vigne Vecchie (oaked 2 yrs). Also v.gd delicately sweet MALVASIA di Casorzo Brigantino.

Adriano, Marco e Vittorio Pie ★★★ High quality, low prices, BARBARESCO full of early appeal. One of great buys in Italian wine.

Aglianico del Taburno Cam DOCG r dr ★ ★★★ Around Benevento. Spicier notes (leather, tobacco) and herbs, higher acidity than other AGLIANICOS. Gd: CANTINA del Taburno, Fontanavecchia, La Rivolta.

Aglianico del Vulture Bas DOC(G) r dr ★→★★★ 11 12 13 15 16 (18) DOC after 1 yr, SUPERIORE after 3 yrs, RISERVA after 5. From slopes of extinct volcano Monte Vulture. More floral (violet), dark fruits (plum), smoke, spice than other AGLIANICOS. V.gd: ELENA FUCCI, GRIFALCO. Also gd: Armando Martino, Basilisco, CANTINA di Venosa, CANTINE del NOTAIO, D'Angelo, Eubea, Madonna delle Grazie, Mastrodomenico, PATERNOSTER, Re Manfredi, Terre dei Re.

Alba Pie Truffles, hazelnuts; PIE's, if not Italy's, most prestigious wines: BARBARESCO, BARBERA D'ALBA, BAROLO, DOGLIANI (DOLCETTO), LANGHE, NEBBIOLO D'ALBA, ROERO.

Albana di Romagna E-R DOCG w dr sw s/sw (sp) ★→★★★ DYA. Italy's 1st white DOCG, justified only by sweet PASSITO; dry and sparkling often unremarkable. Best: *Fattori Zerbina Passito*, Giovanna Madonia, PODERE Morini (Cuore Matto RISERVA Passito), Tre Monti.

Allegrini Ven ★★ Popular VALPOLICELLA producer. Elegant AMARONE is best. Owner of POGGIO al Tesoro in BOLGHERI, Poggio San Polo in MONTALCINO, TUS.

Almondo, Giovanni Pie ★★★→★★★★ Top ROERO estate. Best is Roero ARNEIS Bricco delle Ciliegie; outstanding Freisa too; V.gd Roero Bric Valdiana (r).

Alta Langa Pie DOCG (p) w sp ★★→★★★ 1st METODO CLASSICO made in Italy, since mid-C19 in "underground cathedrals". Vintage only, and simply PINOT N, CHARD. Best: Cocchi, Contratto, Enrico Serafino, ETTORE GERMANO, Ferdinando Principiano, FONTANAFREDDA, GANCIA, RIZZI.

Altare, Elio Pie ★★★ Now run by Silvia, Elio's daughter. Try BAROLOS: Arborina, Cannubi, Cerretta VIGNA Bricco, Unoperuno (selection from Arborina); also Giarborina (NEBBIOLO), Larigi (BARBERA D'ALBA), La Villa (BARBERA/Nebbiolo)

Alto Adige (Südtirol) T-AA DOC r p w dr sw sp Mountainous region with Bolzano its chief city (Austrian until 1919); arguably best Italian whites today but also underrated reds. Germanic vines dominate. GEWÜRZ, KERNER, SYLVANER, but PINOT GRIGIO too. RIES and GRÜNER V hopelessly overrated; probably world's best PINOT BIANCO. PINOT n often overoaked; *Lagrein* in gd yrs.

Alto Piemonte Pie Cradle of PIE quality in C19 (40,000 ha v'yds). Acidic soil, exposure, climate and altitude diversity, ideal for many different NEBBIOLO expressions (here called Spanna). Main DOC(G): BOCA, BRAMATERRA, Colline Novaresi, Coste della Sesia, FARA, GATTINARA, GHEMME, LESSONA, Sizzano, Valli Ossolane. Many outstanding wines. Actually, rarely 100% Nebbiolo: small additions of Croatina, Uva Rara, Vespolina common.

Ama, Castello di Tus ★★★★ If CHIANTI CLASSICO today is well reputed, much credit must go to this top Gaiole estate, among 1st to produce single-v'yd Chianti Classico and push for quality in 80s. Gran Selezione VIGNETO Bellavista is best; La Casuccia close second. V.gd San Lorenzo. MERLOT L'Apparita one of Italy's three best.

Amarone della Valpolicella Ven DOCG r ★★→★★★★ 10 11' 13 15 (16) (17) CLASSICO area (from historic zone), Val d'Illasi and Valpantena (from extended zone) can make unique world-class reds from raisined grapes. Alas, many less than what they should be, despite hype. Choose carefully. (*See also* VALPOLICELLA; box p.150.)

> **Best of Brunello**
>
> Any of the below provide a satisfying BRUNELLO DI MONTALCINO, but we have put a star next to the ones we think are best: Altesino ★, Baricci ★, BIONDI-SANTI ★, Campogiovanni, Canalicchio di Sopra ★, Canalicchio di Sotto, Caparzo, Casanova di Neri ★, CASE BASSE ★, Castelgiocondo, CASTELLO DI Argiano ★, CASTIGLION DEL BOSCO, Ciacci Piccolomini, COL D'ORCIA ★, Collemattoni ★, Colombini, Costanti ★, Cupano ★, Donatella Cinelli, Eredi, Fossacolle, Franco Pacenti ★, FULIGNI ★, GIANNI BRUNELLI ★, Giodo, Il Colle, Il Marroneto ★, Il Paradiso di Manfredi, La Gerla, La Magia, La Poderina, Le Potazzine ★, Le Ragnaie ★, Le Ripi, LISINI ★, Mastrojanni ★, Piancornello ★, PIAN DELL'ORINO ★, Pieri Agostina, Pieve di Santa Restituta, POGGIO ANTICO, POGGIO DI SOTTO★, San Filippo, Salvioni ★, Sesta di Sopra ★, Siro Pacenti, Stella di Campalto ★, TENUTA IL POGGIONE ★, TENUTA di Sesta, Uccelliera, Val di Suga.

Angelini, Tenimenti Tus *See* BERTANI DOMAINS (TOSCANA).

Antinori, Marchesi L&P Tus ★★→★★★★ Since 1385, through 26 generations. This Florentine family were protagonists of C20 renaissance of Italian wine. Top CHIANTI CLASSICO (TENUTE Marchese Antinori and *Badia a Passignano*, latter now elevated to Gran Selezione), Cervaro (Umb *Castello della Sala*), the two excellent SUPER TUSCANS (TIGNANELLO and SOLAIA) and PRUNOTTO BAROLO Bussia (PIE). Also MONTALCINO (Pian delle Vigne), MONTEPULCIANO (La Braccesca), PUG (Tormaresca), TUS MAREMMA (Fattoria Aldobrandesca). RIES-based white from Monteloro n of Florence. Interests in Calfornia, Romania, etc.

Antoniolo Pie ★★★ Age-worthy benchmark GATTINARA. Outstanding Osso San Grato and San Francesco.

Argiano, Castello di Tus ★★★ Distinctive, refined yet flavourful BRUNELLOS from Sesti family estate. Classico Brunello and RISERVA Phenomena equally gd, if different.

Argiolas Sar ★★→★★★ Top producer, native island grapes. Outstanding crus *Turriga* (★★★), Iselis MONICA, Iselis Nasco, *Vermentino* di Sardegna and top sweet Angialis (mainly local Nasco grape). V.gd CANNONAU RISERVA Senes.

Asti Pie DOCG sw sp ★→★★ NV Sparkler from MOSCATO Bianco grapes, inferior to MOSCATO D'ASTI, not really worth its DOCG. Try Bera, Cascina Fonda, Caudrina, Vignaioli di Santo Stefano. Now dry version, Asti Secco.

Avignonesi Tus ★★★ Large bio estate, Belgian-owned since 2007. *Italy's best Vin Santo*. Top is VINO NOBILE single-v'yd Poggetto di Sopra and Grandi Annate, but MERLOT Desiderio, 50&50 (Merlot/SANGIOVESE), CHARD Il Marzocco creditable internationals.

Azienda agricola / agraria Estate (large or small) making wine from its own grapes.

Badia a Coltibuono Tus ★★★ One of top 5–10 CHIANTI CLASSICO producers; every wine worth buying: great terroirs spell non-stop success. Organic certified.

Banfi (Castello or Villa) Tus ★→★★★ Giant of MONTALCINO, at extreme s of zone, but top, limited-production POGGIO all'Oro is a great BRUNELLO. V.gd Moscadello.

Barbaresco Pie DOCG r ★★→★★★★ 10 11 12 13 14 15 16 (18) Often better than BAROLO, Barbaresco's lesser reputation is undeserved. When spot on, the gracefulness, age-worthiness and perfumed intensity is like that of no other wine in Italy – the world, really. Min 26 mths ageing, 9 mths in wood; at 4 yrs becomes RISERVA. (For top crus and producers *see* box, p.130.)

Barbera d'Alba DOC r Unique, luscious, sultry BARBERA, quite different from D'ASTI's more nervy, higher-acid version. GIACOMO CONTERNO (Cascina Francia and Ceretta), CONTERNO FANTINO (Vignota), VIETTI (Scarrone) make benchmarks. Cogno's Pre-Phylloxera is hors classe.

Barbera d'Asti DOCG r Huge denomination, encompassing v. different soil and climatic characteristics. Wines vary, but characterized by high acidity, fruity notes. SUPERIORE: higher quality (often overoaked). Nizza (since 2014; previously, Superiore Nizza) is best: 100% Barbera-only from best sites, needs time. Gd: BRAIDA (Ai Suma, Bricco dell'Uccellone, Bricco della Bigotta), Cascina Castlet, Dacapo, Scarpa (La Bogliona), TENUTA Olim Bauda, *Vietti* (La Crena), Vigne dei Maestri.

Barbera del Monferrato Superiore Pie DOCG r From soils rich in limestone, full-bodied BARBERA with sharpish tannins, gd acidity. Top: Accornero (Bricco Battista and Cima), Iuli (Barabba).

Barberani ★★→★★★ Organic estate on slopes of Lago di Corbara, gd to excellent ORVIETO. Cru Luigi e Giovanna is star; Polvero, Calcaia (noble rot) also excellent.

Bardolino Ven DOC(G) r p DYA Light-bodied fresh red made with VALPOLICELLA's typical grapes; hit with tourists on Lake Garda. Ill-advised plantings of CAB SAUV in order to make bigger wines a tragic idea. Best today are the rosés, called CHIARETTO in this neck of the Italian woods. Look for Cavalchina, *Guerrieri Rizzardi*, Zeni.

Barolo Pie DOCG r 06 07 08 09 10' 11 12 13 15 16' (17) (18) "King of wines and wine of kings" 100% NEBBIOLO. 2000-ha zone. 11 communes incl Barolo itself and 181 MGA (including 11 municipal), the PIE way of Burgundian crus. Traditionally blend of v'yds from different communes, but now most is single-v'yd (*see box, p.142*). Must age 38 mths before release (5 yrs for RISERVA), of which 18 mths in wood. Best are age-worthy, able to join power and elegance, with alluring floral scent and sour red-cherry flavour. (For top crus and producers *see box, below*.)

Bastianich ★★→★★★ FRIULANO Plus one of Italy's 30 best whites; Vespa Bianco not far behind. Reds (Calabrone, Vespa Rosso) can be weedy and tough in poor yrs (air drying unripe grapes is never a gd idea), but in gd vintages can be memorable.

Belisario Mar ★★→★★★ Quality co-op, direct management of v'yds, with gd quality/price. Largest producer of VERDICCHIO DI MATELICA. Top: RISERVA Cambrugiano e Del Cerro. V.gd Meridia, Vigneti B (organic) and SPUMANTE.

Top Barolos

Here are a few top crus and their best producers: **Bricco Boschis** (Castiglione Falletto) CAVALLOTTO (RISERVA VIGNA San Giuseppe); **Bricco delle Viole** (BAROLO) GD VAJRA; **Bricco Rocche** (Castiglione Falletto) CERETTO; **Briccolina** (Serralunga) Enrico Rivetto; **Brunate** (La Morra, Barolo) Ceretto, GIUSEPPE RINALDI, ODDERO, VIETTI; **Bussia** (Monforte) ALDO CONTERNO (Gran Bussia e Romirasco), FENOCCHIO GIACOMO (Riserva 90 Dì), ODDERO (Bussia Vigna Mondoca), PODERI COLLA (Dardi Le Rose); **Cannubi** (Barolo) BREZZA, FENOCCHIO GIACOMO, LUCIANO SANDRONE (Cannubi Boschis), E Pira e Figli – Chiara Boschis; **Cerequio** (La Morra, Barolo) Boroli, ROBERTO VOERZIO; **Falletto** (Serralunga) BRUNO GIACOSA (Riserva Vigna Le Rocche); **Francia** (Serralunga) GIACOMO CONTERNO (Barolo Cascina Francia and Monfortino); **Ginestra** (Monforte) CONTERNO FANTINO (Sorì Ginestra and Vigna del Gris), DOMENICO Clerico (Ciabot Mentin); **Lazzarito** (Serralunga) ETTORE Germano (Riserva), Vietti; **Monprivato** (Castiglione Falletto) GIUSEPPE MASCARELLO (Mauro); **Mongliero** (VERDUNO) CASTELLO DI VERDUNO, COMM. GB BURLOTTO, PAOLO SCAVINO; **Ornato** (Serralunga) PIO CESARE; **Ravera** (Novello) ELVIO COGNO (Bricco Pernice), GD Vajra, Vietti; **Rocche dell'Annunziata** (La Morra) Paolo Scavino (Riserva), Roberto Voerzio, TREDIBERRI; **Rocche di Castiglione** (Castiglione Falletto) BROVIA, Oddero, Vietti; **Vigna Rionda** (Serralunga) Oddero, VR MASSOLINO; **Villero** (Castiglione Falletto) Boroli, Brovia, Fenocchio Giacomo. And the Barolo of Bartolo MASCARELLO blends together Cannubi San Lorenzo, Ruè and Rocche dell'Annunziata.

Bellavista Lom r w sp ★★★ Owned by Francesca Moretti of FRANCIACORTA. Alma Gran Cuvée is flagship. Top: Vittorio Moretti.

Benanti Si r w ★★★ Benanti family turned world on to ETNA. Bianco SUPERIORE *Pietramarina* one of Italy's best whites. Top: Etna red Rovittello and Serra della Contessa. V.gd: mono-variety Nerello Cappuccio.

Berlucchi, Guido Lom sp ★★ Makes millions of bottles of METODO CLASSICO fizz. FRANCIACORTA Brut Cuvée Imperiale is flagship. V.gd: Nature Dosaggio Zero 61.

Bersano Pie ★★→★★★ Large volume but gd quality. V.gd: BARBERA D'ASTI, GRIGNOLINO and Ruchè, all inexpensive, delightful.

Bertani Ven ★★→★★★ Long-est VALPOLICELLA, SOAVE. *See also* BERTANI DOMAINS.

Bertani Domains (Toscana) Tus ★★→★★★ Previously Tenimenti Angelini. Three major wineries: San Leonino, CHIANTI CLASSICO; Trerose, MONTEPULCIANO; Val di Suga, MONTALCINO (esp BRUNELLO Spuntali).

Biondi-Santi Tus ★★★★ Invented BRUNELLO, now sold to Epi group (Piper-Heidsieck): Brunellos and esp RISERVAS high in acid, tannin requiring decades to develop fully. Recently trying more user-friendly style.

Bisol Ven ★★★ Owned by FERRARI'S Lunelli family; quality leader in PROSECCO. CARTIZZE best bubbly, but Crede Brut (with atypical small % PINOT BIANCO) one of Italy's best buys.

Boca Pie DOC r *See* ALTO PIEMONTE. Potentially among greatest reds. NEBBIOLO (70–90%), incl up to 30% Uva Rara and/or Vespolina. Volcanic soil. Needs long ageing. Best: *Le Piane*. Gd: Carlone Davide, Castello Conti.

Priciest Barolo v'yds to buy: Bricco Rocche, Brunate, Bussia Cannubi. €4m/ha.

Bolgheri Tus DOC r p w (sw) Mid-Maremma, on w coast, cradle of many expensive SUPER TUSCANS, mainly French variety-based. Big name and excellent quality: ALLEGRINI (POGGIO al Tesoro), ANTINORI (Guado al Tasso), GAJA (CÀ MARCANDA), Grattamacco. Le Macchiole, SAN GUIDO (SASSICAIA, the original Super Tuscan) and ORNELLAIA (FRESCOBALDI) are best (incl iconic 100% MERLOT Masseto.)

Bolla Ven ★★ Historic Verona firm (owned by GIV) for AMARONE, RECIOTO DELLA VALPOLICELLA, RECIOTO DI SOAVE, SOAVE, VALPOLICELLA.

Borgo del Tiglio F-VG ★★★→★★★★ Nicola Manferrari is one of Italy's top white winemakers. COLLIO FRIULANO RONCO della Chiesa, MALVASIA Selezione, Rosso della Centa, Studio di Bianco esp impressive.

Boscarelli, Poderi Tus ★★★ Small estate of Genovese de Ferrari family with reliably high standard VINO NOBILE DI MONTEPULCIANO, cru Nocio dei Boscarelli, RISERVA Sotto Casa. New single-v'yd Costa Grande (100% SANGIOVESE).

Botte Big barrel, anything from 6–250 hl, usually between 20–50, traditionally of Slavonian but increasingly of French oak. To traditionalists, ideal vessel for ageing wines without adding too much oak smell/taste.

Brachetto d'Acqui / Acqui Pie DOCG r sw (sp) DYA. Pink version of ASTI made with Brachetto grape; similarly undeserving for most part of its DOCG status.

Braida Pie ★★★ If BARBERA D'ASTI is known today, it's thanks to the Bologna family, world ambassadors for this wine. Top: Bricco dell'Uccellone, Bricco della Bigotta and Ai Suma. V.gd Montebruna. GRIGNOLINO d'Asti Limonte one of Italy's five best.

Bramaterra Pie DOC r *See* ALTO PIEMONTE. Gd: Antoniotti Odilio.

Brezza Pie ★★→★★★ Organic certified. Certainty for those who love traditional BAROLOS. Also great value, from famous crus incl Cannubi, Castellero, Sarmassa (esp RISERVA VIGNA Bricco). Range incl BARBERA, DOLCETTO, Freisa (outstanding), plus surprisingly gd CHARD.

Brigaldara Ven ★★★ Elegant but powerful benchmark AMARONE from estate of Stefano Cesari. Top: Case Vecie. Try charming Dindarella Rosato, when made.

Brolio, Castello di Tus ★★→★★★ Since 1141 run by RICASOLI family; historic estate is largest and oldest of CHIANTI CLASSICO. Now with Francesco R, SANGIOVESE has chance to shine again with outstanding Gran Selezione Castello di Brolio and Colledilà. (*See* Chianti Classico box, p.131).

Brovia ★★★→★★★★ Since 1863, classic BAROLOS in Castiglione Falletto. Organic certified. Top: Rocche, Villero and Garblét Sue.

Brunelli, Gianni Tus ★★★ Lovely refined user-friendly BRUNELLOS (top RISERVA) and Rossos from two sites: Le Chiuse di Sotto n of MONTALCINO and Podernovone to s, with views of Monte Amiata.

Brunello di Montalcino Tus DOCG r 07 09 10' 12 13 15' (16) (17) World-famous, but quality all over the shop. When gd, memorable and ageless, archetypal SANGIOVESE. Problems derive mostly from greedily enlarged production area (a ridiculous 2000 ha+), much less than ideal for fickle Sangiovese and world-class wines. Recent push to turn wine into a blend has been successfully stopped (thus far, at least). (For top producers *see* box, p.126).

Bucci Mar ★★★→★★★★ Villa Bucci RISERVA one of Italy's 20 best whites. All wines quasi-Burgundian, especially complex VERDICCHIOS, slow to mature but all age splendidly. V.gd red Pongelli.

Burlotto, Commendatore GB Pie ★★★→★★★★ Fabio Alessandria maintains his ancestor Commander GB Burlotto's (among 1st to make/bottle BAROLO in 1880) high quality and focus. Barolo Monvigliero and superlative Freisa best. Also outstanding: Barolo Cannubi, VERDUNO Pelaverga.

Bussola, Tommaso Ven ★★★★ Self-taught maker of some of the great AMARONES, RECIOTOS, RIPASSOS of our time. The great Bepi QUINTARELLI steered him; he steers his two sons. Top TB selection.

Ca' del Baio Pie ★★★ Family estate; best-value producer in BARBARESCO. Outstanding Asili (RISERVA too) and Pora. V.gd Vallegrande, Autinbej and LANGHE RIES.

Ca' del Bosco Lom ★★★★ Arguably Italy's best METODO CLASSICO sparkling, famous FRANCIACORTA estate owned by giant PINOT GR producer Santa Margherita, still run by ultra-knowledgeable, exuberant founder Maurizio Zanella. Annamaria Clementi (rosé too) both Krug-like and unforgettable. Great Dosage Zero and Dosage Zero Noir Vintage Collection practically as gd; also excellent B'x-style Maurizio Zanella (r), PINOT N, CHARD.

Ca' dei Frati Lom ★★★ Foremost quality estate of revitalized DOC LUGANA, I Frati a fine example at entry level; *Brollettino* a superior cru.

Ca' Marcanda Tus ★★★★ BOLGHERI estate of GAJA. In order of price (high, higher, highest): Promis, Magari, Ca' Marcanda. Grapes mainly international.

Caiarossa Tus ★★★ Dutch-owned (Ch Giscours; *see* B'x) estate, n of BOLGHERI. Excellent Caiarossa Rosso plus reds Aria, Pergolaia.

Calcagno Si r w ★★★→★★★★ Lilliputian size and Brobdingnagian quality from ETNA family estate. Outstanding mineral NERELLO MASCALESE reds. Top (r) Feudo di Mezzo and Arcurìa. V.gd Ginestra (w), Romice delle Sciare (p), Nireddu (r).

Calì, Paolo Si r w ★★★ Passionate Paolo Calì makes numerous wines highly typical of Vittoria's terroir. Top: CERASUOLO DI VITTORIA Forfice and Frappato. V.gd Manene (r), GRILLO Blues (w) and Mood (p sp).

Caluso / Erbaluce di Caluso Pie DOCG w ★→★★★ Can be still, sparkling (dry) and sweet (Caluso PASSITO). Top: Giacometto, Cieck (Misobolo), Favaro (Le Chiusure), Ferrando (Cariola). V.gd: Orsolani, Salvetti (Brut).

Campania Few international varieties cloud native grapes panorama: (w) Biancolella, Caprettone, Coda di Volpe Bianca, FALANGHINAS (more than one), Fenile, FIANO, FORASTERA, GRECO, Grecomusc', Ginestra; (r) AGLIANICO, Piedirosso, Sciascinoso, Tintore di Tramonti are local world-class grapes. FIANO DI AVELLINO may well be Italy's best white zone, TAURASI making better and better reds (but bad oak,

> **Barbaresco Menzioni Geografiche Aggiuntive (MGA)**
> BARBARESCO has four main communes; four distinct styles – **Barbaresco:** most complete, balanced. Growers incl Asili (BRUNO GIACOSA, CA' DEL BAIO, CERETTO, PRODUTTORI DEL BARBARESCO), Martinenga (Marchesi di Gresy), Montefico (Produttori del B, Roagna), Montestefano (Giordano Luigi, Produttori del B, Rivella Serafino), Ovello (CANTINA del Pino, ROCCA ALBINO), Pora (Ca' del Baio, Produttori del B), Rabaja (Bruno Giacosa, BRUNO ROCCA, CASTELLO DI VERDUNO, GIUSEPPE CORTESE, Produttori del B), Rio Sordo (Cascina Bruciata, Cascina delle Rose, Produttori del B), Roncaglie (PODERI COLLA). **Neive:** most powerful, fleshiest. Albesani (Cantina del Pino, Castello di Neive), Basarin (Adriano Marco e Vittorio, Giacosa Fratelli, Negro Angelo, SOTTIMANO), Bordini (La Spinetta), Currà (Bruno Rocca, SOTTIMANO), Gallina (Castello di Neive, Lequio Ugo, ODDERO), Serraboella (Cigliuti). **San Rocco Seno d'Elvio:** readiest to drink, soft. Sanadaive (Adriano Marco e Vittorio). **Treiso:** freshest, most refined. Bernardot (Ceretto), Bricco di Treiso (PIO CESARE), Marcarini (Ca' del Baio), Montersino (ABRIGO ORLANDO, Rocca Albino), Nervo (RIZZI), Pajoré (Rizzi, Sottimano).

and too much of it, plus overripe grapes still a problem). Gd producers: BENITO FERRARA, Caggiano, CANTINE Lonardo, COLLI DI LAPIO, Contrada Salandra, D'AMBRA, De Angelis, Fattoria La Rivolta, *Feudi di San Gregorio*, GALARDI, Guastaferro, La Sibilla, Luigi Maffini, Luigi Tecce, MARISA CUOMO, *Mastroberardino*, Molettieri, Mustilli, Nicola Mazzella, Perillo, Pierlingeri, Pietracupa, QUINTODECIMO, Reale, Rocca del Principe, Sarno 1860, Terredora, Vadiaperti-Traerte. MONTEVETRANO makes world-class international mostly CAB wine.

Canalicchio di Sopra Tus ★★★ Top-ten BRUNELLO estate, with pretty B&B nearby. Owner Francesco Ripaccioli keen observer of terroir and MONTALCINO typicity; RISERVA usually spectacular.

Cantina A cellar, winery or even a wine bar.

Cantina Tramin T-AA ★★★ Quality co-op with benchmark GEWURZ. Outstanding Epokale, Nussbaumer and Terminum. Look for Le Selezioni (single-v'yd) line. V.gd: PINOT GR Unterebner.

Cantine del Notaio Bas ★★→★★★ Organic and bio, specializing in AGLIANICO (p w sp and PASSITO). Star is La Firma, but super-ripe, AMARONE-like Il Sigillo almost as gd.

Capezzana, Tenuta di Tus ★★★ Certified organic production from noble Bonacossi family that made CARMIGNANO's reputation. Now run by founder's children. Excellent Carmignano (Villa di Capezzana and Selezione, Trefiano RISERVA), exceptional VIN SANTO, one of Italy's five best.

Capichera Sar ★★★★ No better VERMENTINO anywhere than that of Ragnedda family, talented enough also to make top reds (Manthenghja CARIGNANO 100%). Outstanding Isola dei Nuraghi Bianco Santigaìni, Vendemmia Tardiva and Vigna'ngena.

Cappellano Pie ★★★ The late Teobaldo Cappellano, a BAROLO hero, devoted part of his cru Gabutti to ungrafted NEBBIOLO (Pie Franco). Son Augusto keeps highly traditional style; also "tonic" Barolo Chinato, invented by an ancestor.

Caprai Umb ★★★→★★★★ MONTEFALCO leader thanks to Marco Caprai and consultant Attilio Pagli. Many outstanding wines eg. 25 Anni. Non-cru Collepiano is better for less oak, while ROSSO DI MONTEFALCO is smooth, elegant. V.gd: GRECHETTO Grecante. Look for Spinning Beauty (10 yrs aged).

Carema Pie DOC r ★★→★★★ 07 08 09 10 11 13 15 (16') (18) 17 ha, n of Turin. Steep terraces, morainic agglomerate soils for light, mineral, intense, outstanding NEBBIOLO. Best: Ferrando (esp Etichetta Nera), Produttori Nebbiolo di Carema.

Carignano del Sulcis Sar DOC r p ★★→★★★ 09 10 12 13 14 15 (16) From SAR's sw, world-class CARIGNANO that ages gracefully but always boasts early appeal, accessibility. Best: ARGIOLAS, CS DI SANTADI (Rocca Rubia, *Terre Brune*), Mesa.

Carmignano Tus DOCG r ★★★ 08 09 10 11 12 13 15 16' (17) Fine SANGIOVESE/B'x-grape blend invented in C20 by late Count Bonacossi of CAPEZZANA. Best: Ambra (Elzana RISERVA), Capezzana (Selezione), Farnete, Piaggia (RISERVA), Le Poggiarelle, Pratesi (Il Circo Rosso).

Carpenè-Malvolti Ven ★★ Huge volume of well-crafted PROSECCO, from 1st to specialize in this fizz (founder Antonio Carpenè considered "father" of Prosecco).

Carpineti, Marco Lat w sw ★★★ Phenomenal bio whites from little-known Bellone and GRECO Moro, Greco Giallo varieties. Benchmark Moro and Ludum, one of Italy's best stickies.

Cartizze Ven ★★★→★★★★ DOCG At 107 ha, this PROSECCO supercru is reportedly the second most expensive v'yd land in Italy, after BAROLO. V. steep hills in heart of VALDOBBIADENE showcase just how great Prosecco DOCG can be. Usually on sweet side due to fully ripe grapes. Best: BISOL, Bortolomiol, Col Vetoraz, Le Colture, NINO FRANCO, Ruggeri.

Case Basse Tus ★★★★ Owner Gianfranco Soldera died 2019: hopefully children will keep up similar lofty level of iconic BRUNELLO. Long-oak-aged. Rare, precious.

Castel del Monte Pug DOC r p w ★→★★ (r) 10 11 13 14 15 (p w) DYA. Dry, fresh, increasingly serious, from mid-PUG DOC. Esp *Bocca di Lupo* from Tormaresca (ANTINORI). Il Falcone RISERVA (Rivera) is iconic.

Castel Juval, Unterortl T-AA ★★★ Owned by mountaineer Reinhold Messner. Distinctive, crystalline. Best: WEISSBURGUNDER, V.gd RIES and PINOT N.

Castellare Tus ★★★ Classy Castellina-in-CHIANTI producer of long standing. 1st-rate SANGIOVESE/MALVASIA Nera I Sodi di San Niccoló and updated CHIANTI CLASSICO, esp RISERVA Il Poggiale. Also POGGIO ai Merli (MERLOT), Coniale (CAB SAUV).

Castell' in Villa Tus ★★★ Individual, traditionalist CHIANTI CLASSICO estate in extreme sw of zone. Wines of class, excellence by self-taught Princess Coralia Pignatelli. V. age-worthy. Top RISERVA.

Castelluccio E-R ★★→★★★ Quality SANGIOVESE from estate run by Vittorio Fiore and son Claudio. IGT RONCO dei Ciliegi and Ronco delle Ginestre stars. Also gd Le More (Romagna DOC).

Castiglion del Bosco Tus ★★★ Ferragamo-owned up-and-coming BRUNELLO producer.

Cataldi Madonna Ab w r ★★★ Organic certified. Top is PECORINO Frontone, from oldest Pecorino vines in Ab. V.gd: Pecorino Supergiulia, CERASUOLO D'ABRUZZO Piè delle VIGNE, MONTEPULCIANO d'Ab Toni and Malandrino.

Caudrina Pie ★★★→★★★★ Romano Dogliotti one of best for MOSCATO D'ASTI. Top: La Galeisa and ASTI La Selvatica. V.gd: La Caudrina.

Who makes really good Chianti Classico?

CHIANTI CLASSICO is a large zone with hundreds of producers, so picking out the best is tricky. Top get a ★: AMA ★, ANTINORI, BADIA A COLTIBUONO ★, BROLIO, Capraia, Casaloste, Casa Sola, CASTELLARE, CASTELL' IN VILLA, CASTELLO DI VOLPAIA ★, FELSINA ★, FONTERUTOLI, FONTODI ★, I Fabbri ★, Il Molino di Grace, ISOLE E OLENA ★, Le Boncie, Le Cinciole, Le Corti, Le Filigare, Lilliano, Mannucci Droandi, Meleto, MONSANTO ★, Monte Bernardi, Monteraponi ★, NITTARDI, NOZZOLE, Palazzino, Paneretta, Poggerino, POGGIOPIANO, QUERCIABELLA ★, Rampolla ★, RIECINE, Rocca di Castagnoli, Rocca di Montegrossi ★, RUFFINO, San Fabiano Calcinaia, SAN FELICE, SAN GIUSTO A RENTENNANO ★, Paolina Savignola, Selvole, Tenuta Perano – FRESCOBALDI, Vecchie Terre di Montefili, Verrazzano, Vicchiomaggio, VIGNAMAGGIO, Villa Calcinaia ★, Villa La Rosa ★, Viticcio.

Cavallotto Pie ★★★ Organic certified. Solid reference for traditional BAROLO, in Castiglione Falletto. Outstanding RISERVA Bricco Boschis VIGNA San Giuseppe, Riserva Vignolo, v.gd LANGHE NEBBIOLO. Surprisingly gd GRIGNOLINO, Freisa too.

Cave Mont Blanc VdA ★★★ Quality co-op at foot of Mont Blanc, with ungrafted indigenous 60–100-yr-old Prie Blanc vines. Organic certified. Outstanding sparkling. Top Blanc de Morgex et de la Salle Rayon and Brut MC Extreme. V.gd Cuvée du Prince.

Cerasuolo d'Abruzzo Ab DOC p ★ DYA ROSATO version of MONTEPULCIANO D'ABRUZZO, don't confuse with red CERASUOLO DI VITTORIA from SI. Can be brilliant; best (by far): CATALDI MADONNA (Pie delle Vigne), EMIDIO PEPE, Praesidium, TIBERIO, VALENTINI. V.gd TENUTA Tre Gemme.

Cerasuolo di Vittoria Si DOCG r ★★ 11 13 14 15 16 17 (18) A blend of Frappato/NERO D'AVOLA. Only SI DOCG, in se, around city of Vittoria, best terroir for Frappato. Try COS, GULFI, OCCHIPINTI ARIANNA, PAOLO CALÌ, PLANETA, Valle dell'Acate.

Ceretto Pie ★★★ Leading producer of BARBARESCO (Asili, Bernadot), BAROLO (Bricco Rocche, Brunate, Prapò, Cannubi San Lorenzo and new Bussia), plus LANGHE Bianco Blange (ARNEIS). Older generation handing over. Organic certified and bio (2015). Wines recently more classic.

Cerruti, Ezio Pie ★★★ Small estate in gd area for MOSCATO: best sweet Sol, naturally dried Moscato and Sol 10 (07 after 10 yrs ageing). V.gd Fol (dr) and (sp) Ri-Fol.

In Asti they drink sparkling Moscato with salami. Try it.

Cesanese del Piglio or Piglio Lat DOCG r ★→★★★ Medium-bodied red, gd for moderate ageing. Best: Petrucca e Vela, Terre del Cesanese. Cesanese di Olevano Romano, Cesanese di Affile are similar.

Chianti Tus DOCG r ★→★★★ Its cheerful fiasco was a wine icon many would like to see come back, though quality is all over the place. But when gd: delightfully delicious, easy-going and food-friendly fresh red. Modern-day production zone covers most of TUS; original, historic Chianti production zone between Florence and Siena now called CHIANTI CLASSICO.

Chianti Classico Tus DOCG r ★★→★★★ 12 13 15 16' 17 (18') No wine in Italy has improved more over last 20 yrs than Chianti Classico, now often 100% SANGIOVESE. More than 500 producers means inconsistent quality, but best are among Italy's greatest. Made in historic (high, rocky) CHIANTI production zone between Florence and Siena in eight communes. Gran Selezione is new top level, above RISERVA. *See also* box, p.131.

Chiaretto Ven Pale, light blush-hued rosé (word means "claret"), produced esp around Lake Garda. *See* BARDOLINO.

Ciabot Berton Pie ★★★ Marco and Paola Oberto, following their father, have turned this La Morra estate into one of best value producers. V.gd blended BAROLO ("1961" and Barolo del Comune di La Morra); crus Roggeri, Rocchettevino have distinctive single-v'yd characters. Try LANGHE NEBBIOLO 3 Utin.

Cinque Terre Lig DOC w dr sw ★★ Dry VERMENTINO-based whites from vertiginous LIG coast. Sweet version: Sciacchetrà. Try Arrigoni, Bisson, Buranco, De Battè.

Ciolli, Damiano Lat ★★★ r One of most interesting wineries in central Italy. Best is Cirsium, 100% Cesanese d'Affile, 80-yr-old vines. V.gd Silene.

Cirò Cal DOC r (p) (w) ★→★★★ Brisk strong red from Cal's main grape, Gaglioppo, or light, fruity white from GRECO (DYA). Best: 'A Vita, Caparra & Siciliani, IPPOLITO 1845, *Librandi* (Duca San Felice ★★★), San Francesco (Donna Madda, RONCO dei Quattroventi), Santa Venere.

Classico Term for wines from a restricted, usually historic and superior-quality area within limits of a commercially expanded DOC. *See* CHIANTI CLASSICO, SOAVE, VALPOLICELLA, VERDICCHIO, numerous others.

Clerico, Domenico Pie ★★★ Influential BAROLO innovator and modernist producer of Monforte d'ALBA, esp crus Ginestra (Ciabot Mentin and Pajana) and Mosconi (Percristina only in best vintages). V.gd Barolo Aeroplanservaj (from Baudana cru).

Coffele Ven ★★★ Sensitive winemaker in up-and-coming SOAVE-land. Gd Soave CLASSICO, crus Ca' Visco and Alzari (100% GARGANEGA), RECIOTO Le Sponde.

Cogno, Elvio Pie ★★★★ Top estate; super-classy, austere, elegant BAROLOS. Best: RISERVA VIGNA Elena (NEBBIOLO Rosè variety), Bricco Pernice, Ravera. V.gd Anas-Cëtta (100% Nascetta) and BARBERA D'ALBA Pre-Phylloxera (100-yr-old vines).

Col d'Orcia Tus ★★★ Top-quality MONTALCINO estate (3rd-largest) owned by Francesco Marone Cinzano. Best: BRUNELLO RISERVA POGGIO al Vento, new Brunello Nastagio. Col d'Orcia is valley between Montalcino and Monte Amiata.

Colla, Poderi Pie ★★★→★★★★ Tino, Federica and Pietro now run this winery based on experience of Beppe Colla. Classic, traditional, age-worthy. Top: BARBARESCO Roncaglie, BAROLO Bussia Dardi Le Rose, LANGHE Bricco del Drago. V.gd Langhe NEBBIOLO and Pietro Colla Extra Brut.

Colli = hills; singular: Colle. **Colline** (singular Collina) = smaller hills. *See also* COLLIO, POGGIO.

Colli di Catone Lat ★→★★★ Top producer of FRASCATI and IGT. Outstanding aged whites from MALVASIA del Lazio (aka Malvasia Puntinata) and GRECHETTO. Look for Colle Gaio or Casal Pilozzo labels.

Colli di Lapio Cam ★★★ Clelia Romano's estate is Italy's **best Fiano** producer. Also v.gd GRECO DI TUFO Alèxandros.

Colli di Luni Lig, Tus DOC r w ★★→★★★ Nr Spezia. VERMENTINO and Albarola whites; SANGIOVESE-based reds easy to drink, charming. Gd: Ottaviano Lambruschi (Costa Marina); Giacomelli (Boboli), La Baia del Sole (Oro d'Isèe); Bisson (VIGNA Erta), Lunae (Numero Chiuso).

Collio F-VG DOC r w ★★→★★★★ Famous white denomination, unfortunately moved steadily to nonsensical Collio Bianco blend rather than highlight terroir differences of its communes, incl coolish San Floriano and Dolegna, warmer Capriva. Happily, Collio boasts glut of talented producers: BORGO DEL TIGLIO, Castello di Spessa, GRAVNER, La Castellada, Livon, MARCO FELLUGA, Podversic, Primosic, Princic, **Radikon**, Renato Keber, RONCO dei Tassi, RUSSIZ SUPERIORE, **Schiopetto**, Venica & Venica, VILLA RUSSIZ.

Colli Piacentini E-R DOC r p w ★→★★★ DYA Light gulping wines, often fizzy, from eg. BARBERA, BONARDA (r), MALVASIA di Candia Aromatica, Pignoletto (w).

Colterenzio CS / Schreckbichl T-AA ★★★ Cornaiano-based main player among ALTO ADIGE co-ops. Whites (SAUV Lafoa, PINOT BIANCO Berg and new LR) tend to be better than reds. V.gd CAB SAUV Lafoa and PINOT N St Daniel.

Conegliano Valdobbiadene Ven DOCG w sp ★→★★ DYA. Name for top PROSECCO, tricky to say: may be used separately or together.

Conero DOCG r ★★→★★★ 15 17 Aka ROSSO CONERO. Small zone making powerful, at times too oaky, MONTEPULCIANO. Try: GAROFOLI (Grosso Agontano), Le Terrazze (Praeludium), Marchetti (RISERVA Villa Bonomi), Moncaro, Monteschiavo (Adeodato), Moroder (Riserva Dorico), UMANI RONCHI (Riserva Campo San Giorgio).

Conterno, Aldo Pie ★★★★ Top estate of Monforte d'ALBA, was considered a traditionalist, esp concerning top BAROLOS Granbussia (outstanding 10), Cicala, Colonello and esp Romirasco. V.gd CHARD Bussiador.

Conterno, Giacomo Pie ★★★★ For many, Monfortino is best wine of Italy. Roberto recently acquired NERVI estate in GATTINARA. Outstanding BARBERAS. Top BAROLO Cascina Francia, Cerretta and now Arione.

Conterno, Paolo Pie ★★→★★★ Family of NEBBIOLO and BARBERA growers since 1886,

ITALY

current *titolare* Giorgio continues with textbook cru BAROLOS Ginestra and Riva del Bric, plus particularly fine LANGHE *Nebbiolo Bric Ginestra*.

Conterno Fantino Pie ★★★ Organic certified. Two families joined to produce excellent modern-style BAROLO crus at Monforte: Ginestra (VIGNA Sorì Ginestra and Vigna del Gris), Mosconi (Vigna Ped) and Castelletto (Vigna Pressenda). Also gd Ginestrino and NEBBIOLO/BARBERA blend Monprà.

Contini Sar ★★★ Benchmark VERNACCIA DI ORISTANO, oxidative-styled whites not unlike v.gd Amontillado or Oloroso. Antico Gregori one of Italy's best whites. Amazing Flor 22 and RISERVA.

Conti Zecca Pug ★★→★★★ SALENTO estate. Donna Marzia line of Salento IGT wines is gd value, as is SALICE SALENTINO Cantalupi. Best-known is Nero (NEGROAMARO/CAB SAUV blend).

Cornelissen, Frank ★★★★ Belgian-owned, today one of ETNA's top producers, esp for red. Practically all outstanding, NERELLO MASCALESE Magma usually SI's most expensive wine. V.gd VA and CS (r).

Correggia, Matteo Pie r ★★★ Organic certified. Leading producer of ROERO (RISERVA Rochè d'Ampsej, Val dei Preti), Roero ARNEIS, plus BARBERA D'ALBA (Marun) and Roero Arneis Val dei Preti 13 (6 yrs aged).

Crivelli Pie ★★★★ Top producer of Ruchè di Castagnole MONFERRATO and of GRIGNOLINO d'Asti. Must-try wines.

CS (Cantina Sociale) Cooperative winery.

Cuomo, Marisa Cam ★★★ Fiorduva is *one of Italy's greatest whites*. V.gd Furore Bianco and Rosso (Costa d'Amalfi).

Cusumano Si ★★→★★★ Recent major player with 500 ha in various parts of SI. Reds from NERO D'AVOLA, CAB SAUV, SYRAH; whites from CHARD, INSOLIA. Gd quality, value. Best from ETNA (Alta Mora).

Dal Forno, Romano Ven ★★★★ V.-high-quality VALPOLICELLA, AMARONE, RECIOTO (latter not identified as such any more); v'yds outside CLASSICO zone but wines great.

D'Ambra Cam r w ★★→★★★ On ISCHIA; fosters rare local native grapes. Best: single-v'yd Frassitelli (w, 100% Biancolella). V.gd Forastera (w) and AGLIANICO-based La VIGNA dei Mille Anni (r).

De Bartoli, Marco Si ★★★★ One of best estates in all Italy. Marco spent his life promoting "real" MARSALA and his have to be tried. Top is 20-yr-old Ventennale and VECCHIO SAMPERI. Delicious table wines (eg. GRILLO Vignaverde and Grappoli del Grillo, ZIBIBBO Pietranera and Pignatello), outstanding sweet Zibibbo di PANTELLERIA *Bukkuram*.

Dei Pie ★★★ Pianist Caterina Dei runs this aristocratic estate in MONTEPULCIANO, making VINO NOBILES with artistry and passion. Her chef d'oeuvre is Bossona and new cru Madonna della Querce.

Derthona Pie w ★→★★★ Wine from Timorasso grapes grown only in COLLI Tortonesi. One of Italy's more *interesting whites*, like v. dry RIES from Rheinhessen. Gd: La Colombera, Mariotto, Mutti, POGGIO Paolo, VIGNETI MASSA, Vigneti Repetto.

Di Majo Norante Mol ★★→★★★ Best-known of Mol with decent Biferno Rosso Ramitello, Don Luigi Molise Rosso RISERVA, Mol AGLIANICO Contado.

DOC / DOCG Quality wine designation: *see* box, p.153.

Dogliani Pie DOCG r ★→★★★ 12 13 15 16 17 DOLCETTO monovariety wine. Some to drink young, some for moderate ageing. Gd: Chionetti, Clavesana, EINAUDI, Francesco Boschis, Marziano Abbona, Pecchenino.

Donnafugata Si r w ★★→★★★ Classy range. Reds Mille e Una Notte, Tancredi, ETNA Rosso Contrada Montelaguardia Fragore; whites Chiaranda, Lighea, Kebir. Also v. fine MOSCATO PASSITO di PANTELLERIA Ben Ryé.

Duca di Salaparuta Si ★★ Top Duca Enrico and everyday Corvo some of Italy's best-known, much-loved wines at home and abroad. White Kados from GRILLO grapes.

Einaudi, Luigi Pie ★★★ 52-ha estate founded late C19 by ex-president of Italy, in DOGLIANI. Solid BAROLOS from Cannubi and Terlo Costa Grimaldi v'yds. Top Dogliani (DOLCETTO) from VIGNA Tecc. Now also in Bussia, Monvigliero.

Elba Tus r w (sp) ★→★★ DYA. Island's white, TREBBIANO/ANSONICA, can be v. drinkable with fish. Dry reds based on SANGIOVESE. Gd sweet white (MOSCATO) and red (*Aleatico Passito DOCG*). Gd: Acquabona, La Mola, Ripalte, Sapereta.

Enoteca Wine library; also shop or restaurant with ambitious wine list. There is a national enoteca at the *fortezza* in Siena.

Est! Est!! Est!!! Lat DOC w dr s/sw ★ DYA. White from Montefiascone, n of Rome.

Etna Si DOC r p w ★★→★★★ (r) 12 13 14 15 16 17 (18) One of hottest areas, remarkable development in last decade. 900 ha of v'yds on n slopes, high-altitude, volcanic soils. Etna Rosso typically 90/10 blend of NERELLOS MASCALESE/Cappuccio, while Etna Bianco can be pure CARRICANTE or incl CATARATTOS Comune or Lucido.

Falchini Tus ★★→★★★ Producer of gd DOCG VERNACCIA DI SAN GIMIGNANO.

Falerno del Massico Cam ★★→★★★ DOC r w ★★ (r) 15 16 (18) Falernian wine, from three portions of hill much like Burgundy's monk-farmed Clos Vougeot, was Antiquity's most famous wine. Today average only. Best are elegant AGLIANICO reds, fruity dry FALANGHINA whites. Try: Masseria Felicia, Villa Matilde.

Fara Pie *See* ALTO PIEMONTE.

Faro Si DOC r ★★★ 12 13 14 15 16 (17) Intense, harmonious red from NERELLO MASCALESE, NERO D'AVOLA and Nocera in hills behind Messina. Top Bonavita. Palari most famous, but Le Casematte just as gd if not better.

Felline Pug ★★★→★★★★ Gregory Perucci was pioneer in rediscovery of PRIMITIVO and Sussumaniello vines. Top PRIMITIVO DI MANDURIA (Giravola and ZIN). V.gd Sum (Sussumaniello).

Felluga, Livio F-VG ★★★ Consistently fine FRIULI COLLI ORIENTALI, esp blends Terre Alte and Abbazia di Rosazzo; Bianco Illivio, *Pinot Gr*, PICOLIT (Italy's best?), MERLOT/REFOSCO blend Sossó.

Felluga, Marco F-VG *See* RUSSIZ SUPERIORE.

Felsina Tus ★★★ CHIANTI CLASSICO estate of distinction in se corner of zone: classic RISERVA Rancia and IGT Fontalloro, both 100% SANGIOVESE. V.gd Gran Selezione Colonia.

Fenocchio, Giacomo Pie ★★★→★★★★ Small but outstanding Monforte d'ALBA-based BAROLO cellar. Traditional style, Crus: Bussia (also RISERVA), Cannubi, Villero Outstanding Freisa, one of Italy's 2–3 best.

Ferrara, Benito Cam ★★★ Maybe Italy's best GRECO DI TUFO producer (VIGNA Cicogna and Terra d'Uva). Talent shows in excellent TAURASI and FIANO too.

Ferrari T-AA sp ★★→★★★★ Trento maker of one of two best Italian METODO CLASSICOS.

Summit of Etna

Some of Italy's most exciting wines come from this famous volcano. Vines grow up to 1000m (3280ft) on si's e coast. *Contrada* is Si way to express cru: differences in soil, altitude and age of lava flows. Top growers incl: Alberelli di Giodo, BENANTI (Rovittello), Calabretta, CALCAGNO (Arcuria, Feudo di Mezzo, w Ginestra), Cottanera (w Calderara, Zottorinoto), Graci (Arcuria, Feudo di Mezzo), Gulfi (Reseca), I Vigneri, Le Vigne di Eli (r w Moganazzi-Volta Sciara), Pietradolce (r w Archineri, Barbagalli, Rampante, Santo Spirito), Santa Maria La Nave (Calmarossa), TASCA D'ALMERITA (Rampante, Sciaranuova, w Tascante), TENUTE Bosco (VIGNA Vico), TENUTE DI FESSINA (w A' Puddara, r w Il Musmeci), TENUTA DELLE TERRE NERE (r w Caldera Sottana, Guardiola, San Lorenzo, w Santo Spirito), VINI FRANCHETTI (Guardiola, Rampante, Sciaranuova). Time to get exploring.

Giulio Ferrari is top. New outstanding Giulio Ferrari Rosè and Perlè Zero. Also gd: CHARD-based Brut RISERVA Lunelli and new Perlè Zero, Perlè Bianco (gd value), PINOT N-based Extra Brut Perlè Nero.

Feudi di San Gregorio Cam ★★→★★★ Much-hyped CAM producer, with DOCGS FIANO DI AVELLINO Pietracalda, GRECO DI TUFO Cutizzi, TAURASI Piano di Montevergine. Also gd: Serpico (AGLIANICO); whites *Campanaro* (Fiano/Greco), FALANGHINA. Feudi is an archaic word for manor or estate.

Feudo di San Maurizio VdA ★★★★ Outstanding wines from rare native grapes CORNALIN, Mayolet and Vuillermin; last two rank among Italy's greatest reds. V.gd Torrette.

Feudo Montoni Si r w ★★★★ Exceptional estate in upland e SI. Best: NERO D'AVOLA Lagnusa and Vrucara. V.gd GRILLO della Timpa (w), CATARRATTO del Masso (w) and Perricone del Core (r).

Fiano di Avellino Cam DOCG w ★★→★★★ 10 12 15 16 Can be either steely (most typical) or lush. Volcanic soils best. Gd: Ciro Picariello, COLLI di Lapio-Romana Clelio, MASTROBERARDINO, Pietracupa, QUINTODECIMO, Tenuta Sarno 1860, Vadiaperti, Villa Diamante.

Fino, Gianfranco Pug ★★★★ Greatest PRIMITIVO, from old, low-yielding bush vines. Outstanding Es among Italy's top 20 reds. V.gd Jo (NEGROAMARO) and Se (PRIMITIVO).

Florio Si Historic quality maker of MARSALA. Specialist in Marsala Vergine Secco. For some reason Terre Arse (= burnt lands), its best wine, doesn't do well in UK.

Between 1954 and 2018 there was only one frost in Franciacorta – in 2017.

Folonari Tus ★★→★★★ Folonari family, ex-RUFFINO, have v'yds in TUS and elsewhere. Estates/wines incl *Cabreo* (CHARD and SANGIOVESE/CAB SAUV), La Fuga (BRUNELLO DI MONTALCINO), NOZZOLE (incl top Cab Sauv Pareto).

Fongaro Ven ★★★★ Classic-method fizz Lessini Durello (Durello = grape). High quality, even higher acidity, age-worthy.

Fontanafredda Pie ★★ Since 1858, former royal estates. Large producer of PIE. V.gd BAROLO La Rosa, Alta Langa Brut Nature VIGNA Gatinera, LANGHE Freisa.

Fonterutoli Tus ★★★ Historic CHIANTI CLASSICO estate of Mazzei family at Castellina. Notable: CHIANTI CLASSICO Gran Selezione and IGT Mix 36. Also owns TENUTA di Belguardo in MAREMMA and Zisola in SI.

Fontodi Tus ★★★★ One of v. best CHIANTI CLASSICOS, also Gran Selezione VIGNA del Sorbo and memorable all-SANGIOVESE Flaccianello. IGT SYRAH Case Via among best of that variety in TUS.

Foradori T-AA ★★★ Much-loved Elizabetta F is "lady of TRENTINO wine" making outstanding *Teroldego* but lovely macerated (too macerated for some), anfora-aged Incrocio Manzoni and Nosiola. Look for TEROLDEGO Morei and Sgarzon, white Nosiola Fontanasanta. Top remains Teroldego-based Granato.

Franciacorta Lom DOCG w (p) sp ★★→★★★★ Italy's zone for top-quality METODO CLASSICO fizz. Best: Barone Pizzini, BELLAVISTA, *Ca' del Bosco*, Cavalleri, UBERTI, Villa. Also v.gd: Bersi Serlini, Contadi Castaldi, Gatti, Monte Rossa, Mosnel, Ricci Curbastro.

Frascati Lat DOC w dr sw s/sw (sp) ★→★★ DYA. Once proud name (Rome's favourite) sadly debased. Large producers continuing to make neutral but mysteriously well-written-about wines leave little hope for change. Buy following small producers only: Borgo del Cedro (SUPERIORE), Castel de Paolis (Superiore), De Sanctis (Abelos Bio), Merumalia (Primo), Villa Simone (RISERVA Filonardi).

Frascole Tus ★★→★★★★ Most n winery of most n CHIANTI RÚFINA zone, small estate run organically by Enrico Lippi, with an eye for typicity. Chianti Rúfina is main driver, but VIN SANTO is to die for.

Frescobaldi Tus ★★→★★★★ Ancient noble family, leading CHIANTI RÚFINA pioneer at NIPOZZANO estate (look for ★★★ *Montesodi*), also BRUNELLO from Castelgiocondo estate, MONTALCINO. Sole owner of LUCE estate (Montalcino), ORNELLAIA (BOLGHERI), new TENUTA Perano (top: CHIANTI CLASSICO Gran Selezione Rialzi). V'yds also in COLLIO, MAREMMA, Montespertoli, Gorgona Island (state prison).

Friuli Colli Orientali F-VG DOC r w dr sw ★★→★★★★ 12 15 16 (18) (Was COLLI Orientali del Friuli.) Hilly area of F-VG next to COLLIO. Unlike latter, not just whites, but outstanding reds and v.gd stickies from likes of r Pignolo, SCHIOPPETTINO, Tazzelenghe and w PICOLIT and VERDUZZO Friulano. Top: Aquila del Torre, Ermacora, Gigante, Grillo Iole, La Sclusa, LE DUE TERRE, LIVIO FELLUGA, Miani, Petrussa, Rodaro, Ronchi di Cialla, VIGNA Petrussa. Ramandolo DOCG is best sweet Verduzzo (look for Anna Berra). Picolit can be Italy's best sweet: Aquila del Torre, Livio Felluga, Marco Sara, Ronchi di Cialla, Vigna Petrussa often amazing.

Friuli Grave F-VG DOC r w ★→★★ Previously Grave del Friuli. Largest DOC of F-VG, mostly on plains, often v. rainy, making quality red production tricky. Mostly big volumes, whites best. Look for Borgo Magredo, Di Lenardo, Le Monde, RONCO Cliona, Villa Chiopris.

Friuli Isonzo F-VG DOC r w ★★★ Used to be just Isonzo. High-alc, luscious, powerful whites from gravel-rich alluvial river plain, a rare flatland high-quality site. Best: LIS NERIS, RONCO DEL GELSO, VIE DI ROMANS. Gd: Borgo Conventi, Pierpaolo Pecorari.

Friuli-Venezia Giulia F-VG Ne region hugging Slovenian border, home to Italy's best whites (along with ALTO ADIGE). Hills to nc give best, but alluvial seaside regions (DOC from Aquileia, Latisana, Annia) improving markedly. DOCs ISONZO, COLLI ORIENTALI, COLLIO and Carso best. All have COLLIO preceded on label by "Friuli".

Frizzante Semi-sparkling, up to 2.5 atmospheres, eg. MOSCATO D'ASTI, much PROSECCO, LAMBRUSCO and the like.

Fucci, Elena Bas ★★★★ AGLIANICO DEL VULTURE Titolo from 55–70-yr-old vines in Mt Vulture's Grand Cru; one of Italy's 20 best. Organic. Seek out o8 11 and outstanding 13 15 (also SUPERIORE RISERVA) and 17 Anniversary.

Fuligni Tus ★★★★ Outstanding producer of BRUNELLO (top RISERVA) and ROSSO DI MONTALCINO.

Gaja Pie ★★★★ Old family firm at BARBARESCO led by eloquent Angelo Gaja; daughter Gaia G following. High quality, higher prices. Top: Barbaresco (Costa Russi, Sorì San Lorenzo, Sorì Tildìn), BAROLO (Conteisa, Sperss). Splendid CHARD (Gaia & Rey). Also owns CA' MARCANDA in BOLGHERI, Pieve di Santa Restituta in MONTALCINO. New acquisition on ETNA (with Graci).

Galardi Cam ★★★ Producer of Terra di Lavoro, much-awarded AGLIANICO/Piedirosso blend, in n CAM. Only one wine.

Gancia Pie Famous old brand of MUSCAT fizz, also ALTA LANGA; still gd.

Garda Ven DOC r p w ★→★★ (r) 15 16 (18) (p w) DYA. Divided between Lom and Ven hugging Italy's largest lake, home to easy-going, early-drinking cheap-and-cheerful r p w. CHIARETTO best. Gd: Cavalchina, Zeni.

Garofoli Mar ★★→★★★ Quality leader in the Mar, specialist in VERDICCHIO (Podium, Serra Fiorese and sparkling Brut RISERVA), CONERO (Grosso Agontano).

Gattinara Pie DOCG r 11 12 13 15 Best-known of a cluster of ALTO PIE DOC(G)s based on NEBBIOLO. Volcanic soil. Suitable for long ageing. Best: ANTONIOLO (Osso San Grato, San Francesco), CANTINA del Signore, Iarretti Paride, NERVI, Torraccia del Piantavigna, TRAVAGLINI (RISERVA and Tre Vigne). *See also* ALTO PIEMONTE.

Gavi / Cortese di Gavi Pie DOCG w ★→★★★ DYA. Overhyped, but at best subtle dry white of Cortese grapes, though much is dull, simple or sharp. Most comes from commune of Gavi, now known as Gavi del Comune di Gavi. Best:

Bruno Broglia/La Meirana, Castellari Bergaglio (Rovereto Vignavecchia and Rolona, Fornaci), Franco Martinetti, La Raia, TENUTA San Pietro, Villa Sparina.

Germano, Ettore Pie ★★★ Small family Serralunga estate run by Sergio and wife Elena. Top BAROLOS: RISERVA Lazzarito and Cerretta. Look for 1st cru VIGNA Rionda. V.gd: LANGHE RIES Herzù, LANGHE NEBBIOLO VR and ALTA LANGA.

Ghemme Pie DOCG r NEBBIOLO (at least 85%), incl up to 15% Uva Rara and/or Vespolina. Top: *Antichi Vigneti di Cantalupo* (Collis Braclemae and Collis Carellae), Ioppa (Balsina). V.gd Torraccia del Piantavigna (VIGNA Pelizzane). *See also* ALTO PIEMONTE.

Giacosa, Bruno Pie ★★★★ Italy's greatest winemaker died in 2018, but wines still top. Now run by daughter Bruna. Splendid traditional-style BARBARESCOS (Asili, Rabajà), BAROLOS (Falletto, Falletto VIGNA Rocche). Top wines (ie. RISERVAS) get famous red label. Amazing METODO CLASSICO Brut, ROERO ARNEIS (w) and Valmaggiore (r).

Giuseppe Cortese Pie ★★★ Traditional producer of outstanding BARBARESCO Rabajà (also RISERVA). Also gd LANGHE NEBBIOLO.

GIV (Gruppo Italiano Vini) Complex of co-ops and wineries, biggest v'yd holders in Italy: Bigi, BOLLA, MELINI, Negri. Also in S: SI, Bas.

Grappa Pungent spirit made from grape pomace (skins, etc., after pressing), can be anything from disgusting to inspirational. What the French call "marc".

Grasso, Elio Pie ★★★ Tiptop BAROLO producer (crus Gavarini VIGNA Chiniera, Ginestra Casa Maté and RISERVA Rüncot); v.gd BARBERA D'ALBA Vigna Martina and DOLCETTO d'Alba.

Gravner, Josko F-VG ★★★→★★★★ Controversial but talented COLLIO producer (unlike some who copy him), vinifies on skins (r w) in buried amphorae without temperature control; long ageing, bottling without filtration. Wines either loved for complexity or loathed for oxidation and phenolic components. Look out for Breg (white blend), RIBOLLA GIALLA (w) and Rosso Breg (Pignolo).

Greco di Tufo Cam DOCG w (sp) DYA. Tannic, oily whites made with local GRECO variety (different to Cal's also outstanding Greco Bianco). Best: Bambinuto (Picoli), BENITO FERRARA (VIGNA Cicogna), Caggiano (Devon), COLLI di Lapio (Alexandros), Donnachiara, FEUDI DI SAN GREGORIO (Cutizzi), Macchialupa, *Mastroberardino* (Nova Serra, Vignadangelo), Pietracupa, QUINTODECIMO, Terredora (Loggia della Serra), Vadiaperti (Tornante).

Grifalco Bas ★★★ Small estate in high-quality Ginestra and Maschito subzone. Best: AGLIANICO DEL VULTURE Daginestra, Damaschito (SUPERIORE DOCG from 15). Look for new Duemila 11.

Grignolino Pie DOC r DYA. Two DOCS: Grignolino d'Asti, Grignolino del MONFERRATO Casalese. At best, light, perfumed, crisp, high in acidity, tannin. D'Asti: BRAIDA, Cascina Tavijin, Crivelli, Incisa della Rocchetta, Spertino, TENUTA Garetto. Monferrato C: Accornero (Bricco del Bosco and Bricco del Bosco Vigne Vecchie: vinified like BAROLO), Bricco Mondalino, Canato (Celio), Il Mongetto, PIO CESARE.

Grosjean VdA r ★★★ Top quality; best CORNALIN, Premetta. Vigne Rovettaz one of Valle's oldest, largest.

Guardiense, La Cam ★★→★★★ Dynamic co-op, decent-value whites (FALANGHINA Senete, FIANO COLLI di Tilio, GRECO Pietralata) and reds (esp I Mille per l'AGLIANICO); World's largest producer of Falanghina, but that in itself means little.

Guerrieri Rizzardi Ven ★★→★★★ Noble family making top-level AMARONE and BARDOLINO. Gd Bardolino CLASSICO Tacchetto, elegant Amarone Villa Rizzardi, but best wine is Amarone cru Calcarole. V.gd SOAVE Classico Costeggiola. Historic garden worth visiting.

Gulfi Si ★★★★ SI's best producer of NERO D'AVOLA, 1st to bottle single-*contrada* (cru) wines. Organic certified. Outstanding: Nerobufaleffj and Nerosanlorè. V.gd:

CERASUOLO DI VITTORIA CLASSICO, Nerobaronj, Neromaccarj, Nerojbleo and Carjcanti (w). Interesting Pinò (PINOT N) and ETNA Rosso Reseca.

Gutturnio dei Colli Piacentini E-R DOC r dr ★→★★ DYA. BARBERA/BONARDA blend from COLLI PIACENTINI; sometimes frothing.

Haas, Franz T-AA ★★★ ALTO ADIGE producer of excellence and occasional inspiration; v.gd PINOT N, LAGREIN (Schweizer), IGT blends, esp Manna (w). Best: MOSCATO Rosa.

Hofstätter T-AA ★★★ Top quality; gd PINOT N. Look for Barthenau VIGNA Sant'Urbano. Also whites, mainly GEWURZ (esp *Kolbehof*, one of Italy's two best). Now owner of Mosel's Dr. Fischer.

IGT (Indicazione Geografica Tipica) Increasingly known as Indicazione Geografica Protetta (IGP). *See* box, p.153.

Ippolito 1845 Cal ★→★★ This CIRÒ Marina-based winery claims to be oldest in Cal. Run quasi-organically, international and indigenous grapes. Top: CIRÒ RISERVA COLLI del Mancuso and Ripe del Falco, Pecorello Bianco and GRECO Bianco (Gemma del Sole).

Ischia Cam DOC (r) w ★→★★ DYA. Island off Naples, own grape varieties (w: Biancolella, Forastera; r. Piedirosso, also found in Cam). Frassitelli v'yd best for Biancolella. Best: Cenatiempo (Kalimera), D'AMBRA (Biancolella Frassitelli, Forastera), Antonio Mazzella (VIGNA del Lume).

Isole e Olena Tus ★★★★ Top CHIANTI CLASSICO estate run by astute Paolo de Marchi, with superb red IGT Cepparello. Outstanding VIN SANTO; v.gd CHIANTI CLASSICO, CAB SAUV, CHARD, SYRAH. Also owns fantastic Proprietà Sperino in LESSONA.

Ripeness levels can vary 1–1.5% alc within same bunch of Sangiovese.

Jermann, Silvio F-VG ★★→★★★ Famous estate with v'yds in COLLIO and ISONZO: top white blend Vintage Tunina and Capo Martino. V.gd Vinnae (mainly RIBOLLA GIALLA) and "Where Dreams ..." (CHARD).

Köfererhof T-AA ★★★ →★★★ Great whites: KERNER, SYLVANER; MÜLLER-T excellent too.

Lacrima di Morro d'Alba Mar DYA. Curiously named aromatic medium-bodied red from small commune in the Mar, no connection with ALBA or La Morra (PIE). Gd: Mario Lucchetti (Guardengo), Marotti Campi (Orgiolo and Rubico), Stefano Mancinelli and Vicari. For PASSITO: Lucchetti, Stefano Mancinelli (Re Sole).

Lacryma (or Lacrima) Christi del Vesuvio Cam r p w dr (sw) (sp) ★→★★ DOC Vesuvio based on Coda di Volpe (w), Piedirosso (r). Despite romantic name Vesuvius comes nowhere nr ETNA in quality stakes. Sorrentino and De Angelis best; Caputo, MASTROBERARDINO, Terredora less inspired.

Lageder, Alois T-AA ★★ →★★★ Famous ALTO ADIGE producer. Most exciting are single-v'yd varietals: *Sauv Bl Lehenhof*, PINOT GR Porer, CHARD Löwengang, GEWURZ Am Sand, PINOT N Krafuss, LAGREIN Lindenberg, CAB SAUV Cor Römigberg. Also owns Cason Hirschprunn for v.gd IGT blends.

Lagrein Alto Adige T-AA DOC r p ★★→★★★ 11 12 13 15 16' 18 Alpine red with deep colour, rich palate (plus a bitter hit at back); refreshing pink *Kretzer rosé* made with LAGREIN. Top ALTO ADIGE: CANTINA Bolzano (Taber), Cantina Santa Maddalena, CS Andriano, CS Tramin (Urban), Elena Walch, LAGEDER, MURI GRIES (cru Abtei), Niedrist Ignaz, Putzenhof, TIEFENBRUNNER. From TRENTINO try Francesco Moser's Deamater.

Lambrusco E-R DOC (or not) r p w dr s/sw ★ →★★★ DYA. 17 different Lambrusco varieties (five most planted by far) make for highly distinct wines, ie. Lambrusco wine does not exist. Plethora of different denominations usually linked to one of five main grapes, so each has its defining characteristics. When gd, delightful fizzy fresh, lively red that pairs divinely with rich, fatty fare. DOCS: L Grasparossa di Castelvetro, L Salamino di Santa Croce, L di Sorbara. Best: [Sorbara] Cavicchioli (Cristo Secco and Cristo Rose), Cleto Chiarli (Antica Modena Premium), Medici

Ermete (Phermento, ancestral method), Paltrinieri. Grasparossa: Cleto Chiarli (Enrico Cialdini), Moretto (Monovitigno and VIGNA Canova), Pederzana (Canto Libero Semi Secco), Vittorio Graziano (Fontana dei Boschi). Maestri: Ceci (Nero di Lambrusco Otello), Dall'Asta (Mefistofele). Salamino: Cavicchioli (Tre Medaglie Semi Secco), Luciano Saetti (Vigneto Saetti), Medici Ermete (Concerto Granconcerto).

Langhe Pie The hills of central PIE, home of BAROLO, BARBARESCO, etc. DOC name for several Pie varietals plus Bianco and Rosso blends. Those wishing to blend other grapes with NEBBIOLO can at up to 15% as "Langhe Nebbiolo" – a label to follow.

Langhe Nebbiolo Pie ★★→★★★ Like NEBBIOLO D'ALBA (Nebbiolo > 85%) but from a wider area: LANGHE hills. Unlike N d'Alba may be used as a downgrade from BAROLO or BARBARESCO. Gd: BURLOTTO, GIACOMO FENOCCHIO, GIUSEPPE RINALDI, PIO CESARE, TREDIBERRI, CIABOT BERTON, *Vajra*.

Le Due Terre F-VG ★★★ Small family-run FRIULI COLLI ORIENTALI estate: Top Sacrisassi Rosso (SCHIOPPETTINO/REFOSCO). V.gd Merlot, (w) Sacrisassi Bianco.

Le Macchiole Tus ★★★★ Organic. One of few native-owned wineries of BOLGHERI; one of 1st to emerge after SASSICAIA, makes *Italy's best Cab Fr* (Paleo Rosso), one of best MERLOTS (Messorio), SYRAHS (Scrio). V.gd Bolgheri Rosso.

Le Piane Pie ★★★ BOCA DOC has resurfaced thanks to Christoph Kunzli. Gd: Maggiorina, Piane (Croatina) and Mimmo (NEBBIOLO/Croatina), plus (w) Bianko (Erbaluce).

Les Cretes VdA ★★★ Costantino Charrère is father of modern VALLE D'AOSTA viticulture and saved many forgotten varieties. *Outstanding Petite Arvine*, two of Italy's best CHARDS; v.gd Fumin, Torrette and Neige d'Or (w blend).

Drink with Arancini Siciliani (rice balls): (w) Carricante or (r) Cerasuolo di Vittoria.

Lessona Pie DOCG r *See* ALTO PIEMONTE. NEBBIOLO (at least 85%). Elegant, age-worthy, fine bouquet, long savoury taste. Best: PROPRIETÀ SPERINO. Gd: Cassina, Colombera & Garella, La Prevostura, TENUTE Sella.

Librandi Cal ★★★ Top producer pioneering research into Cal varieties. V.gd red CIRÒ (*Riserva Duca San Felice* is ★★★), IGT Gravello and Terre Lontane (CAB SAUV/ Gaglioppo blend), Magno Megonio (r) from Magliocco grape, IGT Efeso (w) from Mantonico.

Liguria r w sw ★→★★ Narrow ribbon of extreme mtn viticulture produces memorable (w) VERMENTINO, PIGATO, (r) Rossese di Dolceacqua varieties. Look for Giacomelli, La Baia del Sole, Ottaviano Lambruschi (Vermentino); Bio Vio, Bruna and Lupi (PIGATO). CINQUE TERRE is beautiful and Sciacchetrà (sw) is one of Italy's best stickies; Ormeasco di Pornassio (r) is made with Ligurian biotype of DOLCETTO.

Lisini Tus ★★★→★★★★ Historic estate for some of finest, longest-lasting BRUNELLO, esp RISERVA Ugolaia.

Lis Neris F-VG ★★★ Top ISONZO estate for whites, esp PINOT GR (Gris), SAUV BL (Picol), FRIULANO (Fiore di Campo), plus outstanding blends Confini and Lis. V.gd Lis Neris Rosso (MERLOT/CAB SAUV), sweet white Tal Luc (VERDUZZO/RIES).

Lo Triolet VdA r w ★★★ Top PINOT GR producer, v.gd Fumin, Coteau Barrage (SYRAH/ Fumin), MUSCAT and GEWURZ.

Luce Tus ★★★ FRESCOBALDI's estate. Luce (SANGIOVESE/MERLOT blend for oligarchs), but lovely Luce BRUNELLO DI MONTALCINO too. From 15, Lux Vitis (blend CAB SAUV/Sangiovese).

Lugana DOC w (sp) ★★→★★★ DYA. Much-improved white of s Lake Garda, rivals gd SOAVE next door. Main grape Turbiana (Formerly TREBBIANO di Lugana). Best: CA DEI FRATI (I Frati, esp *Brolettino*), Fratelli Zeni, Le Morette (owned by Valerio ZENATO), Ottella (Brut, Le Crete), ZENATO (oaked).

Lungarotti Umb ★★→★★★ Leading producer of TORGIANO. Star wines DOC Rubesco, DOCG RISERVA *Monticchio*. Gd VIGNA Il Pino, Sangiorgio (SANGIOVESE/CAB SAUV), Giubilante, MONTEFALCO SAGRANTINO.

Maculan Ven ★★★ Quality pioneer of Ven. Excellent CAB SAUV (Fratta, Palazzotto). Best-known for sweet TORCOLATO (esp RISERVA Acininobili).

Malvasia delle Lipari Si DOC w sw ★★★ Luscious sweet, made with one of the many MALVASIA varieties. Best: Capofaro, Caravaglio, Fenech, Lantieri, Marchetta. Gd: Hauner.

Malvirà ★★★→★★★★ Top ROERO producer. Organic certified. Best Roero single-v'yd: (r w) Renesio, Trinità; (w) Saglietto. Waiting for Roero ARNEIS RISERVA Saglietto.

Manduria (Primitivo di) Pug DOC r s/sw ★★→★★★ Manduria is the cradle of PRIMITIVO, alias ZIN, so expect wines that are gutsy, alcoholic, sometimes porty. Best: GIANFRANCO FINO, MORELLA and FELLINE. Gd producers, located in Manduria or not: Cantele, CS Manduria, de Castris, Polvanera.

Manni Nössing T-AA ★★★★ Outstanding KERNER, MÜLLER-T Sass Rigais, SYLVANER. Benchmark wines.

Marchesi di Barolo Pie ★→★★★ Large, historic BAROLO producer, plus other ALBA wines.

Maremma Tus Once malaria-plagued, s TUS coast area boomed in C20 with easygoing, delicious SANGIOVESE-based reds from DOC(G)s: Monteregio, MORELLINO DI SCANSANO, PARRINA, Pitigliano, Sovana.

Marrone, Agricola Pie ★★★ Small estate, low price but gd-quality BAROLO. Top: Bussia. Gd: ARNEIS, BARBERA D'ALBA, BAROLO Pichemej.

Marsala Si DOC w sw sl's once-famous fortified (★→★★★★), created by Woodhouse Bros of Liverpool in 1773. Can be dry to v. sweet; best is bone-dry Marsala Vergine. *See also* DE BARTOLI.

Marzemino Trentino T-AA DOC r ★→★★ Pleasant everyday red. Isera and Ziresi are subzones. Best: Eugenio Rosi (Poiema), Grigoletti. Gd: CANTINA d'Isera, Vivallis.

Mascarello Pie ★★★★ Two top producers of BAROLO: the late Bartolo M, of Barolo, whose daughter Maria Teresa continues her father's highly traditional path (v.gd Freisa); and Giuseppe M, of Monchiero, whose son Mauro makes v. fine, traditional-style Barolo from the great *Monprivato* v'yd in Castiglione Falletto. Both deservedly iconic.

Masi Ven ★★→★★★ Archetypal yet innovative producer of Verona led by inspirational Sandro Boscaini. V.gd Rosso Veronese *Campo Fiorin* and Osar (Oseleta) Top AMARONES Costasera, Mazzon, Campolongo di Torbe. New F-VG Moxxé fizz from PINOT GRIGIO/partly dried VERDUZZO.

Massa, Vigneti Pie ★★★ Walter Massa brought Timorasso (w) grape back from m-extinction. Top: Coste del Vento, Montecitorio, Sterpi. V.gd: Freisa and BARBERA Monleale.

Massolino Vigna Rionda Pie ★★★ One of finest BAROLO estates, in Serralunga. Excellent Parafada, Margheria have firm structure, fruity drinkability; long-ageing VIGNA Rionda best. V.gd Parussi and LANGHE NEBBIOLO.

Mastroberardino Cam ★★★ Historic top-quality producer of mtn Avellino province in CAM. Top *Taurasi* (Historia Naturalis, Radici), also FIANO DI AVELLINO (More Maiorum, Radici and new Stilèma), GRECO DI TUFO Nova Serra.

Melini Tus ★★ Part of GIV. Gd quality/price, esp CHIANTI CLASSICO Selvanella.

Meroi F-VG ★★★ Dynamic estate. Top FRIULANO, MALVASIA Zittelle Durì, RIBOLLA GIALLA, SAUV BL Zitelle Barchetta.

Metodo classico or tradizionale Italian for "Champagne method".

Miani F-VG ★★★★ Enzo Pontoni is Italy's best white winemaker. Top: FRIULANO (Buri and Filip), RIBOLLA GIALLA Pettarin, SAUV BL Zitelle. V.gd: Sauv Bl Saurint, CHARD Zitelle, MERLOT, REFOSCO Buri.

Mogoro, Cantina di Sar ★★→★★★ High-quality co-op. Rare producer of Semidano

> **Barolo MGA**
> The concept of "cru" is gaining acceptance here, although it is not allowed on the label. It is replaced (in BAROLO and BARBARESCO at least) by subzones known officially as MGA (Menzioni Geografiche Aggiuntive). Currently Barolo has 11 village mentions and 170(!) additional geographical mentions. Most celebrated incl (by commune) **Barolo:** Bricco delle Viole, Brunate, Bussia, Cannubi, Cannubi Boschis, Cannubi San Lorenzo, Cannubi Muscatel, Cerequio, Le Coste, Sarmassa; **Castiglione Falletto:** Bricco Boschis, Bricco Rocche, Fiasco, Monprivato, Rocche di Castiglione, Vignolo, Villero; **Cherasco:** Mantoetto; **Diano d'Alba:** La VIGNA, Sorano (partly shared with Serralunga); **Grinzane Cavour:** Canova, Castello; **La Morra:** Annunziata, Arborina, Bricco Manzoni, Bricco San Biagio, Brunate, Cerequio, Fossati, La Serra, Rocche dell'Annunziata, Rocchettevino, Roggeri; **Monforte d'Alba:** Bussia, Ginestra, Gramolere, Mosconi, Perno; **Novello:** Bergera, Ravera; **Roddi:** Bricco Ambrogio; **Serralunga d'Alba:** Baudana, Boscareto, Cerretta, Falletto, Francia, Gabutti, Lazzarito, Marenca, Ornato, Parafada, Prapò, Vignarionda; **Verduno:** Massara, Monvigliero. None of them is likely to be cheap.

(w) variety. Best: Monica San Bernardino and Semidano SUPERIORE Puistéris. V.gd NURAGUS Ajò.

Molettieri, Salvatore Cam ★★★ Outstanding: TAURASI, RISERVA VIGNA Cinque Querce. Gd FIANO DI AVELLINO Apianum.

Monaci Pug r p ★★ →★★★ Part of GIV. Characterful NEGROAMARO Kreos (p), PRIMITIVO Artas and SALICE SALENTINO Aiace (r).

Monferrato Pie Hills between River Po and Apennines. Main cities: Asti, Canelli and Nizza. Some of Italy's most delicious and fairly priced wines from typical local grapes: BARBERA, Freisa, GRIGNOLINO, MALVASIAS (di Casorzo and di Schierano), Ruché.

Monica di Sardegna Sar DOC r ★→★★★ DYA. Delightfully perfumed, medium weight. Best: ARGIOLAS (Iselis), CANTINA DI MOGORO, CANTINA SANTADI (Antigua), CONTINI, Dettori (Chimbanta), Ferruccio Deiana (Karel), Josto Puddu (Torremora).

Monsanto Tus ★★★ Esteemed CHIANTI CLASSICO estate. Best: Il POGGIO RISERVA (1st single-v'yd Chianti Classico). Gd: Chianti Classico (RISERVA) and IGT SANGIOVETO.

Montagnetta, La Pie ★★ ·★★★ One of top producers of Freisas. Best: Freisa d'Asti SUPERIORE Bugianen.

Montalcino Tus Hilltop town in province of Siena, fashionable and famous for concentrated, expensive BRUNELLO and more approachable, better-value ROSSO DI MONTALCINO, both still 100% SANGIOVESE.

Monte Carrubo Si r ★★★ Pioneer Peter Vinding-Diers planted SYRAH on a former volcano s of ETNA. Exciting, complex results.

Montefalco Sagrantino Umb DOCG r dr (sw) ★★★ →★★★★ Once sweet PASSITO only (still best wine of area), drier version is Italy's most powerfully tannic red that requires optimal growing seasons to show best. Gd: Adanti, Antano Milziade, Antonelli, CAPRAI, Colpetrone, LUNGAROTTI, Paolo Bea, Scacciadiavoli, Tabarrini.

Montepulciano d'Abruzzo Ab DOC r p ★★ →★★★ (r) **12 13** 14 15 1st all-region DOC of Italy. Subdenomination: Colline Teramane (now DOCG), wines often tough, charmless despite hype; Controguerra (DOC), usually more balanced. Reds can be either light, easy-going or structured, rich. Look for Cataldi Madonna (Toni, lighter Malandrino), EMIDIO PEPE, Marina Cvetic (S Martino Rosso), Terraviva, TIBERIO (Colle Vota and regular), Torre dei Beati (Cocciapazza, Mazzamurello), Valle Reale, *Valentini* (best, age-worthy), Zaccagnini.

Montevertine Tus ★★★★ Organic certified estate in Radda. Outstanding IGT Le Pergole Torte, world-class, pure, long-ageing SANGIOVESE. V.gd Montevertine.

Montevetrano Cam ★★★ Iconic CAM AZIENDA. Superb IGT Montevetrano (AGLIANICO, CAB SAUV, MERLOT). V.gd Core Rosso (Aglianico).

Morella Pug ★★★ →★★★★ Gaetano Morella and wife Lisa Gilbee make outstanding PRIMITIVO (Old Vines, La Signora, new Mondo Nuovo) from c.90-yr-old vines.

Morellino di Scansano Tus DOCG r ★ →★★★ 11 13 15 16' 17 (18) MAREMMA's famous SANGIOVESE-based red is best when cheerful and light rather than overoaked and gritty. When gd, perfect with grilled foods, pasta, pizza. Best: *Le Pupille*, Mantellassi, Moris Farms, PODERE 414, POGGIO Argentiera, Roccapesta, Terenzi.

Moris Farms Tus ★★★ One of 1st new-age producers of TUS's MAREMMA; Monteregio and *Morellino di Scansano* DOCS, plus VERMENTINO IGT. Top cru is iconic IGT Avvoltore, rich SANGIOVESE/CAB SAUV/SYRAH blend. But *try basic Morellino*.

Moscato d'Asti Pie DOCG w sw sp ★★→★★★ DYA Similar to DOCG ASTI, but usually better grapes; lower alc, lower pressure, sweeter, fruitier, often from small producers. Best DOCG MOSCATO: Bera, Ca' d'Gal, Caudrina (La Galeisa), Elio Perrone, Forteto della Luja, SARACCO, *Vajra*, Vignaioli di Sante Stefano. V.gd: *Braida*, Cascina Fonda, Il Falchetto, L'Armangia, RIZZI, Scagliola.

Muri Gries T-AA ★★→★★★ Monastery in Bolzano suburb of Gries; traditional and still top producer of LAGREIN ALTO ADIGE DOC. Esp cru Abtei-Muri and Klosteranger.

Nals Margreid T-AA ★★★ Small quality co-op making mtn-fresh whites (esp PINOT BIANCO Sirmian). Harald Schraffl is an inspired winemaker.

Nebbiolo d'Alba Pie DOC r dr ★★ →★★★ 11 12 13 14 15 16' (18) (100% NEBBIOLO) Sometimes a worthy replacement for BAROLO/BARBARESCO, though it comes from a distinct area between the two. Gd: BREZZA (VIGNA Santa Rosalia), BRUNO GIACOSA (Valmaggiore), PAITIN, LUCIANO SANDRONE and ORLANDO ABRIGO (Valmaggiore).

Negrar, Cantina Ven ★★ Aka CS VALPOLICELLA. Major producer of Valpolicella, RIPASSO, AMARONE; grapes from various parts of CLASSICO zone. Look for brand name Domini Veneti. So-so quality but affordable.

Nervi Pie ★★★ Historical winery in GATTINARA now owned by Roberto CONTERNO. Best: Valferana and Molsino.

Niedrist, Ignaz I w ★★★ LAGREIN Berger Gei (also RISERVA) is reference. So are RIES, WEISSBURGUNDER (Limes). V.gd BLAUBURGUNDER, SAUV BL, Trias (w blend).

Nino Franco Ven ★★★→★★★★ Owner Primo Franco makes large volumes of top-notch PROSECCO that age surprisingly well. Among finest: Primo Franco Dry, Riva di San Floriano Brut. Excellent CARTIZZE.

Nipozzano, Castello di Tus ★★★→★★★★ FRESCOBALDI estate in RÚFINA, e of Florence, making excellent CHIANTI Rúfina. Top Nipozzano RISERVA (esp Vecchie Viti) and IGT *Montesodi*. V.gd Mormoreto (B'x blend).

World's densest planting: Bonsai v'yd, Montalcino. 62,500 vines/ha. Why?

Nittardi Tus ★★→★★★ Reliable source of quality modern CHIANTI CLASSICO. German owned; oenologist Carlo Ferrini.

Nozzole Tus ★★→★★★ Famous estate in heart of CHIANTI CLASSICO, n of Greve, owned by FOLONARI. V.gd Chianti Classico RISERVA, excellent CAB SAUV Pareto.

Nuragus di Cagliari Sar DOC w ★★ DYA. Lively, easy SAR wine from Nuragus grape, finally gaining visibility. Best: ARGIOLAS (S'Elegas), MOGORO (Ajò), Pala (I Fiori).

Occhio di Pernice Tus "Partridge's eye". A type of VIN SANTO made predominantly from black grapes, mainly SANGIOVESE. *Avignonesi's is definitive*. Also an obscure black variety found in RÚFINA and elsewhere.

Occhipinti, Arianna Si ★★★ Cult producer, deservedly so. Organic certified. Happily, wines now less oxidized and more about finesse than power. Top: Il Frappato. V.gd NERO D'AVOLA Siccagno, Nero d'Avola Frappato SP68 and Bianco SP68.

Oddero Pie ★★★→★★★★ Traditionalist La Morra estate for excellent BAROLO (Brunate, Bussia, Villero, VIGNA Rionda RISERVA now 10 yrs), BARBARESCO (Gallina) crus, plus other serious PIE wines. V.gd value Barolo and RIES. Monvigliero from 2021.

Oltrepò Pavese Lom DOC r w dr sw sp ★→★★★ Multi-DOC, incl numerous varietal and blended wines from Pavia province; best is SPUMANTE. Gd growers: Anteo, Barbacarlo, CS Casteggio, Frecciarossa, Le Fracce, Mazzolino, Monsupello, Ruiz de Cardenas, Travaglino; La Versa co-op.

What to drink with Tuscan ribollita soup? Chianti Classico or Morellino di Scansano.

Ornellaia Tus ★★★★ 10 11 12 13 15 16 Fashionable, indeed cult, estate nr BOLGHERI now owned by FRESCOBALDI. Top wines of B'x grapes/method: Bolgheri DOC Ornellaia, IGT Masseto (MERLOT), Ornellaia Bianco (SAUV BL/VIOGNIER). Gd: Bolgheri DOC Le Serre Nuove and POGGIO alle Gazze (w).

Orvieto Umb DOC w dr sw s/sw ★→★★★ DYA One of few areas of Italy where noble rot occurs spontaneously and often. Sweet late-harvest can be memorable though cheap-and-cheerful dry white v. popular too. Off-dry Amabile less in favour today but delicious. Top: BARBERANI (Luigi e Giovanna) but sweet Calcaia just as gd. Other gd: Bigi, Cardeto, *Castello della Sala*, Decugnano dei Barbi, Palazzone; Sergio Mottura (Lat).

Pacenti, Siro Tus ★★★ Modern-style BRUNELLO, ROSSO DI MONTALCINO from small, caring producer.

Paitin Pie ★★→★★★ Pasquero-Elia family have been bottling BARBARESCO since C19. Today back on track making "real" Barbaresco in large barrels. Sorì Paitin Vecchie Vigne is star.

Paltrinieri ★★→★★★ One of top three LAMBRUSCO producers. Among 1st to produce 100% Lambrusco di Sorbara. Best: Secco Radice; v.gd Leclisse.

Pantelleria Si ★★★ Windswept, black (volcanic) earth SI island off Tunisian coast, famous for superb MOSCATO d'Alessandria stickies. PASSITO versions particularly dense/intense. Try DE BARTOLI (Bukkuram), DONNAFUGATA (Ben Ryé), Ferrandes.

Parrina, La Tus ★★ Popular estate and *agriturismo* on TUS coast dominates DOC Parrina; solid rather than inspired wines.

Passito Tus, Ven One of Italy's most ancient and characteristic styles, from grapes dried briefly under harvest sun (in s) or over a period of weeks or mths in airy attics of winery – a process called *appassimento*. Best-known versions: VIN SANTO (TUS); AMARONE/RECIOTO (Ven), VALPOLICELLA/SOAVE. *See also* Loazzolo, MONTEFALCO, ORVIETO, TORCOLATO, VALLONE. Never cheap.

Paternoster Bas ★★★ Old organic estate now owned by TOMMASI family. Top: Don Anselmo. V.gd AGLIANICO DEL VULTURE Rotondo.

Pepe, Emidio Ab ★★★ Artisanal winery, 15 ha, bio and organic certified. Top MONTEPULCIANO D'ABRUZZO. Gd: TREBBIANO D'ABRUZZO (Old Vines) and PECORINO.

Pian dell'Orino ★★★→★★★★ Small MONTALCINO estate, committed to bio. BRUNELLO seductive, technically perfect, Rosso nearly as gd. Many epic wines.

Picolit F-VG DOCG w sw s/sw ★★→★★★ 12 13 15 16 (18) Potentially Italy's best sweet (most from air-dried grapes, but rare late-harvests even better), but plagued by poor versions that don't speak of the grape. Texture ranges from light/sweet (rare) to super-thick (PASSITO). Best: Aquila del Torre, I Comelli, LIVIO FELLUGA, Marco Sara, Perusini, Ronchi di Cialla, Valentino Butussi, VIGNA PETRUSSA. V.gd: Ermacora, Girolamo Dorigo, Paolo Rodaro.

Piedmont / Piemonte In ne, bordering France to the w. Turin is capital. MONFERRATO, LANGHE, ROERO, ALTO PIEMONTE main areas. With TUS, Italy's most important region for quality (10% of all DOC(G) wines). No IGTS allowed. Grapes incl: BARBERA, Brachetto, Cortese, DOLCETTO, Freisa, GRIGNOLINO, MALVASIA di Casorzo, Malvasia di Schierano, MOSCATO, NEBBIOLO, Ruché, Timorasso. *See also* BARBARESCO, BAROLO.

Pieropan Ven ★★★★ Andrea and Dario, Leonildo's son, now run winery. Cru *La Rocca* still ultimate oaked soave; Calvarino best of all. V.gd AMARONE. Waiting for PASSITO della Rocca.

Pietradolce Si ★★★→★★★★ Faro bros own v'yds in key ETNA crus, often pre-phylloxera vines. Top: Etna Rosso Barbagalli, Rampante. V.gd: Archineri (r w).

Pio Cesare Pie ★★→★★★ Veteran ALBA producer, offers BAROLO, BARBARESCO in modern (barrique) and traditional (large-cask-aged) versions. Particularly gd NEBBIOLO D'ALBA, *a little Barolo at half the price.* New Barolo Mosconi.

Pira & Figli – Chiara Boschis Pie ★★★→★★★★ Organic certified. Must-visit estate. Top: Cannubi, Mosconi, Via Nuova.

Planeta Si ★★★ Leading SI estate with six v'yd holdings in various parts of island, incl Vittoria (CERASUOLO Dorilli), Noto (NERO D'AVOLA Santa Cecilia), most recently on ETNA (Carricante, NERELLO MASCALESE Eruzione 1614). Also gd: Nocera (r), La Segreta (r w), Cometa (FIANO).

Podere Tus Small TUS farm, once part of a big estate.

Poggio Tus Means "hill" in TUS dialect. "**Poggione**" means "big hill".

Poggio Antico Tus ★★★ Paola Gloder looks after this 32-ha estate, one of highest in MONTALCINO at c.500m (1640ft). Style is restrained, consistent, at times too herbal.

Poggio di Sotto Tus ★★★★ Small MONTALCINO estate with a big reputation recently. Has purchased adjacent v'yds. Top BRUNELLO, RISERVA and Rosso of traditional character with idiosyncratic twist.

Poggione, Tenuta Il Tus ★★★ S MONTALCINO estate; consistently excellent BRUNELLO, ROSSO. Top Brunello RISERVA VIGNA Paganelli. V.gd VIN SANTO.

Poggiopiano Tus ★★ Opulent CHIANTI CLASSICO from Bartoli family. Chiantis are pure SANGIOVESE, but SUPER TUSCAN Rosso di Sera incl up to 15% Colorino. V.gd Colorino Taffe Ta'.

Poggio Scalette Tus ★★ Vittorio Fiore and son Jury run CHIANTI organic estate at Greve. Top: Il Carbonaione (100% SANGIOVESE); needs several yrs bottle-age. Above-average CHIANTI CLASSICO and B'x-blend Capogatto.

Poliziano Tus ★★★ MONTEPULCIANO organic estate of Federico Carletti. Superior if often v. dark, herbal VINO NOBILE (esp cru Asinone and new Le Caggiole); gd IGT Le Stanze (CAB SAUV/MERLOT), Cortona In Violas (Merlot).

Pomino Tus DOC r w ★★★ (r) 12 13 15 16 Appendage of RÙFINA, with fine red and white blends (esp Il Benefizio). Virtually a FRESCOBALDI exclusivity.

Poluzzine, Le Tus ★★★★ Organic estate of Gorelli family just s of MONTALCINO. V'yd is quite high. Outstanding BRUNELLOS (also RISERVA) and Rossos, serious and v. drinkable. Try them at family's restaurant in town.

Prà Ven ★★★★ Leading SOAVE CLASSICO producer, csp crus Colle Sant'Antonio, Monte Grande, Staforte v. tasty. Excellent VALPOLICELLA La Morandina and AMARONE.

Coming soon: Valdarno di Sopra (Tus), 1st DOC for organic only.

Produttori del Barbaresco Pie ★★★ One of Italy's earliest co-ops, perhaps best in the world, makes excellent traditional straight BARBARESCO plus crus Asili, Montefico, Montestefano, Ovello, Pora, Rio Sordo. Super values.

Proprietà Sperino Pie ★★★→★★★★ Top estate of LESSONA. One of best of ALTO PIEMONTE run by Luca De Marchi (*see* ISOLE E OLENA). Outstanding: Lessona; V.gd L Franc (one of best Italian CAB FR), Rosa del Rosa (p, NEBBIOLO/Vespolina) and Uvaggio (r).

Prosecco Ven DOC(G) w sp ★→★★ DYA. Prosecco is the wine, GLERA the grape variety with which it is made. Quality is higher in the VALDOBBIADENE. Look for: Adami, Biancavigna, BISOL, Bortolin, Canevel, CARPENÈ-MALVOLTI, Case Bianche, Col Salice, Col Vetoraz, Le Colture, Gregoletto, La Riva dei Frati, Mionetto, NINO FRANCO, Ruggeri, Silvano Follador, Zardetto.

Prunotto, Alfredo Pie ★★★→★★★★ Acquired by ANTINORI in 90s. V.gd BARBARESCO (Bric Turot), BAROLO (Bussia, VIGNA Colonnello), NEBBIOLO (Occhetti), Nizza (Costamiole).

Puglia The "heel" of Italy. Many gd-value reds from likes of Bombino Nero, NEGROAMARO, PRIMITIVO, Susumaniello and Uva di Troia grapes. W Bombino Bianco and aromatic Minutolo most interesting whites, but Verdeca making comeback; by contrast, FIANO is disappointing, best avoided. Dubious winemaking talent and old equipment a real problem. Castel del Monte, Gioia del Colle Primitivo, PRIMITIVO DI MANDURIA, SALICE SALENTINO best denominations.

Querciabella Tus ★★★ Top CHIANTI CLASSICO estate, bio since 2000. Top IGT Camartina (CAB SAUV/SANGIOVESE), Batàr (CHARD/PINOT BL) and new single-commune wines (Greve in CHIANTI, Radda in Chianti, Gaiole). V.gd Chianti Classico (and RISERVA).

Quintarelli, Giuseppe Ven ★★★★ Arch-traditionalist artisan producer of sublime VALPOLICELLA, RECIOTO, AMARONE; plus a fine Bianco Secco, a blend of various grapes. Daughter Fiorenza and sons now in charge, altering nothing, incl the old man's ban on spitting when tasting.

Quintodecimo Cam ★★★ Oenology professor/winemaker Luigi Moio's beautiful estate. Outstanding: TAURASI VIGNA Grande Cerzito and VIGNA Quintodecimo; great AGLIANICO (Terra d'Eclano) and GRECO DI TUFO (Giallo d'Arles).

Ratti, Renato ★★→★★★ Iconic BAROLO estate. Modern wines; short maceration but plenty of substance, esp Barolos Rocche dell'Annunziata and Conca.

Recioto della Valpolicella Ven DOCG r sw (sp) ★★★→★★★★ Sweet wine marketing problems mean this traditional Italian beauty is being made less and less. A shame, esp since always much better than many disappointing overly-sweet, overly tannic AMARONES.

Old Piedmontese name for a wine bar: *piola*. Venetian for a glass of wine: *un ombra*.

Recioto di Soave Ven DOCG w sw (sp) ★★★→★★★★ SOAVE from half-dried grapes: sweet, fruity, slightly almondy; sweetness is cut by high acidity. *Drink with cheese.* Best: Anselmi, COFFELE, Gini, PIEROPAN, Tamellini, often v.gd from Ca' Rugate, Pasqua, PRÀ, Suavia, Trabuchi.

Refosco (dal Peduncolo Rosso) F-VG ★★ 12 13 15 (16) Most-planted native red grape of region. Best from FRIULI COLLI ORIENTALI DOC. Top: Volpe Pasini and Valchiarò. Gd: Ca' Bolani, LIVIO FELLUGA, MIANI, Vignai da Duline.

Ricasoli Tus Historic Tuscan family. First Italian Prime Minister Bettino R devised the classic CHIANTI blend. Main branch occupies medieval Castello di BROLIO.

Riecine Tus ★★★→★★★★ SANGIOVESE specialist at Gaiole since 70s. Riecine di Riecine and La Gioia (100% Sangiovese) potentially outstanding. Gd Tresette (MERLOT).

Rinaldi, Giuseppe Pie ★★★ Beppe R's daughters Marta and Carlotta continue their father's highly traditional path. Outstanding: Brunate and Freisa. Also gd, but overrated, Tre Tine.

Ripasso Ven *See* VALPOLICELLA RIPASSO.

Riserva Wine aged for a statutory period, usually in casks or barrels.

Rivetti, Giorgio (La Spinetta) Pie ★★★ Fine MOSCATO D'ASTI, excellent BARBERA, series of super-concentrated, oaky BARBARESCOS. Also owns v'yds in BAROLO, CHIANTI COLLI Pisane DOCGS and traditional SPUMANTE house Contratto.

Rizzi Pie ★★★→★★★★ Sub-area of Treiso, commune of BARBARESCO, where Dellapiana family look after 35-ha v'yd. Organic. Top cru is Barbaresco Pajore and Rizzi RISERVA Boito. V.gd: Nervo, Rizzi, also gd METODO CLASSICO Pas Dosè, MOSCATO D'ASTI.

Rocca, Bruno Pie ★★★→★★★★ Modern-style BARBARESCO (Rabajà, Currà) and other ALBA wines, also v. fine BARBERA D'ASTI.

Rocca, Albino Pie ★★★→★★★★ A foremost producer of elegant, sophisticated BARBARESCO: top crus Ovello VIGNA Loreto, Ronchi, new Cottà.

Rocca delle Macìe Tus ★★ Large estate in Castellina-in-CHIANTI run by Sergio Zingarelli. Gd quaffing Chianti, plus top wines incl Gran Selezione Sergio Zingarelli and Fizzano. Also Campo Macione estate in Scansano zone.

Roero Pie DOCG r w ★★→★★★ 10 11 13 15 16 (18) Wilder, cooler much sandier soils compared to LANGHE. NEBBIOLO (r) and ARNEIS (w). Best producers: BRUNO GIACOSA ★, Ca' Rossa, Cascina Chicco, Cornarea, GIOVANNI ALMONDO, MALVIRÀ ★, MATTEO CORREGGIA ★, Morra, Negro, Rosso, Taliano, Val del Prete, Valfaccenda.

Romagna Sangiovese Mar DOC r ★★→★★★ At times too herbal and oaky, but often well-made, even classy SANGIOVESE red. Gd: Ca' di Sopra, Cesari, Condello, Drei Donà, FATTORIA ZERBINA, Nicolucci, Papiano, Paradiso, Tre Monti, Trere, Villa Venti (Primo Segno). Seek also IGT RONCO delle Ginestre, Ronco dei Ciliegi from CASTELLUCCIO.

Ronco Term for a hillside v'yd in ne Italy, esp F-VG.

Ronco del Gelso F-VG ★★★ →★★★★ Tight, pure ISONZO: PINOT GR Sot lis Rivis, FRIULANO Toc Bas and MALVASIA VIGNA della Permuta are regional benchmarks. V.gd Latimis (w blend).

Rosato General Italian name for rosé. Other rosé names incl CHIARETTO from Lake Garda; CERASUOLO from Ab; Kretzer from ALTO ADIGE.

Rossese di Dolceacqua or Dolceacqua Lig DOC r ★★ →★★★ Interesting reds. Intense, salty, spicy; greater depth of fruit than most. Gd: Ka Mancine, Maccario-Dringenberg (Posaù), Poggi dell'Elmo, TENUTA Anfosso, Terre Bianche.

Rosso Conero Mar *See* CONERO.

Rosso di Montalcino Tus DOC r ★★→★★★ 11 12 13 15 16 18 DOC for earlier-maturing wines from BRUNELLO grapes, usually from younger or lesser v'yd sites, but bargains exist.

Rosso di Montefalco Umb DOC r ★★ 12 13 15 16 18 SANGIOVESE/SAGRANTINO blend, often with a splash of softening MERLOT. *See* MONTEFALCO SAGRANTINO.

Rosso di Montepulciano Tus DOC r ★ 13 15 16 Junior version of VINO NOBILE DI MONTEPULCIANO, growers similar.

Rosso Piceno / Piceno Mar DOC r ★ 13 15 (18) Blend of MONTEPULCIANO (>35%) and SANGIOVESE (>15%). SUPERIORE means it comes only from far s of region. Gd: Boccadigabbia, BUCCI, GAROFOLI, Moncaro, Monte Schiavo, Saladini Pilastri, Santa Barbara, TENUTA di Tavignano, Velenosi.

Ruffino Tus ★→★★★ Venerable CHIANTI firm, at Pontassieve nr Florence, now owned by US giant Constellation Brands, produces reliable wines such as CHIANTI CLASSICO RISERVA Ducale and Ducale Oro.

Rùfina Tus ★★ →★★★ E of Florence, the most n sub-zone of CHIANTI grows SANGIOVESE at highest altitudes, meaning tight, refined, age-worthy wines. Gd to outstanding: CASTELLO DI NIPOZZANO (FRESCOBALDI), Castello del Trebbio, Colognole, Frascole, Grati/Villa di Vetrice, I Veroni, Lavacchio, SELVAPIANA, TENUTA Bossi, Travignoli. Don't confuse with RUFFINO, which happens to have HQ in Pontassieve, Rúfina's main town.

Russiz Superiore F-VG ★★→★★★ LIVIO FELLUGA's brother, Marco, est v'yds in various parts of F-VG. Now run by Marco's son Roberto. Wide range; best is PINOT GRIGIO, COLLIO Bianco blend Col Disôre. V.gd PINOT BIANCO RISERVA.

Salento Pug Home to Italy's best rosé from NEGROAMARO (along with Abruzzo's CERASUOLO made from MONTEPULCIANO). But also v.gd Negramaro, with a bit of help from MALVASIA Nera, and now increasingly monovarietal reds from local Sussumaniello. *See also* PUG, SALICE SALENTINO.

Salice Salentino Pug DOC r ★★ →★★★ 11 13 15 16 17 Best-known of Salento's too many NEGROAMARO-based DOCs. RISERVA after 2 yrs. Gd: Vallone (Vereto Riserva), Cantele, Leone de Castris (Riserva), Mocavero, Conti Zecca (Cantalupi), Cosimo Taurino.

Salvioni Tus ★★★★ Aka La Cerbaiola; iconic small, highest-quality MONTALCINO estate run by father and daughter team. BRUNELLO, ROSSO DI MONTALCINO among v. best available and worthy of their high prices.

Sandrone, Luciano Pie ★★★→★★★★ Modern-style ALBA wines. Deep BAROLOS: Aleste (formerly Cannubi Boschis), Le VIGNE. Also gd NEBBIOLO D'ALBA Valmaggiore.

San Felice Tus ★★★ Important historic TUS grower, owned by Gruppo Allianz, run by Leonardo Bellaccini. Fine CHIANTI CLASSICO and RISERVA POGGIO Rosso from estate in Castelnuovo Berardenga. Gd too: IGT *Vigorello* (1ST SUPER TUSCAN, from 1968), BRUNELLO DI MONTALCINO Campogiovanni.

San Gimignano Tus TUS medieval town known for its towers and dry (w) VERNACCIA DI SAN GIMIGNANO DOCG. Top: Cesani, FALCHINI, Guicciardini Strozza, Il Colombaio, Il Palagione, Montenidoli, Mormoraia, Panizzi, Pietrafitta.

San Giusto a Rentennano Tus ★★★★ Top CHIANTI CLASSICO estate. Outstanding SANGIOVESE IGT Percarlo, sublime VIN SANTO (Vin San Giusto), MERLOT (La Ricolma); v.gd Chianti Classico, RISERVA Le Baroncole.

San Guido, Tenuta Tus *See* SASSICAIA.

San Leonardo T-AA ★★★★ Top TRENTINO estate of Marchesi Guerrieri Gonzaga. Main wine is B'x blend, *San Leonardo*, Italy's most claret-like wine. V.gd CARMENÈRE and Villa Gresti (MERLOT/Carmenère).

San Michele Appiano T-AA ★★★ Historic co-op. *Mtn-fresh whites*, brimming with varietal typicity, drinkability, are speciality. Seek out wines of The Wine Collection and Appius (selected by Hans Terzer). Top PINOT BL Schulthauser and Sanct Valentin line.

Santadi Sar ★★★ SAR'S, and one of Italy's, best co-ops, esp for CARIGNANO-based reds *Terre Brune*, Grotta Rossa, Rocca Rubia RISERVA (all DOC CARIGNANO DEL SULCIS). V.gd MONICA DI SARDEGNA Antigua.

Santa Maddalena / St-Magdalener T-AA DOC r ★→★★★ DYA. Teutonic-style red from SCHIAVA grapes from v. steep slopes behind ALTO ADIGE capital Bolzano. Notable: CS St-Magdalena (Huck am Bach), CS Tramin, Gojer, Hans Rottensteiner (Premstallerhof), Untermoserhof, Waldgries.

Sant'Antimo Tus DOC r w sw ★★ Lovely little Romanesque abbey gives its name to this catch-all DOC for (almost) everything in MONTALCINO zone that isn't BRUNELLO DOCG or Rosso DOC.

Saracco, Paolo Pie ★★★★ Italy's best MOSCATO D'ASTI producer. V.gd: LANGHE RIES, PINOT N and new BARBERA D'ALBA "Per Sbaglio".

Sardinia / Sardegna Italy's 2nd-largest island is home to world-class whites and reds. Look for VERMENTINO DI GALLURA DOCG and VERMENTINO DI SARDEGNA (fruitier and less mineral), Sherry-like VERNACCIA DI ORISTANO and NURAGUS among whites, late-harvest sweet Nasco; forgotten Semidano deserves much better. CANNONAU (GARNACHA) and CARIGNANO most famous reds, but Bovale Sardo and Pascale just as gd.

17 different vines in Italy called Malvasia: v. fashionable wine in C17.

Sassicaia Tus DOC r ★★★★ 04' 05 06 07' 08 09 10 13 15' 16 Italy's sole single-v'yd DOC (BOLGHERI), a CAB (SAUV/FR) made on First Growth lines by Marchese Incisa della Rocchetta at TENUTA SAN GUIDO. More elegant than lush, made for age – and often bought for investment, but hugely influential in giving Italy a top-quality image. 16 extremely elegant, one of best recent yrs.

Satta, Michele Tus ★★★ Virtually the only BOLGHERI grower to succeed with 100% SANGIOVESE (Cavaliere). Also Bolgheri DOC red blends Piastraia and SUPERIORE I Castagni.

Scavino, Paolo Pie ★★★ Modernist BAROLO producer of Castiglione Falletto, esp crus Rocche dell'Annunziata, Bric del Fiasc, Cannubi, Monvigliero. V.gd new Barolo crus: Ravera and Prapò.

Schiava Alto Adige T-AA DOC r ★ DYA. Schiava (VERNATSCH in German) gives practically tannin-free, easy-glugging red from most s territory of German-speaking world. Sadly disappearing from Tyrolean v'yds.

Schioppetto, Mario F-VG ★★★ Legendary late COLLIO pioneering estate now owned by Rotolo family. V.gd DOC SAUV BL, *Pinot Bl*, RIBOLLA GIALLA, FRIULANO, IGT blend Blanc des Rosis, etc.

Sella & Mosca Sar ★★ Major SAR grower and merchant with v. pleasant white Torbato (esp Terre Bianche) and light, fruity VERMENTINO Cala Viola (DYA). Gd Alghero DOC Marchese di Villamarina (CAB SAUV) and Monteluce (Nasco). Also interesting Port-like Anghelu Ruju.

Bolzano, source of juicy red Lagrein and cool mtn whites – summer as hot as Palermo.

Selvapiana Tus ★★★★ CHIANTI RÚFINA organic estate among Italian greats. Best: RISERVA Bucerchiale, IGT Fornace; but even *basic Chianti Rúfina is a treat.* Also fine red Petrognano, POMINO, Riserva VIGNETO Erchi.

Settesoli, CS Si ★ → ★★ Co-op giving SI gd name with reliable, gd-value wines.

Sforzato / Sfursat Lom ★★★ DOCG Sforzato di VALTELLINA is made AMARONE-like, from air-dried NEBBIOLO grapes. Ages beautifully.

Sicily The Med's largest island, modern source of exciting original wines and value. Native grapes (r. Frappato, NERO D'AVOLA, NERELLO MASCALESE; w: CATARRATTO, Grecanico, GRILLO, INZOLIA), plus internationals. V'yds on flatlands in w, hills in centre, volcanic altitudes on Mt Etna.

Soave Ven DOC w (sw) ★ → ★★★ Famous, hitherto underrated, Veronese white. CHIARD, GARGANEGA, TREBBIANO di Soave. Wines from volcanic soils of CLASSICO zone can be intense, saline, v. fine, quite long-lived.

Solaia Tus r ★★★★ 06 07 08 09 10 11 12 13 15 16 Occasionally magnificent CAB SAUV/SANGIOVESE blend by ANTINORI, made to highest B'x specs; needs yrs of laying down.

Sottimano Pie ★★★ → ★★★★ Family estate. One of most inspired in BARBARESCO (crus: Basarin, Cottà, Currà, Fausoni, Pajorè). V.gd DOLCETTO D'ALBA, BARBERA D'ALBA.

Speri Ven ★★★ VALPOLICELLA family estate, organic certified (2015), gd sites incl outstanding Monte Sant'Urbano. Traditional style. Top: AMARONE Sant'Urbano.

Spumante Sparkling.

Südtirol T-AA German name for ALTO ADIGE.

Superiore Wine with more ageing than normal DOC and 0.5 1% more alc. May indicate a restricted production zone, eg. ROSSO PICENO Superiore.

Super Tuscan Tus Wines of high quality and price developed in 70s/80s to get round silly laws then prevailing. Now, esp with Gran Selezione on up, scarcely relevant. Wines still generally considered in Super Tuscan category, strictly unofficially: CA' MARCANDA, Flaccianello, Guado al Tasso, Messorio, ORNELLAIA, Redigaffi, SASSICAIA, SOLAIA, TIGNANELLO.

Sylla Sebaste Pie ★★★ Illustrates merits of rare NEBBIOLO Rosé variety: lighter, v. perfumed BAROLO. A beauty.

Tasca d'Almerita Si ★★★ New generation of Tasca d'Almeritas runs historic, still prestigious ETNA estate. High-altitude v'yds; balanced IGTS under its old Regaleali label. Top: NERO D'AVOLA-based *Rosso del Conte* and Etna wines Tascante Sciaranuova and Rampante. V.gd MALVASIA delle Lipari Capofaro, GRILLO Mozia TENUTA Whitaker.

Taurasi Cam DOCG r ★★★ 09 10 11 12 13 15 16 S Italy's 1st DOCG. Best AGLIANICO of CAM: none so potentially *complex, demanding, ultimately rewarding.* 17 communes, four sub-zones (nw, w, Taurasi, s). Top: Contrade di Taurasi (Vigne d'Alto, Coste), Guastaferro (Primum), MASTROBERARDINO (Radici), MOLETTIERI SALVATORE (VIGNA

CinqueQuerce), QUINTODECIMO (VIGNA Gran Cerzito). V.gd: FERRARA BENITO (Vigna Quattro Confini), FEUDI DI SAN GREGORIO, Perillo.

Tedeschi Ven ★★★ Bevy of v. fine AMARONE, VALPOLICELLA. Amarone Capitel Monte Olmi and RECIOTO Capitel Monte Fontana best.

Tedeschi, Fratelli Ven ★★→★★★ One of original quality growers of VALPOLICELLA when zone was still ruled by mediocrities.

Tenuta An agricultural holding (*see* under name – eg. SAN GUIDO, TENUTA).

Tenuta di Fessina Si ★★★→★★★★ Silvia Maestrelli has one of youngest and best estates in Rovittello area of ETNA. Elegant wines. Top Il Musmeci (r w). V.gd: Erse.

Terlano T-AA w ★★→★★★ DYA. ALTO ADIGE Terlano DOC applies to one white blend and eight white varietals, esp PINOT BL, SAUV BL. Can be v. fresh, zesty or serious and surprisingly long-lasting. Best: CS Terlano (esp Pinot Bl Vorberg), Maninor, MURI GRIES.

Teroldego Rotaliano T-AA DOC r p ★★→★★★ TRENTINO's best local grape makes seriously tasty wine on flat Campo Rotaliano. *Foradori* is tops. Gd: Dorigati, Endrizzi, MEZZACORONA'S RISERVA Nos, Zeni.

Terre da Vino Pie ★ →★★ Co-op in BAROLO; millions of bottles of Barolo and other LANGHE; remarkably consistent quality. Look for brand Vite Colte.

Terre Nere, Tenuta delle Si ★★★★ Marc de Grazia shows great wine can be made from NERELLO and CARRICANTE grapes, on coveted n side of Mt Etna. Top: Cuvèe delle Vigne Niche wines, Guardiola and pre-phylloxera La VIGNA di Don Peppino. V.gd Le Vigne di Eli.

Terriccio, Castello del Tus ★★★ Large estate s of Livorno: excellent, v. expensive B'x-style IGT Lupicaia; v.gd IGT Tassinaia, Terriccio (mainly Rhône grapes blend).

Tiberio ★★★★ Outstanding TREBBIANO D'ABRUZZO Fonte Canale (60-yr-old vines) one of Italy's best whites and new MONTEPULCIANO D'ABRUZZO Colle Vota; CERASUOLO D'ABRUZZO, PECORINO also exceptional.

Tiefenbrunner T-AA ★★★→★★★★ Grower-merchant in Teutonic castle (*Turmhof*) in s ALTO ADIGE. Wide range of mtn-fresh white and well-defined red varietals, esp 1000m-high MÜLLER-T *Feldmarschall*, one of Italy's best whites. Look for new CAB SAUV VIGNA Toren.

Tignanello Tus r ★★★★ 07′ 08 09 10 11 12 13 15 16 SANGIOVESE/CAB SAUV blend, barrique-aged, 1st made in 1971 as CHIANTI CLASSICO; with 1975 moved to international/native blend we know today. Such a large volume of such high quality speaks of ANTINORI family talent.

Tommasi Ven ★★★ 4th generation now in charge. Top: AMARONE (RISERVA Ca' Florian, new RISERVA De Buris), VALPOLICELLA Rafael. Other estates in Bas (PATERNOSTER), OLTREPÒ PAVESE (TENUTA Caseo), PUG (Masseria Surani), Ven (Filodora).

Torcolato Ven Sweet white from BREGANZE in Ven; Vespaiolo grapes laid on mats or hung up to dry for mths, as nearby RECIOTO DI SOAVE. Best: CS Beato Bartolomeo da Breganze, MACULAN, Miotti.

Torgiano Umb DOC r p w (sp) ★★ and **Torgiano, Rosso Riserva** DOCG r ★★ →★★★

Valpolicella: the best

Time to take gd VALPOLICELLA more seriously. AMARONE DELLA VALPOLICELLA and RECIOTO DELLA VALPOLICELLA are now DOCG, while RIPASSO has new rules. Following producers make gd to great wine: ALLEGRINI ★, Begali, BERTANI, BOLLA, Boscaini, Brigaldara ★, BRUNELLI, BUSSOLA ★, Ca' la Bianca, Ca' Rugate, Campagnola, CANTINA Valpolicella, Castellani, Corteforte, Corte Sant'Alda, CS Valpantena, DAL FORNO ★, GUERRIERI-RIZZARDI ★, Le Ragose, Le Salette, MASI ★, Mazzi ★, Nicolis, PRÀ, QUINTARELLI ★, Roccolo Grassi ★, Serego Alighieri ★, SPERI ★, Stefano Accordini ★, TEDESCHI ★, TOMMASI ★, Valentina Cubi, Venturini, Viviani ★, ZENATO, Zeni.

08 09 10 11 12 13 15. Gd-to-excellent red from Umb, virtually an exclusivity of LUNGAROTTI. Top: *Vigna Monticchio* Rubesco RISERVA. Keeps many yrs.

Torrette VdA DOC r ★→★★★ Blend based on Petit Rouge and other local varieties. Best: Torrette SUPERIEUR. Gd: Anselmet, D&D, Didier Gerbelle, GROSJEAN, FEUDO DI SAN MAURIZIO, LES CRETES, Elio Ottin.

Travaglini Pie ★★★ Solid producer of n PIE NEBBIOLO, with v.gd GATTINARA RISERVA, Gattinara Tre VIGNE, pretty-gd MC Nebolè (Nebbiolo) and Coste della Sesia.

Trebbiano d'Abruzzo Ab DOC w ★→★★★★ DYA. Generally crisp, simple, but VALENTINI's and Tiberio's Fonte Canale are *two of Italy's greatest* whites.

Trediberri Pie ★★★ Dynamic estate, top BAROLO Rocche dell'Annunziata (best value); v.gd LANGHE NEBBIOLO, BARBERA D'ALBA.

Trentino T-AA DOC r w dr sw ★→★★★ Varietally named DOC wines; best are perfumed, flavourful, inexpensive. Less successful ones dilute, boring, neutral. Best: GEWURZ, Nosiola, MÜLLER-T, SCHIAVA, MARZEMINO, TEROLDEGO. **Trento DOC** is name of potentially high-quality METODO CLASSICO wines.

Trinoro, Tenuta di Tus ★★★★ Individualist TUS red estate, pioneer in DOC Val d'Orcia between MONTEPULCIANO and MONTALCINO. Heavy accent on B'x grapes in flagship TENUTA di Trinoro, also in Palazzi, Camagi, Tenagli and Magnacosta. *See also* VINI FRANCHETTI (ETNA).

Tua Rita Tus ★★★★ 1st producer, as new BOLGHERI in 90s, of possibly Italy's greatest MERLOT in Redigaffi, also outstanding B'x blend *Giusto di Notri*, SYRAH Per Sempre. *See* VAL DI CORNIA.

Tuscany / Toscana Home of world's top SANGIOVESE. CHIANTI CLASSICO, BRUNELLO DI MONTALCINO and Chianti RÚFINA best, but BOLGHERI just as gd and world-class for international grapes (esp CAB FR, MERLOT), though SASSICAIA (CAB SAUV) is most famous.

Uberti Lom ★★★ →★★★★ Historical estate, excellent interpreter of FRANCIACORTA's terroir. Outstanding DeQuinque (blend of 10 yrs) and Comarì del Salem. V.gd Dosaggio Zero Sublimis and Francesco I.

Umani Ronchi Mar ★★→★★★ Leading Mar producer, esp for VERDICCHIO (Casal di Serra, Plenio), CONERO Cumaro, IGTS Le Busche (w), Pelago (r).

Vajra, GD Pie ★★★ →★★★★ Leading quality BAROLO producer in Vergne. Outstanding Bricco delle Viole and LANGHE Freisa Kyè. Gd: Langhe RIES Petracine, Barolo Albe, DOLCETTO Coste & Fossati; and Serralunga's Luigi Baudana Barolos and new Coste di Rose.

Val di Cornia Tus DOC(G) r p w ★★→★★★ 10 11 12 13 15 16 Quality zone s of BOLGHERI. MERLOT, MONTEPULCIANO, SYRAH. Look for: Bulichella, Gualdo del Re, Casadei, Terricciola, TUA RITA.

Valentini, Edoardo Ab r w ★★★★ Collectors seek out his MONTEPULCIANO D'ABRUZZO, TREBBIANO D'ABRUZZO, among Italy's v. best. Traditional, age-worthy.

Valle d'Aosta DOC r p w ★★→★★★ Italy's smallest region makes some its best reds and whites, though hard to find. DOCS mostly varietal; famous names incl NEBBIOLO-based Arnad Montjovet, Donnas; Torrette (mostly Petit Rouge), Blanc de Morgex, made with Prié (w); Chambave (made with local biotype of white MUSCAT), Nus MALVOISIE (made with PINOT GR), Premetta a lovely light red.

Valle Isarco T-AA DOC w ★★ DYA. ALTO ADIGE DOC for seven Germanic varietal whites made along the Isarco (Eisack) River ne of Bolzano. Gd GEWURZ, MÜLLER-T, RIES, SILVANER. Top: Abbazia di Novacella, Eisacktaler, KÖFERERHOF, Kuenhof, Manni Nossing.

Valpolicella Ven DOC(G) r ★ →★★★★ Light, easy-going red (Valpolicella), medium-bodied to powerfully alc, rich, tannic (AMARONE) and supersweet (RECIOTO). Popular RIPASSO in between Valpolicella and Amarone, but few noteworthy. *See* box, left.

Valpolicella Ripasso Ven DOC r ★★→★★★ 10 11 12 13 15 (16) (18) In huge demand, so changes from 2016. Used to be only from VALPOLICELLA SUPERIORE re-fermented (only once) on RECIOTO or AMARONE grape-skins to make a more age-worthy wine. Now can blend 10% Amarone with standard Valpolicella and call it Ripasso. Best: BUSSOLA, Castellani, DAL FORNO, QUINTARELLI, ZENATO.

Valtellina Lom DOC/DOCG r ★→★★★ A rare e-w valley, just s of Swiss border. Home of CHIAVENNASCA (NEBBIOLO). Best labelled Valtellina SUPERIORE (5 subzones: Grumello, Inferno, Maroggia, Sassella, Valgella), *see* SFORZATO. Top: Fay (Valgella Carteria), AR.PE.PE. (Sassella: RISERVA VIGNA Regina and Riserva Rocce Rosse), Mamete Prevostini (Sassella Sommarovina). For *Sforzato*: Fay (RONCO del Picchio), Mamete Prevostini (Albareda) and Nino Negri (Cinque Stelle).

Wines from Arneis, Ribolla Gialla – Sem too – all acquire saffron notes with age.

Vecchio Samperi Si *See* DE BARTOLI.

Verdicchio dei Castelli di Jesi Mar DOC w (sp) ★★→★★★ DYA. Versatile white from nr Ancona on Adriatic; light and quaffable or sparkling or structured, complex, long-lived (esp RISERVA DOCG, min 2 yrs old). Also CLASSICO. Best: Andrea Felici, *Bucci* (Riserva), Casalfarneto, Colognola, Coroncino (Gaiospino e Stracacio), Fazi Battaglia (Riserva San Sisto), GAROFOLI (Podium), La Staffa, Marotti Campi (Salmariano), MONTECAPPONE (FEDERICO II), MONTE SCHIAVO (Le Giuncare), Santa Barbara, Sartarelli (Balciana, rare late-harvest, and Tralivio), TENUTA di Tavignano (Misco and Riserva), UMANI RONCHI (Plenio and Casal di Serra).

Verdicchio di Matelica Mar DOC w (sp) ★★→★★★ DYA. Similar to last, smaller, more inland, higher, so more acidic, so longer-lasting though less easy-drinking young. RISERVA is likewise DOCG. Esp Belisario, Bisci, Borgo Paglianetto, Collestefano, La Monacesca (Mirum).

Verduno Pie DOC r ★★★ DYA. Berry and herbal flavours. Top producers: CASTELLO DI VERDUNO (Basadone), Fratelli Alessandria (Speziale), GB BURLOTTO. Gd: Bel Colle and Reverdito.

Verduno, Castello di Pie ★★★ Husband/wife team, v.gd BARBARESCO Rabaja and Rabajà Bas, BAROLO Monvigliero, VERDUNO Basadone.

Verduzzo F-VG DOC (Friuli Colli Orientali) w dr sw s/sw ★★→★★★ Full-bodied white from local variety. Ramandolo (DOCG) is well-regarded subzone for sweet wine. Top: Marco Sara, Scubla, I Clivi.

Vermentino di Gallura Sar DOCG w ★★→★★★ DYA. VERMENTINO makes gd light wines in TUS, LIG and all over SAR, but best, most intense in ne corner of island, under this its DOCG name. Try Capichera, CS del Vermentino, CS di Gallura, Depperu, Masone Mannu, Mura, Siddura, Zanatta.

Vermentino di Sardegna Lig DOC w ★★ DYA. From anywhere on SAR; generally fails to measure up to VERMENTINO DI GALLURA for structure, intensity of flavour. Gd producers incl: ARGIOLAS, Deiana, Mora e Memo, Quartomoro, *Santadi*, *Sella & Mosca*.

Vernaccia di Oristano Sar DOC w dr ★→★★★★ Flor-affected wine, similar to light Sherry, a touch bitter, full-bodied. Delicious with *bottarga* (dried, salted fish roe). Must try. Top: CONTINI (Flor 22). Gd: Orro, Serra, Silvio Carta.

Vernaccia di San Gimignano Tus *See* SAN GIMIGNANO.

Vie di Romans F-VG ★★★→★★★★ Gianfranco Gallo has built up his father's ISONZO estate to top FV-G status. Outstanding Isonzo PINOT GR Dessimis, SAUV BL Piere and Vieris (oaked), Flors di Uis blend and MALVASIA. V.gd Pinot Gr Dessimis.

Vietti Pie ★★★★ Organic estate at Castiglione Falletto owned by Krause Group but still run by Luca Currado and Mario Cordero. Characterful. Textbook BAROLOS: Lazzarito, Ravera, Brunate, Rocche di Castiglione, Villero. V.gd BARBARESCO Masseria, BARBERA D'ALBA Scarrone, BARBERA D'ASTI la Crena.

Vignamaggio Tus ★★→★★★ Historic, beautiful CHIANTI CLASSICO estate, nr Greve. Leonardo da Vinci painted the *Mona Lisa* here. RISERVA is called – you guessed it.

Vigna (or vigneto) A single v'yd, generally indicating superior quality.

Vigna Petrussa F-VG ★★★ Small family estate, with high-quality Schioppettino and PICOLIT.

Villa Russiz F-VG ★★★ Historic estate for DOC COLLIO. V.gd SAUV BL and MERLOT (esp "de la Tour" selections), CHARD, FRIULANO, PINOT BL, PINOT GR.

Vin Santo / Vinsanto / Vin(o) Santo T-AA, Tus DOC w sw s/sw ★★→★★★★ Sweet PASSITO, usually TREBBIANO, MALVASIA and/or SANGIOVESE in TUS (VIN SANTO), Nosiola in TRENTINO (Vino Santo). Tus versions extremely variable, anything from off-dry and Sherry-like to sweet & v. rich. May spend 3–10 unracked yrs in small barrels called *caratelli. Avignonesi's is legendary;* plus CAPEZZANA, FELSINA, FRASCOLE, ISOLE E OLENA, Rocca di Montegrossi, SAN GIUSTO A RENTENNANO, SELVAPIANA, Villa Sant'Anna, Villa di Vetrice. *See also* OCCHIO DI PERNICE.

Vini Franchetti Si ★★★→★★★★ ETNA estate (former Passopisciaro) run by Franchetti (*see* TENUTA DI TRINORO), contributor to fame of Etna. Outstanding Rosso Franchetti (PETIT VERDOT/Cesanese d'Affile) and NERELLO MASCALESE single-*contrada*. Best: Contrada G, Contrada S and Contrada C. V.gd Contrada R.

Vino Nobile di Montepulciano Tus DOCG r ★★→★★★ 10 11 12 13 15 16 (17) (18') 1st Italian DOCG (1980), Prugnolo Gentile (SANGIOVESE)-based, from TUS town MONTEPULCIANO (distinct from grape). Recent focus on single-v'yd wines. Complex, long-lasting Sangiovese expression, often tough with drying tannins. Top: AVIGNONESI, BOSCARELLI, DEI, La Braccesca, POLIZIANO, Salcheto. Also gd: Bindella, Canneto, Contucci, Fattoria del Cerro, Fattoria della Talosa, Montemercurlo, Romeo, Valdipiatta, Villa Sant'Anna. RISERVA after 3 yrs.

Voerzio, Roberto Pie ★★★→★★★★ BAROLO modernist: concentrated, tannic. More impressive/expensive than delicious. Range incl Brunate, Cerequio, La Serra, RISERVA 10 yrs Fossati Case Nere, Rocche dell'Annunziata-Torriglione, Sarmassa and Torriglione.

Volpaia, *Castello di* Tus ★★→★★★ V.gd CHIANTI CLASSICO estate at Radda. Organic certified. Top Chianti Classico RISERVA, Gran Selezione Coltassala (SANGIOVESE/Mammolo), Balifico (Sangiovese/CAB SAUV).

Zenato Ven ★★★ V. reliable, sometimes inspired GARDA wines, also AMARONE, LUGANA, SOAVE, VALPOLICELLA. Look for labels RISERVA Sergio Zenato.

Zerbina, Fattoria E-R ★★★ Leader in Romagna; best sweet ALBANA DOCG (Scacco Matto and AR), v.gd SANGIOVESE (Pietramora); barrique aged IGT Marzieno.

Zibibbo Si ★★★ dr sw Alluring sweet MUSCAT d'Alessandria, most associated with PANTELLERIA and extreme w SI. Dry version exemplified by DE BARTOLI.

Zonin Ven ★→★★★ One of Italy's biggest estate owners, based at Gambellara in Ven, but also big in F-VG, TUS, PUG, SI and elsewhere in world (eg. Virginia, US).

What do the initials mean?

DOC (Denominazione di Origine Controllata) Controlled Denomination of Origin, cf. AOC in France.

DOCG (Denominazione di Origine Controllata e Garantita) "G" = "Guaranteed". Italy's highest quality designation. Guarantee? It's still caveat emptor.

IGT (Indicazione Geografica Tipica) "Geographic Indication of Type". Broader and more vague than DOC, cf. Vin de Pays in France.

DOP/IGP (Denominazione di Origine Protetta/Indicazione Geografica Protetta) "P" = "Protected". The EU's DOP/IGP trump Italy's DOC/IGT.

MGA (Menzione Geografica Aggiuntiva) or Additional Geographical Definitions, eg. subzones, cf. crus in France.

Germany

Abbreviations used in the text:

Bad	Baden
Frank	Franken
Hess	Hessische-Bergstrasse
M-M	Mittelmosel
M Rh	Mittelrhein
Mos	Mosel
Na	Nahe
Pfz	Pfalz
Rhg	Rheingau
Rhh	Rheinhessen
Sa-Un	Saale-Unstrut
Sachs	Sachsen
Würt	Württemberg

More heavily shaded areas are the wine-growing regions.

The realities of modern German wine have changed. Helped by climate change, it has a radically new profile as a full-on food wine. Red as well as white: Germany's mastery of the ticklish Pinot Noir is remarkable. Riesling, though, is still the greatest grape in most regions, making wines of depth and tension, with no shortage of ripeness either. Climate change is undoubtedly benefiting German wine; there hasn't been a truly bad vintage since the 80s. But unpredictability is becoming an issue for winemakers. The 19 vintage, for example, was a rollercoaster, the most nerve-wracking harvest in years, though many 19s are surprisingly good. In the cold 70s growers would have cheerfully welcomed a vintage like this. But yields of good grapes in 19 are down 20–40%. And the meticulous viticulture that years like this need is expensive. Strict sorting of the bunches is costly too. Consumers want growers to go organic, with good reason. Some of the best estates already lead the way. But we have to be willing to pay for it. Great German wine is plentiful, but it can't, shouldn't, be a cheap option.

Recent vintages

Mosel

Mosels (incl Saar and Ruwer wines) are so attractive young that their capabilities for developing are not often enough explored. But fine Kabinetts can gain from at least 5 yrs in bottle and often much more: Spätlese from 5–20, and Auslese and BA anything from 10–30 yrs. "Racy" is their watchword. Dry Mos used to be pretty mean; climate change is rounding them out. Saar and Ruwer make leaner wines than Mos, but surpass the whole world for elegance and thrilling, steely "breed".

2019 Mixed bag, summer too dry, autumn wet. Sound quality, but low quantity (-25%, Ruwer -40%: frost).
2018 Powerful, at times really big wines, but balanced; better acidity than 2003.
2017 Low yield (frost) = high extract. Brilliant Kabinett, Spätlese with steely acidity.
2016 Balanced wines of textbook raciness.
2015 Warm yr, rich Trocken and Spätlesen, Auslesen to keep.
2012 Classic, discreet wines, might turn out to be long-lived.
2011 Brilliant vintage, particularly Saar, Ruwer, sensational TBAs.
2009 Magnificent Spätlesen, Auslesen, gd balance. Keep.
2008 Kabinetts, Spätlesen can be fine, elegant. Now perfect to drink.
Fine vintages: 07 05 04 03 01 99 97 95 94 93 90 89 88 76 71 69 64 59 53 49 45 37 34 21.

Rheinhessen, Nahe, Pfalz, Rheingau, Ahr

Apart from Mos, Rhg wines tend to be longest lived of all German regions, improving for 15 yrs or more, but best wines from Rhh, Na and Pfz can last as long. Modern dry wines such as Grosses Gewächs (GG) are generally intended for drinking within 2–5 yrs, but the best undoubtedly have the potential to age. The same holds for Ahr Valley reds (and their peers from Bad and other regions of the s): their fruit makes them attractive young, but best wines can develop for 10 yrs and longer. Who will give them a chance?

2019 Complicated yr, drought and heatwaves in July, sunburn (esp Ahr), rain in August and Sept. High acidity; rain affected concentration.
2018 Record summer, powerful wines. Growers allowed to acidify.
2017 Roter Hang and Mittelhaardt outstanding: rare combination of freshness, extract.
2016 Quality, quantity mixed, well-balanced Spätburgunder.
2015 Hot, dry summer. Rhg excellent, both dry and nobly sweet.
2014 Complicated, mixed results, best now mature.
2013 Much variation; best in s Rhh, Franken, Ahr Valley.
2012 Quantities below average, but v.gd, classical at every level.
2011 Fruity wines, harmonious acidity, now fully mature.
2010 Uneven quality: some v.gd Spätburgunder; dry whites drink now.
2009 Excellent wines, esp dry. Some acidification needed.
2008 Ries of great raciness, ageing well.
Fine vintages: 05 03 02 01 99 98 97 96 93 90 83 76 71 69 67 64 59 53 49 45 37 34 21.

Adams, Weingut Rhh ★★→★★★ Simone A proves why in C19 PINOT N from INGELHEIM was among Germany's best.

Adelmann, Weingut Graf Würt ★★→★★★ Young Count Felix Adelmann raises quality at idyllic Schloss Schaubeck. V.gd PINOT GR 17, Süssmund RIES 17 and LEMBERGER GG 16.

Ahr ★★→★★★★ Small river valley s of Bonn, elegant, fruit-driven PINOT N from slate. Best: Adeneuer, Bertram, Brogsitter, Deutzerhof, Kreuzberg, MEYER-NÄKEL, Nelles, Riske, Schumacher, STODDEN.

Aldinger, Gerhard Würt ★★★→★★★★ Family estate of great versatility. Whites and reds with density and tension, sensational Brut Nature SEKT (5 yrs lees ageing) 09′ 10 11′ 12′.

Alte Reben Old vines. But no min age.

Alter Satz Frank Wines from old co-planted (different varieties all mixed up) v'yds, esp in FRANK, many of them being more than 100 yrs old and ungrafted. Try (w) Otmar Zang, Scheuring, Scholtens – or Stritzinger (r).

Amtliche Prüfungsnummer (APNr) Official test number, on every label of a quality wine. Useful for discerning different lots of AUSLESE a producer has made from the same v'yd.

Assmannshausen Rhg ★★→★★★★ 05′ 09 10 13 15′ 16 17 18′ 19 The only RHG village with almost no RIES – but tradition for *Spätburgunder*. Most fame has GROSSE LAGE Höllenberg (45 ha on slate), but wines from neighbouring plots (Frankenthal, Hinterkirch) can be outstanding too. Growers: BISCHÖFLICHES WEINGUT RÜDESHEIM, CHAT SAUVAGE, KESSELER, König, KRONE, KÜNSTLER, Schloss Reinhartshausen, SOLVEIGS.

Auslese Wines from selective picking of super-ripe bunches affected by noble rot (*Edelfäule*). Unctuous, but – traditionally – elegant rather than super-concentrated. 99% are sweet, but specialists (JB BECKER, Koehler-Ruprecht) show Auslese TROCKEN can be elegant too.

Ayl Mos ★→★★★ All v'yds known since 1971 by name of historically best site: Kupp. Such are German wine laws. Growers: BISCHÖFLICHE WEINGÜTER TRIER, *Lauer*, Vols, ZILLIKEN.

BA (Beerenauslese) Luscious sweet wine from exceptionally ripe, individually selected berries concentrated by noble rot. Rare, expensive.

Bacharach M Rh ★→★★★ Small, idyllic Rhine-side town; centre of M RH RIES. Classified GROSSE LAGE: Hahn, Posten, Wolfshöhle. Growers incl: Bastian, JOST, Kauer, RATZENBERGER.

Baden Huge sw region and former Grand Duchy, 15,000 ha stretch over 300 km (186 miles), best-known for PINOT N, GRAU- and WEISSBURGUNDER and pockets of RIES, usually dry. Two-thirds of crop goes to co-ops.

Bassermann-Jordan Pfz ★★★ Famous DEIDESHEIM estate with stake in FORST's acclaimed Kirchenstück and Jesuitengarten v'yds. *Majestic dry Ries,* but also a tradition for TBA.

On schist

The magic of many German RIES (and some German PINOT N) is largely due to the schist in the subsoil. Schist is formed when a parent rock's inner structure is transformed under unimaginably high pressure. The resulting "parallel structure of flat planes", as geologists put it, will split (Greek *skhistos* = schist = split) along these planes, offering vine roots the chance to burrow deep for moisture, which it will find there even if the topsoil has already dried out. Highly adapted microbes are also involved, which help to make nutrients available to the vine roots via ion exchange. Life down there in the rock remains a secret, but taste any fine MOS Ries to sense it.

Battenfeld-Spanier Rhh ★★★ Leading estate with v'yds at Hohen-Sülzen, Mölsheim and in neighbouring ZELLERTAL, bio, led by passionate HO Spanier (*see also* KÜHLING-GILLOT). Brilliant Brut Nature 09'.

Becker, Friedrich Pfz ★★→★★★★ Outstanding SPÄTBURGUNDER (Heydenreich, Kammerberg, Sankt Paul, Res) from most s part of PFZ; some v'yds actually lie across border in Alsace. Wines need 5–10 yrs cellaring. Gd whites (CHARD, RIES, PINOT GR) too.

Becker, JB Rhg ★★→★★★ 10 11 12 15' 16 17 18' Delightfully old-fashioned, cask-aged (and long-lived) dry RIES, SPÄTBURGUNDER at WALLUF and Martinsthal. Outstanding 15 18, mature vintages (back to 90s) great value.

Bercher Bad ★★★ KAISERSTUHL family estate, known for barrique-aged PINOTS (1 W) from Burkheim.

Bergdolt Pfz ★★★ Organic estate at Duttweiler, known for food-friendly, age-worthy WEISSBURGUNDER GG Mandelberg. Gd RIES, SPÄTBURGUNDER too.

Bernkastel M-M ★★→★★★★ Senior wine town of the M-M, known for timbered houses and flowery, balsamic RIES. GROSSE LAGE: BADSTUBE, DOCTOR, Graben, Johannisbrünnchen, Lay. Top growers: Kerpen, JJ PRÜM, Lauerburg, LOOSEN, MOLITOR, SCHLOSS LIESER, Studert-Prüm, THANISCH (both estates), WEGELER. Kurfürstlay GROSSLAGE name is a deception: avoid.

Bernkasteler Ring Mos One of two MOS growers' associations organizing an auction every yr mid–end Sept. Other is GROSSER RING.

Bertram, Julia Ahr ★★★ Young Julia B and partner Benedikt Baltes produce Burgundy-inspired PINOT N from prime AHR valley sites, new cellar at Dernau to follow.

Bischöfliches Weingut Rüdesheim Rhg ★★★ 8 ha of best sites in ASSMANNSHAUSEN, JOHANNISBERG, RÜDESHEIM; vault cellar in Hildegard von Bingen's historic monastery. Peter Perabo (ex-KRONE) is *Pinot N specialist*, RIES also v.gd.

Bischöfliche Weingüter Trier Mos ★★ 130 ha of potentially 1st-class historical donations. Not v. reliable; do not buy without prior tasting.

Bocksbeutel Frank Belly-shaped bottle dating back to C18, today only permitted in FRANK and village of Neuweier, BAD. New: modernized, stackable "Bocksbeutel PS" – at last.

Bodensee Bad Idyllic district of s BAD and on Bavarian shore of Lake Constance, at altitude: 400–580m (1312–1903ft). Dry MÜLLER-T with elegance, light but delicate SPÄTBURGUNDER. Top villages: Hagnau, Meersburg, Nonnenhorn, Reichenau.

Doppard M Rh ★→★★★ Wine town of M RH with GROSSE LAGE Hamm, an amphitheatre of vines. Growers: Heilig Grab, Lorenz, M Müller, Perll, WEINGART. Unbeatable *value*.

Braunebeg M-M ★★★→★★★★ Top village on M-M; excellent full-flavoured RIES of great raciness. GROSSE LAGE v'yds Juffer, Juffer-SONNENUHR. Growers: *F Haag*, KESSELSTATT, MF RICHTER, Paulinshof, SCHLOSS LIESER, THANISCH, *W Haag*.

Bremer Ratskeller Town-hall cellar in n Germany's commercial town of Bremen, founded in 1405, UNESCO World Heritage Site. Oldest wine is a barrel of 1653 RÜDESHEIMER Apostelwein.

Breuer Rhg ★★★→★★★★ Exquisite RIES from RAUENTHAL, RÜDESHEIM. Berg Schlossberg 89 90' 96' 97' 08' 10 11 12' 13' 14 15' has depth at 12% alc, Nonnenberg transforms austerity into age-worthiness. Exciting experiments with forgotten historic grape Gelber Orleans.

Buhl, Reichsrat von Pfz ★★★ Historic PFZ estate at DEIDESHEIM. Ex-Bollinger cellarmaster Mathieu Kauffmann came in 2013, but left in 2019 before the fruit of his work became clear.

Bunn, Lisa Rhh ★★ →★★★ Shooting star in NIERSTEIN, refined Hipping and Oelberg RIES, stylish Res CHARD.

Bürgerspital zum Heiligen Geist Frank ★★★ Ancient charitable estate with great continuity: only six directors in past 180 yrs. Traditionally made whites (*Silvaner*, RIES) from best sites in/around WÜRZBURG. SILVANER GG from monopole Stein-Harfe 15' 16' 17' 18' is a monument.

Bürklin-Wolf, Dr. Pfz ★★→★★★★ 30 ha of best MITTELHAARDT v'yds incl important holdings of FORST's Kirchenstück, Jesuitengarten, Pechstein. Estate own classification since 1994, bio farming (incl horses). Wines made to age.

Initiative to make Germany's viticulture UNESCO Intangible World Heritage.

Busch, Clemens Mos ★★★→★★★★ Steep Pündericher Marienburg farmed by hand, bio, for seven GGS from different parcels. Best usually: Felserrasse (mineral, deep), Raffes (power, balance), Rothenpfad (silky, balsamic). Now also Res line: 2 yrs barrel-ageing. Breathtaking 18' AUSLESE GOLDKAPSEL.

Castell'sches Fürstliches Domänenamt Frank ★★→★★★ Superb dry SILVANER, RIES from monopoly v'yd *Casteller Schlossberg*, occasionally also TBA (outstanding Ries 06) or BA (Silvaner 67 08).

Chat Sauvage Rhg ★★★ Created out of nothing in 2000, Burgundian approach to RHG PINOT N. Sensational 15s (ASSMANNSHAUSEN Höllenberg, Clos de Schultz, LORCH Schlossberg). Some CHARD too, and delicate SEKT.

Christmann Pfz ★★★ VDP President Steffen Christmann, MITTELHAARDT bio pioneer, now joined by daughter Sophie. Best known is RIES Königsbacher Idig, but other GGS fine too, eg. Mandelgarten Meerspinne. Since Mathieu Kauffmann left VON BUHL, promising new Christmann & Kauffmann SEKT project.

Clüsserath, Ansgar Mos ★★★ Tense TRITTENHEIMER Apotheke RIES. KABINETTS delicious.

Corvers-Kauter Rhg ★★★ 31-ha-organic estate at Mittelheim making a name for textbook mineral RÜDESHEIM RIES. Now taken over most of former LANGWERTH v'yds, incl MARCOBRUNN, RAUENTHAL Baiken.

Crusius, Dr. Na ★★→★★★ Family estate at TRAISEN. Vivid, age-worthy RIES from sun-baked Bastei and Rotenfels of Traisen and SCHLOSSBÖCKELHEIM. Daughters Judith, Rebecca bring fresh air.

Deidesheim Pfz ★★→★★★★ Central MITTELHAARDT village and series of GROSSE LAGE v'yds: Grainhübel, Hohenmorgen, Kalkofen, Kieselberg, Langenmorgen. Top growers: BASSERMANN-JORDAN, Biffar, BUHL, BÜRKLIN-WOLF, CHRISTMANN, Fusser, MOSBACHER, Seckinger, Siben, Stern, VON WINNING. Gd co-op.

Deinhard, Dr. Pfz ★★→★★★ Since 2008, a brand of the VON WINNING estate, used for wines with no oak influence.

Diel, Schlossgut Na ★★★→★★★★ Caroline Diel follows her father: exquisite *GG Ries* (best usually Burgberg of Dorsheim). Magnificent SPÄTLESEN, serious *Sekt*.

Doctor M-M Emblematic steep v'yd at BERNKASTEL, the place where TBA was invented (1921, THANISCH). Only 3.2 ha, and five owners: both Thanisch estates, WEGELER (1.1 ha), Lauerburg and local Heiligen Geist charity (0.26 ha, leased until 2024 to MARKUS MOLITOR, SCHLOSS LIESER). RIES of extraordinary depth and richness, but pricey – up to a record-breaking 1100€/bottle for Molitor's 18 TROCKEN.

Dönnhoff Na ★★★→★★★★ NA superstar. Cornelius D continuing success of father Helmut, with splendid series of GGS from Roxheim (*Höllenpfad*), NIEDERHAUSEN (Hermannshöhle), *Norheim (Dellchen)* and SCHLOSSBÖCKELHEIM (Felsenberg). Dazzling EISWEIN – and tiny quantities of Blanc de Noirs SEKT.

Durbach Bad ★★→★★★ ORTENAU village for full-bodied RIES, locally called Klingelberger. Growers: Graf Metternich, LAIBLE (both), Männle (both), MARKGRAF VON BADEN. Reliable co-op.

Egon Müller zu Scharzhof Mos ★★★★ 59 71 83 90 03 15 16 17 18 Legendary SAAR family estate at WILTINGEN with a treasury of old vines. Racy RIES from SCHARZHOFBERGER is among the world's greatest wines: sublime, vibrant and immortal. *Kabinetts* feather-light and long-lived. New (from 18 on): a TROCKEN version of Scharzhofberg.

Einzellage Individual v'yd site. Never to be confused with GROSSLAGE.

Eiswein Made from frozen grapes with the ice (ie. water content) discarded, thus v. concentrated: of BA ripeness or more. Outstanding Eiswein vintages: 98 02 04 08. Less and less produced in past decade: climate change is Eiswein's enemy.

Emrich-Schönleber Na ★★★ Werner Schönleber and son Frank make precise RIES from Monzingen's classified Frühlingsplätzchen and Halenberg v'yds.

Erden M-M ★★★→★★★★ Village on red slate soils; noble AUSLESEN and TROCKEN RIES with rare delicacy. GROSSE LAGE: Prälat and Treppchen. Growers incl: BREMER RATSKELLER, JJ Christoffel, LOOSEN, MARKUS MOLITOR, MERKELBACH, Mönchhof, Rebenhof, Schmitges.

Erste Lage Classified v'yd, 2nd-from-top level, similar to Burgundy's Premier Cru, only in use with VDP members outside AHR, M RH, MOS, NA, RHH.

Erstes Gewächs Rhg "First growth". Previous name for classified dry wines from RHG v'yds; VDP members changed to GG in 2012, non-VDP estates followed in 2018.

Erzeugerabfüllung Bottled by producer. Incl the guarantee that only own grapes have been used. May be used by co-ops also. GUTSABFÜLLUNG is stricter, applies only to estates.

Escherndorf Frank ★★★ Village with steep GROSSE LAGE Lump ("scrap" – as in tiny inherited parcels). Marvellous *Silvaner* and RIES (dr sw) Growers: Fröhlich, H SAUER, R SAUER, Schäffer, zur Schwane.

Feinherb Imprecisely defined traditional term for wines with around 10–25g sugar/litre, not necessarily tasting sweet. More flexible than HALBTROCKEN. I often choose Feinherbs.

Forst Pfz ★★→★★★★ Outstanding MITTELHAARDT village, a mosaic of individualistic GROSSE LAGE v'yds giving RIES of terroir expression, longevity. Most famous: Kirchenstück, Jesuitengarten, Pechstein, but Ungeheuer and Freundstück not far behind. ORTSWEIN usually excellent value. Top growers: Acham-Magin, BASSERMANN-JORDAN, BÜRKLIN-WOLF, H Spindler, MOSBACHER, VON BUHL, VON WINNING, WOLF. Gd co-op.

Franken / Franconia Region of distinctive dry wines, esp *Silvaner*, often bottled in round-bellied flasks (BOCKSBEUTEL). Centre is WÜRZBURG. Top villages: Bürgstadt, ESCHERNDORF, IPHOFEN, Klingenberg, RANDERSACKER.

Fricke, Eva Rhg ★★→★★★ Born nr Bremen without viticultural background, now rising star in RHG: expressive, taut RIES (13 ha, in conversion to organic) from KIEDRICH, LORCH.

Fuder Traditional German cask, sizes 600–1800 litres depending on region, traditionally used for fermentation and (formerly long) ageing.

Kabinett prices are rising: for back vintages, Kabinett retails at Spätlese prices.

Fürst, Weingut Frank ★★★→★★★★ Sebastian F continues – and tops – his father Paul's work. *Spätburgunders* 05' 09 10 15' 16 17 18' of great finesse from red sandstone (most dense: Hundsrück, most powerful: Schlossberg, most typical: Centgrafenberg, best value: Bürgstadter Berg). Newly planted Hundsrück parcel, 17,000 vines/ha.

Gallais, Le Mos 2nd estate of EGON MÜLLER ZU SCHARZHOF with 4-ha-monopoly Braune Kupp of WILTINGEN. Soil is schist with more clay than in SCHARZHOFBERG; AUSLESEN can be exceptional.

Geisenheim Rhg Town primarily known for Germany's university of oenology and viticulture (can be controversial: 3rd-party-funded research, and technocratic winemaking). GROSSE LAGEN: Kläuserweg, Mäuerchen, Rothenberg (best).

> **Grosse Lage / Grosslage: spot the difference**
> *Bereich* means district within an *Anbaugebiet* (region). *Bereich* on
> a label should be treated as a flashing red light; the wine is a blend
> from arbitrary sites within that district. Do not buy. The same holds
> for wines with a GROSSLAGE name, though these are more difficult to
> identify. Who could guess if "Forster Mariengarten" is an EINZELLAGE
> or a Grosslage? (It's Gross.) But now, from the 12 vintage, it's even
> more tricky. Don't confuse Grosslage with GROSSE LAGE: the latter
> refers to best single v'yds, Germany's "Grands Crus" according to the
> classification set up by wine-grower's association VDP. They weren't
> thinking about us poor consumers.

GG (Grosses Gewächs) "Great/top growth". The top dry wine from a VDP-classified GROSSE LAGE.

Goldkapsel / Gold Capsule Mos, Na, Rhg, Rhh Designation (and bottle seal) mainly for AUSLESE and higher. V. strict selection of grapes, which should add finesse and complexity, not primarily weight and sweetness. Lange Goldkapsel (Long Gold Capsule) is even better. Not a legal term.

Graach M-M ★★★→★★★★ Small village between BERNKASTEL and WEHLEN. GROSSE LAGE v'yds: Domprobst, Himmelreich, JOSEPHSHÖFER. Growers: *JJ Prüm*, Kees-Kieren, LOOSEN, MARKUS MOLITOR, SA PRÜM, SCHAEFER, *Selbach-Oster*, Studert-Prüm, WEGELER.

Griesel & Compagnie Hess ★★★ SEKT startup at Bensheim, top Prestige series (Rosé Extra Brut, PINOT Brut Nature). Since 2016, also excellent still wine under Schloss Schönberg label.

Grosse Lage Top level of VDP's classification, but only for VDP members. Dry wine from a Grosse Lage site is called GG. *NB* Not on any account to be confused with GROSSLAGE. Stay awake, there.

Grosser Ring Mos Group of top (VDP) MOS estates, whose annual Sept auction at TRIER sets world record prices.

Grosslage Term destined, maybe even intended, to confuse. Introduced by the disastrous 1971 wine law. A collection of secondary v'yds with supposedly similar character – but no indication of quality. Not on any account to be confused with GROSSE LAGE. Repeat after me.

Gunderloch Rhh ★★★→★★★★ Historical NACKENHEIM estate portrayed in Carl Zuckmayer's play *Der fröhliche Weinberg* (1925). Known for nobly sweet RIES and culinary *Kabinett Jean-Baptiste* from prime ROTER HANG sites. Recently increasing emphasis on TROCKEN.

Gut Hermannsberg Na ★★★ Former state domain at NIEDERHAUSEN. Some of densest RIES GGS of NA, eg. Kupfergrube, Felsberg from SCHLOSSBÖCKELHEIM and Hermannsberg at Niederhausen.

Gutsabfüllung Estate-bottled, and made from own grapes.

Gutswein Wine with no v'yd or village designation, but only the producer's name: entry-level category. Ideally, Gutswein should be an ERZEUGERABFÜLLUNG (from own grapes), but is not always the case.

Haag, Fritz Mos ★★★★ BRAUNEBERG'S top estate; Oliver Haag follows footsteps of father Wilhelm, but wines more modern in style. *See also* SCHLOSS LIESER.

Haag, Willi Mos ★★→★★★ BRAUNEBERG family estate, led by Marcus Haag. Old-style RIES, mainly sweet, rich but balanced, and inexpensive.

Haart, Julian M-M, Mos ★★→★★★ Talented nephew of Theo H making a name for dense, spontaneously fermented RIES.

Haart, Reinhold M-M ★★★→★★★★ Best estate in PIESPORT with important holding in famous Goldtröpfchen ("gold droplet") v'yd, RIES SPÄTLESEN, AUSLESEN and higher PRÄDIKAT wines are *racy, copybook Mosels* – with great ageing potential.

Haidle Würt ★★★ Family estate now led by young Moritz Haidle, using the cool climate of Remstal area for RIES, LEMBERGER of distinctive freshness.

Halbtrocken Medium-dry with 9–18g unfermented sugar/litre, inconsistently distinguished from FEINHERB (which sounds better).

Hattenheim Rhg ★★→★★★★ Town famous for classic RHG RIES from GROSSE LAGEN Nussbrunnen, Hassel, STEINBERG, Wisselbrunnen. Estates: Barth, HESSISCHE STAATSWEINGÜTER, Kaufmann, Knyphausen, Ress, Schloss Reinhartshausen, SPREITZER. The *Brunnen* ("well") v'yds lie on a rocky basin that collects water, protection against drought.

Heger, Dr. Bad ★★★ KAISERSTUHL estate known for dry parcel selections from volcanic soils in Achkarren and IHRINGEN, esp Vorderer Berg (PINOTS N/GR/BL, RIES) from steepest Winklerberg terraces, and Häusleboden Pinot N from old Clos de Vougeot (*see* France) cuttings planted in 1956.

Heitlinger / Burg Ravensburg Bad ★★→★★★ Two leading estates of KRAICHGAU, under same ownership: Heitlinger more elegant, Burg Ravensburg full-bodied.

Hessische Bergstrasse ★→★★★ Germany's smallest wine region (only around 440 ha), n of Heidelberg. Best: Bergsträsser Winzer co-op, GRIESEL (SEKT), HESSISCHE STAATSWEINGÜTER, Schloss Schönberg, Simon-Bürkle, Stadt Bensheim.

Hessische Staatsweingüter Hess, Rhg ★★→★★★★ State domain of big – and historical – dimensions: 220 ha in ASSMANNSHAUSEN, HATTENHEIM (monopoly STEINBERG), HOCHHEIM, *Rauenthal*, RÜDESHEIM, and along HESS. Superb C12 Cistercian abbey KLOSTER EBERBACH has vinothèque, 12 historical presses. The new oenologist Kathrin Puff made wine in Thailand before she returned to Germany.

Heyl zu Herrnsheim Rhh ★★→★★★ Historic NIERSTEIN estate, bio, part of ST-ANTONY estate, now to be repositioned.

Heymann-Löwenstein Mos ★★★ Pacemaker downstream on Mosel at WINNINGEN nr Koblenz. Spontaneously fermented RIES from steep terraces: intense, individual.

Hochgewächs Designation for a MOS RIES that obeys stricter requirements than plain QbA, today rarely used. A worthy advocate is *Kallfelz* in Zell-Merl.

Hochheim Rhg ★★→★★★★ Town e of main RHG, on River Main. Rich, earthy RIES from GROSSE LAGE v'yds: Domdechaney, Hölle, Kirchenstück, KÖNIGIN VICTORIABERG, Reichestal. Growers: *Domdechant Werner*, Flick, HESSISCHE STAATSWEINGÜTER, Himmel, *Künstler*.

Hock Traditional English term for Rhine wine, derived from HOCHHEIM.

Hövel, Weingut von Mos ★★★ Fine SAAR estate, now bio, with v'yds at Oberemmel (Hütte is 4.8-ha monopoly), at KANZEM (Hörecker) and in SCHARZHOFBERG. Magnificent 18s.

Huber, Bernhard Bad ★★★→★★★★ Young Julian H has a vision of BAD as Germany's burgundy: Wildenstein SPÄTBURGUNDER 16' ranks among best reds ever produced by this fine estate. Equally outstanding tight, demanding CHARDS (eg. ALTE REBEN 17').

Ihringen Bad ★→★★★ Village in KAISERSTUHL known for fine SPÄTBURGUNDER, GRAUBURGUNDER (historically also for SILVANER) on steep volcanic Winklerberg. Top growers: DR. HEGER, Konstanzer, Michel, Stigler.

Ingelheim Rhh ★★→★★★ RHH town across Rhine from RHG, with limestone beds under v'yds; historic fame for SPÄTBURGUNDER being reinvigorated by ADAMS, Arndt F Werner, Bettenheimer, Dautermann, NEUS, Schloss Westerhaus, Wasem.

Iphofen Frank ★★→★★★ STEIGERWALD village with famous GROSSE LAGE Julius-Echter-Berg. Rich, aromatic, well-ageing SILVANER from gypsum soils. Growers: Arnold, Emmerich, JULIUSSPITAL, RUCK, Seufert, VETTER, Weigand, WELTNER, *Wirsching*, Zehntkeller.

Jahrgang Year – as in "vintage".

Johannisberg Rhg ★★→★★★★ RHG village best known for berry- and honey-scented RIES from SCHLOSS JOHANNISBERG. GROSSLAGE (avoid!): Erntebringer.

Johannishof (Eser) Rhg ★★→★★★ Family estate with v'yds at JOHANNISBERG, RÜDESHEIM. Johannes Eser makes RIES with perfect balance of ripeness and steely acidity.

Josephshöfer Mos ★★→★★★ GROSSE LAGE v'yd at GRAACH, the sole property of KESSELSTATT. Harmonious, berry-flavoured RIES.

Jost, Toni M Rh ★★★ Leading estate in BACHARACH with monopoly Hahn, now led by Cecilia J. Aromatic RIES with nerve, and recently remarkable PINOT N 15' 17. Family also run estate at WALLUF (RHG).

Germany's biggest wine city? Stuttgart. 600,000 inhabitants, 420 ha vines.

Jülg Pfz ★★★ Shooting star at Schweigen, dense PINOT N and sharply mineral CHARD, SAUV BL from limestone soils on both sides of the PALATINATE-Alsace border.

Juliusspital Frank ★★★ Ancient WÜRZBURG charity with top v'yds all over FRANK known for *dry Silvaners* that age well. Recently less opulence and more structure – GGS now cellared 1 yr more before sale.

Kabinett *See* box, p.165. Germany's unique featherweight contribution, but with climate change ever more difficult to produce.

Kaiserstuhl Bad Outstanding district and low mtn range nr Rhine with notably warm climate and volcanic soil. Altitudes up to 400m (1312ft). Renowned for SPÄTBURGUNDER, GRAUBURGUNDER.

Kanzem Mos ★★★ SAAR village with steep GROSSE LAGE v'yd Altenberg (slate and weathered red rock). Growers: BISCHÖFLICHE WEINGÜTER TRIER, *Van Volxem*, VON OTHEGRAVEN.

Karthäuserhof Mos ★★★★ Outstanding RUWER estate, now led by Mathieu Kauffmann (ex-Bollinger). Characteristic neck-only label stands for refreshing dry and sublime sweet wines.

Keller, Franz Bad *See* SCHWARZER ADLER.

Keller, Klaus Peter Rhh ★★★→★★★★ Star of RHH, cultish for ALTE REBEN RIES G-Max from undisclosed parcel, and GGS Hubacker, Morstein. Also Ries from NIERSTEIN (Hipping, Pettenthal) and M-M (PIESPORT Schubertslay).

Kesseler, August Rhg ★★★→★★★★ August K's long-time employees have taken over – and continue to produce refined SPÄTBURGUNDERS from ASSMANNSHAUSEN, RÜDESHEIM – and purist RIES (dr sw).

Kesselstatt, Reichsgraf von Mos ★★→★★★★ Annegret Reh-Gartner's successors maintain high levels of quality. 35 ha top v'yds on Mosel and both tributaries, incl remarkable stake in SCHARZHOFBERG.

Kiedrich Rhg ★★→★★★★ Top RHG village, almost a monopoly of the WEIL estate, other growers (eg. FRICKE, Knyphausen, PRINZ VON HESSEN) own only small plots. Famous church and choir.

Kloster Eberbach Rhg Glorious C12 Cistercian abbey in HATTENHEIM with iconic STEINBERG, domicile of HESSISCHE STAATSWEINGÜTER.

Klumpp, Weingut Bad ★★★ Rising star in Kraichgau. SPÄTBURGUNDER LEMBERGER of depth, elegance. Markus Klumpp is married to Meike Näkel of MEYER-NÄKEL.

Knipser, Weingut Pfz ★★★→★★★★ N PFZ family estate, barrique-aged SPÄTBURGUNDER (iconic RdP 15') straightforward RIES (GG Steinbuckel), Cuvée X (B'x blend). Many specialities, incl historical bone-dry Gelber Orleans.

Königin Viktoriaberg Rhg ★★→★★★ Historic v'yd at HOCHHEIM, 4.5 ha along shores of River Main, known for textbook RHG RIES, today run by Flick estate of Wicker. After 1845 visit, Queen Victoria granted owner right to rename v'yd as "Queen-Victoria-mountain".

Kraichgau Bad Small district se of Heidelberg. Top growers: HEITLINGER/BURG RAVENSBURG, Hoensbroech, Hummel, KLUMPP.

Krone, Weingut Rhg ★★★ Famous SPÄTBURGUNDER estate with old v'yds in ASSMANNSHAUSEN's steep Höllenberg (slate), run by WEGELER. Fine Rosé SEKT too.

Kühling-Gillot Rhh ★★★★ Top bio estate, run by Caroline Gillot and husband HO Spanier. Best in already outstanding range of ROTER HANG RIES: GG Rothenberg Wurzelecht from ungrafted, 70 yrs+ vines.

Kühn, Peter Jakob Rhg ★★★ Excellent estate in OESTRICH led by PJ Kühn and son. Obsessive bio v'yd management and long macerations shape *nonconformist but exciting* RIES. Res RPJK Unikat aged 4 yrs in cask.

Kuhn, Philipp Pfz ★★★ Reliable estate of great versatility. Rich RIES (eg. SAUMAGEN, Schwarzer Herrgott), succulent barrel-aged SPÄTBURGUNDER, specialities (FRÜHBURGUNDER, SAUV BL, SEKT).

Künstler Rhg ★★★ Superb dry RIES at HOCHHEIM (Hölle), Kostheim (Weiß Erd), and other side of RHG at RÜDESHEIM (Rottland, Schlossberg).

Kuntz, Sybille Mos ★★★ Progressive organic 12-ha estate at Lieser, esp Niederberg-Helden v'yd. MOS TROCKEN pioneer; intense wines, one of each ripeness category, intended for gastronomy, listed in many top restaurants.

Laible, Alexander Bad ★★→★★★ DURBACH estate of ANDREAS LAIBLE's (*see* next entry) younger son; aromatic dry RIES, SCHEUREBE, WEISSBURGUNDER.

Laible, Andreas Bad ★★★ Crystalline dry RIES from DURBACH's Plauelrain v'yd exhibiting typicity of Ries on granite soil. Gd SCHEUREBE, GEWÜRZ too.

Landwein Technically "ggA" (*see* box, p.165), meant to label wines with only broadly defined origin. But now popular among ambitious growers to avoid official quality testing (eg. because of spontaneous fermentations, or low sulphur levels). Best known (all from BAD): Brenneisen, Enderle & Moll, Forgeurac, Höfflin, WASENHAUS, ZIEREISEN.

Langwerth von Simmern Rhg Once famous, but now history. In 2018, von Langwerth family sold their manor house in Eltville and leased their v'yds to the CORVERS-KAUTER estate.

Lauer Mos ★★★ Fine, precise SAAR RIES: tense, poised. Parcel selections from huge Ayler Kupp v'yd. Best. Kern, Schonfels, Stirn.

Leitz, Josef Rhg ★★★ RÜDESHEIM-based family estate known for rich but elegant single-v'yd RIES (dr sw) and for inexpensive but reliable Eins-Zwei-Dry label.

Liebfrauenstift, Weingut Rhh Owner of best plots of historical LIEBFRAUENSTIFT-KIRCHENSTÜCK v'yd. Formerly linked to a merchant house, but now autonomous. Katharina Prüm (of JJ PRÜM) consults. *See* next entry.

Liebfrauenstift-Kirchenstück Rhh A walled v'yd in city of Worms producing flowery RIES from gravelly soil. Producers Gutzler, Schembs (or' almost youthful in 2017), WEINGUT LIEBFRAUENSTIFT. Not to be confused with Liebfraumilch, a cheap and tasteless imitation.

Loewen, Carl Mos ★★★ RIES of elegance, tension, complexity. Best v'yd Longuicher Maximin Herrenberg (planted 1896, ungrafted). Entry-level Ries Varidor excellent *value*.

Loosen, Weingut Dr. M-M ★★→★★★★ Charismatic Ernie L produces fine traditional RIES from old vines in BERNKASTEL, ERDEN, GRAACH, ÜRZIG, WEHLEN. Erdener Prälat

---

Teasing the neighbours

The famous MARCOBRUNN v'yd takes its name from a well on the border between the villages of Erbach and HATTENHEIM. When, in 1810, Erbachers proudly engraved: "Marcowell – commune Erbach" onto the well housing, it didn't take long for Hattenheimers to reply: "*So ist es richtig/so soll es sein./Für Erbach das Wasser/und für Hattenheim den Wein*" (This is correct/this is just fine./For Erbach the water/and for Hattenheim the wine).

AUSLESE cultish for decades, dry Prälat Res (2 yrs cask-ageing, 1st vintage 2011) about to follow. Dr. L Ries, from bought-in grapes, is reliable. *See also* WOLF (PFZ), Ch Ste Michelle (Washington State).

Lorch Rhg ★ →★★★ Village in extreme w of RHG, conditions more M-RH-like than Rhg-like. Sharply crystalline wines, both RIES and PINOT N, now re-discovered. Best: CHAT SAUVAGE, FRICKE, KESSELER, SOLVEIGS, von Kanitz. Soon to come: BREUER.

Löwenstein, Fürst Frank, Rhg ★★★ Princely estate with holdings in RHG, FRANK. Classic Rhg RIES from HALLGARTEN, unique *Silvaner, Ries* from ultra-steep v'yd *Homburger Kallmuth.*

Lützkendorf, Weingut Sa-Un ★★→★★★ Quality leader in SA-UN. 18 brought record ripeness, but wines have balance.

Marcobrunn Rhg Historic 7-ha v'yd in Erbach, GROSSE LAGE. Potential for rich, long-lasting RIES. Growers: CORVERS-KAUTER, HESSISCHE STAATSWEINGÜTER, Knyphausen, Schloss Reinhartshausen, Von Oetinger.

Markgraf von Baden Bad ★★→★★★ Important noble estate (135 ha) at Salem castle (BODENSEE) and Staufenberg castle (ORTENAU).

Markgräflerland Bad Up-and-coming district s of Freiburg, cool climate due to breezes from Black Forest, and limestone soils. Typical GUTEDEL a pleasant companion for local cuisine. Climate change making PINOT varieties successful.

Maximin Grünhaus Mos ★★★★ Supreme RUWER estate led by Carl von Schubert, who also presides over GROSSER RING (VDP MOS). V. traditional winemaking shapes herb-scented, *delicate, long-lived Ries.* Astonishing WEISSBURGUNDER, PINOT N.

Merkelbach, Weingut M-M ★★→★★★ Tiny Estate at URZIG, 2 ha. Brothers Alfred and Rolf (both c.80), inexpensive MOS made not to sip, but to drink. Superb list of old vintages.

Meßmer, Weingut Pfz ★★→★★★ Brothers Gregor and Martin M have gd chunk of Burrweiler's Schäwer v'yd, one of the few pockets of schist in PFZ. V. consistent quality, many specialities.

Meyer-Näkel Ahr ★★★→★★★★ Werner Näkel and daughters make the most fruit-driven, refined AHR Valley SPÄTBURGUNDER. Freshness of wines obscures how hard it is to produce them: eg. Walporzheimer Kräuterberg has ten steep terraces on a surface equal to football penalty area. All must be done by hand. Also in S Africa (Zwalu, together with Neil Ellis) and Portugal (Quinta da Carvalhosa).

Mittelhaardt Pfz 08' 09 10 11 12 15 16 17' 18 19 N-central and best part of PFZ, incl DEIDESHEIM, FORST, RUPPERTSBERG, WACHENHEIM; largely planted with RIES.

Mittelmosel Central and best part of MOS, a RIES Eldorado, incl BERNKASTEL, BRAUNEBERG, GRAACH, PIESPORT, WEHLEN, etc.

Mittelrhein ★★→★★★ Dramatically scenic Rhine area nr tourist-magnet Loreley. Best villages: BACHARACH, BOPPARD. Delicate yet *steely Ries, underrated* and underpriced. Longtime decline in production now finally halted.

Molitor, Markus M-M, Mos ★★★ Growing estate (120 ha in more than 180 parcels throughout M-M, SAAR) led by perfectionist Markus M. Styles, v'yds and vintages in amazing depth. Dry DOCTOR (since 16) fetches €1000+/bottle at auction.

The Mosel bridge

For almost 10 yrs growers like Dr. Manfred PRÜM, Ernie LOOSEN, MARKUS MOLITOR and WILLI SCHÄFER (with supporters such as HJ) have been fighting against it: the giant 160m (525ft)-high Mosel bridge whose supports rest in famous v'yd sites (eg. ZELTINGER Himmelreich). At the end of 2019 the bridge was officially inaugurated. 1st inspection shows practically no traffic. And workers continue to pump concrete into one of the supports. From the v. beginning, geologists had warned that the slope under the pillars might be sliding.

Germany's quality levels

The official range of qualities and styles in ascending order is (take a deep breath):

1 Wein: formerly known as Tafelwein. Light wine of no specified character, mostly sweetish.

2 ggA (geschützte geographische Angabe): or Protected Geographical Indication, formerly known as LANDWEIN. Dryish Wein with some regional style. Mostly a label to avoid, but some thoughtful estates use the Landwein, or ggA designation to bypass official constraints.

3 gU (geschützte Ursprungsbezeichnung): or protected Designation of Origin. Replacing QUALITÄTSWEIN. So far, only two: Bürgstadter Berg, and Uhlen of Winningen.

4 Qualitätswein: dry or sweetish wine with sugar added before fermentation to increase its strength, but tested for quality and with distinct local and grape character. Don't despair.

5 Kabinett: dry/dryish natural (unsugared) wine of distinct personality and distinguishing lightness. Can occasionally be sublime – esp with a few yrs' age.

6 Spätlese: stronger, sweeter than KABINETT. Full-bodied (but no botrytis). Dry SPÄTLESE (or what could be considered as such) is today mostly sold under Qualitätswein designation.

7 Auslese: sweeter, stronger than Spätlese, often with honey-like flavours, intense and long-lived. Occasionally dry and weighty. The lower the alc (read the label), the sweeter the wine.

8 Beerenauslese (BA): v. sweet, dense and intense, but seldom strong in terms of alc. Can be superb.

9 Eiswein: from naturally frozen grapes of BA/TBA quality: concentrated, sharpish and v. sweet. Some examples are extreme, unharmonious.

10 Trockenbeerenauslese (TBA): intensely sweet and aromatic; alc slight. Extraordinary and everlasting.

Mosbacher Pfz ★★★ Some of best GG RIES of FORST: refined rather than massive. Traditional ageing in big oak casks. Excellent SAUV BL too ("Fumé").

Mosel Wine-growing area formerly known as Mosel-Saar-Ruwer, 8800 ha in total, 62% RIES. Conditions on the RUWER and SAAR tributaries are v. different from those along the Mosel (aka Moselle in French).

Moselland Mos, Na, Pfz, Rhh Huge MOS co-op, at BERNKASTEL, after mergers with co-ops in NA, RHH, PFZ; 2000 members, 1900 ha. Little is above average.

Nackenheim Rhh ★→★★★★ NIERSTEIN neighbour with GROSSE LAGE Rothenberg on red shale, famous for *Rhh's richest Ries*, superb TBA. Top growers: *Gunderloch*, KÜHLING-GILLOT.

Nahe Tributary of the Rhine and dynamic region; dozens of lesser-known producers, excellent value. Great soil variety; best RIES from slate has almost MOS-like raciness.

Naturrein "Naturally pure": designation on old labels (pre-1971), indicating as little technical intervention as possible, esp no chaptalizing (sugar added at fermentation). Should be brought back.

Neipperg, Graf von Würt ★★★ LEMBERGER and SPÄTBURGUNDER of grace and purity, and v. fine sweet TRAMINER. Count Karl-Eugen von Neipperg's younger brother Stephan makes wine at Canon la Gaffelière in St-Émilion and elsewhere.

Neus Rhh ★★★ Revived historic estate at INGELHEIM, excellent PINOT N (best: Pares).

Niederhausen Na ★★→★★★★ Village of the middle NA Valley. Complex RIES from famous GROSSE LAGE Hermannshöhle and neighbouring steep slopes. Growers: CRUSIUS, *Dönnhoff*, GUT HERMANNSBERG, J Schneider, Mathern.

Nierstein Rhh ★→★★★★ Important wine town (c.800 ha) with accordingly variable wines. Best are rich, tense, eg. GROSSE LAGE v'yds Brudersberg, Hipping, Oelberg, Orbel, Pettenthal. Growers: BUNN, Huff (both), Gehring, GUNDERLOCH, Guntrum, HEYL ZU HERRNSHEIM, KELLER, KÜHLING-GILLOT, Manz, Schätzel, ST-ANTONY, Strub. But *beware Grosslage Gutes Domtal*: a supermarket deception.

Ockfen Mos ★★→★★★ Village with almost atypical powerful SAAR RIES from GROSSE LAGE v'yd Bockstein. *M Molitor*, OTHEGRAVEN, SANKT URBANS-HOF, WAGNER, *Zilliken*.

Odinstal, Weingut Pfz ★★→★★★ Highest v'yd of PFZ, 150m (492ft) above WACHENHEIM. Bio farming and low-tech vinification bring pure RIES, SILVANER, GEWÜRZ. Harvest often extends into Nov.

Germans drink on average 102 litres beer/yr and 24 litres wine (incl sparkling).

Oechsle Scale for sugar content of grape juice. Until 90s, more Oechsle meant better wine. But global warming has changed game.

Oestrich Rhg ★★→★★★ Exemplary steely RIES and fine AUSLESEN from GROSSE LAGE v'yds: Doosberg, Lenchen. Top growers incl: A Eser, KÜHN, Querbach, SPREITZER, WEGELER.

Ökonomierat Rebholz Pfz ★★★ Top SÜDLICHE WEINSTRASSE estate: bone-dry, zesty and reliable RIES GGS, best usually Kastanienbusch from red schist 07' 11' 15 16 17 18. Also gd CHARD, SPÄTBURGUNDER.

Oppenheim Rhh ★→★★★ Town s of NIERSTEIN, GROSSE LAGE Kreuz, Sackträger. Growers: Guntrum, Kissinger, KÜHLING-GILLOT, Manz. Spectacular C13 church.

Ortenau Bad ★★→★★★ District around and s of city of Baden-Baden. Mainly Klingelberger (RIES) and SPÄTBURGUNDER from granite soils. Top villages: DURBACH, Neuweier, Waldulm.

Ortswein Second rank up in VDP's pyramid of qualities: a village wine, often from single-v'yd grapes. Many bargains.

Othegraven, von Mos ★★★ Fine SAAR estate with superb GROSSE LAGE Altenberg of KANZEM, as well as parcels in OCKFEN (Bockstein) and Wawern (Herrenberg). Since 2010 owned by TV star (and von Othegraven family member) Günther Jauch.

Palatinate English for PFALZ.

Pfalz The 2nd-largest German region, 23,350 ha, balmy climate, Lucullian lifestyle. MITTELHAARDT RIES best; s Pfalz (SÜDLICHE WEINSTRASSE) is better suited to PINOT varieties. ZELLERTAL now fashionable: cool climate.

Piesport M-M ★→★★★★ M-M village for rich, aromatic RIES. GROSSE LAGE v'yds Domherr, Goldtröpfchen. Growers: Grans-Fassian, Joh Haart, JULIAN HAART, Hain, KESSELSTATT, *Reinhold Haart*, SANKT URBANS-HOF. Avoid GROSSLAGE Michelsberg.

Piwi ★→★★ Crossings of European and American vines, for fungal resistance ("Pilz-Widerstandsfähigkeit"). Most widely planted: Cabernet Blanc, Solaris (w), Regent, Pinotin (r).

Prädikat Legally defined special attributes or qualities. *See* QMP.

Prinz, Weingut ★★★ Distinctly fresh and elegant RIES from Hallgarten's altitude v'yds, organic.

Prinz von Hessen Rhg ★★★ Glorious wines from historic JOHANNISBERG estate, esp at SPÄTLESE and above, and mature vintages.

Prüm, JJ Mos ★★★★ 59 71 76 83 90 03 15 16 17 18 Legendary WEHLEN estate; also BERNKASTEL, GRAACH. Delicate but extraordinarily long-lived wines with finesse and distinctive character.

Prüm, SA Mos ★★→★★★ Quality revolution since Saskia A Prüm took over in 2017. Brilliant 18 BERNKASTEL Badstube SPÄTLESE and WEHLEN SONNENUHR AUSLESE, TROCKEN RIES also much improved.

QbA (Qualitätswein bestimmter Anbaugebiete) "Quality Wine", controlled as to area, grape(s), vintage. May add sugar before fermentation (chaptalization). Intended as middle category, but now VDP obliges its members to label their best dry wines (GGS) as QbA. New EU name gU is scarcely found on labels (*see* box, p.165).

QmP (Qualitätswein mit Prädikat) Top category, meant to replace the NATURREIN designation: no sugaring of must. Apart from that, less strict. Six levels according to ripeness of grapes: KABINETT to TBA.

Randersacker Frank ★★→★★★ Village s of WÜRZBURG with GROSSE LAGE: Pfülben. Top growers: BÜRGERSPITAL (remarkable RIES), JULIUSSPITAL, Schmitt's Kinder, STAATLICHER HOFKELLER, Störrlein & Krenig.

Ratzenberger M Rh ★★→★★★ Family estate producing racy RIES from BACHARACH and v. fine SEKT. Bought 10 ha steep slope Oberdiebacher Fürstenberg (2017), saving it from becoming fallow.

Rauenthal Rhg ★★→★★★★ Once RHG's most expensive RIES. *Spicy, austere but complex* from inland slopes. Baiken, Gehrn and Rothenberg v'yds contain GROSSE LAGE and ERSTE LAGE parcels, while neighbouring Nonnenberg (monopole of BREUER) is unclassified, despite its equal quality. Top growers: A Eser, Breuer, CORVERS-KAUTER, Diefenhardt, HESSISCHE STAATSWEINGÜTER.

Raumland Rhh ★★★ SEKT expert with deep cellar and full range of fine, balanced cuvées. Best: CHARD Brut Nature (disgorged after 10 yrs), Cuvée Triumvirat, MonRose. 1st SEKT-only estate to become VDP member.

Restsüsse Unfermented grape sugar remaining in (or in cheap wines added to) wine to give it sweetness. Can range from 1g/l in a TROCKEN wine to 300g in a TBA.

Rheingau ★★→★★★★ Birthplace of RIES. Historic s- and sw-facing slopes overlooking Rhine between Wiesbaden and RÜDESHEIM. Classic, substantial Ries, famous for steely backbone, and small amounts of delicate SPÄTBURGUNDER. Also centre of SEKT production.

Rheinhessen ★→★★★★ Germany's largest region by far (26,750 ha and rising), between Mainz and Worms. Much dross, but also treasure trove of well-priced wines from gifted young growers.

Richter, Max Ferd M-M ★★→★★★ Reliable estate. Esp gd RIES KABINETT, SPÄTLESEN: full and aromatic. Round, pretty Brut (EISWEIN dosage). Thoughtful winemaking.

Riffel Rhh ★★★ Organic family estate with holdings in Bingen's once famous Scharlachberg (red soils). RIES Turm has class. Now also pét-nat (cloudy SEKT) and barrel-fermented SILVANER.

Rings, Weingut Pfz ★★★→★★★★ Brothers Steffen and Andreas have made a name for dry RIES (esp Kallstadt SAUMAGEN), and precise SPÄTBURGUNDER (Saumagen, Felsenberg im Berntal).

Roter Hang Rhh ★★→★★★★ 08 11 12 15 16 17' 18 Leading RIES area of RHH (NACKENHEIM, NIERSTEIN, OPPENHEIM). Name ("red slope") refers to red shale soil.

Ruck, Johann Frank ★★★ Spicy, age-worthy SILVANER, RIES, SCHEUREBE and TRAMINER from IPHOFEN.

Bio growers lament cowhorn shortage: organic dairy farmers prefer hornless breeds.

Rüdesheim Rhg ★★→★★★★ Most famous RHG RIES on slate, best GROSSE LAGE v'yds (Kaisersteinfels, Roseneck, Rottland, Schlossberg) called Rüdesheimer Berg. Full-bodied but never clumsy wines, floral, esp gd in off-yrs. Best growers: BISCHÖFLICHE WEINGÜTER RÜDESHEIM (not to be confused with BISCHÖFLICHE WEINGÜTER TRIER), *Breuer*, CHAT SAUVAGE, CORVERS-KAUTER, HESSISCHE STAATSWEINGÜTER, *Johannishof*, KESSELER, KÜNSTLER, LEITZ, Ress.

Ruppertsberg Pfz ★★→★★★ MITTELHAARDT village known for elegant RIES. Growers: BASSERMANN-JORDAN, Biffar, BUHL, BÜRKLIN-WOLF, CHRISTMANN, VON WINNING.

Ruwer Mos ★★→★★★★ Tributary of MOS nr TRIER, higher in altitude than M-M. Quaffable light dry and intense sweet RIES. Best growers: Beulwitz, Karlsmühle, KARTHÄUSERHOF, KESSELSTATT, MAXIMIN GRÜNHAUS.

Saale-Unstrut ★→★★★ A ne region around confluence of these two rivers nr Leipzig. Terraced v'yds have Cistercian origins. Quality leaders: Böhme, Born, Gussek, Hey (VDP member), Kloster Pforta, LÜTZKENDORF (VDP), Pawis (VDP).

Saar Mos ★★→★★★★ Tributary of Mosel, bordered by steep slopes. Most austere, steely, *brilliant Ries* of all, consistency favoured by climate change. Villages incl: AYL, KANZEM, OCKFEN, SAARBURG, Serrig, WILTINGEN (SCHARZHOFBERG).

Saarburg Mos SAAR Valley small town. Growers: WAGNER, ZILLIKEN. GROSSE LAGE: Rausch.

KP Keller (Rhh) is growing Ries in Norway. 76 Oechsle, so Norwegian "Kabinett".

Sachsen ★→★★★ Region in Elbe Valley around Meissen and Dresden. Characterful dry whites. Best growers: Aust, Richter, *Schloss Proschwitz*, SCHLOSS WACKERBARTH, Schuh, Schwarz (try co-fermented RIES/TRAMINER), ZIMMERLING.

St-Antony Rhh ★★→★★★ NIERSTEIN estate with exceptional v'yds, known for sturdy ROTER HANG RIES, complex BLAUFRÄNKISCH. New winemaker (2018).

Salm, Prinz zu Na, Rhh ★★→★★★ Owner of Schloss Wallhausen in NA and v'yds there and at BINGEN (RHH); ex-president of VDP.

Salwey Bad ★★★→★★★★ Leading KAISERSTUHL estate. Konrad S picks early for freshness. Best: GGS Henkenberg and Eichberg GRAUBURGUNDER, Kirchberg SPÄTBURGUNDER and WEISSBURGUNDER.

Sankt Urbans-Hof Mos ★★★ Large family estate led by Nik Weis based in Leiwen, v'yds along M-M and SAAR. Limpid RIES, impeccably pure, racy, age well. Also runs a nursery with unique collection of genetically varied Ries.

Sauer, Horst Frank ★★★ Finest exponent of ESCHERNDORF's top v'yd Lump. Racy, straightforward *dry Silvaner* and RIES, sensational TBA.

Sauer, Rainer Frank ★★★ Top family estate producing seven different dry SILVANERS from ESCHERNDORF's steep slope Lump. Best: GG am Lumpen, ALTE REBEN and L 99' 03' 07'.

Saumagen Popular local dish of PFZ: stuffed pig's stomach. Also one of best v'yds of region: a calcareous site at Kallstadt producing excellent RIES, PINOT N.

Schaefer, Willi Mos ★★★ Willi S and son Christoph finest in GRAACH (but only 4 ha). MOS RIES at its best: pure, crystalline, feather-light, rewarding at all levels.

Schäfer-Fröhlich Na ★★★ Ambitious NA family estate known for spontaneously fermented RIES of great intensity, *GG* incl Bockenau Felseneck and Stromberg.

Scharzhofberg Mos ★★→★★★★ Superlative SAAR v'yd: rare coincidence of micro-climate, soil, human intelligence to bring about perfection of RIES. Top estates: BISCHÖFLICHE WEINGÜTER TRIER, EGON MÜLLER, KESSELSTATT, VAN VOLXEM, VON HÖVEL.

Schloss Johannisberg Rhg ★★→★★★ Historic RHG estate and Metternich mansion, 100% RIES, owned by Henkell (Oetker group). Usually v.gd SPÄTLESE Grünlack ("green sealing-wax"), reliable GUTSWEIN (Gelblack).

Schloss Lieser M-M ★★★★ Thomas Haag (elder son of FRITZ HAAG estate) produces painstakingly elaborate RIES both dry and sweet from Lieser (Niederberg Helden), BRAUNEBERG, WEHLEN, PIESPORT. Now also small (leased) plot in BERNKASTEL's DOCTOR. Hotel Lieser Castle has no ties to wine estate.

Schloss Proschwitz Sachs ★★ Prince Lippe's resurrected estate at Meissen in SACHS; pioneering E German viticulture after the wall came down, esp with *dry Weissburgunder*, GRAUBURGUNDER. 80 ha, S African winemaker.

Schloss Vaux Rhg ★★→★★★ SEKT house best known for single-v'yd RIES Sekt (eg. MARCOBRUNN, RÜDESHEIMer SCHLOSSBERG).

Schloss Vollrads Rhg ★★→★★★ One of greatest historic RHG estates; owned by a bank.

Schloss Wackerbarth ★★→★★★ Saxon state domaine on outskirts of Dresden, showing remarkable progress recently, esp 18 Goldener Wagen RIES (TROCKEN, AUSLESE), TRAMINER SPÄTLESE.

Schlossböckelheim Na ★★→★★★★ Village with GROSSE LAGE v'yds Felsenberg,

Kupfergrube. Firm RIES that needs ageing. Top growers: C Bamberger, CRUSIUS, DÖNNHOFF, GUT HERMANNSBERG, Kauer, SCHÄFER-FRÖHLICH.

Schnaitmann Würt ★★→★★★★ Excellent barrel-aged reds from Fellbach (nr Stuttgart). Whites (eg. RIES, SAUV BL), SEKT (Evoé!), and wines from lesser grapes (SCHWARZRIESLING, TROLLINGER) tasty too.

Schneider, Markus Pfz ★★ Shooting star in Ellerstadt, PFZ. A full range of soundly produced, trendily labelled wines.

Schoppenwein Café (or bar) wine, ie. wine by the glass.

Schwarzer Adler Bad ★★★→★★★★ French restaurant (Michelin star since 1969) at Oberbergen, KAISERSTUHL, and top wine estate led by Fritz Keller (also elected president of German Football Federation DFB in 2019) and son Friedrich Keller. PINOTS (r w) have class.

Schwegler, Albrecht Würt ★★★→★★★★ 7 ha, now led by young Aaron S. Red blends Beryll, Saphir, Granat have ultra-pure fruit. Top selection Solitär only produced once in a decade.

Seeger Bad ★★★ Best producer of the Badische Bergstrasse area s of Heidelberg, known for clever barrel-ageing. Reds and whites equally gd.

Sekt ★→★★★★ German sparkling. Avoid cheap offers: bottle fermentation is not mandatory. Serious Sekt producers are making spectacular progress, eg. ALDINGER, Bardong, Barth, BATTENFELD-SPANIER, DIEL, GRIESEL & COMPAGNIE, JÜLG, Leiner, Melsheimer, RAUMLAND, Reinecker, Schembs, SCHLOSS VAUX, SCHWARZER ADLER, Solter, S Steinmetz, Strauch, WAGECK, WEGELER, Wilhelmshof, ZILLIKEN. A VDP classification of Sekt v'yds is underway.

Selbach-Oster M-M ★★★ Scrupulous ZELTINGEN estate with excellent v'yd portfolio, known for classical style and focus on sweet PRÄDIKAT wines.

Solveigs Rhg ★★→★★★★ PINOT N specialist with v'yds on red slate at ASSMANNSHAUSEN and LORCH, only 2 ha, organic viticulture, minimal winemaking for depth, longevity. Best: single-plots Micke 06' 13' 15'; Present 95' 99' 03' 04 06 09' 12 15'.

Sonnenuhr M-M Sundial. Name of GROSSE LAGE sites at BRAUNEBERG, WEHLEN, ZELTINGEN.

Sorentberg M-M ★★★ V'yd in a side valley of M-M nr Reil, fallow for 50 yrs (except a tiny plot of 1000 vines), now replanted by young Tobias Treis and partner from South Tyrol. Red slate, cool climate.

Spätlese Late-harvest. One level riper and potentially sweeter than KABINETT. Gd examples age at least 7 yrs. Spätlese TROCKEN designation was abandoned by VDP members: a shame.

Spreitzer Rhg ★★★ Brothers Andreas and Bernd S produce deliciously *racy, harmonious* RIES from v'yds in HATTENHEIM, OESTRICH, Mittelheim. Mid-price range ALTE REBEN a bargain.

Staatlicher Hofkeller Frank ★★★ Bavarian state domain; 120 ha of fine FRANK v'yds, spectacular cellars under great baroque Residenz at WÜRZBURG. Three directors in past 5 yrs, now 1st signs of improvement.

Staatsweingut / Staatliche Weinbaudomäne State wine estates or domains exist in BAD (IHRINGEN, Meersburg), WÜRT (Weinsberg), RHG (HESSISCHE STAATSWEINGÜTER), RHH (OPPENHEIM), PFZ (Neustadt), MOS (TRIER), SACHS (WACKERBARTH) and SA-UN (Kloster Pforta). Some have been privatized in recent yrs, eg. at Marienthal (AHR), NIEDERHAUSEN (NA).

The steam locomotive lesson

There have always been terroir sceptics who scoff at geology. At WÜRZBURG, for instance, people used to say that the typical smoky flavour of the STEINWEIN was due to the railway at the bottom of the slope. But today there are no more steam trains. And what does Steinwein smell like? Smoky. It's the *Muschelkalk* (shell limestone), not the trains.

Steigerwald Frank District in e FRANK. V'yds at considerable altitude. Best: Castell, Hillabrand, Roth, RUCK, VETTER, WELTNER, *Wirsching*.

Steinberg Rhg ★★★ Walled-in v'yd above HATTENHEIM, est by Cistercian monks 700 yrs ago: German Clos de Vougeot. Monopoly of HESSISCHE STAATSWEINGÜTER. Classified parcels (14 ha of 37) have unique soil (clay with fragments of decomposed schist in various colours). Fascinating old vintages (eg. NATURREIN 43, TBA 59).

53% employees in German viticulture are seasonal workers, 33% family members.

Steinwein Frank Wine from WÜRZBURG's best v'yd, Stein. Goethe's favourite. Only six producers: BURGERSPITAL, JULIUSSPITAL, L Knoll, Meinzinger, Reiss, STAATLICHER HOFKELLER. The author once tasted the 1540 vintage.

Stodden Ahr ★★★→★★★★ AHR SPÄTBURGUNDER with a Burgundian touch, delicately extracted and subtle. Best usually ALTE REBEN and Rech Herrenberg. Pricey – but production is tiny.

Südliche Weinstrasse Pfz District in s PFZ, known esp for PINOT varieties. Best growers: BECKER, JÜLG, Leiner, Minges, Münzberg, Ö REBHOLZ, Siegrist, WEHRHEIM.

Taubertal Bad, Frank, Würt ★→★★★ Cool-climate district along River Tauber, divided by Napoleon into BAD, FRANK and WÜRT sections, SILVANER from limestone soils, local red Tauberschwarz. Frost a problem. Growers: Hofmann, Schlör, gd co-op at Beckstein.

TBA (Trockenbeerenauslese) Sweetest, most expensive category of German wine, extremely rare, viscous and concentrated with dried-fruit flavours. Made from selected dried-out grapes affected by noble rot (botrytis). Half bottles a gd idea.

Thanisch, Weingut Dr. M-M ★★★ BERNKASTEL estate, founded 1636, famous for its share of the DOCTOR v'yd. After family split-up in 1988 two homonymous estates with similar qualities: Erben (heirs) Müller-Burggraef and Erben Thanisch.

Trier Mos The n capital of ancient Rome, on the Mosel, between RUWER and SAAR. Big charitable estates have cellars here among awesome Roman remains.

Trittenheim M-M ★★→★★★ Racy, textbook M-M RIES if from gd plots within extended GROSSE LAGE v'yd Apotheke. Growers: A CLÜSSERATH, Clüsserath-Weiler, E Clüsserath, FJ Eifel, Grans-Fassian, Milz.

Trocken Dry. Used to be defined as max 9g/l unfermented sugar. Today height of gastronomic fashion. Generally, further s in Germany, the more Trocken wines.

Ürzig M-M ★★★→★★★★ Mosel village on red sandstone and red slate, famous for ungrafted old vines and *unique spicy Ries*. GROSSE LAGE v'yd: Würzgarten. Growers: Berres, Christoffel, Erbes, *Loosen*, MERKELBACH, *M Molitor*, Mönchhof, Rebenhof.

Van Volxem Mos ★★★ Historical SAAR estate revived by obsessive Roman Niewodniczanski. Low yields from top sites (KANZEM Altenberg, OCKFEN Bockstein, SCHARZHOFBERG, WILTINGEN Gottesfuss) bring about monumental (mostly dry) RIES. Spectacular castle-like new cellar building in a Saar loop nr Wiltingen – and capacity to age selected wines 5 yrs+ in tank.

VDP (Verband Deutscher Prädikatsweingüter) Influential association of 200 premium growers setting highest standards. Look for its eagle insignia on wine labels, and for GROSSE LAGE logo on wines from classified v'yds. A VDP wine is usually a gd bet. President: Steffen CHRISTMANN.

Vetter, Stefan Frank ★★→★★★ Natural wine (*see* A Little Learning at back of book): SILVANER fermented on skins. To watch.

Vollenweider, Daniel Mos ★★★ A Swiss in M-M: excellent RIES in v. small quantities from Wolfer Goldgrube v'yd nr Traben-Trarbach.

Wachenheim Pfz ★★★ Celebrated village with, according to VDP, no GROSSE LAGE v'yds. See what you think. Top growers: Biffar, BÜRKLIN-WOLF, Karl Schäfer, ODINSTAL, WOLF, Zimmermann (bargain).

Wageck Pfz ★★→★★★ MITTELHAARDT estate for unaffected, brisk CHARD (still and sparkling) and PINOT N of great finesse.

Wagner, Dr. Mos ★★→★★★ Estate with v'yds in OCKFEN and Saarstein led by young Christiane W. SAAR RIES with purity, freshness.

Wagner-Stempel Rhh ★★★ Seriously crafted RHH wines from Siefersheim nr NA border. Best usually RIES GGS Heerkretz (porphyry soil).

Walluf Rhg ★★★ Underrated village, 1st with important v'yds as one leaves Wiesbaden going w. GROSSE LAGE v'yd: Walkenberg. Growers: *JB Becker, Jost.*

Wasenhaus Mos ★★★ Young Alexander Götze and Christoph Wolber produce Burgundy-inspired PINOT N Bellen and PINOT BL Möhlin from limestone sites in MARKGRÄFLERLAND, labelled as LANDWEIN, only 2 ha, all farmed by hand.

Wegeler M-M, Rhg ★★→★★★★ Important family estates in OESTRICH and BERNKASTEL (both in top form) plus a stake in the famous KRONE estate of ASSMANNSHAUSEN. Geheimrat J blend maintains high standards, single-v'yd RIES usually outstanding value. Old vintages available ("vintage collection").

Wehlen M-M ★★★→★★★★ Wine village with legendary steep SONNENUHR v'yd expressing RIES from slate at v. best: rich, fine, everlasting. Top growers: JJ PRÜM, Kerpen, KESSELSTATT, LOOSEN, MARKUS MOLITOR, RICHTER, SA PRÜM, SCHLOSS LIESER, SELBACH-OSTER, Studert-Prüm, THANISCH, WEGELER.

Wehrheim, Weingut Dr. Pfz ★★★ Top organic estate of SÜDLICHE WEINSTRASSE. V. dry, culinary style, esp white PINOT varieties.

Weil, Robert Rhg ★★★→★★★★★ 17 37 59 90 01 05 09 12 15 16 17 18 (19) Outstanding estate in KIEDRICH with classified v'yds Gräfenberg (steep slope on phyllite schist), Klosterberg, Turmberg. Superb sweet KABINETT to TBA (sensational 18 GOLDKAPSEL) and EISWEIN, gd GG.

Weingart M Rh ★★★ Outstanding estate at Spay, v'yds in BOPPARD (esp Hamm Feuerlay). Refined, taut RIES, low-tech in style, superb value.

Weingut Wine estate.

Weissherbst Pale-pink wine, made from a single variety, often SPÄTBURGUNDER. V. variable quality.

Weltner, Paul Frank ★★★ STEIGERWALD family estate. Densely structured, age-worthy SILVANER from underrated Rödelseer Küchenmeister v'yd.

Wiltingen Mos ★★→★★★★ Heartland of the SAAR. SCHARZHOFBERG crowns a series of GROSSE LAGE v'yds (Braune Kupp, Braunfels, Gottesfuss, Kupp). Top growers: BISCHÖFLICHE WEINGÜTER TRIER, EGON MÜLLER, KESSELSTATT, LE GALLAIS, SANKT URBANS-HOF, VAN VOLXEM, Vols.

Winning, von Pfz ★★★→★★★★ DEIDESHEIM estate with prime v'yds there and at FORST. *Ries of great purity*, terroir expression, slightly influenced by fermentation in new FUDER casks.

German vintage notation

The vintage notes after entries in the German section are mostly given in a different form from those elsewhere in the book. If the vintages of a single wine are rated, or are for red wine regions, the vintage notation is identical with the one used elsewhere (*see* front jacket flap). But for regions, villages or producers, two styles of vintage are indicated:

Bold type (eg. **16**) indicates classic, ripe vintages with a high proportion of SPÄTLESEN and AUSLESEN; or, in the case of red wines, gd phenolic ripeness and must weights.

Normal type (eg. 17) indicates a successful but not outstanding vintage. Generally, German white wines, esp RIES, can be drunk young for their intense fruitiness, or kept for a decade or even two to develop their potential aromatic subtlety and finesse.

> **EU terminology**
> Germany's part in new EU classification involves, firstly, abolishing
> the term Tafelwein in favour of plain **Wein** – this is, up to now, the
> only visible change on labels. LANDWEIN is still called Landwein, even
> if its bureaucratic name would be **geschützte geographische Angabe
> (ggA)**, or Protected Geographical Indication. Brussels generally allows
> continued use of est designations. **Geschützte Ursprungsbezeichnung
> (gU)**, or Protected Designation of Origin, was expected to replace
> QUALITÄTSWEIN and QUALITÄTSWEIN MIT PRÄDIKAT. But as it's hard and
> time-consuming (4 yrs) to get recognition for a village- or v'yd-specific
> gU, only two gUs were in place by end of 2019: Bürgstadter Berg
> (see WEINGUT FÜRST) and WINNINGEN Uhlen (see HEYMANN-LÖWENSTEIN).
> WÜRZBURGER Steinberg (for historical extension of STEIN) is likely to follow.
> The existing predicates – SPÄTLESE, AUSLESE and so on (see box, p.165) –
> stay in place; the rules for these styles hasn't changed, and isn't going to.

Winningen Mos ★★→★★★ Lower MOS town nr Koblenz; powerful dry RIES. GROSSE LAGE v'yds: Röttgen, Uhlen. Top growers: HEYMANN-LÖWENSTEIN, Knebel, Kröber, Richard Richter.

Wirsching, Hans Frank ★★★ Renowned estate in IPHOFEN known for classically structured dry RIES and *Silvaner*. Andrea W extends range with spontaneously fermented Ries Sister Act and kosher SILVANER. Plus excellent SCHEUREBE (ALTE REBEN).

Wittmann Rhh ★★★ Philipp W has propelled this bio estate to the top ranks. Crystal-pure, zesty dry RIES GG (Morstein 05 07' 08 11 12' 15 16 17 18).

Wöhrle Bad ★★★ Organic pioneer at Lahr (25 yrs+), son Markus a PINOT expert, excellent GGs.

Wolf JL Pfz ★★→★★★ WACHENHEIM estate, leased by Ernst LOOSEN of BERNKASTEL. Dry PFZ RIES (esp FORSTER Pechstein), sound and consistent rather than dazzling.

Württemberg Würt Formerly mocked as "TROLLINGER republic", but today dynamic, with many young growers eager to experiment. Best usually LEMBERGER, SPÄTBURGUNDER. Only 30% white varieties. RIES needs altitude v'yds.

Würzburg Frank ★★→★★★★ Great baroque city on the Main, centre of FRANK wine. Classified v'yds: Innere Leiste, Stein (STEINWEIN), Stein-Harfe.

Zell Mos ★→★★★ Best-known lower MOS village, notorious for GROSSLAGE Schwarze Katz (Black Cat) – avoid.

Zellertal Pfz ★★→★★★ Area in n PFZ, high, cool, recent gold-rush: BATTENFELD-SPANIER, KÜHN have bought in Zellertal's best RIES v'yd Schwarzer Herrgott or neighbouring RHH plot Zellerweg am Schwarzen Herrgott. Gd local estates: Bremer, Janson Bernhard, Klosterhof Schwedhelm.

Zeltingen M-M ★★→★★★ Top MOS village sometimes overshadowed by neighbour WEHLEN. GROSSE LAGE v'yd: SONNENUHR. Top growers: JJ PRÜM, MARKUS MOLITOR, SELBACH-OSTER.

Ziereisen Bad ★★→★★★ Outstanding estate in MARKGRÄFLERLAND, advocating LANDWEIN, mainly PINOTS and GUTEDEL. Best are SPÄTBURGUNDERS from small plots: Rhini, Schulen, Talrain. Jaspis = old-vine selections.

Zilliken, Forstmeister Geltz Mos ★★★→★★★★ 93 94 95 96 97 99 01 03 04 05 07 08 09 10 11 12 14 15 16 17 SAAR family estate: intense racy/savoury *Ries from Saarburg Rausch* and OCKFEN Bockstein, incl superb long-lasting AUSLESE, EISWEIN. V.gd SEKT too (and Ferdinand's gin).

Zimmerling, Klaus Sachs ★★★ Small, perfectionist estate, one of 1st to be est after Berlin wall came down. Best v'yd is Königlicher Weinberg (King's v'yd) at Pillnitz nr Dresden. RIES, sometimes off-dry, can be exquisite.

Luxembourg

The tiny Duchy of Luxembourg lies on the River Moselle/Mosel's left bank. Annual wine consumption is 61 litres/capita, more than France and Italy, and only surpassed by the Vatican (71 litres/year). The wines are different to those of the Middle Mosel or nearby Saar valley. The soil is limestone, and has more in common with Chablis or Champagne than Piesport. Only 11% is Riesling. The big ones are Müller-Thurgau (aka Rivaner), Auxerrois, Pinots Blanc and Gris. Crémant fizz is a speciality. Climate change has been kind: 18 brought the most powerful Pinots of Luxembourg's history. But 19 showed painfully that frost is still a danger (as did 17 16). Most whites have strong acidity and some sweetness. A common term (but of little significance) is "Premier Grand Cru". More reliable: groups of winemakers who come together to promote their high standards: Domaine et Tradition (eight producers) has most credibility.

Alice Hartmann ★★→★★★★ Luxembourg's best RIES v'yd, Koeppchen (Les Terrasses, 70-yr-old vines, Au Coeur de la Koeppchen, limestone). Also in Burgundy (St-Aubin), Mittelmosel (Trittenheim) and a plot in Scharzhofberg.

Aly Duhr ★★→★★★ Barrique (PINOT BL/AUXERROIS), refined RIES Ahn Palmberg, Ahn Nussbaum PINOT GR (18'). V.gd Crémant Grande Cuvée.

Bastian, Mathis ★★→★★★ Substantial whites (v.gd 17 Domaine et Tradition RIES Wellenstein Foulschette).

Bernard-Massard ★→★★★ Big producer, esp Crémant. Top: Ch de Schengen and Clos des Rocher. Makes Sekt in Germany too.

Château Pauqué ★★★ ★→★★★★ Passionate Abi Duhr bridges gap between Burgundy and Germany: top RIES. Clos de la Falaise is outstanding barrel-fermented CHARD.

Duhr Frères / Clos Mon Vieux Moulin ★★→★★★ Classically built whites (mineral RIES Ahn Palmberg).

Gales ★★→★★★ Reliable. Best: Crémant, Domaine et Tradition. Old labyrinth cellar.

Schumacher-Knepper ★★→★★★ Some v.gd: Ancien Propriété Constant Knepper.

Sunnen-Hoffmann ★★★ Full range of textbook whites, best usually RIES Wintrange Felsberg VV Domaine et Tradition from a v'yd planted in 1943.

Other good estates: Cep d'Or, Fränk Kayl, Paul Legill, Ruppert, Schmit-Fohl, Stronck-Pinnel. Domaines Vinsmoselle is a union of co-ops.

Belgium

The Kingdom of Belgium is better known for beer and chocolates, yet more and more interesting wines originate from its c.500 ha of vineyards. Most wineries are small, with only a dozen exceeding 10 ha, but number and acreage are rising. Belgium's cold climate shows in its wines: fresh, lively, food-friendly, with 80% white or sparkling. Quality, quantity and age-ability vary between vintages and the best wines come from vineyards with a favourable microclimate. Many winemakers swear by classics: Chardonnay, Pinot Noir, Pinot Gris, Auxerrois or Pinot Blanc; others advocate disease-resistant varieties, eg. Johanniter, Regent, Solaris.

Best producers: Aldeneyck, Bon Baron, Chant d'Eole, Clos d'Opleeuw, Crutzberg, Entre-Deux-Monts, Genoels-Elderen, Gloire de Duras, Hoenshof, Kitsberg, Kluisberg, Meerdael, Pietershof, Schorpion, Vandeurzen, Vignoble des Agaises, Vin de Liège, Waes.

Spain

Abbreviations used in the text:

PORTUGAL		Jum	Jumilla
Alen	Alentejo	La M	La Mancha
Alg	Algarve	Mad	Madeira
Bair	Bairrada	Mall	Mallorca
Bei Int	Beira Interior	Man	Manchuela
Dou	Douro	Mén	Méntrida
Lis	Lisboa	Mont-M	Montilla-Moriles
Min	Minho	Mont	Montsant
Set	Setúbal	Mur	Murcia
Tej	Tejo	Nav	Navarra
Vin	Vinho Verde	Pen	Penedès
		Pri	Priorat
SPAIN		P Vas	País Vasco
Alel	Alella	R Bai	Rías Baixas
Alic	Alicante	Rib del D	Ribera
Ara	Aragón		del Duero
Bier	Bierzo	Rio	Rioja
Bul	Bullas		
Cád	Cádiz		
Can	Canary Islands		
C-La M	Castilla-	R Ala	Rioja Alavesa
	La Mancha	R Alt	Rioja Alta
C y L	Castilla y León	R Or	Rioja Oriental
Cat	Catalonia	Rue	Rueda
Cos del S	Costers del Segre	Som	Somontano
Emp	Empordà	U-R	Utiel-Requena
Ext	Extremadura	V'cia	Valencia
Gal	Galicia		

The warmer wine countries always make their name for red wines before they master the freshness essential for white. Now quietly, almost surreptitiously, Spain's whites have come to take a starring role. Galicia has led the way with an exceptional range. The region's long isolation from the rest of Spain, tucked away above Portugal, has made it come as a surprise. What Galicia has is fresh, unoaked, aromatic Atlantic whites: notably Albariños on the coast in Rías Baixas. But travel inland to Ribeiro and further still to Valdeorras, and there are whites from warmer climates. Treixadura, Loureiro and above all Godello are the varieties to watch. Nor is it just Galicia. Viura, also known as Macabeo, has for years been a workhorse producing frankly drab whites. Suddenly it seems, Viura can be fresh too. Rioja's Viuras are much more interesting now, with young whites joining the famously venerable wines of López de Heredia and Marqués de Murrieta's Castillo Ygay. In Valdejalón, in Aragón, newcomer Frontonio proves that Macabeo from old bush vines can make fine wine. Further east, the tiny DO of Terra Alta produces one-third of the world's Garnacha Blanca, from high-altitude vineyards. Also in Cataluña, the Cava grape Xarel·lo is winning recognition as a table wine. It has had a long history in Cava, but in this guise it's a welcome

Portugal & Spain

newcomer. These wines are unoaked, or only subtly so: the varieties sing out as the fresh voices Spain has long needed. Meanwhile, the greatest of all Spain's white wines, Sherry, continues to diversity with new releases.

Recent Rioja vintages

2019 Uneven ripening demanded careful selection. Quality varies.

2018 Rain at right meant super-abundant harvest, excellent quality to match.

2017 Dramatic frost led to much-reduced harvest. What was left is v.gd.

2016 Difficult spring, v. hot summer, rain at harvest. Pick your producer.

2015 Reliably back to form. Top wines are as gd as 2010.

2014 After two small vintages, a return to quality, quantity.

2013 Cool yr, small harvest, with some gd wines.

2012 Gd. One of lowest yields for two decades.

2011 Officially *excelente*; not as gd as 2010, perhaps, but still v.gd.

2010 *Excelente*. Perfect yr. Wines to enjoy now. Best have plenty of time.

Aalto Rib del D r ★★★→★★★★ Big, polished, structured wines: Aalto; and flagship PS (from 200 small plots) now entering 3rd decade. MARIANO GARCÍA, ex-VEGA SICILIA, builds wines for cellaring that blossom after 10–20 yrs. Co-owns with Masaveu, owners of Enate (SOM), Fillaboa (R BAI), Murúa (RIO).

Abadía de Poblet Pri, C de Bar r w ★★ Exciting project within Cistercian monastery of Poblet, burial place of Ara kings. SCALA DEI winemaker working with local varieties esp (r) Trepat.

Abadía Retuerta C y L r w ★★★ Height of luxury: Michelin-starred restaurant, glam hotel, winery. Just outside RIB DEL D. V.gd white DYA Le Domaine. Single-v'yd international reds, eg. Pago Garduña SYRAH, PV PETIT VERDOT, Pago Valdeballón CAB SAUV. Novartis-owned.

Abel Mendoza R Ala r w ★★→★★★ For knowledge of RÍO villages and varieties Abel and Maite Mendoza have few equals. Discover no fewer than five varietal whites. Grano a Grano are only-the-best-berry-selected TEMPRANILLO and GRACIANO.

Alexander Jules Sherry w ★★→★★★ US-based négociant bottling selected BUTTS of distinctive Sherries.

Algueira Rib Sac r w ★★→★★★ Exceptional producer in RIBEIRA SACRA, expert in its extreme viticulture. Fine selection of elegant wines from local varieties. Outstanding is Merenzao (aka Jura's Trousseau), almost burgundian in style.

Alicante r w sw ★→★★★ Spiritual home of MONASTRELL; spicy reds and rare traditional fortified *Fondillón*. Heritage of old bush-vines. Top: ARTADI, ENRIQUE MENDOZA.

Almacenista Sherry, Man A Sherry stockholding cellar; provides wines for BODEGAS to increase or refresh stocks. Important in MANZANILLA production. Can be terrific. Few left; many now sell direct to consumers, eg. GUTIÉRREZ COLOSÍA, EL MAESTRO SIERRA. Often source for individual négociant bottlings.

Alonso, Bodegas Man w ★★★→★★★★ Part of revival of SANLÚCAR DE BARRAMEDA BODEGAS. Asencio brothers own Dominio de Urogallo (Asturias). Bought exceptional stock of Pedro Romero, incl v. fine SOLERAS of Gaspar Florido. Excellent if super-priced four-bottle collection. More accessibly priced is Velo Flor, 9–10-yr-old MANZANILLA.

Alonso del Yerro Rib del D, Toro r ★★→★★★ Stéphane Derenoncourt (B'x consultant) entices elegance from extreme continental climate of RIB DEL D. Transformation since 2016, altogether more delicate. Family business, estate wines. Top wine: María, inky but not overblown. Paydos is its TORO.

Alta Alella Cat r w sp ★★→★★★ With toes in the Med and just up the coast from Barcelona, the BODEGA welcomes visitors warmly. Renowned for CAVAS. Delightful sweet red Dolç Mataró from MONASTRELL. Organic.

Álvaro Domecq Sherry, Man w ★★→★★★ SOLERAS drawn from former ALMACENISTA Pilar Aranda. Polished wines. Gd FINO La Janda. Excellent 1730 VORS series.

Alvear Mont-M, Ext r w ★★→★★★★ Historic Alvear has superb array of PX wines in MONT-M. Gd FINO CB and Capataz, lovely sweet SOLERA 1927, unctuous DULCE Viejo. V. fine vintage wines. Also Palacio Quemado BODEGA in Ext. Imaginative appointment of ENVINATE team has brought innovative thinking, eg. 3 Miradas series – lower alc wines with FLOR and focus on terroir.

Añada Vintage.

Arrayán, Bodegas Mén r w ★★ Winemaker Maite Sánchez is restoring reputation of MÉN with fine GARNACHA and Albillo Real (w).

Artadi Alic, Nav, P Vas, Ala r w sp ★★→★★★★ Formed in 1985 from a growers' co-op around Laguardia in R Ala. Juan Carlos López de Lacalle has led it to outstanding success. Left RIO DO end 2015, believing it failed to defend quality. Former Rios now called Álava. Focus on single v'yds: luxuriant La Poza de Ballesteros; dark, stony El Carretil; outstanding El Pisón. Also in ALIC (v.gd r El Sequé), NAV (r Artazuri, p DYA). Latest project Izar-Leku, sparkling TXAKOLI (w) from Getaria.

Artuke Rio r ★★★ Arturo and Kike de Miguel, 5th generation of growers, lend their names to this. Specializes in small, single, old v'yds, notably La Condenada.

"Atlantic" wines Gal, Rio, P Vas r p w Unofficial collective term for bright, often unoaked style, with firm acidity. Increasingly used to describe crisp, delicate reds, esp in R BAI; or Cantabrian Sea: the TXAKOLIS. Also used to describe cool climatic influences, eg. inland GAL DOS, and specific vintages in R Ala, R Alt.

Ausàs Rib del D r ★★ Look out for Interpretación, from eponymous project of Xavier A, former winemaker at VEGA SICILIA.

Baigorri R Ala r w ★★→★★★ Wines as glamorous as BODEGA's glassy architecture. Gravity-fed, producing bold and modern RIO. Gd restaurant, tasting menus and v'yd views.

Barbadillo Man ★→★★★★ Cathedral-like grand cellars dominate SANLÚCAR's upper town. Montse Molina manages wines from supermarket to superb. Pioneer of MANZANILLA EN RAMA. Top of the range Reliquía wines unbeatable, esp AMONTILLADO, PALO CORTADO. Outstanding century-old Versos Amontillado. Sherry guru Armando Guerra advises on adventurous new releases (Nude is Beaujolais-style Tintilla de Rota) and returning to traditional practices, eg. Mirabras, unfortified PALOMINO. Worth following: this once-slumbering giant is now wide awake. Also owns Vega Real (RIB DEL D), BODEGA Pirineos (SOM).

Belondrade C y L, Rue r w ★★→★★★ VERDEJO as it should be but so rarely is. Didier B was early (1994) exponent of finesse in RUE, and lees-ageing. Quinta Apollonia (w) and light, summery, Quinta Clarisa TEMPRANILLO (r), both C Y L.

Bentomiz, Bodegas Mál r p w sw ★★→★★★ Sweet wine experts, Dutch by birth, Spanish by adoption: Clara and André are in Axarquía, inland from MÁLAGA. Sweet MOSCATEL and MERLOT. Also rare dry Romé (p). Visit restaurant.

Beronia Rio r p w ★→★★★ Popular RIO BODEGA owned by GONZÁLEZ BYASS. RES v. reliable. Award-winning new winery.

Ribei, Dominio do Gal r w ★★ One of stars of RIBEIRA SACRA DO. Lapena is GODELLO grown on schist; Llama is spicy MENCÍA blend.

New Sherry-based Vermouth: Barbadillo, Fernando de Castilla, González B, Lustau.

Bierzo r w ★→★★★ Isolated DO in nw Spain shot to international fame with two producers, Raúl Pérez and Ricardo Pérez Palacios (no relation). Key variety is red MENCÍA. Grown on slate soils it makes perfumed crunchy *Pinot-like red*. Best sites are high-altitude. Bier quality is uneven. Best advice: follow the producer. Look for DESCENDIENTES DE J PALACIOS, RAÚL PÉREZ, plus Dominio de Tares, Losada, Luna Berberide, Mengoba. Also fine GODELLO (w).

Bilbao, Ramón R Bai, Rib del D, Rio, Rue r p w sp ★→★★ Major RIO producer making strides in quality. BODEGAS also in R BAI (Mar de Frades, in blue bottles), RUE and RIB DEL D (Cruz de Alba). Delivering fresh GARNACHA at altitude in RIO, and Provençal-style pale ROSADO Lalomba. Owned by Diego Zamora, producer of Spain's bestselling Licor 43.

Bodega A cellar; a wine shop; a business making, blending and/or shipping wine.

Butt Sherry 600-litre barrel of long-matured American oak used for Sherry. Filled 5/6 full, allows space for FLOR to grow. Trend for whites aged in ex-FINO butts – CVNE's Monopole Clasico, BARBADILLO Mirabras.

Calatayud Ara r p w ★→★★★ Old-bush-vine GARNACHA grown at 700–900m (2297–2953ft) has put Calatayud on map as part of heritage of Ara. Still best known for cheap co-op wines. Best: EL ESCOCÉS VOLANTE, San Alejandro co-op.

Callejuela Sherry, Man ★★→★★★ Blanco brothers are growers with v'yds in some of Sherry's most famous PAGOS. In 2005 launched 1st Callejuela Sherries as part of return to recognizing terroir in Sherry. V.gd aged MANZANILLA, AMONTILLADO, OLOROSO; vintage releases of Manzanilla.

Campo de Borja Ara r p w ★★★★ Self-proclaimed "Empire of GARNACHA". Heritage of old vines, plus young v'yds = 1st choice for gd-value Garnacha, now starting to show serious quality: BODEGAS Alto Moncayo, Aragonesas, Borsao, FRONTONIO.

Campo Viejo Rio r p w sp ★→★★★ RIO's biggest brand. In addition to value RES, GRAN

RES, has varietal GARNACHA, and adds TEMPRANILLO Blanco to white RIO. V.gd top Res Dominio. Part of Pernod Ricard (also owns much-improved YSIOS winery in Rio).

Canary Islands r p w ★→★★ Seven main islands, nine DOS. TENERIFE has five DOs. Plenty of dull wine for tourists. Seek unusual varieties, old vines, distinct microclimates, volcanic soils. Dry white LISTÁN (aka PALOMINO) and Marmajuelo, black Listán Negro, Negramoll (TINTA NEGRA), Vijariego offers *enjoyable original flavours*. Gd dessert MOSCATELS, MALVASÍAS, esp fortified El Grifo from Lanzarote. Top: Borja Pérez, ENVINATE, SUERTES DEL MARQUÉS. In La Palma, Victoria Pecis TORRES.

Cangas ★→★★ Isolated DO in wild Asturias beginning to export. Isolation means unique vines: fresh Albarín Blanco, firm reds from Albarín Negro, Verdejo Negro, and most promising, Carrrasquín. Producers: Dominio de Urogallo (owned by BODEGAS ALONSO), Monasterio de Corias, VidAs.

Capçanes, Celler de Mont r p w sw ★→★★ One of Spain's top co-ops. Great-value, expressive wines from MONT. Kosher specialist esp Peraj Ha'abib.

Cariñena Ara r p w ★→★★ The one DO that is also the name of a grape variety. DO, formerly co-op country, is not exciting, but gd value. Jorge Navascués, winemaker at RIO's CONTINO, makes own wines at Navascués Enologia. Also consults at VINO DE PAGO FINCA Ayles.

Casa Castillo Jum r ★★→★★★ Proves JUM can be tiptop. Family business high up in Jum *altiplano*. V. fine Las Gravas single v'yd blend. Showcase for MONASTRELL, esp PIE FRANCO (plot escaped fairly recent phylloxera).

Castell d'Encús Cos del S r p w ★★→★★★ CAT wineries are searching for cool sites. Raül Bobet (also of PRI Ferrer-Bobet) has all the cool climates he wants at 1000m (3281ft), for *superbly fresh, original wines*. Ancient meets modern: grapes fermented in stone *lagares*, while winery is up to date. Acusp PINOT, Ekam RIES, Thalarn SYRAH have become classics.

Castilla y León r p w ★→★★★ Spain's largest region. Plenty to enjoy: new projects, those who prefer to be outside DO rules. DOs in region: Arlanza, Arribes, BIER, Cigales, RUE, Sierra de Salamanca (one to watch: r Rufete variety), Tierra de León, Tierra del Vino de Zamora, TORO, Valles de Benavente, Valtiendas. Catch-all DO is Vino de la Tierra de Castilla y León. Top: ABADÍA RETUERTA, MAURO, MARQUÉS DE RISCAL VERDEJO, Ossian, Prieto Pariente – either geographically outside a smaller DO or want to avoid regulations or poor reputation of a particular DO.

Castillo de Cuzcurrita R Alt r ★★ Walled v'yd, top consultant Ana Martín, v. fine RIO.

Castillo Perelada Emp, Nav, Pri r p w sp ★→★★★ Glamorous estate and tourist destination. Vivacious CAVAS, esp Gran Claustro; modern red blends. Rare 12-yr-old, SOLERA-aged GARNATXA de l'EMPORDÀ. V. fine Casa Gran del Siurana from PRI. Has purchased CHIVITE group.

Catalonia r p w sp Vast DO, covers whole of Cat: seashore, mtn, in-between. Top chefs and top BODEGAS (eg. TORRES). Yet actual DO is just umbrella, too large to have identity, an excuse for characterless cross-DO blends.

Cava p w ★→★★★ Could Cava be starting to get its house in order? Producers who lost faith in the DO have formed several: CLÀSSIC PENEDÈS, Conca del Riu Anoia, CORPINNAT; each with tighter quality regulations. At last Cava intends to follow suit, with proposals to extend min ageing, organic viticulture and more. The vast majority of Spain's traditional-method sparkling is made in PEN (in or around San Sadurní d'Anoia), but also Ext, RIO, V'CIA. Local grapes back in favour: MACABEO (VIURA of Rio), PARELLADA, XAREL·LO (best for ageing). Best can age 10 yrs. Highest quality category is CAVA DE PARAJE CALIFICADO.

Cava de Paraje Calificado Cava Launched 2017 by CAVA CONSEJO as top category of single-v'yd Cava with stringent rules. Single estate; low or no dosage; min 36 mths age, most exceed that. Currently ten wines, produced by six companies: Alta Alella Mirgin, CODORNÍU, FREIXENET, JUVÉ Y CAMPS, Pere Ventura, Vins el Cep.

Cebreros r w ★→★★ Brand-new DO n of Madrid with heritage of old-vine GARNACHA. Most of Sierra de Gredos is within boundaries.

Celler del Roure V'cia r p w ★→★★ Remarkable BODEGA in s V'CIA. A *masía* (country house) with traditional underground cellar filled with vast amphoras buried up to neck. V.gd, fresh, elegant; Cullerot, Parotet, Safrà, from local grape varieties.

César Florido Sherry ★→★★★ Master of MOSCATEL, since 1887. Explore gloriously scented, succulent trio: Dorado, Especial, Pasas.

Chacolí *See* TXAKOLI.

Chipiona Sherry's MOSCATEL grapes come from this sandy coastal zone. CÉSAR FLORIDO is a classic. Best are floral delicacies, far less dense than PX.

Chivite Nav r p w sw ★★→★★★ Great name in NAV. Colección 125, *top Chard* (the late Denis Dubordieu consulted), one of Spain's greatest whites. Gd late-harvest MOSCATEL. Sold to CASTILLO PERELADA, providing welcome investment.

Clàssic Penedès Pen Category of DO PEN for traditional-method fizz, more strict rules than Cava. Min 15 mths ageing, organically grown grapes. Members incl Albet i Noya, Colet, LOXAREL, Mas Bertran.

Clos Mogador Pri r w ★★★ René Barbier was one of PRI's founding quintet and mentor to many. Still commands respect. One of 1st wineries to gain a VI DE FINCA designation. Son René Jr in charge, works with partner Sara Pérez of MAS MARTINET.

Codorníu Raventós Cos del S, Pen, Pri, Rio r p w sp ★→★★★★ Historic art nouveau CAVA winery worth a visit. V. fine single-v'yd, single-variety CAVAS DE PARAJE CALIFICADO trio; plus 456, blend of three v'yds, most expensive Cava ever. Elsewhere in the group, Legaris in RIB DEL D has v.gd village wines; Raimat in COS DEL S still finding its way. BODEGA Bilbaínas in RIO has bestseller VIÑA Pomal, back on form. Part owns outstanding PRI SCALA DEI. Latest project ABADÍA DE POBLET.

Conca de Barberà Cat r p w Small CAT DO once a feeder of fruit to large enterprises, now some excellent wineries, incl ABADÍA DE POBLET, TORRES.

Consejo Regulador Organization that controls a DO – each DO has its own. Quality as inconsistent as wines they represent: some bureaucratic, others enterprising.

Contador Cat, R Alt r w ★★→★★★ Benjamín Romeo known for TEMPRANILLO: Contador, El Bombón; wines to age. Try whites: Pirata (Vino de Mesa), Predicador (RIO).

Contino R Ala r p w ★★→★★★★ Estate incl one of RIO's great single v'yds. Promising developments under winemaker Jorge Navascués. CVNE-owned.

Corpinnat Cat sp Group of producers of traditional-method sparkling. More stringent quality controls. Formed 2018, left CAVA DO 2019. Members: GRAMONA, Huguet-Can Feixes, Júlia Vernet, Llopart, Mas Candí, Nadal, RECAREDO, Sabaté i Coca and Torelló.

Costers del Segre r p w sp ★→★★★ Geographically divided DO combines mountainous CASTELL D'ENCÚS and lower-lying Castell del Remei, Raimat.

Cota 45, Bodegas Sherry ★→★★ From SANLÚCAR Sherry star Ramiro Ibáñez. Ube brand is PALOMINO from different famous PAGOS, eg. Carrascal, Miraflores. Unfortified but aged in Sherry BUTTS to give FINO character. Reveals strong terroir differences.

Crianza Label term, indicates ageing of wine, not quality. New or unaged wine is *sin crianza* (without oak) or *joven*. In general Crianzas must be at least 2 yrs old (with 6 mths to 1 yr in oak) and must not be released before 3rd yr. *See* RES.

Cusiné, Tomás Cos del S r w ★★→★★★ Innovative winemaker, now returned to CASTELL DEL REMEI; group incl Cara Nord, Cérvoles, FINCA Racons, Vilosell. Individual, modern, always interesting; incl multi-variety Auzells (w).

CVNE R Ala, R Alt r p w ★★→★★★★ One of RIO's great names, based in Haro. Pronounced "coo-nee", Compañía Vinícola del Norte de España, founded 1879. Four Rio wineries: CONTINO, CVNE, Imperial, VIÑA Real. Most impressive at top end. Great wines, long-lived **64 70**. Recent purchases in RIB DEL DUERO, R BAI.

Delgado Zuleta Man ★→★★ Oldest (1744) SANLÚCAR firm. Flagship is 6/7-yr-old

SPAIN

> **Terroir I**
> The geological base of Spanish terroirs: PRI has its *costers* (slopes), of friable *llicorella* slate which forces the roots to dig deep for water, and which heats up the grapes in summer. R BAI, esp in subzone of Salnés, has granite soils, but also the Atlantic ocean close by. Never too cold in winter, nor too hot in summer, maintaining aromatics, but with regular rain risking rot. The Sierra de Gredos has high, convoluted slopes of granite sand where every site has a different aspect, a slightly different climate. And as RIB DEL D and TORO modify their habit of overextraction, the differences of altitude and soil in their v'yds are starting to sing.

La Goya MANZANILLA PASADA, served at wedding of King Felipe VI of Spain; also 10-yr-old Goya XL EN RAMA. Impressively aged 40-yr-old Quo Vadis? AMONTILLADO.

Díez-Mérito Sherry ★→★★★ Reliable Bertola range; fine VORS Sherries: AMONTILLADO *Fino Imperial*, Victoria Regina OLOROSO, Vieja SOLERA PX.

Dinastía Vivanco R Alt r p w sw ★→★★ In Briones, *outstanding Vivanco wine museum*.

DO / DOP (Denominación de Origen / Protegida) Has replaced the former Denominación de Origen (DO) category.

Domaines Lupier Nav r ★★★ Lupier has rescued scattered v'yds of old GARNACHA to create two exceptional wines: floral La Dama, bold El Terroir. Bio.

Dulce Sweet. Spain has a tempting selection of sweet wines (r w). Some late-harvest, others botrytis, or fortifed. Seek out treasures: Alta Alella, BENTOMIZ, GUTIÉRREZ DE LA VEGA, OCHOA, TELMO RODRÍGUEZ, TORRES. Also EMPORDÁ, MÁLAGA, TXAKOLÍ, YECLA.

El Puerto de Santa María Sherry One of three towns forming the "Sherry Triangle". Production in decline; few BODEGAS remain incl GUTIÉRREZ COLOSÍA, OSBORNE, Terry. Puerto FINOS are prized as less weighty than JEREZ, not as "salty" as SANLÚCAR. Taste Lustau's EN RAMA trio to understand differences of Sherries aged in the three towns.

Emilio Hidalgo Sherry w ★★★→★★★★ Outstanding family BODEGA. All wines (except PX) start by spending time under FLOR. Excellent unfiltered 15-yr-old La Panesa FINO, thrilling 50-yr-old AMONTILLADO Tresillo 1874, rare Santa Ana PX 1861.

Emilio Rojo Gal w ★★★ Rojo's eponymous wine is Treixadura/LOUREIRO/ALBARIÑO/ Lado/TORRONTÉS/GODELLO. Superb; thrilling freshness. Star of RIBEIRO. Purchased by RIB DEL D PAGO de Carrovejas 2019; Rojo remains involved in winemaking.

Empordà Cat r p w sw ★→★★ One of number of centres of creativity in CAT. Best: CASTILLO PERELADA, Celler Martí Fabra, Pere Guardiola, Vinyes dels Aspres. Quirky, young Espelt grows 17 varieties: try GARNACHA/CARIGNAN Sauló. Sumptuous natural sweet wine from Celler Espolla: SOLERA GRAN RES.

En rama Sherry bottled from butt; v. low filtration, to capture max freshness.

Enrique Mendoza Alic r w sw ★→★★ Pepe, son of Enrique, serious winemaker and lively host, cheerleader for ALIC and its MONASTRELL. Top wines from dry inland *altiplano*: vibrant Tremenda, single-v'yd Las Quebradas. His personal project Casa Agrícola, for Med wines, is one to follow.

Envínate Quartet of winemakers casting original light on lesser-known regions, incl Almansa, RIBEIRA SACRA, TENERIFE. Also with ALVEAR at Palacio Quemado, Ext and at MONT-M making amphora wines.

Epicure Wines Cat, Pri r w sp ★★ Sommelier Franck Massard building characterful portfolio from DOS across Spain, incl CAVA, MONT, RIBEIRA SACRA, TERRA ALTA, VALDEORRAS. Lively ROSADO Mas Amor.

Equipo Navazos Sherry ★★★→★★★★ Academic Jesús Barquín and Sherry winemaker Eduardo Ojeda pioneered négociant approach to Sherry, bottling individual BUTTS. Colet-Navazos (sparkling: uses Sherry in *liqueur d'expédition*), Navazos-Palazzi (brandy), Pérez Barquero (MONT-M). Early adopters of unfortifed PALOMINO.

Escocés Volante, El Gal, Ara r w ★→★★★ The Scot Norrel Robertson MW was a flying winemaker in Spain, hence the brand. In CALATAYUD focuses on old-vine GARNACHA grown at altitude, often blending in local varieties. Individual, characterful wines, part of movement transforming Ara. Consultancies incl ALBARIÑO in R BAI, GODELLO in MONTERREI.

Espumoso Sparkling, but confusing: incl cheap, injected bubble wine as well as traditional method, like CAVA.

Fernando de Castilla Sherry ★★→★★★★ Gloriously consistent quality. Seek out Antique Sherries; all qualify as VOS or VORS, but label doesn't say so. Youngest of these, Antique FINO, is fascinating, complex, fortified to historically correct 17% alc. New, v. gd OLOROSO and PX Singular. Also v.-fine brandy, vinegar. Favoured supplier to EQUIPO NAVAZOS.

Finca Farm or estate (eg. FINCA SANDOVAL).

Finca Allende R Alt r w ★★→★★★★ Top (in all senses) RIO BODEGA at BRIONES in merchant's house with tower looking over town to v'yds, run by irrepressible Miguel Ángel de Gregorio. Single v'yd incl mineral Calvario and Mingoritz, grown on limestone. Splendid aromatic Martires (w).

Finca Sandoval Man r ★★→★★★ Winery founded by Victor de la Serna, leading wine writer/critic, in DO MANCHUELA. New investors suggest renewed energy.

Flor Sherry Spanish for "flower": refers to the layer of *Saccharomyces* yeasts that grow and live on top of FINO/MANZANILLA Sherry in a BUTT 5/6 full. Flor consumes oxygen and other compounds ("biological ageing") and protects wine from oxidation. Traditional AMONTILLADOS begin as Finos or Manzanillas before the flor dies naturally or with addition of fortifying spirit. It grows a thicker layer nearer the sea at EL PUERTO and SANLÚCAR, hence finer character of Sherry there. Trend to market unfortified Palomino aged for a short time with flor, a Sherry style with lower alc. Growing interest in creating flor wines: Spain, Jura, Argentina, NZ.

Fondillón Alic sw ★→★★★ Fabled unfortified *rancio* semi-sweet wine from overripe MONASTRELL grapes, made to survive sea voyages. Now oak matured for min 10 yrs; some SOLERAS of great age. Sadly shrinking production: Brotons, GUTIÉRREZ DE LA VEGA, Primitivo Quiles. MG Wines with Bodegas Monovár reissuing v. old wines.

Freixenet Pen, Cava r p w sp ★→★★★ Biggest CAVA producer. Best-known for black-bottled Cordón Negro. Casa Sala is CAVA DE PARAJE CALIFICADO. Other Cava brands: Castellblanch, Conde de Caralt, Segura Viudas. Plus: Morlanda (PRI), Solar Viejo (RIO), Valdubón (RIB DEL D), Vionta (R BAI). Also Gloria Ferrer (US), Katnook (Australia), FINCA Ferrer (Argentina). Owned by sparkling wine giant Henkell.

Terroir II

Terroir in Sherry? Of course; in v'yds and in BODEGAS; terroir for ageing and terroir for growing. For buildings where wine is aged, aspect is key: BUTTS in different parts of same bodega will age differently: it's all to do with heat, humidity and whether a cooling breeze blows through the building. BARBADILLO has two MANZANILLAS, from the same SOLERA (Solear) and the same bodega building, from butts 200m (656ft) apart. Poniente (w) is cooler and Levante (e) hotter part of bodega and resulting Sherries are subtly different. In SANLÚCAR, some soleras are humid and lie below sea level, such as the Napoléon at HIDALGO-LA GITANA. Up on the hill of JEREZ there are drying winds, which explains why the style of FINO aged there is richer and sometimes heavier than Manzanilla. Sherry lovers learn that the bright-white soils of Jerez are *albariza*, chalk with limestone, clay, sand (with some darker *barros* and *arenas* soils). WILLY PÉREZ and Ramiro Ibáñez now communicating the complexities that growers always knew, revealing subtle differences between v'yds of Albariza with boxes of soil samples. Their message: Sherry has terroir too.

Frontonio Ara Fernando Mora MW making waves seeking out old-vine GARNACHA, GARNACHA BLANCA. Also v. gd MACABEO, from El Jardin de la Iguales v'yd. Cuevas de Arom is 2nd project made in old underground cellars in DO CAMPO DE BORJA.

Fundador Pedro Domecq Sherry Former Domecq BODEGAS were sliced up through Sherry's multiple mergers. VORS wines now owned by OSBORNE; *La Ina, Botaina, Rio Viejo*, VIÑA 25 by LUSTAU. Andrew Tan of Emperador, world's largest brandy company, bought remainder, focus on Fundador brandy. Group also incl Terry Centenario brandy, Harvey's, famed for Bristol Cream and v. fine VORS, and Garvey, known for *San Patricio* FINO.

Galicia r w (sp) Isolated nw corner of Spain, destination of pilgrims walking the Camino de Santiago; home to many of Spain's best whites (*see* MONTERREI, R BAI, RIBEIRA SACRA, RIBEIRO, VALDEORRAS), and bright crunchy reds. Isolation ensures number of rare varieties.

Garcia, Mariano Fixture in n and nw Spain. For many yrs his life was VEGA SICILIA, where he was winemaker until 1998. He co-founded AALTO, and launched MAURO (TORO). He and sons Eduardo and Alberto also run San Román (Toro), and Garmón (RIB DEL D), specializing in v'yd selection.

Genéricos Rio If there's no category shown on the bottle – such as RES – then it's a *genérico*. *Genéricos* need not follow all DO rules on ageing. An unattractive name that gives no guidance, but they can be v.gd or outstanding. Here it is a case of needing to know the producer.

Gómez Cruzado R Alt Boutique BODEGA between MUGA and LA RIOJA ALTA. Wines being revived by dynamic duo. V.gd Montes Obarenes (w) blend, Pancrudo GARNACHA.

González Byass Sherry, Cád r w sw ★★→★★★★ Family business (1845). Cellarmaster Antonio Flores is a debonair, poetic presence. From the *Tío Pepe* SOLERA Flores continues to extract *glories: En Rama and the Palmas Finos*. Consistently polished VIÑA AB AMONTILLADO, Matúsalem OLOROSO, Noë PX. Boutique hotel soon to open inside BODEGA site. Other wineries: BERONIA (RIO), Pazos de Lusco (R BAI), Vilarnau (CAVA), VIÑAS del Vero (SOM); lastest acquisition O Fournier winery in RIB DEL D; plus (not so gd, but popular) Croft Original Pale Cream Sherry. FINCA Moncloa, close to JEREZ, produces still reds; also Tintilla de Rota (sweet red fortified).

Gramona Cat, Pen r w sw sp ★★→★★★★ Cousins make impressively long-aged traditional-method sparkling, esp Enoteca, *III Lustros, Celler Batlle*. Founder members of CORPINNAT producer group. Inspiring hive of research, incl bio; sweet incl Icewines, experimental wines, table wines.

Grandes Pagos Network of mainly family-owned estates across Spain who work together for collective marketing. Easily confused with VINO DE PAGO. Some are Vinos de Pago but not all.

Gran Reserva In RIO Gran Res spends min 2 yrs in 225-litre barrique, 3 yrs in bottle. Seek out superb old Rio vintages, often great value. Age does not always equal beauty, thus many recent examples less exciting.

Guita, La Man ★→★★★ Classic *Manzanilla* distinctive for sourcing its fruit from v'yds close to maritime SANLÚCAR. Grupo Estévez-owned (also VALDESPINO).

Divine (Ardennes horse) ploughs at Gramona. But has to be addressed in French.

Gutiérrez Colosía Sherry ★→★★★ Rare remaining riverside BODEGA in EL PUERTO. Family business. Former ALMACENISTA. Excellent old PALO CORTADO.

Gutiérrez de la Vega Alic r w sw ★★→★★★ Remarkable BODEGA specializing in sweet wine. In ALIC, but no longer in DO, after disagreement over regulations. Expert in MOSCATEL and FONDILLÓN.

Hacienda Monasterio Rib del D r ★★★ PETER SISSECK co-owns and consults here, where he 1st started in RIB DEL D. More accessible in price, palate than DOMINIO DE PINGUS.

Haro R Alt Town at heart of R Alt, reputation made when railway enabled exports to

B'x during phylloxera. Top producers (seven of them) of RIO are clustered around station: BODEGAS BILBAÍNAS (CODORNÍU), CVNE, GÓMEZ CRUZADO, LA RIOJA ALTA, LÓPEZ DE HEREDÍA, MUGA, RODA. Annual open house day for public.

Harvey's Sherry ★→★★★ Once-great Sherry name, famed for Bristol Cream. Now owned by Emperador, owners of FUNDADOR PEDRO DOMECQ. V.gd VORS Sherries.

Hidalgo-La Gitana Man ★★ →★★★★ Historic (1792) SANLÚCAR firm. MANZANILLA La Gitana a classic. Finest Manzanilla is single-v'yd *Pastrana Pasada*, verging on AMONTILLADO maturity. Outstanding VORS, incl Napoleon Amontillado, Triana PX, Wellington *Palo Cortado*.

Oldest remains of winemaking in W Europe? Iberi people, C7 BC, U-R DO.

Jerez de la Frontera Sherry Capital of Sherry region, between Cádiz and Seville. "Sherry" is corruption of C8 "Sherish", Moorish name of city. Pronounced "hereth". In French, Xérès. Hence DO is Jerez-Xérès-Sherry. MANZANILLA has own DO: Manzanilla-SANLÚCAR DE BARRAMEDA.

Joven Young, unoaked wine.

Juan Carlos Sancha Rio r w ★★ Professor of oenology turned winemaker, understands soils, traditions of RIO. Works with lesser-known varieties (Tempranillo Blanco, Maturana Tinta, Maturana Blanca, Monastel) as well as GARNACHA.

Juan Gil Family Estates Jum r w ★ →★★★ Family BODEGA; has helped transform reputation of JUM. Gd young MONASTRELLS (eg. 4 Meses); long-lived top Clio, El Nido. Other wineries incl Ateca (CALATAYUD), Can Blau (MONT), Shaya (RUE).

Jumilla Mur r (p) (w) ★→★★★ Arid v'yds in mtns n of Mur with heritage of old MONASTRELL vines. Top: CASA CASTILLO, JUAN GIL. Also: Agapito Rico, Carchelo.

Juvé y Camps Pen, Cava w sp ★★→★★★ Consistently gd CAVA. RES de la Familia is stalwart, La Capella is CAVA DE PARAJE CALIFICADO.

La Mancha C-La M r p w ★→★★ Don Quixote country, but Spain's least impressive (except for its size) wine region, s of Madrid. Key sources of grapes for distillation to brandy. Too much bulk wine, yet excellence still possible: JUAN GIL Volver, MARTÍNEZ BUJANDA's FINCA Antigua, PESQUERA's El Vínculo.

Landi, Daniel Mén Leader in new generation of GARNACHA producers.

León, Jean Pen r w ★★ →★★★ Part of TORRES family of wineries, run by Mireia Torres. Getting better all the time; some classics but also taking a modern approach.

López de Heredia R Alt r p w ★★ →★★★★ Haro's oldest (1877) family business with wines that have become a cult. Take a look at its "Txori-toki" tower and Zaha Hadid designed shop. See how RIO was made (as it still is, here). Cubillo is younger range with GARNACHA; darker Bosconia; delicate, ripe *Tondonia*. Whites have seriously long barrel-and-bottle-age; GRAN RES ROSADO is like no other.

Loxarel Pen r p w sp ★★ Josep Mitjans is passionately committed to his terroir and to XAREL·LO variety. (Loxarel is anagram). Range incl skin contact and amphora wines. Cora is fun, fresh (w). Cent Nou 109 Brut Nature RES is quirky treat: traditional-method fizz, but lees never disgorged. Complex, cloudy, unsulphured, v. youthful after 109 mths. Bio.

Lustau Sherry ★★★ →★★★★ Launched original ALMACENISTA collection. Sherries from JEREZ, SANLÚCAR, EL PUERTO. Only BODEGA to produce EN RAMA from three Sherry towns. Emilín is superb MOSCATEL, VORS PX is outstanding, carrying age and sweetness lightly. New cellarmaster but quality continues unchanged.

Maestro Sierra, El Sherry ★★★ Discover how a JEREZ cellar used to be. Run by Mari-Carmen Borrego, following on from her mother, the redoubtable Pilar Plá. Fine AMONTILLADO 1830 VORS, FINO, OLOROSO 1/14 VORS. *Brilliant quality wines.*

Málaga r w sw ★→★★★ MOSCATEL-lovers should explore hills of Málaga. TELMO RODRÍGUEZ revived ancient glories with subtle, sweet *Molino Real*. Barrel-aged No 3 Old Vines Moscatel from Jorge Ordóñez is gloriously succulent. BENTOMIZ has impressive portfolio. Sierras de Málaga DO for dry table wines; Ordóñez' Botani is delicately aromatic dry Moscatel.

Mallorca r p w ★→★★★ Uneven quality, some v.gd, some simply prestige projects. Can be high-priced and hard to find off island. Incl 4 Kilos, Án Negra, Biniagual, Binigrau, Hereus de Ribas, Son Bordils. Reds blend traditional varieties (Callet, Fogoneu, Mantonegro) plus CAB, SYRAH, MERLOT. Whites (esp CHARD) improving fast. DOS: Binissalem, Pla i Llevant.

Manchuela r w sw ★→★★ Traditional region for bulk wine, showing promise with Bobal, MALBEC, PETIT VERDOT. Leaders: Alto Landón, FINCA SANDOVAL, Ponce.

Marqués de Cáceres R Alt r p w ★→★★ Important contributor to RIO in 70s, introducing French winemaking techniques. Fresh white, rosé. Gaudium is modern top wine; GRAN RES traditional classic. Owns Deusa Nai in R BAI.

Marqués de Murrieta R Alt r p w ★★★→★★★★ Between them, the marqueses of RISCAL and Murrieta launched RIO. At Murrieta, step change in quality continues with new BODEGA on estate. Two styles, classic and modern: Castillo Ygay GRAN RES is one of Rio's traditional greats. Latest release of Gran Res Blanco is **86**, and Gran Res Tinto **75**. Dalmau is impressive contrast, glossy modern Rio, v. well made. *Capellania* is fresh, taut, complex white, one of Rio's v. best; ROSADO, v. pale, unusual Primer Rosé from Mazuelo; v.gd Pazo de Barrantes ALBARIÑO (R BAI).

Marqués de Riscal R Ala, C y L, Rue r (p) w ★★→★★★★ Riscal is living history of RIO, able to put on a tasting of every vintage going back to its 1st in 1862. Take your pick of styles: reliable RES, modern FINCA Torrea, balanced GRAN RES. Powerful *Barón de Chirel Res*. The marqués discovered and launched RUE (1972) and makes vibrant DYA SAUV BL, VERDEJO and v.gd Barón de Chirel Verdejo, though now chooses to put wines in C Y L not Rue. Eye-popping Frank Gehry hotel attached to Rio BODEGA.

Pinot N in sunny Spain? Yes: at Cortijo Los Aguilares in Málaga mtns, it shines.

Mas Doix Pri ★★→★★★ Brand new BODEGA has arrived in Poboleda, Pri, enabling Mas Doix to expand, funded by new joint owners Lede Family Winery of California. Doix's treasures are superb CARIÑENA (grape), all blueberry and velvet, astonishingly pure, named after yr v'yd was planted: *1902*; old-vine GARNACHA (*1903*) Latest release is Murmuri, DYA GARNACHA BLANCA.

Mas Martinet Pri r ★★→★★★ Sara Pérez is daughter of one of original PRI quintet. She's the most passionate of Pri's 2nd generation, fermenting freshly picked grapes in vats in v'yds, and in TINAJAS. Venus La Universal is MONT project with partner René Barbier Jr of CLOS MOGADOR. Also consults on projects across Spain.

Mauro Rib del D, C y L, Toro r w ★★→★★★ Founded by MARIANO GARCÍA, formerly of VEGA SICILIA. Mauro wines typically García, full-bodied, built to age.

Méntrida C-La M r p ★→★★ Former co-op country s of Madrid, now being put on map by ARRAYÁN, Canopy, DANIEL LANDI with Albillo, GARNACHA grapes.

Monterrei Gal r w ★→★★★ Small DO on Portuguese border, with trace of Roman winemaking. Discovering its potential. Best: Quinta da Muradella: fascinating parcels of unusual vines.

Montilla-Moriles ★→★★★ Andalucian DO nr Córdoba. Hidden treasure, unfairly regarded as JEREZ's poor relation. Makes dry to sweetest wines all with PX. Shop nr top end for superbly rich treats, some with long ageing in SOLERA. Top: ALVEAR, PÉREZ BARQUERO, TORO ALBALÁ. Important source of PX for use in Jerez DO.

Montsant Cat r (p) w ★→★★★ Tucked in around PRI, plenty to discover. Fine GARNACHA BLANCA, esp Acústic. Dense reds: Alfredo Arribas, Can Blau, CAPÇANES, Domènech, Espectacle, Joan d'Anguera, Mas Perinet, Masroig, Venus la Universal.

Muga R Alt r p w (sp) ★★→★★★★ Muga brothers and cousin are the friendly giants of HARO, producing some of RIO's finest reds. New-wave pale ROSADO, FLOR de Muga;

lively traditional-method sparkling; classic reds delicately crafted. Best: classical GRAN RES *Prado Enea;* modern, powerful *Torre Muga;* expressive, complex Aro.

Mustiguillo V'cia r w ★★→★★★ Dynamic BODEGA has led renaissance of unloved Bobal, also reviving Merseguera (w) variety. V.gd GARNACHA. FINCA El Terrerazo is VINO DE PAGO with top wine Quincha Corral. New: Hacienda Solana (RIB DEL D).

Navarra r p (w) sw ★→★★★ Next door to RIO and always in its shadow. Early focus on international varieties confused identity. Best: old-vine GARNACHA (DOMAINES LUPIER). Also CHIVITE, Nekeas, OCHOA, Otazu, Tandem, VIÑA ZORZAL. MOSCATEL (sw).

Numanthia Toro r ★★→★★★★ One of TORO's heavyweights. Founded by the Egurens of SIERRA CANTABRIA who sold to LVMH. Exceptional, powerful wines. Top Termanthia comes round with 10 yrs of age.

Ochoa Nav r p w sw sp ★→★★★ Ochoa *padre* led modern growth of NAV, daughters now carry the torch. Winemaker Adriana O calls her range 8a, incl Mil Gracias GRACIANO, fun, sweet, Asti-like sparkling MdO; classic MOSCATEL.

Osborne Sherry ★★→★★★★ Historic Sherry BODEGA, treasure trove of richer styles incl AOS AMONTILLADO, PDP PALO CORTADO. Owns former Domecq VORS incl 51–1a Amontillado. Based in EL PUERTO; its FINO Quinta and mature Coquinero Fino typical of town. Wineries in RIO, RUE, RIB DEL D and super-succulent 5 Jotas jamón.

Pago, Vinos de Officially, top category of DOP; actually, not always. Currently fewer than 20, typically in less famous zones. Obvious regions (RIO, PRI, RIB DEL D) absent.

Pago de Carraovejas Rib del D Strikingly situated v'yds on slopes below Peñafiel's romantic castle. Major improvements here with a focus on single v'yds. Further investment in v. fine properties, EMILIO ROJO (RIBEIRO) and Ossian, top-quality VERDEJO producer in CYL. V.gd restaurant, wine tourism.

Pago de los Capellanes Rib del D ★★→★★★★ V. fine estate, once belonging to church as name suggests, founded 1996. All TEMPRANILLO. El Nogal has plenty of yrs ahead; top El Picón reveals best of RIB DEL D.

Palacio de Fefiñanes Gal w ★★★→★★★★ Standard DYA *R Bai* one of finest ALBARIÑOS. Two superior styles: barrel-fermented 1583 (yr winery was founded, oldest winery of DO); super-fragrant, lees-aged "III". Visit palace/winery at Cambados.

Palacios, Álvaro Bier, Pri, Rio r ★★★→★★★★ Eloquent ambassador who has built global reputation of Spanish wine by obsession with quality. One of quintet who revived PRI. FINCA Dofí mainly GARNACHA, superbly aromatic; Les Aubaguetes, from Bellmunt, boosted by 26% CARIÑENA. L'Ermita is powerful, from low-yielding Garnacha. Also at PALACIOS REMONDO in RIO, restoring reputation of R Baj and its Garnachas, and with nephew Ricardo at DESCENDIENTES DE J PALACIOS in BIER.

Palacios, Descendientes de J Bier r ★★★→★★★★ Superb wines, MENCÍA at its best. Ricardo Pérez Palacios, Álvaro's nephew, grows old vines on steep slate. Sadly not all BIER lives up to this promise. Gd-value, floral *Pétalos* and Villa de Corullón; Las Lamas and Moncerbal are v. different soil expressions, one more clay, the other rocky. Exceptional single-v'yd La Faraona (but only one barrel), grows on complex tectonic fault. Bio.

Palacios, Rafael Gal w ★★★→★★★★ Rafael P can't put a foot wrong in VALDEORRAS. Singular focus on GODELLO across many tiny v'yds over more than a decade. Lovely Louro do Bolo; As Sortes, a step up; *Sorte O Soro*, surely Spain's best white. Latest wine is Sorte Antiga from tiny v'yd, v. delicate orange wine.

Palacios Remondo R Baj r w ★★→★★★ ÁLVARO PALACIOS has put deserved spotlight on R Baj and its GARNACHAS. Complex Plácet (w) originally created by brother RAFAEL PALACIOS. Reds: organic, Garnacha-led, red-fruited La Montesa; big, mulberry-flavoured, old-vine Propriedad. Top wine is Quiñón de Valmira from slopes of Monte Yerga.

Pariente, José Rue w ★★→★★★ Victoria P makes VERDEJOS of shining clarity. Cuvée Especial is fermented in concrete eggs; silky late-harvest Apasionado. Daughter

Martina also runs Prieto Pariente with brother Ignacio, working in C Y L and with GARNACHA in Sierra de Gredos.

Pazo Señorans Gal w ★★★ Consistently excellent ALBARIÑOS from glorious R BAI estate. Outstanding Selección de AÑADA, proof v. best Albariños age beautifully.

Penedès Cat r w sp ★→★★★★ Region w of Barcelona, with v. varied styles. Best: Agustí Torelló Mata, Alemany i Corrio, Can Rafols dels Caus, GRAMONA, JEAN LEÓN, Parés Baltà, TORRES.

Pérez, Raúl Bier One of Spain's stars, but avoids celebrity, renowned for finesse and elegance. Specialist in nw. Terroir-driven; always interesting wines. Provides generous house-room for new winemakers in his cellar in BIER. Magnet for visiting (eg. Spanish and Argentine) winemakers.

Pérez, Willy Jer, Sherry Effective communicator, cheerleader for debate about transformation of Sherry with colleague and winemaker Ramiro Ibañez. Researching and reviving practices, traditions, terroir of JEREZ is their aim. At tastings, matches samples of v'yd soils to v'yd wines, an entirely new experience in marketing in Jerez. La Barajuela wines are unfortified vintage PALOMINOS.

Pérez Barquero Mont-M ★→★★★ Part of revival of MONT-M PX. GD Gran Barquero FINO, AMONTILLADO, OLOROSO; La Cañada PX. Supplier to EQUIPO NAVAZOS.

Pesquera, Tinto Rib del D r ★★ Alejandro Fernández put RIB DEL D on map with simply named but majestic Tinto Pesquera. Family divisions mean business is now under different management as Familia Fernández Rivera. Wineries also at Condado de Haza, Dehesa La Granja (C Y L), El Vínculo (LA MANCHA).

Pie franco Ungrafted vine, on own roots. Typically on sandy soils where phylloxera could not penetrate. Some are well over a century old. Lots in TORO.

At long last, Rib del D allowing producers to make white wines. Grape: Albillo.

Pingus, Dominio de Rib del D r ★★★★ One of RIB DEL D's greats. Tiny bio winery of Pingus (PETER SISSECK's childhood name), made with old-vine Tinto Fino, shows refinement of variety in extreme climate. *Flor de Pingus* from younger vines; Amelia is single barrel named after his wife. PSI uses grapes from growers, long-term social project to encourage them to preserve vines and viticultural practices and stay on land. Sisseck now has a Sherry BODEGA.

Priorat r w ★★★★ Some of Spain's finest wines. Named after former monastery tucked under craggy cliffs. The key is the slate soil – known as *llicorella*. Best wines (eg. *Palacios, Barbier*) show purity, finesse, sense of place and drink smoothly at a young age. Pri has pioneered classification pyramid rising from village wines through Vi de Finca to Gran Vi de Vinya.

Raventós i Blanc Cat r p w ★→★★★ Pepe R led legendary family business out of CAVA. Created Conca del Riu Anoia DO for high-quality sparklings with strict controls. Wines incl De Nit ROSADO, Mas del Serral, ringingly pure Textures de Pedra. Can Sumoi is estate for still; Naturals Pepe Raventós, low-intervention range.

Recaredo Pen w sp ★★→★★★ Outstanding producer of traditional-method sparkling, small family concern. Few wines, all outstanding. Hand-disgorges all bottles. Tops is characterful, mineral *Turó d'en Mota*, from vines planted 1940, ages brilliantly, and RES Particular. Bio. Member of CORPINNAT.

Remelluri, La Granja Nuestra Señora R Ala r w ★★→★★★ TELMO RODRÍGUEZ's family property. Makes his original multi-varietal white here. Renewed focus on exceptional old GARNACHA v'yds. Some ethereal wines.

Reserva (Res) Has actual meaning in RIO: aged min 3 yrs, of which 1 yr is in oak of 225 litres. Many producers now prefer to follow own rules. *See* GENÉRICOS.

Rías Baixas Gal (r) w ★★→★★★ Atlantic DO, ALBARIÑO in five subzones, mostly DYA. Best: Forjas del Salnés, Gerardo Méndez, Martín Códax, PALACIO DE FEFIÑANES, Pazo de Barrantes (MARQUÉS DE MURRIETA), *Pazo de Señorans*, Terras Gauda, ZÁRATE.

Many small producers to discover. Until recently Spain's premier DO for whites, now at risk of overproduction. Influential new generation of consultants, eg. Dominique Roujou de Boubée (Adega Pombal), RAÚL PÉREZ (Sketch).

Ribeira Sacra Gal r w ★★→★★★ Magical DO with v'yds running dizzyingly down to River Sil. Increasingly fashionable, esp for fresh reds. Top: Adegas Moure, ALGUEIRA, DOMINIO DO BIBEI, FINCA Viñoa, Guímaro.

Ribeiro Gal (r) w sw ★→★★★ Historic region, famed in Middle Ages for Tostado (sw). Deserving rediscovery, with textured whites made from GODELLO, LOUREIRO, Treixadura. Top: Casal de Armán, Coto de Gomariz, EMILIO ROJO, FINCA Viñoa.

Ribera del Duero r p (w) ★ →★★★★ Ambitious DO with great appeal in Spain, created 1982. Anything that incl AALTO, HACIENDA MONASTERIO, PESQUERA, PINGUS, VEGA SICILIA has to be serious, but consistency hard to find. Too many v'yds planted in wrong places, plus domestic demand for oaky, concentrated wines. At last, elegance breaking through. Wineries in Soria (to the e) provide most delicate wines. Try ALONSO DEL YERRO, *Pago de los Capellanes*, PAGO DE CARRAOVEJAS. Also of interest: Arzuaga, Bohórquez, Cillar de Silos, Dominio de Atauta, Dominio del Aguila, Garmón, Hacienda Solano, Pérez Pascuas, Tomás Postigo. *See also* C Y L neighbours ABADÍA RETUERTA, MAURO.

Rioja r p w sp ★→★★★★ Spain's most famous wine region. Three sub-regions: R Ala, R Alt and R Or (meaning e-facing; formerly named R Baja). Much-debated new regulations allow producers to name villages and "singular" v'yds, and make sparkling RIO. Rioja is composed of two provinces: La Rioja and Álava, or the Basque country. Growing political differences between the provinces are also reflected in movement by Alavesa producers to separate from the DO.

Rioja Alta, La R Ala, R Alt r ★★→★★★★ For lovers of classic RIO, a favourite choice. *Gran Res 904* and GRAN RES *890* are stars. But rest of range from *Ardanza*, down to Arana, Alberdi each carry classic house style; all of them qualify as Gran Res. Also owns R Ala modern-style Torre de Oña, R BAI Lagar de Cervera, RIB DEL D Áster. Hard to fault.

Rioja 'n' Roll Rio Something new for RIO. New generation of winemakers formed a network for fun and for marketing. All small production, with serious focus on v'yds. Seek them out: Alegre & Valgañón, ARTUKE, Barbarot, Exopto, Macrobert & Canals, Olivier Rivière, Sierra de Toloño.

Roda Rib del D, R Alt r ★★→★★★ One of HARO's seven "station quarter" wineries. TEMPRANILLO specialists: Roda, Roda I, Cirsión, approachable Sela. Also RIB DEL D BODEGAS La Horra, *Corimbo* (w), Corimbo I.

Rosado Rosé. NAV dark rosados were defeated by Provence pinks. Spain has fought back with pale hues, CSP: SCALA DEI's Pla dels Àngels (PRI), MARQUÉS DE MURRIETA's Primer Rosé (RIO), Dominio del Águila Pícaro Clarete (RIB DEL D).

Rueda C y L w ★→★★★ Spain's response to SAUV BL: zesty VERDEJO. Mostly DYA. Too much poor quality. Best: *Belondrade*, JOSÉ PARIENTE, Naia. Pálido is flor-aged Verdejo, with 3 yrs in oak.

Saca A withdrawal of Sherry from the SOLERA (oldest stage of ageing) for bottling. For EN RAMA wines most common *sacas* are in *primavera* (spring) and *otoño* (autumn), when FLOR is richest, most protective.

Sánchez Romate Sherry ★★→★★★ Old (1781) BODEGA with wide range, also sourcing and bottling rare BUTTS for négociants and retailers. 8-yr-old *Fino Perdido*, nutty AMONTILLADO NPU, PALO CORTADO Regente, excellent VORS AMONTILLADO and OLOROSO La Sacristía de Romate, unctuous Sacristía PX.

Sandeman Sherry ★→★★★ More famous for its Port than its Sherry. Interesting VOS wines: Royal Esmeralda AMONTILLADO, *Royal Corregidor* Rich Old OLOROSO.

Sanlúcar de Barrameda Sherry, Man Sherry-triangle town (with JEREZ, EL PUERTO) at mouth of River Guadalquivír. Port where Magellan, Columbus and the admiral of the Armada set sail. Humidity in low-lying cellars encourages FLOR. Sea air said to encourage "saltiness". Wines aged in Sanlúcar BODEGAS qualify for DO MANZANILLA-Sanlúcar de Barrameda.

Scala Dei Pri r p w ★★★ Tiny v'yds of "stairway to heaven" cling to craggy slopes. Managed by part-owner CODORNÍU. Winemaker Ricard Rofes returning to the old ways, eg. fermenting in stone *lagares*. Focus on local varieties, esp GARNACHA and now CARIÑENA. Single-v'yds Sant'Antoni and Mas Deu show terroir. Also rare GARNACHA BLANCA/CHENIN BL blend.

Sierra Cantabria R Ala, Toro r w ★★★ Exceptional family business. The Egurens specialize in elegant, single-v'yd, low-intervention wines. Organza (w). Reds, all TEMPRANILLO. At Viñedos de Paganos, superb El Puntido; powerful, structured La Nieta. Other properties: Señorío de San Vicente in RIO and Teso la Monja in TORO, where Alabaster is the star. 150th anniversary in 2020.

Sisseck, Peter Sherry, Rib del D Dane who attracted world interest to RIB DEL D with DOMINIO DE PINGUS. With 2018 purchase of JEREZ BODEGA and v'yd in Balbaina zone, he should work same magic for Sherry. Also at Ch Rocheyron, B'x.

Solera System for blending Sherry and, less commonly, Madeira (*see* Portugal). Consists of topping up progressively more mature BUTTS with younger wines of same sort from previous stage, or *criadera*. With FINOS, MANZANILLAS it maintains vigour of FLOR. For all wines gives consistency and refreshes mature wines.

Proportion of legally ungrafted vines in Toro: c.60%. Illegally ungrafted: c.10%+.

Somontano r p w ★→★★ DO in Pyrénéan foothills still searching for an identity, growing international varieties. Opt for GEWURZ – rare for Spain. Try Enate, VIÑAS del Vero (owned by GONZÁLEZ BYASS) – its high-altitude Secastilla has gd old-vine GARNACHA, GARNACHA BLANCA.

Suertes del Marqués Can r w ★→★★ Rising star in TENERIFE. Works with LISTÁN Blanco, Listán Negro, Vijariego, Tintilla, making vibrant village and single-v'yd wines. Exceptional v'yds, with unique local *trenzado* – plaited – vines.

Telmo Rodríguez, Compañía de Vinos Rio, Mál, Toro r w sw ★★ →★★★ Groundbreaking winemaker Rodríguez has returned to REMELLURI in RIO but continues his pioneering business: in MÁLAGA (*Molino Real* MOSCATEL), ALIC (Al-Murvedre), RUE (Basa), TORO (Dehesa Gago), Cigales (Pegaso), *Valdeorras* (DYA Gaba do Xil GODELLO). Return to Rio and BODEGA Lanzaga has led to work on recuperating old v'yds, esp exceptionally pure La Estrada, Las Beatas, Tabuérniga.

Tenerife Can Rising star of CAN. Top: Borja Pérez, ENVINATE, SUERTES DEL MARQUÉS.

Terra Alta Cat Up-and-coming inland DO neighbouring PRI. GARNACHA territory, esp Bárbara Forés, Celler Piñol, Edetària, Lafou. 90% of Catalan GARNACHA BLANCA v'yds, 75% of Spain's.

Tinaja Aka amphora. Huge clay pots used in revival of traditional winemaking. Found across Spain, incl ALVEAR, CELLER DEL ROURE, LOXAREL, MAS MARTINET.

Toro r ★→★★★ ★ Small DO w of Valladolid famed for rustic reds from Tinta del Toro (TEMPRANILLO). Today best more restrained, but still firm tannic grip. Dense old-vine San Román. Glamour from VEGA SICILIA-owned Pintia, and LVMH property NUMANTHIA. Also: Las Tierras de Javier Rodríguez, Paydos, Teso la Monja.

Toro Albalá Mont-M ★→★★★★ From young dry FINOS to glorious sweet wines, a triumph for MONT-M. Among them lively AMONTILLADO Viejísimo. Seek out remarkable, sumptuous Don PX Convento Selección 1931.

Torres Cat, Pri, Rio r p w sw ★★ →★★★★ Celebrated 150 yrs in 2020. Miguel Jr runs business, sister Mireia is technical director and runs JEAN LEÓN, Miguel Sr is busy on many fronts. Top wines: outstanding, elegant B'x-blend *Res Real*, top PEN CAB *Mas la Plana*; CONCA DE BARBERÀ duo (Burgundy-like *Milmanda*, one of Spain's finest CHARDS, *Grans Muralles* blend of local varieties) is oustanding. Also gd-value portfolio; lovely MOSCATEL. Newer, improving wineries in RIB DEL D (Celeste), PRI (Perpetual) and RIO (Ibéricos). Newest launch in Pri is single-v'yd Mas de la Rosa. Pioneer in Chile. Marimar T a star in Sonoma.

Tradición Sherry ★★→★★★★ BODEGA assembled by the great José Ignacio Domecq from exceptional selection of SOLERAS. Based on oldest-known Sherry house (1650). Glorious VOS, VORS Sherries, also a 12-yr-old FINO. Outstanding art collection, and archives of Sherry history.

Txakolí / Chacolí P Vas (r) (p) w (sw) ★→★★ Wines from Basque country DOS in Getaria, Bizkaya and Álava. Many v'yds face Atlantic winds and soaking rain, hence acidity of pétillant whites, esp in Getaria where DYA Txakolí is poured into tumblers from a height to add to spritz. Bizkaya wines, with less exposed v'yds, can have depth and need not be DYA. Top: Ameztoi, Astobiza, Doniene Gorrondona, Txomín Etxaníz. Also Gorka Izagirre, with Michelin three-star restaurant Azurmendi, nr Bilbao airport.

Utiel-Requena r p (w) ★→★★ Marriage of two towns, slowly forging identity with Bobal grape. Try: Bruno Murciano, Caprasia, Cerrogallina.

Valdeorras Gal r w ★→★★★ Warmest, most inland of GAL'S DOS, named after gold Romans found in valleys. Exceptional GODELLO, potentially more interesting than ALBARIÑO. Best: Godeval, RAFAEL PALACIOS, TELMO RODRÍGUEZ, Valdesil.

Valdepeñas C-La M r (w) ★→★★ Large DO S of LA MANCHA. Historic favourite for cheap reds. ARA reds now offer best quality/value.

Valdespino Sherry ★★→★★★★ Home to Inocente FINO from top Macharnudo single-v'yd, rare oak-fermented Sherry (EN RAMA bottled by EQUIPO NAVAZOS). Terrific dry AMONTILLADO Tío Diego; outstanding 80-yr-old **Toneles** MOSCATEL, JEREZ'S v. best. Winemaker Eduardo Ojeda also experimenting. Grupo Estévez (LA GUITA) own.

Valencia r p w sw ★→★★ Known for bulk wine, cheap MOSCATEL and oranges. Slowly on move with higher-altitude old vines and min-intervention winemaking: eg. Aranleon, Baldovar 923, CELLER DEL ROURE, El Angosto, Los Frailes.

VDT (Vino de la Tierra) Table wine usually of superior quality made in a demarcated region without DO. Covers immense geographical possibilities; category incl many prestigious producers, non-DO by choice to be freer of inflexible regulation and use varieties they want. (*See* Super Tuscan, Italy).

Vega Sicilia Rib Del D r ★★★★ Spain's "First Growth" acquired a new winemaker, only the 3rd in 40 yrs. Wines take yrs, so any change of style will be slow. Único, 6 yrs in oak; *Valbuena* outstanding. Flagship: RES Especial, NV blend of three vintages, with up to 10 yrs in barrel: v. fine. Neighbouring Alión shows modern take on RIB DEL D at last coming in its own. Pintia (TORO) also transformed, joint-venture project Macán (RIO) with Rothschild, plus Oremus in Tokaji (Hungary).

Vendimia Harvest.

Viña Literally, a v'yd.

Viña Zorzal Nav, Rio Family business. Entrepreneurial new generation – young, gd-value wines, eg. GRACIANO. Restoring old-vine NAV GARNACHA eg. Malayeto.

Williams & Humbert Sherry ★→★★★★ Winemaker Paola Medina transforming historic BODEGA. Initially famed for eg. Dry Sack, Winter's Tale AMONTILLADO, *As You Like It* sweet OLOROSO. Now pioneering specialities such as organic Sherry, vintage Sherries, incl FINO. One of new leaders.

Ximénez-Spinola Sherry V. fine small producer of PX. Grows PX in JEREZ (v. rare); most source from MONT-M. Exceptional Harvest, from overripe PX, unfortified.

Yecla Mur Traditional bulk wine country, but changing. Drivers are Castaño family: MONASTRELLS (eg. Hécula) and blends (GSM). Castaño Dulce a modern classic.

Ysios Rio r ★→★★ Famous winery for its Calatrava architecture and undulating roof. Wines now growing to match building's reputation.

Yuste Sherry, Man ★★→★★★ Bodega with a growing collection of SOLERAS. Home to MANZANILLAS Aurora and La Kika. Acquired Herederos de Argüeso, bringing with it v.gd San León, dense and salty San León RES and youthful Las Medallas. Conde de Aldama label has AMONTILLADO and PALO CORTADO both over 100 yrs old.

Zárate Gal (r) w ★★→★★★ BODEGA in R BAI. Elegant ALBARIÑOS with long lees-ageing. El Palomar is from centenarian v'yd, one of DO's oldest, on own rootstock, aged in *foudre*. Ethereal. Owner/winemaker Eulogio Pomares one of key figures in GAL.

Sherry styles

Manzanilla: lightest of Sherries, v. dry, matured by the sea at SANLÚCAR where FLOR grows thickly and wine grows salty. Usually a mere 15% alc. Serve cool with almost any food, esp crustaceans. Drink up once open, eg. I Think (EQUIPO NAVAZOS), Deliciosa (VALDESPINO), LA GUITA, La Gitana.

Manzanilla Pasada: mature, where flor is fading, turning into AMONTILLADO; v. dry, complex. Eg. LUSTAU's ALMACENISTA Cueva Jurado.

Fino: dry, biologically aged in JEREZ or EL PUERTO; weightier than MANZANILLA; min age is 2 yrs (as Manzanilla) but don't drink so young. Trend for mature FINOS aged 8 yrs+, eg. FERNANDO DE CASTILLA Antique, GONZÁLEZ BYASS Palmas range.

Amontillado: Fino in which layer of protective flor has died. Oxygen gives more complexity. Naturally dry. Eg. LUSTAU Los Arcos. Many brands are sweetened: look for "medium" on label.

Oloroso: not aged under flor. Heavier, often less brilliant when young, matures to nutty intensity. Naturally ultra-dry, even fierce. May be sweetened and sold as CREAM. Eg. EMILIO HIDALGO Gobernador (dr), Old East India (sw). Keeps well.

Palo Cortado: a cult. Traditionally wine that had lost flor – between Amontillado and v. delicate OLOROSO. Difficult to identify with certainty. Refined, complex: worth looking for. Eg. BARBADILLO Reliquía, Fernando de Castilla Antique. Drink with meat or cheese.

Cream: blend sweetened with grape must, PX and/or MOSCATEL for a commercial medium-sweet style. Can be gd with ice cream.

En Rama: new craze, bottled from BUTT with low or no filtration or cold stabilization to reveal full character of wine. Typically Manzanilla or Fino. More flavoursome, said to be less stable. *Saca* or withdrawal is typically when flor is most abundant, in spring.

Pedro Ximénez (PX): raisined sweet, dark, from partly sun-dried PX grapes (grapes mainly from MONT-M; wine matured in Jerez DO). Unctuous, decadent, bargain. Sip with ice-cream. Tokaji Essencia apart, world's sweetest wine. Eg. Emilio Hidalgo Santa Ana 1861, Lustau VORS.

Moscatel: aromatic appeal, around half sugar of PX. Eg. Lustau Emilín, Valdespino Toneles. Now permitted to be called "Jerez".

VOS / VORS: age-dated: some of treasures of Jerez BODEGAS. V. necessary move to raise perceived value of Sherry. Wines assessed by carbon dating to be 20 yrs old+ called VOS (Very Old Sherry/Vinum Optimum Signatum); 30 yrs+ are VORS (Very Old Rare Sherry/Vinum Optimum Rare Signatum). Also 12-yr, 15-yr examples. Applies only to Amontillado, Oloroso, PALO CORTADO, PX. Eg. VOS Hidalgo Jerez Cortado Wellington. Some VORS wines are softened with PX: sadly producers can overdo the PX. VORS with more than 5 g/l residual sugar are labelled Medium.

Añada "Vintage": Sherry with declared vintage. Runs counter to tradition of vintage blended SOLERA. Formerly private bottlings now winning public accolades. Eg. WILLIAMS & HUMBERT series, Lustau Sweet Oloroso Añada 1997.

Portugal

There's little overlap in style between Portugal and Spain. Maritime, mountainous Portugal has 250 grapes of its own and historic local cultures to go with them; and it is discovering itself with modern, polished wines. The fine table wines of the Douro Valley are an obvious example. Each of its classic regions has a new take on tradition and new energy. Plus, the country is tackling climate change with aplomb. Its deeply rooted old vines are pretty resistant to heat and drought, but savvy producers are also heading for the hills or the seaside for cooler sites. The Symingtons (Douro) and Esporão (Alentejo) have new high-altitude projects in the upper Alentejo, and Quinta Nova (Douro) has set up in mountainous Dão. Sogrape has a new Lisbon estate facing the sea, and Cortes de Cima has new vineyards on the coast. Bairrada and Vinho Verde (especially in Monção e Melgaço) are more dynamic than ever.

Recent Port vintages

A vintage is "declared" when a wine is outstanding by shippers' highest standards. In gd but not quite classic yrs (increasingly in top yrs too by single-estate producers) shippers use the names of their quintas (estates) for single-quinta wines of real character but needing less ageing in bottle, and there are more limited-production (often single-v'yd) Vintage Ports. The vintages to drink now are 63 66 70 77 80 83 85 87 92 94 00 03 04 05 07 though v. young Vintage Port is an unconventional delight, esp with chocolate cake.

2019 Exceptionally long harvest after a cool summer. Gd quality for some.
2018 Heavy spring rain. Absurdly low yields. Gd quality, declaration for some.
2017 Superlative yr, widely declared. V. hot and dry growing cycle often compared to the historic 1945.
2016 Classic yr, widely declared. Great structure and finesse after challenging harvest.
2015 V. dry, hot. Controversial yr. Declared by many (top-quality Niepoort, Noval), but not Fladgate, Symingtons or Sogrape.
2014 Excellent from v'yds that ducked September's rain; production low.
2013 Single-quinta yr; mid-harvest rain. Stars: Vesuvio, Fonseca Guimaraens.
2012 Single quinta yr. V. low yielding, drought afflicted. Stars: Noval, Malvedos.
2011 Classic yr, widely declared. Considered by most on par with iconic 1963. Inky, outstanding concentration, structure. Stars: Dow, Noval Nacional, Vargellas Vinha Velha, Fonseca. You can even start on them now.
2010 Single-quinta yr. Hot, dry but higher yields than 2009. Stars: Vesuvio, Senhora da Ribeira.
2009 Controversial yr. Declared by Fladgate, but not Symingtons or Sogrape. Stars: Taylor, Niepoort, Fonseca, Warre.
2008 Single-quinta yr. Low-yielding, powerful wines. Stars: Noval, Vesuvio.
Fine vintages: 07 03 00 97 94 92 91 87 83 80 77 70 66 63 45 35 31 27.

Recent table wine vintages

2019 No rain, cool summer. V.gd quality all around. Keep.
2018 Heavy rains. V. low yields. Aromatic whites, concentrated reds.
2017 3rd consecutive fine vintage. V.gd quality all around. Keep for yrs.
2016 V.gd quality for those who had patience. Keep for yrs.
2015 Fine yr on quality, quantity. Aromatic, balanced reds drinking specially well. Keep.

See Portugal map p.174.

2014 Rainy winter, cool summer. Fresh whites, bright reds (picked before rain). Drink now.

2013 Great whites, balanced reds (picked before rain). Keep/drink.

2012 Forward, scented reds, elegant whites. For early-drinking.

Açores / Azores (r) w sw ★→★★★ Mid-Atlantic archipelago of nine volcanic islands with DOCS Pico, Biscoitos and Graciosa for whites and traditional *licoroso* (late-harvest/fortified). Pico landscape, incl vine-protecting *currais* (pebble walls), is UNESCO World Heritage Site. Dynamic winemakers, thrilling volcanic-soil, sea-threatened whites from indigenous varieties Arinto dos Açores, Terrantez do Pico, VERDELHO. Watch: Azores Wine Company, Pico Wines, Biscoitos.

Adega A cellar or winery.

Alentejo r (w) Reliably warm popular central region, divided into subregional DOCS Borba, Granja-Amareleja, PORTALEGRE, Évora, Moura, Redondo, Reguengos, Vidigueira (known for quality white). Atlantic-influenced Costa Vicentina area making fresh white and red (watch: CORTES DE CIMA, Vicentino). Ancient clay amphora technique Vinho de Talha seeing a comeback. More liberal VR Alentejano preferred by many top estates. Rich, ripe reds, esp from ALICANTE BOUSCHET, SYRAH, TRINCADEIRA, TOURIGA N. New classics incl CARTUXA, ESPORÃO, JOÃO PORTUGAL RAMOS, JOSÉ DE SOUSA, Malhadinha Nova, MOUCHÃO, which have potency, style. Watch: Aldeia de Cima, Coelheiros, Fonte Souto (SYMINGTON-owned), do Peso, do Rocim, Fita Preta, MONTE DE RAVASQUEIRA, SUSANA ESTEBAN, Terrenus.

Algarve r p w sp S coast producing mostly VR, national and international varieties. Wines progressing but still fall short of famous beaches, Michelin-starred gastronomy. Barranco Longo, QUINTA dos Vales honourable mentions. Watch: AVELEDA's new Villa Alvor project.

Aliança Bair r p w sp ★ →★★★ Large firm with gd reds and *sparkling*. Art and wines at Aliança Underground Museum. Interests in ALEN (da Terrugem, Alabastro), DÃO (da Garrida), DOU (dos Quatro Ventos). Owner of popular Casal Mendes brand.

Ameal, Quinta do Vin w sw sp ★★★ Now owned by ESPORÃO. Superior, age-worthy, organic LOUREIRO incl oaked Escolha and, in top yrs (11 14) low-yield, low-intervention Solo. Gd to visit.

Andresen Port ★★ ·★★★★ Portuguese-owned house with excellent wood-aged Ports, esp 20-yr-old TAWNY. Outstanding *Colheitas* 1900' 1910' (bottled on demand) 68' 80' 91' 03'. Pioneered age-dated WHITE PORTS 10-, 20-, v.gd 40-yr-old.

Aveleda, Quinta da Vin r p w ★→★★ DYA Home of Casal García, biggest VIN seller (since 1939). Regular range of estate-grown wines. Owns DOU's QUINTA DO VALE DONA MARIA and new ALGARVE project Villa Alvor. Visitor centre 30 mins from Porto.

Bacalhôa Vinhos Alen, Lis, Set r p w sw sp ★★ ·★★★ Principal brand and HQ of billionaire art-lover José Berardo's group. Also owns National Monument QUINTA da Bacalhôa (v.gd B'x blend incl CAB SAUV 1st planted 1974, also used in iconic red Palácio da Bacalhôa), sparkling estate Quinta dos Loridos. TOP MOSCATEL DE SETÚBAL barrels, incl rare Roxo. Owner of historic Quinta do Carmo ALEN brand making v.gd reds, and a botanic garden in Funchal. Modern, well-made brands: Serras de Azeitão, Catarina, Cova da Ursa (SET), TINTO da Ânfora (Alen).

Baga Friends Bair Group of BAGA producers. Bágeiras, BUÇACO, FILIPA PATO, LUIS PATO, NIEPOORT'S QUINTA de Baixo, Quinta da Vacariça, Sidonio de Sousa. High quality.

Best talha / amphora producers

Ancestral *talha* (clay amphorae) winemaking is spreading from original ALEN. Best: Anta de Cima, Bojador, CORTES DE CIMA, ESPORÃO, JOSÉ MARIA DA FONSECA, Rocim. Watch: Murças (DOU), field-blend-based NIEPOORT (Dou), Quinta de Santiago (VIN), SUSANA ESTEBAN (Alen).

Bairrada r p w sw sp ★★ →★★★★ Atlantic influenced DOC and Beira Atlântico VR also famous for roast suckling pig. Age-worthy, structured BAGA reds (often from old vines), v.gd sparklings (new Baga BAIR designation for best). Top Baga specialists: Bágeiras, FILIPA PATO, LUÍS PATO, Casa de Saima, CAVES SÃO JOÃO, Sidónio de Sousa. Watch: NIEPOORT'S QUINTA de Baixo, Vadio, V Puro. *See* BAGA FRIENDS.

Sweetness of Vintage Port dropped in recent yrs: better balance now, drier finish.

Barbeito Mad ★★→★★★★ Leading innovative MAD producer with noticeable labels. Unique, single-v'yd, single-cask COLHEITAS. Outstanding 20-, 30-, 40-yr-old MALVASIAS. New 40-yr-old Boal "Vinho do Embaixador". Excellent Ribeiro Real range with 20-yr-old BOAL, Malvasia, SERCIAL, VERDELHO, with dash of 50S TINTA NEGRA. 96 Colheita pioneered mention of Tinta Negra on front label. Historic Series: MAD most coveted wine in US in C18 and C19. Also *Rainwater* and new v.gd, salty Verdelho and RES table wine.

Barca Velha Dou r ★★★★ Portugal's iconic red; 1st bottled in 1952 forging DOU's reputation for world-class table wine. Aged several yrs pre-release and launched only in exceptional yrs: 91' 95' **99 00 04 08' (11').** V.gd second label in v.gd yrs when no BV, from CASA FERREIRINHA's best barrels, *Res Especial* 89' 94' 97' 01' **07 09.** Both last decades. Arguably 89' 94' 97' 01' 09 could have been BV.

Barros Port ★★→★★★ Founded 1913, Sogevinus-owned since 2006, maintains substantial stocks of aged TAWNY and COLHEITA. V.gd Colheitas from the 30s on, and 63 66' 74' 78 80' 97'. V.gd 20-, 30-, 40 yr old Tawny. VINTAGE PORT: 87 95 05 07 11 16 17.

Barros e Sousa Mad ★★ →★★★ Old lodge now new visitor centre after acquisition by neighbour PEREIRA D'OLIVEIRA VINHOS. Look for rare Bastardo Old RES.

Beira Interior r p w ★ →★★ Distinctive DOC with huge potential. Some of the highest mtns in Portugal, between DÃO and Spanish border. V.-old, high (up to 750m/2461ft) v'yds gd for white Fonte Cal, Siria. Gd-value: ANSELMO MENDES, Beyra, do Cardo, dos Currais, dos Termos.

Blandy Mad ★★→★★★★ Historic MAD family firm run by dynamic CEO Chris Blandy. *Funchal lodges* showcase history, incl vast library of FRASQUEIRA (BUAL 1920' 1957' **1966'**, MALMSEY **1988'** **1977' 1981'**, SERCIAL 1968' 1975' 1980' 1988', VERDELHO **1979'**, Terrantez 1980'). V.gd 20-yr-old Terrantez and COLHEITAS (Bual 1996 2008, Malmsey 1999 2004, Verdelho 2000 2008, Sercial 2002). Superb 50-yr-old Malmsey; new expensive, rare MCDXIX blends, 11 yrs between 1863 and 2004, celebrate Madeira's 600 yr history. Also RAINWATER, and Atlantis table wine: Verdelho (w), TINTA NEGRA (p).

Borges, HM Mad ★→★★★ Sisters Helena and Isabel Borges hold tiny amounts of fine Terrantez 1877 demi-john from founding yr. V.gd 30-yr-old MALVASIA incl wine from 1932. V.gd SERCIAL 1990.

Branco White.

Bual (or Boal) Mad Classic MAD grape: medium-rich (sweet), tangy, smoky wines; less rich than MALVASIA. Perfect with harder cheeses and lighter desserts. Tends to be darkest in colour.

Buçaco Bei At r w ★★★ Manueline-Gothic monument *Bussaco Palace hotel* lists its classic, austere, age-worthy wines back to 40s. Blends of two regions. R: BAGA (BAIR), TOURIGA N (DÃO). W: ENCRUZADO (Dão), MARIA GOMES, Bical (Bair). Barriques, new oak since 2000 slightly modernized style (esp w). Member of BAGA FRIENDS.

Bucelas Lis w sp ★★ Tiny LIS DOC set for comeback (ROMEIRA now owned by SOGRAPE). Makes gd-value dry, racy, ARINTO-based whites, fizz. Watch: de Pancas.

Burmester Port ★→★★★ Sogevinus-owned Port house making elegant, wood-aged, gd-value Ports, esp 20-, 40-yr-old TAWNY, 1890 1900' 37' 52' 55' 57' COLHEITAS. Age-dated WHITE PORTS, incl fine 30-, 40-yr-old. Gd VINTAGE PORT.

Cálem Port ★→★★★ Sogevinus-owned Port house; lodge in Gaia with over 300,000 visitors/yr. Popular entry-level Velhotes. Best: COLHEITAS 61', 10-, 40-yr-old TAWNY.

Canteiro Mad Method of naturally cask-ageing finest MAD in warm, humid lodges for greater subtlety/complexity than ESTUFAGEM.

Carcavelos Lis br sw ★★★ Unique, mouthwatering, gripping, off-dry fortified. New Villa Oeiras breathed life into v. old, tiny, ailing 12.5 ha seaside DOC.

Cartuxa, Adega da Alen r w sp ★★→★★★★ Historic ALEN estate with old cellars, new restaurant, modern art centre. Flagship Pêra Manca red draws connoisseurs. Consistent best-buy Cartuxa RES. Scala Coeli is single variety (changes every yr). Gd-value volume Vinea and EA (organic version available) reds.

Planting Bastardo grape is gd way to become poor: old Portuguese saying.

Carvalhais, Quinta dos Dão r p w sp ★→★★★ SOGRAPE-owned boutique DAO estate. V.gd, consistent, age-worthy range, esp oak-aged ENCRUZADO, RES (r w), Alfrocheiro, TOURIGA N, TINTA RORIZ and top wine Único. Unusual BRANCO Especial (w) blends Encruzado yrs. Home of popular Duque de Viseu and Grão Vasco.

Castro, Álvaro de Dão ★★ →★★★★ Emblematic producer; gd-value Saes and superior Pellada. Superb Primus (w), *Pape* (r), Carrocel (TOURIGA N) released in great yrs.

Cello, Casa de Dão, Vin ★★ Family-run, characterful two region project. Unique QUINTA de San Joanne (VIN) age-worthy whites incl outstanding Superior. V.gd Escolha, gd-value Terroir Mineral. Distinctive classic Quinta da Vegia (DÃO) range (r), esp RES, Superior.

Chaves, Tapada do Alen ★★★ Historic property with unique v.old, high-altitude v'yds. Gd white, v.gd age-worthy reds, esp VINHAS VELHAS. Owned by CARTUXA.

Chocapalha, Quinta de Lis r p w ★★★ Family-run estate blending mostly native with some international varieties. Winemaker Sandra Tavares da Silva (WINE & SOUL). *Among Lisboa's best reds*, esp QUINTA, CASTELÃO, CAB SAUV and flagship Vinha Mãe and TOURIGA N CH. Vibrant, fresh whites, esp v.gd RES (CHARD, ARINTO) and old-v'yd Arinto CH.

Chryseia Dou r ★★★ B'x's Bruno Prats and SYMINGTON FAMILY ESTATES DOU partnership. Polished TOURIGA-driven (Nacional and Franca) red. Fresher, finer since sourced from QUINTA de Roriz. Gd-value second label *Post Scriptum*. Also Prazo de Roriz.

Churchill Dou, Port r p w sw ★★★ Family-run Port house est 1981 by John Graham. V.gd DRY WHITE PORT (10 yrs old), 20-, 30-yr-old (new), unfiltered LBV, VINTAGE PORT 82 85 91 94 97 00 03 07' 11' 16'. QUINTA da Gricha is source of old-vine, grippy Single-Quinta Vintage Port and v.gd single-v'yd DOU red. Gd Churchill's Estates label (esp TOURIGA N).

Cockburn's Port ★★→★★★ SYMINGTON-owned Port house, now back on form. New Gaia visitor centre incl Symington's cooperage tour. Drier, fresher style of VINTAGE PORT in 11' 15' 16' 17'. Extraordinary 08' 27' 34 63 67 70'. Consistently gd Special RES aged longer in wood than others. Vibrant LBV aged 1 yr less. V.gd single-QUINTA dos Canais.

Colares Lis r w ★★★ Unique, historic coastal DOC (1908). Windswept ungrafted vines on sand produce Ramisco *tannic reds*, MALVASIA fresh, salty whites. Revitalized Casal Santa Maria bring modern flair to traditional style of ADEGA Regional de Colares and Viúva Gomes. Old vintages not hard to find.

Colheita Port, Mad Crowd-pleasing vintage-dated TAWNY Port or MAD of a single yr. Cask-aged: min of 7 yrs for Port (often 50 yrs+, some 100 yrs+); min 5 yrs for Mad. Bottling date shown on label. Serve chilled.

Cortes de Cima Alen r w ★★★ Sustainable pioneer built from scratch by Danish/Californian couple. V.gd, now more elegant, top red Incógnito 11' 12' 14'. Consistent range, esp (r w) Cortes de Cima, RES, varietals (PINOT N, ARAGONEZ, Syrah, TRINCADEIRA). V.gd whites from new coastal v'yds incl ALVARINHO,

SAUV BL. Gd-value Dois Terroirs (r w) blends coastal, inland v'yds. Leading amphora/*talha* producer.

Cossart Gordon Mad ★★★ MADEIRA WINE COMPANY-owned brand. Drier style than BLANDY eg. bracing BUAL 1962 is bottled electricity.

Côtto, Quinta do Dou r ★★★ Historic DOU table-wine pioneer making auspicious comeback. V.gd iconic Grande Escolha 15' (only in best yrs), single-old-v'yd Vinha do Dote, rare limited-production Bastardo. Gd-value red.

Covela, Quinta de Vin w ★★ Impressive VIN/DOU border property making v.gd age-worthy whites incl single-variety Edição Nacional (Avesso, ARINTO), RES (Avesso), Escolha (Avesso/CHARD), Res (Avesso/Chard/Arinto oak-aged). Gd rosé.

Crasto, Quinta do Dou, Port r w ★★★ →★★★★ Reputed DOU's family-run estate; striking hilltop location. Jewels in crown are v.old, field-blend, single-v'yd reds Vinha da Ponte and Vinha Maria Teresa. New Honore, a "super-blend" of both (also name of exquisite 100-yr-old+ TAWNY). Gd-value old-v'yd RES. Superb single-variety TINTA RORIZ. Great TOURIGA N. V'yds in Dou Superior give value wines, incl attractive red, innovative acacia-aged white and SYRAH with VIOGNIER dash. Gd VINTAGE PORT and unfiltered LBV. Member of DOURO BOYS.

Croft Port ★★ →★★★ FLADGATE-owned historic shipper with visitor centre in glorious v'yds at Pinhão. Sweet, fleshy VINTAGE PORT 75 77 82 85 91 94 00 03' 07 09' 11' 16' 17'. *Quinta da Roêda* Vintage Port incl v.gd value 07 08' 09 12' 15' and superlative new, old-vine Sērikos 17'. Popular: Indulgence, Triple Crown, Distinction and Pink ROSÉ PORT.

Crusted (Port) Port's hidden gem. Gd-value, fine, rare, traditional NV Port style. Blend of two or more vintage quality yrs, aged up to 4 yrs in casks and 3 yrs in bottle. Unfiltered, forms deposit ("crust") so decant. Look for DOW, FONSECA, GRAHAM'S, NIEPOORT, NOVAL.

Dão r p w sp ★★ →★★★ Revived historic DOC makes age-worthy reds and textured, tasty whites (ENCRUZADO is king). Modern pioneers Álvaro de Castro, Cabriz, CARVALHAIS, Casa de Santar, Falorca, Maias, Roques, Vegia. 2nd wave incl Caminhos Cruzados, CASA DA PASSARELLA, CASA DE MOURAZ, Julia Kemper. To watch: António Madeira, Conciso (NIEPOORT-owned), (outstanding) Druida, Lemos, MOB, Paço dos Cunhas, Ribeiro Santo. Top Dão Nobre ("noble") designation now used. Superb, v.gd-value GARRAFEIRAS. Often known as Portugal's Burgundy.

DOC / DOP (Denominação de Origem Controlada / Protegida) Quality-oriented protected designation of origin controlled by a regional commission. Similar to France's AOC/AC. *See also* VR.

Doce (vinho) Sweet (wine).

Alicante Bouschet was planted in Dou for colour, to replace elderberries in Port.

Douro r p w sw ★ →★★★★ World's 1st demarcated and regulated wine region (1756), high up the eponymous river. Dramatic UNESCO World Heritage Site. Once inaccessible, now wine-tourism ready. Famous for Port, also quality table wine (Dou DOC). Three subregions (Baixo Corgo, Cima Corgo, fast-expanding Dou Superior); great diversity of terroir. Over 100 native varieties (often planted together, 80 yrs+) unforgiving schist terraces. Powerful, increasingly elegant, age-worthy reds; fine, characterful, high-altitude whites. Best: ALVES DE SOUSA, BARCA VELHA, *Casa Ferreirinha*, *Chryseia*, CRASTO, da Boavista, das Carvalhas, DO VESÚVIO, MARIA IZABEL, Muxagat, *Niepoort*, POEIRA, QUINTA NOVA, RAMOS PINTO, *Vale D. Maria*, *Vale Meão*, Vallado, WINE & SOUL. To watch: Costa Boal, CÔTTO, Esmero, Ferradosa, Lavradores de Feitoria, Murças, NOVAL, POÇAS, Quanta Terra, Pessegueiro, Pôpa, S José, REAL COMPANHIA VELHA. VR is Duriense.

Douro Boys Dou Group of DOU producers: CRASTO, NIEPOORT, VALE DONA MARIA, VALE MEÃO, VALLADO. All old friends, but different styles.

Dow Port ★★★★ Historic, reputed SYMINGTON-owned house. Drier-style VINTAGE PORT 85' 94' 00' 07' 11' 16' 17'. Single-QUINTAS do Bomfim and Senhora da Ribeira (v.gd 15) in non-declared vintage yrs. Beautiful Bomfim visitor centre in Pinhão.

Duorum Dou, Port r w ★★ →★★★ Consistent DOU Superior project of JOÃO PORTUGAL RAMOS and ex-FERREIRA/BARCA VELHA José Maria Soares Franco. Top-notch, undervalued O. Leucura, fine RES, gd-value fruity, entry-level *Tons*, COLHEITA. V.gd dense, pure-fruited VINTAGE PORT 07 11' 15' from 100-yr-old vines. Fine second label Vinha de Castelo Melhor and gd-value LBV.

Esporão Alen r w ★★ →★★★ Dynamic eco-conscious group with organic landmark ALEN estate. Attractive visitor centre. High-quality, fruit-focused, modern. Gd-value entry-level Monte Velho, reputed RES (r w). V.gd single-v'y'd/variety range. Sophisticated GARRAFEIRA-like Private Selection and rare Torre do Esporão 07' 11. New *talhas* (clay amphorae) red, white using old ALEN tradition. Auspicious DOU project (QUINTA dos Murças) for elegant single-v'y'd reds. New owner of VIN, LOUREIRO-focused, organic AMEAL. New restaurants in Porto, Lisbon.

Espumante Sparkling. Best: DOU's Vértice, BAIR (esp Bágeiras, Colinas São Lourenço, Kompassus, São Domingos. Lookout for BAGA Bair quality designation. Gd-value Távora-Varosa (esp MURGANHEIRA), VIN (esp SOALHEIRO).

Esteban, Susana Alen r w ★★→★★★ ALEN rising star. Flagship Procura (r w), from PORTALEGRE's v.-old low-yield v'yds (red adds ALICANTE BOUSCHET from Évora). V.gd amphora/*talha* wine. Value second label: Aventura. Innovative: Sidecar (invites other winemakers), Sem Vergonha (elegant, fresh, single-variety CASTELÃO made with Dirk NIEPOORT).

Estufagem Mad Tightly controlled "stove" process of heating MAD for min 3 mths for faster ageing, characteristic scorched-earth tang. Used mostly on entry-level wines. Finer results with external heating jackets and lower max temperature (45°C/113°F).

Falua Tej r p w ★ →★★ Value-driven producer, now owned by French group Roullier. Well-made export-focused Tagus Creek blends native, international grapes. Gd-value entry-level Conde de Vimioso (RES a step up). New gd Falua Res (r w) range.

Ferreira Port ★★ →★★★ Historic Port house owned by SOGRAPE. Stand out 11' 16' vintages, v.gd-value QUINTA do Porto 17'. Winemaker Luis Sottomayor (BARCA VELHA) reckons *LBV* now as gd as last decade's VINTAGE PORT; both on the up here. V.gd-value spicy TAWNY incl Dona Antonia RES, 10-, 20-yr-old Tawny (Quinta do Porto, *Duque de Bragança*).

Ferreirinha, Casa Dou r w ★★ →★★★★ Remarkable range of (SOGRAPE-owned) age-worthy DOU wines. Gd-value entry-level, popular Callabriga, Esteva, Papa Figos, Vinha Grande. Well made QUINTA da Lêda, Tinta Francisca, Antónia Adelaide Ferreira (r w). Rarely released RES Especial and (iconic) BARCA VELHA.

Fladgate Port Important independent family-owned partnership. Owns leading Port houses (CROFT, KROHN, FONSECA, TAYLOR) and interests in tourism incl hotels: Infante Sagres (Porto), The Yeatman (VILA NOVA DE GAIA), Vintage House (Pinhão) and new ambitious World of Wine visitor centre set to open in Porto 2020.

Encruzado is a hidden jewel

Mostly planted in DÃO, ENCRUZADO is one of Portugal's top white grapes. It's easy to see why: it gives gd yields, great natural balance (making fresh, thrilling, structured wines) and has great ageing potential: 60s wines can still be drinking well. Best producers (super-best marked with ★), incl v.gd, expensive, burgundian, oak-aged versions: Antonio Madeira, Boas QUINTAS, Caminhos Cruzados, Carlos Lucas, Casa da Passarella ★, Druida ★, João Paulo Gouveia, Quintas da Fata, da Pellada ★, das Marias, DOS CARVALHAIS, dos Roques.

Fonseca Port ★★★ ·★★★★ FLADGATE-owned Port house, founded 1815. Gd-value Bin 27. V.gd 20-,40-yr-old TAWNY. Excellence in VINTAGE PORT 63' 70' 85' 94' 00' 03' 11' 16' 17'. Superb second label: Fonseca Guimaraens. Single-QUINTA Panascal.

Fonseca, José Maria da Alen, Set r p w sw sp ★·★★★★ 200-yr-old, 7th-generation producer with extensive portfolio, visitor centre in Azeitão, attractive Lisbon wine bar. LANCERS, PERIQUITA incl RES are value brands. Jewel in crown is fortified **Moscatel de Setúbal**, which mines aged stock to great effect esp great-value 20-yr-old Alambre. Remarkable Superior 55' 66 71 and limited-release Roxo Superior 18'. Owner of historic, amphora-based, great-value ALEN JOSÉ DE SOUSA estate.

Pigs like cork-oak forests: 18kg of cork-oak acorns/11kg ilex acorns = 1kg pig-weight.

Frasqueira Mad Top MAD category. Also called Vintage. Single-yr, single-noble-variety aged min 20 yrs in wood, usually much longer. Date of bottling required. Highly respected, sought-after and dear.

Garrafeira Label term for superior quality. Traditionally a merchant's "private RES", often v.gd value. Must be aged for min 2 yrs in cask and 1 yr in bottle (often much longer). Whites need 6 mths in cask, 6 mths in bottle.

Global Wines Bair, Dão r w sp ★★ ·★★★ Also known as DAO Sul. One of Portugal's biggest producers, Dão-based, with estates in many other regions. Great-value popular brands Cabriz (esp RES) and Casa de Santar (esp Res, superb Nobre). Classy Paço dos Cunhas single-v'yd Vinha do Contador. Modern wines, striking architecture, visitor centre at BAIR'S QUINTA do Encontro. Other brands: Grilos, Encostas do Douro (DOU), Monte da Cal (ALEN), Quinta de Lourosa (VIN).

Graham's Port ★★★ ·★★★★ Highly reputed SYMINGTON-owned Port house. Top-notch VINTAGE PORT 85' 91' 94' 97 00' 03' 07' 11' 16' 17', superlative Stone Terraces 11' 15' 16' 17'. Gd-value RES RUBY Six Grapes. V.gd-value single-QUINTA dos Malvedos, attractive 20-, 30-, 40-yrs-old TAWNY, LBV. Fine Single-Harvest (COLHEITAS), esp 40' 52' 61' 63' 69' 72' 82 94' 03. Top-notch Ne Oublie V. Old Tawny, one of three 1882 casks, is a stunner.

Gran Cruz Port ★·★★★ Port's largest brand (Porto Cruz) owned by French group La Martiniquaise, focused on volume and cocktails. VILA NOVA DE GAIA museum, popular rooftop terrace bar, new Porto hotel. Dalva brand has outstanding TAWNY stocks incl COLHEITAS and stunning golden white 52' 63' 71'. Gd Pinhão-based QUINTA de Ventozelo wines, new hotel.

Henriques & Henriques Mad ★★ ·★★★★ MAD shipper owned by rum giant La Martiniquaise. Best are 20 yr old MALVASIA and Terrantez, 15-yr-old (NB *Sercial*), Single Harvest (aged in old bourbon barrels, 1997' 1998'), Vintage (VERDELHO 1957, Terrantez 1954', SERCIAL 1971'). V.gd new TINTA NEGRA 50-yr-old.

Justino Mad ★·★★★ Largest MAD shipper, owned by rum giant La Martiniquaise, makes Broadbent label. Fairly large entry-level range. Some jewels: Terrantez Old Res (NV, probably around 50-yrs-old), Terrantez 1978' (oldest in cask), MALVASIA 1964' 1968' 1988'.

Kopke Port ★·★★★★ Oldest Port house, est 1638, now Sogevinus owned. Well-known for v.gd spicy, structured COLHEITAS 35' 41' 57' 64' 65' 66' 78 80' 84 87 00 02 05 07 09, unique WHITE PORT range, esp now-rare 1935' and 30-, 40-yr-olds. Gd old-vines DOU red and new ARINTO/Rabigato white.

Krohn Port ★·★★★ Now FLADGATE-owned. Exceptional stocks of aged TAWNY, now source of TAYLOR's Single Harvest range. V.gd, rich COLHEITAS: 10-, 20-yr-old. Gd, elegant VINTAGE PORT 16' 17'.

Lancers p w sp ★ JOSÉ MARIA DA FONSECA's semi-sweet, semi sparkling, ROSADO, now white, fizzy (p w) and alc-free versions.

Lavradores de Feitoria Dou r w ★★ ·★★★ Innovative collaboration of 15 DOU producers. Gd-value whites, esp SAUV BL, Meruge (100% oak-aged old-vines

Viosinho). Great-value reds incl Três Bagos RES. V.gd Grande Escolha (esp long-aged Estágio Prolongado), QUINTA da Costa das Aguaneiras, elegant Meruge (mostly TINTA RORIZ from n-facing 400m/1312ft v'yd).

LBV (Late Bottled Vintage) Port Splendid, accessible, affordable alternative to VINTAGE PORT. A single-yr wine, aged 4–6 yrs in cask (twice as long as Vintage Port). V.gd, age-worthy, unfiltered versions eg. DE LA ROSA, FERREIRA, NIEPOORT, NOVAL, RAMOS PINTO, SANDEMAN, WARRE. Best decanted.

Lisboa r p w sp sw ★→★★★ Large region n of capital making generally gd blends of local and international grapes. Best-known DOCS: microclimatic Alenquer (age-worthy reds from boutique great-value DE CHOCAPALHA, SYRAH pioneer MONTE D'OIRO) and traditional BUCELAS, COLARES. Crisp whites growing in strength, esp from limestone, coastal/elevated v'yds, eg. ADEGA Mãe (Viosinho), Casal Figueira (Vital), Casal Sta Maria (Colares), QUINTA DE SANT'ANA (PINOT N, RIES), do Pinto (blends), da Serradinha (natural), Vale da Capucha (organic). Watch: Boa Esperança, Hugo Mendes.

Maçanita, Antonio Alen, Dou r p w ★★→★★★ Unstoppable winemaker/consultant, interests in many regions. ALEN (fashionable Sexy, characterful Fita Preta, esp v.gd Palpite, gd-value Touriga vai Nua), AZORES (started revival of volcanic Pico island wines: exquisite Terrantez do Pico, v.gd ARINTO do Açores), DOU (gd-value r w with sister Joana incl Arinto, As Olgas, Cima Corgo, Gouveio, TOURIGA N).

Madeira r w ★→★★★★ Island and DOC, famous for fortifieds. Modest table wines. VERDELHO best. Look for Atlantis, Barbeito, Barbusano, Moledo, Palmeira, Primeira Paixão, Terras do Avô and innovative TINTA NEGRA-based Ilha.

"Montado" landscape: cork oaks, holm oaks, open, undulating. UNESCO-protected.

Madeira Wine Company Mad Association of all 26 British MAD companies, est 1913. Owns BLANDY, COSSART GORDON, Leacock, Miles and accounts for over 50% of bottled Mad exports. Since Blandy family gained control, almost exclusively focused on promoting Blandy brand.

Malvasia (Malmsey) Mad Sweetest and richest of traditional MAD noble grape varieties, yet with Mad's unique sharp tang. Delightful with rich fruit, chocolate puddings or just dreams.

Maria Izabel, Quinta Dou r p w ★★★ Dynamic, new-wave rising star. Consistent range made with Dirk NIEPOORT's overlook, esp QUINTA, old-vine Vinhas da Princesa. Elegant, limited-edition Sublime; single-variety, rare, expensive Bastardo.

Mateus Rosé p (w) sp ★ Bestselling, medium-dry, lightly carbonated rosé. Also ARAGONEZ, fizzy (p w).

Mendes, Anselmo Vin r w sw sp ★★★ VIN winemaker star. Several benchmark, age-worthy ALVARINHOS, incl v.gd-value (aged on lees) Contacto, excellent skin-contact, voluptous Curtimenta, superb single-v'yd Parcela Única, classy Muros de Melgaço and vibrant new Expressões. Gd LOUREIRO, silky, modern red Vin (Pardusco), surprising orange Tempo. Watch out for new BEI INT, DÃO, DOU wines.

Minho Vin River (and province) between n Portugal and Spain, also VR covering same region as VIN. Some leading Vin producers prefer to use VR Minho label (eg. AMEAL).

Monte de Ravasqueira Alen r p w ★★→★★★ Family-owned estate with experienced winemaker, v.gd terroir (high amphitheatre, clay-limestone, granite) and precision viticulture. V.gd Premium range, esp ALICANTE BOUSCHET. Gd single-v'yd Vinha das Romãs.

Monte d'Oiro, Quinta do Lis r p w ★★→★★★ Family estate started with Hermitage vines from Chapoutier. Savoury red (SYRAH), fine Res (Syrah/ VIOGNIER), age-worthy Têmpera (TINTA RORIZ); crisp white (Viognier/MARSANNE/ARINTO), RES (Viognier). V.gd Ex-Aequo, Bento & Chapoutier Syrah/TOURIGA N.

Moscatel de Setúbal Set sw ★★★ One of Portugal's fortified treasures. Exotic sweet MOSCATEL incl rare Roxo and Superior. Best: BACALHÔA VINHOS, Horácio Simões, JOSÉ MARIA DA FONSECA (oldest stocks incl 100-yr-old Torna-Viagem). Value: ADEGA DE PEGÕES, Casa Ermelinda Freitas, do Piloto, SIVIPA.

Moscatel do Douro Dou High Favaios region; surprisingly fresh, fortified MOSCATEL Galego Branco (MUSCAT Blanc à Petits Grains). Try: ADEGA de Favaios, POÇAS, Portal.

Mouchão, Herdade de Alen r w sw ★★★ Historic family-run ALICANTE BOUSCHET pioneer estate. V.gd historic estate red (1st bottled 1949), COLHEITAS Antigas (cellar releases) 02' 03', iconic *Tonel 3–4* 05' 08 11' 13', fortified *licoroso*. Gd-value Ponte das Canas (incl TOURIGA N/TOURIGA FRANCA/SYRAH), Dom Rafael (old vines).

Mouraz, Casa de Dão r w ★★ Organic pioneer run by winemaker Antonio Lopes Ribeiro. Modern but characterful (esp Elfa) wines from family-owned v'yds at 140–400m (459–1312ft). Lost cellar, some v'yds in forest fires. AIR label from bought-in ALEN, DOU, VIN organic grapes.

Murganheira, Caves sp ★★ Gd ESPUMANTE producer; owns Raposeira. Blends and single-varietal (native, French grapes) fizz: Vintage, Grande RES, Czar rosé.

Niepoort Bair, Dão, Dou r p w ★★★→★★★★ Port and DOU pioneer owned by larger-than-life Dirk Niepoort. Now with many interests. Gd-value, globetrotter Diálogo/Fabelhaft. Fine Dou range, esp *Redoma* (r p w, Res w), Batuta, top-notch Coche (w), iconic Charme and unique 130-yr-old single-v'yd Turris. Exciting Projectos cross-region/winemaker wines incl BUÇACO, Gonçalves Faria, Spanish partnerships Ladredo (Ribeira Sacra) and Navazos (Jerez). Dirk's vision incl BAIR's Quinta de Baixo (esp GARRAFEIRA, Poeirinho, VV), DÃO (esp Conciso) and VIN. Port highlights: VINTAGE PORT 15' 17', CRUSTED, unique demijohn-aged GARRAFEIRA and single v'yd Bioma. V.gd TAWNY, esp elegant bottle-aged COLHEITAS. Lalique-bottled 1863 Port is world's most expensive Port sold at auction (€100k+).

Noval, Quinta do Dou, Port r w ★★★→★★★★ Historic DOU estate owned by AXA since 1993. Consistent, fine VINTAGE PORT 97' 00' 03' 07' 08' 11' 12' 13' 15' 16' 17'. Superlative *Nacional* 63' 66' 94' 96' 97' 00' 01' 03' 04' 11' 16' 17' from 2.5 ha ungrafted vines is pricey jewel in crown. Superb COLHEITAS, 20-, 40-yrs-old, unfiltered LBV. Gd-value Cedro (native/SYRAH r, Viosinho/Gouveio w), v.gd RES, single-variety PETIT VERDOT, TOURIGA N.

Offley Port ★→★★ Old house now owned by SOGRAPE. Gd recent fruity VINTAGE PORT, unfiltered LBV, TAWNY. Aperitif/cocktail styles: Cachuca RES WHITE PORT, ROSÉ PORT.

Palmela Set r w ★→★★★ CASTELÃO-focused DOC. Best: Pegos Claros, Horácio Simões, QUINTA do Piloto. To watch

Passarella, Casa da Dão r p w ★★→★★★ Historic estate leading DÃO revival. V.gd range, esp flagship Villa Oliveira: Encruzado, TOURIGA N (old field-blend v'yd), single-v'yd Pedras Altas (r), Vinha do Província (w), 1ª Edição (five vintages ENCRUZADO blend), top-notch 125 Anos (r). V.gd boutique Fugitivo range esp Enólogo, Enxertia (Jaen), Vinhas Centenárias (red blend of 100-yr-old vines), single-variety Tinta Pinheira, Uva Cão. Excellent (gd-value) GARRAFEIRA (w).

Pato, Filipa Bair r w sp sw ★★→★★★ Star BAIR winemaker, daughter of LUIS P. Defends old-vines-driven "wines with no make-up". V.gd pre-phylloxera, 130-yr-vine Nossa Missão, 90-yr-vine Nossa Calcario. Silky, perfumed BAGA (r), complex Bical (w). V.gd old-vine, oak-*lagares* fermented Territorio Vivo. Tests boundaries esp with thrilling amphora-aged Post Quer**s (r w).

Pato, Luís Bair r w sw sp ★★→★★★★ BAIR star; *seriously age-worthy, single-v'yd Baga* (Vinhas Barrio, Barrosa, Pan) and two Pé Franco (ungrafted) wines (superb sandy-soil QUINTA do Ribeirinho, chalky-clay Valadas). Ready to drink, gd value: VINHAS VELHAS (r w), BAGA Rebel, wacky red FERNÃO PIRES (fermented on Baga skins). V.gd whites incl Vinhas Velhas (single-v'yd Vinha Formal), fizzy MARIA GOMES Método Antigo, early-picked (Informal). Daughter is FILIPA P.

Península de Setúbal Set Atlantic-facing region s of Lisbon making value-driven wines. VR wines mostly from chalky or sandy banks of Sado and Tagus Rivers. Popular: ADEGA de Pegões, BACALHÔA VINHOS, Casa Ermelinda Freitas, JOSÉ MARIA FONSECA. Watch: QUINTA do Piloto, Herdade do Portocarro.

Pereira d'Oliveira Vinhos Mad ★★→★★★★ Family-run producer with vast stocks (1.6 million litres) of bottled-on-demand old FRASQUEIRA, many available to taste at emblematic 1619 cellar door. Best incl stunning C19 vintages (MOSCATEL 1875, SERCIAL 1875, Terrantez 1880) and rare Bastardo 1927.

Periquita Grape also known as CASTELÃO. Also trademark of JOSÉ MARÍA DA FONSECA's successful brand.

Just over half the mtn v'yds in the world are in the Douro Valley.

Poças Dou, Port ★★→★★★ 100-yr-old family-owned Port and DOU firm with increasingly gd table wine range. V.gd Símbolo (with Hubert de Bouard of B'x), Branco da Ribeira (w). Gd RES (r), v.gd-value Vale de Cavalos (r w). New Fora da Série range incl innovative orange, amphora. Old stocks allow for fabulous 90-yr-old+ 1918 Very Old TAWNY, outstanding 20-, 30-, 40-yr-old Tawny, COLHEITAS 67 92 00 01 03 07. V.gd VINTAGE PORT 15' 16' 17'.

Poeira, Quinta do Dou r w ★★★ Consultant Jorge Moreira's own project. Cool n-facing slopes make intense yet softly spoken wines, esp red. V.gd CAB SAUV blend, single-v'yd Ímpar. Taut, keen, oaked rare DOU ALVARINHO. Classy, gd-value Pó de Poeira (r w).

Portalegre Alen r p w ★ →★★★ Most n subregion of ALEN undergoing a revival. SYMINGTON and SOGRAPE's acquisition of cooler, high-altitude (often v. old field-blend) v'yds bodes well. Revival pioneers incl ESPORÃO, Fonte Souto (Symingtons), Cabeças do Reguengo, QUINTA do Centro (now Sogrape), RUI REGUINGA (Terrenus), SUSANA ESTEBAN, Tapada do Chaves, Vitor Claro. To watch.

Quinta Portuguese for "estate". "Herdade" in ALEN. "Single-quinta" denotes single-estate VINTAGE PORTS made in non-declared yrs (increasingly made in top yrs too by single-estate producers).

Quinta Nova Dou r p w ★★→★★★★ (r) Hilltop-located, historic 250-yr-old DOU estate revamped by Amorim family. Incl hotel, auspicious new DÃO project. Winemaker Jorge Alves makes consistent quality-driven range incl top-notch oak-aged Mirabilis (r w), Referência (TINTA RORIZ), Grande RES (TOURIGA N), homage Aeternus (100-yr vines). V.gd Rosé, Res. Gd-value (r w) Grainha, Pomares. Aldeia de Cima is Luisa Amorim's new ALEN project.

Rainwater Mad Light style of MADEIRA, popular in US, great as apéritif, with food.

Ramos, João Portugal Alen r w ★ →★★★ ALEN's pioneer, top winemaker, also in VIN and other brands (DUORUM, Foz de Arouce, QUINTA da Viçosa, Vila Santa). Success due to gd-value, true-to-region wines with commercial appeal incl Marquês de Borba (r, RES r), old vines (r w). V.gd single-v'yds São Lázaro, Jeremias. Estremus is top wine.

Ramos Pinto Dou, Port ★★★ Port and DOU pioneer owned by Champagne Roederer. Gd, consistent range esp Duas Quintas RES (r), Res Especial (mainly TOURIGA N from Bom Retiro). V.gd age-worthy VINTAGE PORT incl single-QUINTA Vintage (de Ervamoira). Complex single-quinta TAWNY 10-yr-old (de Ervamoira) and best-in-class 20-yr-old (Bom Retiro). Gd 30-yr-old incl a dash of centenarian Tawny.

Raposeira Dou sp ★★ MURGANHEIRA-owned. Classic-method fizz. Flagship Velha RES, CHARD/PINOT N lees-aged 4 yrs.

Real Companhia Velha Dou, Port r p w sw ★ →★★★ Family-run, historic (1756) DOU company with new Port museum and wine bar in Gaia. Silva Reis family renewing Port (incl Royal Oporto and Delaforce) and DOU portfolio with precision viticulture (540 ha) and winemaking (led by POEIRA's Jorge Moreira). Grandjó is best late-harvest in Portugal. V.gd old-vine flagship QUINTA das Carvalhas (r w), VINTAGE PORT, 20-yr old TAWNY, Quinta de Cidrô (v.gd CAB SAUV/TOURIGA N, Rufete). Value brands incl Aciprestes, Evel. New, thrilling whites from Quinta do Síbio incl rare Samarrinho, v.gd ARINTO.

Reserve / Reserva (Res) Port Higher quality than basic or aged before being sold (or both). In Port, bottled without age indication (used in RUBY, TAWNY). In table wines, ageing rules vary between regions. Apply pinch of salt.

Romeira, Quinta da r p w sp ★→★★ SOGRAPE-owned, historic BUCELAS estate. Extended ARINTO-focused v'yds (75 ha). V.gd, oaked Morgado Sta Catherina RES. Gd-value Prova Regia, Res.

Rosa, Quinta de la Dou, Port r p w ★★★ Family-run Pinhão estate with Port and DOU range ever better under winemaker Jorge Moreira (POEIRA). V.gd VINTAGE PORT, LBV, new 30-yr old TAWNY. Rich but elegant wines, esp RES (r w). V.gd, age-worthy white TIM. Generous gd-value Passagem label.

Rosado Rosé. Growing category. Best incl Colinas São Lourenço Tête de Cuvée, CORTES DE CIMA, DE COVELA, Monte da Ravasqueira, QUINTA NOVA (esp RES), SOALHEIRO (sp), VÉRTICE (sp).

Rosé Port Port Pioneered by CROFT'S Pink (2005) now made by other shippers (eg. POÇAS). Quality variable. Serve chilled, on ice, or, if you must, in a cocktail.

Rozès Port ★★★ Port shipper owned by Vranken-Pommery. VINTAGE PORT, incl LBV, sourced from DOU Superior QUINTAS (Grifo, Anibal, Canameira). Terras do Grifo Vintage is blend of all three; v.gd LBV (from Grifo).

Ruby Port Most simple, young, cheap (sw Port style. Can still be delicious. RES a step up.

Rui Reguinga Alen, Tej ★★★ Consultant winemaker with own projects. ALEN: v.gd old-vines Terrenus range, incl single-v'yd, 100-yr-old vines Vinha da Serra (w). TEJO: v.gd Rhône-inspired SYRAH/GRENACHE/VIOGNIER Tributo. Also in Argentina.

Sandeman Port ★★→★★★ Historic house, SOGRAPE-owned, famous for its caped man (the don) image. Striking new bottle, glass closure, v.gd-value 20-, 30-, 40-yr-old TAWNY. Gd unfiltered LBV. Great VINTAGE PORT 07' 11' 16' bring back quality. Superb Very Old Tawny Cask 33.

Sant'Ana, Quinta de Lis r w ★★ Historic idyllic family-run LIS estate. Consultant ANTONIO MAÇANITA works national, international varieties. V.gd Atlantic-influenced RIES, fresh PINOT N. Age-worthy red, RES (TOURIGA N/ARAGONEZ/MERLOT), Homenagem. Gd-value (r w), VERDELHO.

São João, Caves Bair r w sp ★★→★★★ Established firm known for gd old-fashioned red/white, esp *Frei João*, Poço do Lobo (BAIR), Porta dos Cavaleiros (DÃO). Rare gd-value museum releases from vast stock (back to 1963). Gd ARINTO/CHARD white, sparkling blends.

Seabra, Luis Dou, Vin r w ★★★ New-wave winemaker, min intervention. V.gd, fresh, elegant single-v'yd DOU (r w) and VIN (w) from indigenous varieties. To watch.

Sercial Mad White grape. Makes driest MAD. *Supreme apéritif;* gd with gravadlax or sushi. *See* Grapes chapter.

World leaders: Portuguese wine consumption is 62 litres/head/yr.

Smith Woodhouse Port ★★★ SYMINGTON-owned small Port firm est 1784. Gd unfiltered LBV; some v.gd drier VINTAGE PORT 83 85 91 94 97 00' 03 07 11' 16. Single-QUINTA da Madelena.

Soalheiro, Quinta de Vin r p w sp ★★→★★★ Leading Monção e Melgaço (VIN subregion) ALVARINHO specialist. V.gd, age-worthy range incl mineral Granit, subtly barrel-fermented old-vine Primeiras Vinhas, oak-aged RES, chestnut barrel/partial malolactic fermentation Terramatter, unfiltered Pur Nature. 1st red, Oppaco, is unique Vinhão/Alvarinho blend. V.gd PINOT N/Alvarinho blend rosé. Great *fizz* (p w).

Sogrape Alen, Dou, Vin ★→★★★★ Portugal's most successful firm, global interests (Portugal, Argentina, NZ, Spain). Popular Port brands: FERREIRA, OFFLEY, SANDEMAN. MATEUS ROSÉ and prestigious BARCA VELHA. Estates by region: ALEN (value Herdade do Peso), DÃO (boutique CARVALHAIS), DOU (popular CASA FERREIRINHA, reputed Legado), LIS (ENCRUZADO-focused QUINTA DA ROMEIRA), VIN (value Azevedo).

Sousa, Alves de Dou, Port r w ★★→★★★ Family-run DOU pioneer. Characterful Baixo Corgo range esp age-worthy *Quinta da Gaivosa*, unique late-released RES Pessoal. V.gd old-vine field-blends Abandonado, Vinha de Lordelo. Expanding Port range incl elegant VINTAGE PORT, v.gd 20-yr-old TAWNY.

Sousa, José de Alen r (w) ★→★★★ Historic, prestigious estate keeping Ancient Roman tradition alive with use of 114 (Portugal's largest collection) ceramic amphorae/*talhas*. Superb J de José de Sousa. Great-value Mayor and classic José de Sousa. Puro Talha range is 100% fermented in *talha*.

Symington Family Estates Dou, Port r w ★★→★★★★ DOU's biggest landowner now also in ALEN. Family-run, owns clutch of top Port houses incl COCKBURN, DOW, GRAHAM, VESUVIO, WARRE. Classy Dou range (incl top-notch CHRYSEIA and VESÚVIO) and gd-value Altano range incl RES r w). New upper Alen range (r w) from Fonte Souto estate incl v.gd Vinha do Souto ALICANTE BOUCHET/SYRAH.

Tawny Port Crowd-pleasing wood-aged Port. RES, age-dated (10-, 20-, 30-, 40-yr-old) wines go up in complexity, price. 20-yr-old often best balance between ages. Single-year COLHEITAS can cost gd deal more than VINTAGE PORT. Luscious Very Old Tawny Ports (min 40 yrs old, most much older) can be expensive, give v. different pleasure from Vintage. Best: 1900, 1910 (ANDRESEN) 1918 (POÇAS), 5G (WINE & SOUL), Honore (QUINTA DO CRASTO), Ne Oublie (GRAHAM'S), Scion (TAYLOR), Tributa (Vallado), VV (NIEPOORT).

Taylor Port ★★→★★★★ Historic Port shipper, FLADGATE's jewel in the crown. Imposing VINTAGE PORTS 63' 66' 83 70' 77' 92' 94 97 00' 03' 07' 09' 11' 16' 17', incl single-QUINTAS (Terra Feita, Vargellas), rare, stunning Vargellas VINHA VELHA from 70-yr-old+ vines. Market leader for TAWNY incl v.gd, year-released, 50-yr-old COLHEITAS 68' 69' and luscious 1863' Scion Very Old Tawny.

Tejo r w ★→★★ Region surrounding River Tagus (Tejo) n of Lisbon. Looking for identity, quantity-to-quality shift. Solid: da Alorna, da Lagoalva, da Lapa, FALUA. More ambitious: Casal Branco, RUI REGUINGA/Tributo show potential of v.gd terroir. Old vines produce gd results with stalwart CASTELÃO, FERNÃO PIRES.

Tinto Red.

Trás-os-Montes Mountainous inland DOC just n of DOU; promising. Valle Pradinhos is reference. Watch: Encostas de Sonim, Sobreiró de Cima, Valle de Passos.

Vale Dona Maria, Quinta do Dou, Port r p w ★★ DOU table wine pioneer, owned by AVELEDA. Plush reds, sometimes burly, incl single-parcel Vinha do Rio, Vinha da Francisca. Smoky, oaky but brisk whites made with bought-in fruit, incl single-parcel Vinha do Martim, flagship CV, new VVV. VINTAGE PORT 15' 16' 17. Member of DOURO BOYS.

Vale Meão, Quinta do Dou r w ★★★ Family-run DOU Superior estate; once source of BARCA VELHA. Fine, age-worthy, elegant top red. Gd-value second label Meandro

---

Stay in a wine hotel
Take the train, or hire a car and visit the wine country. Best places to stay (★ for v.gd food): Casa da Calçada ★ (Amarante), Casas do Côro ★ (Marialva), Convento do Espinheiro (Evora), DE LA ROSA (Pinhão), Herdade Dos Grous (Albernoa), Malhadinha Nova (Albernoa), Monverde ★ (Amarante), Pacheca (Lamego), QUINTA NOVA ★ (Pinhão), São Lourenço do Barrocal (Monsaraz), Six Senses ★ (Lamego), VALLADO (Régua), Vila Vita Parc ★ (Porches), Yeatman ★ (Porto).

(r w). Gd single-QUINTA VINTAGE PORT. V.gd single-varietal (BAGA/TOURIGA N/TINTA RORIZ) Monte Meão range. Member of DOURO BOYS.

Vallado, Quinta da Dou r p w ★★→★★★ Family-owned Baixo Corgo estate, modern hotel/winery. DOU Superior QUINTA do Orgal (with boutique hotel Casa do Rio) makes fresh organic red. Range of Dou (r w), incl RES field blend, worth following. Gd 10-, 20-, 30-, 40-yr-old TAWNY. Adelaide is top Dou red, VINTAGE PORT, thrilling Tributa Very Old (pre-phylloxera) Tawny. Member of DOURO BOYS.

Vasques de Carvalho Dou, Port ★★★ New producer (something rare in Port), est 2012 by António Vasques de Carvalho (inherited family cellars, stock, v'yd) and business partner Luís Vale (capital). Top-notch, stylish 10-, 20-, 30- and 40-yr-old TAWNY. New boutique in Gaia.

Verdelho Mad Grape of medium-dry MAD; pungent but without spine of SERCIAL. Gd apéritif or pair with pâté.

Vértice Dou sp ★★★ DOU-fizz producer, often considered Portugal's best. V.gd Gouveio. Superb, high-altitude, 84-mth-aged PINOT N.

Vesuvio, Quinta do Dou, Port ★★★★ Magnificent riverside QUINTA making v.gd, age-worthy, old-vine, high-altitude Vesuvio (r), v.gd-value second label Pombal do Vesuvio (r) and Port on par with best. 07' 08' 11' 13' 15' 16' 17'. Still foot-trodden by people (not robots).

Vila Nova de Gaia Dou, Port Historic home of major Port shippers, across River Douro from Oporto. Hotels, restaurants, tour boats, classy lodges are tourist attractions (CÁLEM, COCKBURN'S, GRAHAM'S, POÇAS, SANDEMAN, TAYLOR). Now home of FLADGATE's World of Wine museum.

Vinhas Velhas Old vines; but old changes by region. N (60–120-yrs) to s (30–40-yrs).

Vinho Verde r p w sp ★ →★★★ Portugal's biggest region in rainy, verdant nw. Signs of a renaissance with gd-value, fresh, elegant whites. Best: high-end ALVARINHO from Monçâo e Melgaço subregion (ANSELMO MENDES, Regueiro, SOALHEIRO, Vale dos Ares), LOUREIRO from Lima (eg. AMEAL, Aphros), Avesso from Baião (COVELA). Red Vinhão grape getting a makeover by leading players (eg. Anselmo Mendes, Aphros, Soalheiro). Large brands (ADEGA de Monção, AVELEDA, Azevedo) often slightly fizzy (Casal Garcia, Gazela, Muralhas); DYA. Watch: LUIS SEABRA, San Joanne, Santiago, Vale dos Ares, Vilacetinho.

Vintage Port Port Classic vintages are best wines declared in exceptional yrs by shippers. Bottled without filtration after 2 yrs in wood, mature v. slowly in bottle, throwing a deposit – always decant. Modern vintages broachable earlier (and hedonistic young) but best will last more than 50 yrs. Single-QUINTA Vintage Ports also drinking earlier, but can last 30 yrs+.

VR / IGP (Vinho Regional / Indicação Geográfica Protegida) Same status as French Vin de Pays. More leeway for experimentation than DOC/DOP.

Warre Port ★★★ →★★★★ 1st and oldest of British Port shippers (1670), now owned by SYMINGTON FAMILY ESTATES. Rich, long-aging VINTAGE 83 85 91 94 97 00' 03 07' 09' 11' 16' 17' and unfiltered LBV. Elegant Single-QUINTA and 10-, 20-yr-old TAWNY Otima reflect Quinta da Cavadinha's cool elevation.

White Port Port from white grapes. Ranges from dry to sweet (*lágrima*); mostly off-dry and blend of yrs. Apéritif straight or drink iced with tonic and fresh mint. Growing, high-quality, niche: age-dated (10-, 20-, 30-, or 40-yr-old), eg. ANDRESEN, KOPKE, QUINTA de Santa Eufemia; rare COLHEITAS eg. Dalva, Kopke.

Wine & Soul Dou, Port r w ★★★ →★★★★ Family-owned DOU estate run by winemaking couple Sandra Tavares and Jorge Serôdio Borges. V.gd, oak-aged Guru (w), superb old-vine QUINTA da Manoella VINHAS VELHAS and complex, dense (80-yr vine) Pintas. V.gd-value Pintas Character and second-label Manoella (r w). V.gd Pintas VINTAGE PORT. Oustanding 5G (120-yr-old barrel kept from five generations) Very Old TAWNY.

PORTUGAL

Switzerland

Abbreviations used in the text:

Aar	Aargau
Ber	Bern
Gris	Grisons
Luc	Lucerne
Neu	Neuchâtel
Schaff	Schaffhausen
Thur	Thurgau
Tic	Ticino
Val	Valais
Vd	Vaud
Zür	Zürich

Swiss wine is full of paradoxes: while Burgundy's wine prices are exploding, Swiss Pinot Noir is considered expensive even though prices haven't gone up much for almost two decades. And Swiss Pinot is better than ever, be it in the fruit-driven and tannic style of Graubünden and other German Swiss cantons, or in the finely scented, elegant manner of Neuchâtel and its neighbours. Swiss growers are obsessed and passionate. How else could they get something serious out of Chasselas? How else could they have preserved two dozen local varieties? The only thing they are not good at is making themselves known abroad. In the long term, a little international fame could be helpful, if only to reassure a new generation of domestic consumers about these wines' excellence.

Recent vintages

2019 Rain at harvest time, esp e Switzerland; Vaud and Vaiais better.

2018 Powerful, round wines all over the country.

2017 Frost; some cantons have only 20% of a normal crop. V.gd quality.

2016 Frost in April, rainy summer then sun: mostly mid-weight wines.

2015 Great vintage, ripe fruit, perfectly balanced acidity.

2014 Yr of mid-weight, classically structured wines.

2013 V. small crop; e Switzerland outstanding, great freshness, purity.

Fine vintages: 10 (Pinot N) 09 05 (all) 99 (Dézaley) 97 (Dézaley) 90 (all).

Aigle Vd ★★→★★★ Commune for CHASSELAS best known for BADOUX Les Murailles. Try Terroir du Crosex Grillé.

AOC Equivalent of France's Appellation Contrôlée, 62 AOCs countrywide.

Bachtobel, Schlossgut Thur ★★★ Since 1784 owned by Kesselring family, known for refined PINOT N from slopes nr Weinfelden. Estate's own clonal selection, in early C20 dismissed because of high acidity, now rediscovered.

Bad Osterfingen Schaff ★★★ Restaurant and wine estate in historical baths (est 1472). Michael Meyer is a PINOT specialist with a famous line in Spätzle (noodles). Co-producer of ZWAA.

Badoux, Henri Vd ★★ CHASSELAS AIGLE les Murailles (classic lizard label) most popular Swiss brand. Ambitious Lettres de Noblesse series has gd barrel-aged YVORNE.

Baumann, Ruedi Schaff ★★★ Leading estate at Oberhallau; berry-scented, age-able PINOT N, esp -R-, Ann Mee. ZWAA a collaboration with BAD OSTERFINGEN estate.

Bern Capital and canton. Villages Ligerz, Schafis, Twann (Lake Biel), Spiez (Lake Thun). Mainly CHASSELAS, PINOT N. Top: Andrey, Johanniterkeller, SCHLÖSSLI, STEINER.

Besse, Gérald et Patricia Val ★★★ Leading VAL family estate; Gérard and Patricia B and daughter Sarah. Mostly steep terraces up to 600m (1969ft); intense old-vines *Ermitage Les Serpentines* 10' 13' 15 16 (MARSANNE from granite soils, planted 1945).

Bonvin Val ★★→★★★ Old name of VAL, recently much improved, esp local grapes: *Nobles Cépages* series (eg. HEIDA, PETITE ARVINE, SYRAH).

Bovard, Louis Vd ★★→★★★★ Family estate (ten generations) famous for textbook DÉZALEY La Médinette 99' 05' 12' 15 16 17 18; old vintages available from domaine.

Bündner Herrschaft Gris ★★→★★★★ 13' 15' 16 17 18' PINOT N with structure, fruit and great capacity to age. Only four villages: FLÄSCH, Jenins, Maienfeld, MALANS. Climate balanced between mild s winds and coolness from nearby mtns.

Calamin Vd ★★★ GRAND CRU of LAVAUX, 16 ha of deep, calcareous soils on a landslide, tarter CHASSELAS than neighbour DÉZALEY, growers incl BOVARD, Dizerens, DUBOUX.

Chablais Vd ★★→★★★ Wine region at upper end of Lake Geneva around AIGLE and YVORNE. Name is from Latin *caput lacis*, head of the lake.

Chanton Val ★★★ *Terrific Valais spécialités*; v'yds up to 800m (2625ft): Eyholzer Roter, Gwäss, HEIDA, Himbertscha, Lafnetscha, Resi.

Chappaz, Marie-Thérèse Val ★★★→★★★★ Small bio estate, famous for magnificent sweet wines of local grape Petite ARVINE and Ermitage (MARSANNE). Hard to find.

Colombe, Domaine La Vd ★★→★★★ Family estate of FÉCHY, LA CÔTE, 15 ha, bio. Best-known for range of age-able CHASSELAS, eg. La Brez.

Cortaillod Neu Village on shores of Lake NEUCHÂTEL renowned for refined PINOT N. Eponymous, low-yielding local clone.

Côte, La Vd ★→★★★ 2000 ha w of Lausanne on Lake Geneva, mainly CHASSELAS of v. light, commercial style. Villages incl FÉCHY, Mont-sur-Rolle, Morges.

Cruchon Vd ★★★ Bio producer of LA CÔTE, now led by young Catherine C, lots of

SWITZERLAND

The geology of Dézaley

DÉZALEYS grow on a geological macédoine. Over millions of yrs rivers have carried stones and sand down from the arising Alps. Around 26–23 million yrs ago layers of so-called *Molasse* formed, each one a mix of particles from different parent rocks. 15 million yrs later, when the European tectonic plate further underrode the African plate, five existing layers of *Molasse* tilted so that all of them came to the surface within a few miles. Hence the great variety of subsoils in the LAVAUX. Then around 25,000 yrs ago the VAL glacier moved along and eroded some of the softer *Molasse* layers. Particularly hard *Molasse* is called *Herrgottsbeton* – God's concrete; round pebbles cemented by a matrix of finer components (mostly sand, marl, lime). That's the subsoil on which the finest Dézaleys grow.

SPÉCIALITÉS (eg. outstanding Altesse). Top growth: PINOT N Raissennaz. Sublime Coeur de Cuvée Sparkling (5 yrs on lees).

Dézaley Vd ★★★ Celebrated LAVAUX GRAND CRU on steep slopes of Lake Geneva, 50 ha; planted in C12 by Cistercian monks. Potent CHASSELAS develops with age (7 yrs+). Best: DUBOUX, *Fonjallaz*, *Louis Bovard*, Monachon, Ville de Lausanne. Tiny red production too, mostly blends.

Dôle Val ★→★★ VAL's answer to Burgundy's Passetoutgrains: PINOT N plus GAMAY. Lightly pink Dôle Blanche pressed straight after harvest. Rarely exciting.

Donatsch, Thomas Gris ★★★ Barrique pioneer (1974) at MALANS. Now son Martin in charge: 13' PINOT N Res Privée fetched 1075 CHF/bottle at auction.

Duboux, Blaise Vd ★★★ 5-ha family estate in LAVAUX. Outstanding DÉZALEY vieilles vignes Haut de Pierre (v. rich, mineral), CALAMIN Cuvée Vincent.

Einsiedeln, Kloster Schw ★★ Benedictine abbey, founded 934, famous for its black Madonna. 8 ha of vines at Lake Zürich, v.gd Konvent PINOT N.

Epesses Vd ★→★★★ Well-known LAVAUX AOC, 130 ha surrounding GRAND CRU CALAMIN: sturdy, full-bodied whites. Growers incl BOVARD, DUBOUX, Fonjallaz, Luc Massy.

Féchy Vd ★→★★★ Famous though unreliable AOC of LA CÔTE, mainly CHASSELAS.

Federweisser / Weissherbst German-Swiss pale rosé or Blanc de Noirs (BLAUBURGUNDER).

Fendant Val ★→★★★ Full-bodied VAL CHASSELAS, ideal for fondue or raclette. Try BESSE, Cornulus, GERMANIER, PROVINS, SIMON MAYE. Name derived from *se fendre* (to burst): ripe grapes of local Chasselas clone crack open if pressed between fingertips.

Fläsch Gris ★★★→★★★★ Village of BÜNDNER HERRSCHAFT known for PINOT N from schist and limestone. Lots of gd estates, esp members of Adank, Hermann, Marugg families. *Gantenbein* is outstanding.

Flétri / Mi-flétri Late-harvested grapes for sweet/slightly sweet wine.

Fribourg Fri 115 ha on shores of Lake Murten (Mont Vully): round CHASSELAS, elegant TRAMINER. Best: Chervet, Cru de l'Hôpital.

Fromm, Georg Gris ★★★ 05' 13' 15' 16 17 18 Top grower in MALANS, known for subtle single-v'yd PINOT N (Fidler, Selfi/Selvenen, Spielmann, Schöpfi).

Gantenbein, Daniel & Martha Gris ★★★★ 09' 10' 13' 15' 16 17 18 Star growers, based in FLÄSCH. Top PINOT N (DRC clones, *see* France), exceptional v. limited CHARD.

Geneva 1400 ha of vines remote from the lake (v'yds there belong mainly to neighbouring canton VD). Wide range of varieties. Growers incl Balisiers, Grand'Cour, Les Hutins, Novelle.

Germanier, Jean-René Val ★★→★★★ Big estate (150 ha), Reliable FENDANT Les Terrasses, elegant SYRAH Cayas, nobly sweet AMIGNE Mitis from schist at Vétroz.

Glacier, Vin du (Gletscherwein) Val ★★★ Fabled oxidized, (larch)-wooded white from rare Rèze grape of Val d'Anniviers. Find it at the Rathaus of Grimentz. A sort of sharp Alpine Sherry.

Grain Noble ConfidenCiel Val Quality label for authentic sweet wines, eg. CHAPPAZ, DOMAINE DU MONT D'OR, Dorsaz (both estates), GERMANIER, Philippe Darioli, PROVINS.

Grand Cru Inconsistent term, in use in VAL (commune Salgesch for PINOT N) and in VD (as "Premier Grand Cru" for a wide range of single-estate wines). Switzerland has only two Grands Crus in sense of classification of v'yd sites: CALAMIN, DÉZALEY.

Grisons (Graubünden) Gris Mtn canton, German-speaking. PINOT N king. *See* BÜNDNER HERRSCHAFT. Best growers in other areas: Manfred Meier, VON TSCHARNER.

Huber, Daniel ★★→★★★ Pioneer who reclaimed historical sites from fallow in 1981. Partly bio (2003). Son Jonas taken over. Top: Montagna Magica (MERLOT/CAB FR).

Johannisberg Val VAL name for SILVANER, often off-dry or sweet; great with fondue. Excellent: *Domaine du Mont d'Or*.

Joris, Didier Val ★★★→★★★★ Only 3 ha, but a dozen varieties bringing wines of depth and complexity, incl outstanding (and rare) MARSANNE. Now also reviving nearly extinct local white grape Diolle.

Lavaux Vd ★★→★★★★ 30 km (19 miles) of steep s-facing terraces e of Lausanne; UNESCO World Heritage site. Uniquely rich, mineral CHASSELAS. GRANDS CRUS DÉZALEY, CALAMIN, several village AOCS.

Litwan, Tom Aar ★★★ Passionate bio grower at Schinznach. 3 ha. Delicate, fine-grained PINOT N Auf der Mauer ("On Top of the Wall") and Chalofe ("Lime Kiln").

Malans Gris ★★→★★★★ Village in BÜNDNER HERRSCHAFT. Top PINOT N producers incl DONATSCH, FROMM, Liesch, Studach, Wegelin. Late-ripening local grape Completer gives a long-lasting phenolic white. Adolf Boner 01' 05' is keeper of the Grail.

Maye, Simon et Fils Val ★★★ 11-ha family estate. Dense SYRAH Vieilles Vignes perhaps best in Switzerland; spicy, powerful Païen (HEIDA), FENDANT v.gd too.

Mémoire des Vins Suisses Union of 56 leading growers in effort to create stock of Swiss icon wines, to prove their ageing capacities. Oldest wines from 1999.

Mercier Val ★★★→★★★★ SIERRE family estate, now young Madeleine M in charge, meticulous v'yd management produces dense, aromatic reds, eg. rare CORNALIN 05' 09' 10' 11 15 16 17 18 and SYRAH.

Mont d'Or, Domaine du Val ★★→★★★★ Emblematic VAL estate for nobly sweet wines, esp JOHANNISBERG Saint-Martin. Recently, more emphasis on dry wines.

Neuchâtel ★→★★★ 600 ha around city and lake on calcareous soil. Slightly sparkling CHASSELAS, exquisite PINOT N from local clone (CORTAILLOD). Best: Ch d'Auvernier, Domaine de Chambleau, La Maison Carrée, PORRET, TATASCIORE.

Oeil de Perdrix "Partridge's eye": PINOT N Rosé, originally from NEU, now elsewhere.

Pircher, Urs Zür ★★★→★★★★ Top estate at Eglisau, steep s-facing slope overlooking Rhine. Outstanding PINOT N Stadtberger Barrique 15' 16 17 18 from old Swiss clones. Whites of great purity.

Few wine regions anywhere can beat Vd for exquisite food. Hotel de Villle, Crissier, nr Lausanne: a foodie icon for generations.

Porret Neu ★★→★★★★ Leading family estate at CORTAILLOD with burgundian approach to CHARD and PINOT N (best: Cuvée Elisa, made for decades to age). Also v'yds in Aloxe-Corton (see France).

Provins Val ★→★★★ Co-op with 4000+ members, Switzerland's biggest producer, 1500 ha, 34 varieties. Sound entry level, v.gd oak-aged Maître de Chais range.

R3 Zür ★★★ 08' 09 12' 17' 18 Räuschling (grape) collaboration of three leading ZÜR growers: Rütihof, Lüthi, SCHWARZENBACH. Soils: sandstone, limestone, clay.

Rodeline, Domaine La Val ★★★ VAL family estate known for local varieties from prime terraced single v'yds at Fully and Leytron, eg. Les Claives MARSANNE.

Rouvinez Vins Val ★→★★★ Famous producer at SIERRE, best known for cuvées La Trémaille (w) and Le Tourmentin (r). Controls also BONVIN, Caves Orsat, Imesch.

Ruch, Markus Schaff ★★★ Excellent PINOT N from Hallau (Chölle from 60-yr-old vines, Haalde from steep slope) and Gächlingen (Schlemmweg on limestone). Amphora-fermented MÜLLER-T. Only 3 ha.

St. Jodern Kellerei Val ★★→★★★ VISPERTERMINEN co-op famous for **Heida Veritas** from ungrafted old vines: superb reflection of Alpine terroir.

St-Saphorin Vd ★→★★★ Neighbour AOC of DÉZALEY, lighter, but equally delicate. Best: Leyvraz, Monachon.

Schaffhausen ★→★★★ Canton/town on Rhine with famous falls, BLAUBURGUNDER stronghold. Best-known village is Hallau, but be v. careful. Top growers: BAD OSTERFINGEN, BAUMANN, RUCH, Strasser.

Schenk SA Vd ★→★★★ Wine giant with worldwide activities, based in Rolle, founded 1893. Classic wines (esp VD, VAL); substantial exports.

Schlössli, Weingut Ber ★★ →★★★ Top producer, excellent Le Grand Pinot (PINOT N).

Schwarzenbach, Hermann Zür ★★★ Leading family estate on Lake Zürich, many SPECIALITIES: Completer, Freisamer, outstanding Räuschling Seehalden 15' 17 18'.

Sierre Val ★★→★★★ VAL town on six hills, home of rich, luscious wines. Best-known: des Muses, Imesch, MERCIER, ROUVINEZ, Zufferey.

Sion Val ★★→★★★ Capital/wine centre of VAL, domicile of big producers: *Charles Bonvin* Fils, Gilliard, PROVINS, Varone.

Spécialités / Spezialitäten Quantitatively minor grapes producing some of best Swiss wines, eg. Räuschling, GEWURZ or PINOT GR in German Switzerland, or local varieties (and grapes like JOHANNISBERG, MARSANNE, SYRAH) in VAL.

Sprecher von Bernegg Gris ★★★ Historic estate at Jenins, BÜNDNER HERRSCHAFT, esp PINOT N: Lindenwingert, vom Pfaffen/Calander.

Steiner, Weingut Ber ★★★ Family estate at Ligerz. Since young Sabine S took over, organic – and rising quality. Outstanding Clos au Comte CHARD.

Stucky-Hügin Tic ★★★→★★★★ MERLOT pioneer Werner Stucky and son Simon plus Jürg Hügin. Best: Temenos (Completer/SAUV BL), Soma (Merlot/CAB FR), Conte di Luna (Merlot/CAB SAUV).

Tatasciore, Jacques Neu ★★★★ Refined (and rare) NEU PINOT N.

Ticino ★→★★★ Italian-speaking. MERLOT (leading grape since 1948) in a taut style. Best: Agriloro, Castello di Morcote, Gialdi, HUBER, Klausener, Kopp von der Crone Visini, STUCKY, Tamborini, Valsangiacomo, Vinattieri, ZÜNDEL.

Tscharner, Gian-Battista von ★★★ Family estate, father Gian-Battista and son Johann-Baptista produce tannin-laden PINOT N (Churer Gian-Battista, Jeninser Alte Reben) to age. Churer Johann-Baptista is more delicate, approachable.

Switzerland grows 250 varieties+, 21 of them indigenous.

Valais (Wallis) Largest wine canton, in dry, sunny upper Rhône Valley. Local varieties outstanding; best MARSANNE, SYRAH rival French legends. Top: BESSE, CHANTON, CHAPPAZ, Cornulus, Darioli, des Muses, Dorsaz, GERMANIER, JORIS, MAYE, MERCIER, MONT D'OR, PROVINS, ROUVINEZ, ST. JODERN K, Zufferey (both). Youngsters: Mathilde Roux, Cave de l'Orlaya (Fully); Romain Cipolla (Raron); Sandrine Caloz (Miège).

Vaud (Waadt) 2nd largest wine canton. Important: Bolle, Hammel, Obrist, SCHENK. CHASSELAS is main grape, but only gd terroirs justify growers' loyalty.

Visperterminen Val ★→★★★ Upper VAL v'yds, esp for HEIDA. One of highest v'yds in Europe (at 1000m/3281ft+; called Riben). Try CHANTON, ST. JODERN KELLEREI.

Yvorne Vd ★★→★★★ CHABLAIS village with v'yds on detritus of 1584 avalanche, eg. BADOUX, Ch Maison Blanche, Commune d'Yvorne, Domaine de l'Ovaille.

Zündel, Christian Tic ★★★→★★★★ German Swiss geologist. Bio farming, wines of purity and finesse, esp MERLOT/CAB SAUV Orizzonte and CHARDS Velabona, Dosso.

Zürich Largest wine-growing canton in German Switzerland, 610 ha. Mainly BLAUBURGUNDER, Räuschling a local SPECIALITY. Best growers: Gehring, Lüthi, PIRCHER, SCHWARZENBACH, Zahner.

Zürichsee Zür, Schw Dynamic AOC uniting v'yds of cantons ZÜR and Schwyz on shores of Lake Zürich. Best: Bachmann, Diederik, E Meier, Höcklistein, Kloster KLOSTER EINSIEDELN, Lüthi, Rütihof, Schipf, Schnorf, SCHWARZENBACH. Räuschling (autochthonous), MÜLLER-T, PINOT N, SPEZIALITÄTEN. But also a proof of *Kantönligeischt* (small-minded canton spirit): St Gallen abutters of lake refused to join.

Zur Metzg, Winzerei Zür ★★→★★★ Banker turned winemaker, bought-in grapes vinified in old butcher's shop, eg. Kirschberg PINOT N.

Zwaa Schaff ★★★ Collaboration: BAUMANN (calcareous, deep soil) and BAD OSTERFINGEN (light, gravelly). PINOT N 94' 09' 13' 15' 16 17 18; PINOT BL/CHARD equally long-lasting.

Wine regions
Switzerland has six major wine regions: VAL, VD, GENEVA, TIC, Trois Lacs (NEU, Bienne/BER, Vully/FRIBOURG) and German Switzerland (AAR, GRIS, SCHAFF, St Gallen, Thur, ZÜR and some smaller wine cantons).

Austria

Abbreviations used in the text:

Burgen	Burgenland
Carn	Carnuntum
Kamp	Kamptal
Krems	Kremstal
Nied	Niederösterreich
Stei	Steiermark
S Stei	Südsteiermark
Therm	Thermenregion
Trais	Traisental
V Stei	Vulkanland Steiermark
Wach	Wachau
Wag	Wagram
Wein	Weinviertel
W Stei	Weststeiermark

F ew European countries can match Austria's dynamism and success over the past decade. None produce such reliably savoury and refreshing wines in what amounts to a national style. Producers here want to express site, soils, variety and vintage as clearly as possible: if you want wines of crystalline purity and tense energy, this is the place to look. Stylistically anything goes: from the soaring purity of Riesling fruit via the savoury richness of Austria's flagship variety Grüner Veltliner to salty, skin-fermented earthiness in orange wines. Some 14% of the vineyard is certified organic, and a further 7% certified sustainable; most top growers are bio. Provenance and regional typicity are moving to centre stage with the DAC system, which started in 2002: there are now 15 DACs, the latest being Carnuntum and Wachau. The classification of single vineyards now includes sites in Vienna and Carnuntum. Austria's pristine, expressive wines have seen growth in export markets year after year, and no wonder.

Recent vintages

2019 Dream vintage boasting both ripeness and freshness.

2018 The heatwave yr: ripe wines. Gd and plentiful.

2017 Gd juicy, rounded wines.

2016 Lovely fruit expression, fine freshness, but choose growers carefully.

2015 V.gd quality. Full-bodied, ripe, with the stuffing to age.

2014 Difficult, cooler yr. Tread carefully; thrilling freshness where selection was stringent. Slender but charming reds.

2013 Slender, v. fresh but expressive whites. Taut, crunchy reds.

Don't hesitate to try more mature vintages from known producers.

Achs, Paul Burgen r (w) ★★★ 13 15 16 Sumptuous, cherry-scented BLAUFRÄNKISCH.

Allram Kamp w ★★★ 13 14 15 Consistently savoury GRÜNER V (esp Renner, Gaisberg), RIES Heiligenstein.

Alzinger Wach w ★★★★ 10 13 14 15 16 17 Quiet genius of dazzling, ethereal RIES, spicy GRÜNER V, in Steinertal and Loibenberg.

Ausbruch Quality/style designation for Prädikat wine; restricted to RUST and botrytized, dried grapes. Min must weight 27°KMW or 138.6°Oechsle.

Austria's entire v'yd amounts to only 41% of the B'x wine region.

Ausg'steckt ("Hung out") Greenery and branches hung prominently outside traditionally signal that a HEURIGE or Buschenschank is open.

Bauer, Anton Wag r w ★★★★ Intense, expressive, savoury GRÜNER V: single-v'yds Rosenberg, Spiegel. Age-worthy, profound reds, purist PINOT N.

Braunstein, Birgit Burgen r w ★★★ 13 15 16 17 Intuitive LEITHABERG bio-winemaker. Gorgeous reds. Try amphora-aged series Magna Mater.

Bründlmayer, Willi Kamp r w sw sp ★★★★ 10 13 15 17 Longstanding KAMP icon. Stellar GRÜNER V, thrilling RIES, esp Heiligenstein Lyra and Alte Reben. Also fine Sekt and PINOT N.

Burgenland r (w) Federal state and wine region bordering Hungary. Warmer than NIED, hence reds like BLAUFRÄNKISCH, ST-LAURENT, ZWEIGELT prevalent. Shallow NEUSIEDLERSEE, eg. at RUST, creates ideal botrytis conditions.

Carnuntum Nied r w Previously unsung region se of VIENNA now blossoming with accomplished, fresh reds, esp ZWEIGELT marketed as Rubin Carnuntum. Look out for BLAUFRÄNKISCH from Spitzerberg. Best: G Markowitsch, MUHR, NETZL, TRAPL.

Christ Vienna r w ★★★ VIENNA stalwart and HEURIGE. Leading light for GEMISCHTER SATZ and unusual red blends.

DAC (Districtus Austriae Controllatus) Provenance- and quality-based appellation system denoting regionally typical wines and styles. Creation of 1st DAC 2002, WEIN, prompted regional quality turnaround. Currently 15 DACs: EISENBERG, KAMP, KREMS, LEITHABERG, MITTELBURGENLAND, NEUSIEDLERSEE, Rosalia, S STEI, TRAIS, V STEI, Wein, Wiener GEMISCHTER SATZ, W STEI. Latest are CARN and WACH.

Domäne Wachau Wach w ★★★ Co-op with unparalleled quality focus and top execution of fine site portfolio. Whistle-clean GRÜNER V, RIES, esp Achleiten, Kellerberg. Always great value.

Ebner-Ebenauer Wein w r sp ★★★ Talented, ambitious couple setting new WEIN standards with expressive GRÜNER V, sinuous ST LAURENT, vivid RIES and age-worthy PINOT N. Gorgeous Blanc de Blancs Brut Nature.

Eichinger, Birgit Kamp w ★★★ Est address for stellar RIES, esp Heiligenstein, always savoury GRÜNER V, esp Hasel.

Eisenberg Burgen Small DAC (since 2009) restricted to BLAUFRÄNKISCH from local slate soil. Powerful but elegant.

Erste Lage Single-v'yd quality designation in CARN, KAMP, KREMS, TRAIS, VIENNA, WAG: 62 member estates with 81 sites classified as Erste Lage, ongoing process. *See* ÖTW.

Esterhazy Burgen r (w) ★★★ Historic schloss where Josef Haydn composed in centre of Eisenstadt (BURGEN). Solid reds, bright whites.

Federspiel Wach VINEA WACHAU middle category of ripeness, min 11.5%, max 12.5% alc. Understated, gastronomic wines as age-worthy as SMARAGD.

Feiler-Artinger Burgen r w sw ★★★→★★★★ Leading producer of exquisite AUSBRUCH and elegant reds in historic town centre of RUST.

Gemischter Satz Vienna Revived historic concept of co-planted and co-fermented field-blend of white varieties. Complex, "winey" wines. Prevalent in WEIN and VIENNA: determined producers achieved DAC status in 2013 for Vienna. No variety to exceed 50%. Look for CHRIST, GROISS, LENIKUS, WIENINGER.

Geyerhof Krems r w ★★→★★★ Superstar bio estate with top RIES and great, expressive entry-level Stockwerk.

Gritsch Mauritiushof Wach w ★★★ Est quality producer in Spitz. Speciality: high-altitude plantings of GRÜNER V. Also RIES from 1000-Eimberberg v'yd.

Groiss, Ingrid Wein w ★★★ Young WEIN star with reputation for reviving old v'yds, making salty GEMISCHTER SATZ and peppery GRÜNER V.

Grosse Lage Stei Highest v'yd classification in STEI; work still in progress along Danube (*see* ERSTE LAGE).

Gruber-Röschitz Wein w ★★★ Family-run WEIN estate consistently raising bar. Racy RIES from granite soils, fine GRÜNER V, esp Mühlberg.

Gumpoldskirchen Therm Once famed, still popular HEURIGEN village s of VIENNA. Home to white rarities ZIERFANDLER, *Rotgipfler.*

Gut Oggau Burgen r w ★★→★★★ Unconventional, hip but solid bio estate.

Hager, Matthias Kamp w sp ★★★ Bio producer of expressive GRÜNER V Mollands, fun pét-nat.

Harkamp S Stei w sp ★★★ STEI's traditional-method stalwart, long-aged, creamy and brisk Sekt.

Hartl, Heinrich Therm r w ★★★ Ambitious young winemaker with increasing following for elegant PINOT N.

Heinrich, Gernot Burgen r w dr sw ★★★ PANNOBILE member and BLAUFRÄNKISCH specialist, branching out into skin-fermented and natural whites.

Heinrich, J Burgen r w ★★★ 10 13 15 BLAUFRÄNKISCH specialist. Fresh-faced, snappy entry-level; layered, complex single-v'yd, esp Goldberg.

Heuriger Wine of most recent harvest. **Heurige**: homely tavern where growers serve own wines with rustic, local food often in the open air – integral to Austrian culture. Called Buschenschank outside Vienna. *See* AUSG'STECKT.

Hiedler Kamp w sw ★★★ Great address for GRÜNER V, esp Thal and Kittmannsberg.

Hirsch Kamp w ★★★ 13 15 17 Outstanding RIES, GRÜNER V from Heiligenstein, Lamm. Great value entry-level Grüner V Hirschvergnügen.

Hirtzberger, Franz Wach w ★★★ 08 10 13 14 15 WACH fixture in Spitz known for powerful styles, now reintroducing subtlety. Dazzling RIES, GRÜNER V, esp Honivogl, Singerriedel.

Huber, Markus Trais w ★★★ Encapsulates delicacy of TRAIS limestone soils with RIES, GRÜNER V, and branching into PINOT N

Illmitz Burgen sw SEEWINKEL town on NEUSIEDLERSEE, famous for BA, TBA (*see* Germany). Best from *Kracher*, Opitz.

Jäger Wach w ★★★ Great GRÜNER V, RIES, esp Achleiten, Klaus.

Jalits Burgen ★★★ Est EISENBERG estate, bold but elegant BLAUFRÄNKISCH, esp single-v'yds Szapary and Diabas.

Jamek, Josef Wach w ★★★→★★★★ Danube-facing WACH icon and famed *restaurant* at Joching, now returned to top tier. RIES, GRÜNER V, esp Achleiten, Klaus.

830-km (516-mile) Niederösterreich Wine Route connects all eight growing regions.

Johanneshof Reinisch Therm r w ★★★ Estate at Tattendorf with deserved reputation for long-lived PINOT N and ST-LAURENT, also leading advocate of local vines ROTGIPFLER, ZIERFANDLER.

Jurtschitsch Kamp w sp ★★★ Exemplary, enterprising bio estate enriching KAMP with energy and verve; impressive across board.

Kamptal Nied (r) w Wine region along Danube tributary Kamp n of WACH; rounder style, lower hills. Top v'yds: Heiligenstein, Lamm. Best: BRÜNDLMAYER, EICHINGER, HIEDLER, HIRSCH, JURTSCHITSCH, LOIMER, SCHLOSS GOBELSBURG. DAC for GRÜNER V, RIES.

Klosterneuburg Wag r w Wine town in WAG, seat of 1860-founded viticultural college and research institute. *See* next entry.

KMW Abbreviation for KLOSTERNEUBURGER Mostwaage ("must level"), Austrian unit denoting must weight, ie. sugar content of grape juice. 1°KMW = 4.86°Oe (*see* Germany). 20°Bx = 83°Oe.

Knoll, Emmerich Wach w ★★★★ 05 06 07 08 10 Defining WACH producer of intense, *long-lived Ries*, GRÜNER V at any level. Notable Auslese. Best when mature.

Kollwentz Burgen r w ★★★ 10 11 Andi K is a national champ, equally famous for accessible CHARD, BLAUFRÄNKISCH and CAB blends.

So ingrained are wine and song in Viennese life: Heurigen Lieder a distinct genre.

Kracher Burgen sw ★★★★ 05 07 08 10 15 Botrytis specialist in ILLMITZ with world fame for TBA (*see* Germany) from various varieties. Nouvelle Vague series is oak-matured.

Kremstal (r) w Wine region and DAC for GRÜNER V, RIES. Top: MALAT, MOSER, NIGL, SALOMON-UNDHOF, STIFT GÖTTWEIG, WEINGUT STADT KREMS.

Krutzler Burgen r ★★★ 11 13 15 Powerful reds from BLAUFRÄNKISCH. Icon: Perwolff.

Lagler Wach w ★★★ Precise RIES, GRÜNER V, rare NEUBURGER SMARAGD.

Laurenz V Kamp ★★★ GRÜNER V specialist in four distinct styles aimed at international market.

Leithaberg Burgen Important DAC on n shore of NEUSIEDLERSEE, limestone and schist soils. Red restricted to BLAUFRÄNKISCH, some of Austria's best, whites can be GRÜNER V, PINOT BL, CHARD or NEUBURGER.

Lenikus Vienna w ★★ Impressive VIENNA newcomer with gd GEMISCHTER SATZ from Bisamberg.

Lesehof Stagård Krems w ★★★ Young winemaker; stellar range of single v'yd RIES.

Loimer, Fred Kamp (r) w sp ★★★★ 10 11 13 Pioneer of bio: evolving, expressive RIES, GRÜNER V, esp single-v'yds Heiligenstein, Steinmassl. Now gaining a name for PINOT N and *lovely sparkling*.

Malat Krems w ★★★→★★★★ Always excellent, thrilling RIES, GRÜNER V, esp single-v'yds Gottschelle, Silberbichl.

Mantlerhof Krems w ★★★ Rich aromatic GRÜNER V from bio-farmed loess soils.

Mayer am Pfarrplatz Vienna (r) w ★★ VIENNA institution, out in Heiligenstadt village. HEURIGE where Beethoven wrote his 3rd symphony, now tourist heaven. Wine/food both gd.

Mittelburgenland Burgen r DAC (since 2005) on Hungarian border: structured, age-worthy BLAUFRÄNKISCH. Producers: GESELLMANN, J HEINRICH, Kerschbaum, WENINGER.

Moric Burgen ★★★ 10 11 13 15 Groundbreaking producer with deserved cult following for BLAUFRÄNKISCH; note single-v'yds Neckenmarkt, Lutzmannsburg.

Moser, Lenz Krems ★→★★ Austria's largest producer and négociant.

Muhr Carn r w ★★★ Chief revivalist of dolomite limestone Spitzberg slope. Poetic, fragrant BLAUFRÄNKISCH. Entry-level Samt & Seide, age-worthy Spitzberg and Liebeskind.

Netzl, Franz & Christine Carn ★★★ Father-daughter team specializing in top ZWEIGELT, esp Haidacker. Gd red blends and value Rubin CARN Zweigelt.

Neumayer Trais w ★★★ Est quality producer of graceful, concentrated GRÜNER V, RIES.

Neumeister V Stei ★★★★ 12 15 Stellar estate for SAUV BL, esp single-v'yds Klausen, Moarfeitl. Look for Stradener Alte Reben. Also notable GEWÜRZ.

Neusiedlersee (Lake Neusiedl) Burgen Largest European steppe-lake and nature reserve on Hungarian border. Lake mesoclimate and humidity key to botrytis development. Eponymous DAC limited to ZWEIGELT.

Niederösterreich (Lower Austria) Northeast region comprising three parts: Danube (KAMP, KREM, TRAIS, WACH, WAG), WEIN (ne) and CARN, THERM (s). 59% Austria's v'yds.

Nigl Krems w ★★★★ Stylish, juicy, taut RIES, GRÜNER V, esp Privat bottlings.

Nikolaihof Wach w ★★★★ 08 10 13 15 Historic estate with pioneering bio ethos in

Mautern on s bank of Danube. Always penetrating, thrilling, pure RIES, GRÜNER V, esp late Vinothek releases.

Nittnaus, Anita & Hans Burgen r w sw ★★★★ Perennial bio performer with fragrant, powerful but sinuous red. Single-v'yd BLAUFRÄNKISCH from Tannenberg, Jungenberg. Lovely entry-level Kalk & Schiefer. Note PANNOBILE blend, Comondor.

Nittnaus, Hans & Christine Burgen ★★★ Fine, elegant reds, esp ZWEIGELT Heideboden, BLAUFRÄNKISCH Edelgrund; red blend Nit'ana. Super TBA, Eiswein.

Ott, Bernhard Wag w ★★★★ Invariably impressive, salty, savoury GRÜNER V from bio wunderkind, esp Rosenberg, Spiegel, Stein v'yds.

ÖTW (Österreichische Traditionsweingüter) Kamp, Krems, Trais, Wag Private association working on v'yd classification. Currently 62 member estates and 81 classified sites. *See* ERSTE LAGE, GROSSE LAGE; excludes WACH.

Pannobile Burgen Association of nine progressive NEUSIEDLERSEE quality growers centred on Gols. Pannobile bottlings may only use indigenous reds (ZWEIGELT, BLAUFRÄNKISCH, ST-LAURENT), whites only PINOTS BL, GR, CHARD. Members: ACHS, BECK, HEINRICH, NITTNAUS, PITTNAUER, PREISINGER.

Pfaffl Wein r w ★★→★★★ 13 14 15 16 Large, enterprising négociant and grower. Gd RES wines but famed for ultra-successful brand The Dot Austrian Pepper, Austrian Cherry, etc.

Pichler, Franz X Wach w ★★★★ 08 10 13 15 Storied estate of incisive, long-lived *Ries*, GRÜNER V, from top WACH sites. Cult RIES Unendlich.

Pichler, Rudi Wach w ★★★★ 10 13 15 17 Magnificent, clear-cut RIES, GRÜNER V from top sites Achleiten, Steinriegl.

Pichler-Krutzler Wach w ★★★ Reliably energetic and thrilling *Ries* from WACH single v'yds, esp In der Wand, Kellerberg.

Pittnauer, Gerhard Burgen r ★★★→★★★★ Constantly evolving, cutting-edge talent for reds. World-class ST-LAURENT, MashPitt orange wine and fun pét-nat.

Prager, Franz Wach w ★★★★ 10 13 14 15 Visionary, long-est philosopher-winemaker taking WACH to the future with thrilling RIES, GRÜNER V.

Preisinger, Claus Burgen r ★★★ Unconventional, trend-setting PANNOBILE member of rare talent from easy entry-level to complex single-v'yd.

Prieler Burgen r w ★★★★ Famed for powerful, long-lived BLAUFRÄNKISCH requiring bottle-age, esp Marienthal, Goldberg; leading light for equally long-lived PINOT BL.

Proidl, A&F Krems w ★★★ Reliably brilliant RIES, GRÜNER V, both from Ehrenfels. Look out for library Ries releases.

Rebenhof S Stei w ★★★ Exciting, vivid skin-fermented whites from bio luminary Hartmut Aubell.

Reserve (Res) Attribute for min 13% alc and prolonged (cask) ageing.

Ried V'yd. As of 2016 compulsory term for single-v'yd bottlings.

Rust Burgen r w dr sw Well-preserved fortified C17 town on NEUSIEDLERSEE. Noisy nesting storks. Famous Ruster AUSBRUCH. FEILER-ARTINGER, SCHRÖCK, TRIEBAUMER.

Sabathi, Hannes S Stei w ★★★→★★★★ Expressive, age-worthy SAUV BL. Notable single-v'yd Pössnitzberg.

Salomon-Undhof Krems w ★★★→★★★★ Always stylish, elegant, slender GRÜNER V, RIES, single-v'yds Kögl, Pfaffenberg, Wachtberg. Fun pét-nat, summery pink fizz.

Gelber Muskateller is best base for G'spritzter, an ideal thirst-quencher.

Sattlerhof S Stei w ★★★ 10 12 15 World-class, long-lived, creamy SAUV BL, MORILLON, esp single-v'yds Kranachberg, Sernauberg.

Sax Kamp w r sp ★★★ Winemaking twins make uncommonly savoury GRÜNER V.

Schauer S Stei w ★★★ Exacting, filigree whites. Notable PINOT BL Höchtemmel.

Schiefer & Domaines Kilger Burgen r ★★★ Savoury, unusual, individualistic BLAUFRÄNKISCH in EISENBERG, esp Szapary.

Schilcher W St Racy, peppery rosé of local importance from indigenous Blauer Wildbacher grape, speciality of w STEI. Also try sparkling version.

Schilfwein (Strohwein) Sweet wine made from grapes dried on reeds from NEUSIEDLERSEE. *Schilf* = reed, *Stroh* = straw.

Schloss Gobelsburg Kamp r w dr sp sw ★★★★ 13 14 15 16 Cistercian-founded mansion and estate making exquisite RIES, GRÜNER V under quality-champion Michael Moosbrugger. Notable Tradition series and single-v'yd Gaisberg, Heiligenstein, Lamm, Renner. Fine sparkling. Also elegant ZWEIGELT, PINOT N.

Schlumberger sp C19 Sekt pioneer. Today high-volume, value producer of traditional-method fizz. Try sparkling GRÜNER V.

Schmelz Wach w ★★★ Exquisite, authentic wines strangely below the radar.

Schröck, Heidi Burgen (r) w sw ★★★ World-class RUST producer of exquisitely concentrated AUSBRUCH, also notable dry FURMINT.

Seewinkel Burgen ("Lake corner") nature reserve and region e of NEUSIEDLERSEE; ideal conditions for botrytis.

Smaragd Wach Ripest category of VINEA WACHAU, min 12.5% alc but can exceed 14%, dry, potent, age-worthy. Often botrytis-influenced but dry. Named after emerald (=Smaragd) lizard.

Spätrot-Rotgipfler Therm Blend of ROTGIPFLER/Spätrot (ZIERFANDLER). Aromatic, weighty, textured. Typical for GUMPOLDSKIRCHEN. *See* Grapes chapter.

Spitz an der Donau Wach w Bijou town at cool, w end of WACH on Danube. Famous v'yds Singerriedel, 1000-Eimerberg. GRITSCH MAURITIUSHOF, HIRTZBERGER, LAGLER.

Stadlmann Therm r w sw ★★→★★★ Exquisite ZIERFANDLER/ROTGIPFLER, esp single-v'yds Tagelsteiner, Mandelhöh. Fragrant, evocative PINOT N.

Steiermark (Styria) Most s region of Austria, known for aromatic, expressive dry whites, esp SAUV BL. *See* S STEI, V STEI, W STEI.

Steinfeder Wach Lightest VINEA WACHAU category for dry wines of max 11.5% alc. Named after fragrant Steinfeder grass. Increasingly difficult/impossible to produce in warming conditions.

Stift Göttweig w ★★→★★★ Prominent hilltop Benedictine abbey surrounded by v'yds; quality ethos, crystalline RIES, GRÜNER V from single-v'yds Gottschelle, Silberbichl.

Strobl, Clemens Wag r w Quality-focused newcomer making slender wine on WAG loess and gravel. Expressive PINOT N.

Südsteiermark (South Styria) STEI region close to Slovenian border, famed for light but highly aromatic MORILLON, MUSKATELLER, SAUV BL from steep slopes. DAC (2018). Best growers: SABATHI, SATTLERHOF, TEMENT, WOHLMUTH.

Tement, Manfred S Stei w ★★★ 12 13 15 17 Benchmark, long-lived SAUV BL, MORILLON. Top sites Grassnitzberg, Zieregg. Look out for Zieregg RES SAUV BL made only in exceptional vintages.

Thermenregion Nied r w Spa region e of VIENNA centred on GUMPOLDSKIRCHEN, home to indigenous ZIERFANDLER, ROTGIPFLER; historic PINOT N hotspot. Producers: Alphart, HARTL, JOHANNESHOF REINISCH, STADLMANN.

Tinhof, Erwin Burgen r w ★★★ Quiet but stellar bio LEITHABERG estate with exquisite

---

Styrian high jinks

Despite lying in the deep s and on the border to Slovenia, the STEI is known as "the green heart of Austria". Deservedly so: the lushness of vegetation in these often steep hills is striking. So is the climate: cooler, more marginal than rest of Austria and therefore prime country for world-beating aromatic wines, esp SAUV BL. The oak-aged single-vineyard wines have more in common with Pessac-Léognan than with Marlborough but are far more subtle, fresh. If you haven't tasted Stei Sauv Bl, do so. Best: NEUMEISTER, SATTLERHOF, SCHAUER, TEMENT, WOHLMUTH.

BLAUFRÄNKISCH, esp Gloriette; ST-LAURENT, esp Feuerstieg. Nutty, rounded whites from Golden Erd v'yd: PINOT BL, NEUBURGER.

Traisental Nied Tiny district s of KREMS. Notable for prevalence of limestone soils lending finesse. Top: HUBER, NEUMAYER.

Trapl, Johannes Carn r ★★→★★★ Star youngster with rare talent for poised, floral BLAUFRÄNKISCH, esp Sitzerberg and Pinot-esque ZWEIGELT.

Triebaumer, Ernst Burgen r (w) (sw) ★★★★ 08 09 10 12 Iconic RUST producer, blazed trail for indigenous reds, esp BLAUFRÄNKISCH in 80s (Mariental). V.gd AUSBRUCH.

Roast chestnuts are traditional pairing to young, fresh Schilcher. Cools your fingers...

Tschida Burgen r w Cult star in natural wine circles. Notable ZWEIGELT/CAB SAUV Himmel auf Erden.

Umathum, Josef Burgen r w dr sw ★★★★ 13 15 Bio pioneer, now a legend. Exceptionally elegant reds. Probably Austria's best ZWEIGELT: single-v'yd Hallebühl. Also BLAUFRÄNKISCH Kirschgarten.

Velich w sw ★★★ SEEWINKEL producer of Austria's cult CHARD Tiglat. Great sweet too.

Veyder-Malberg Wach ★★★ Rigorous boutique producer of pure RIES, GRÜNER V.

Vienna (Wien) (r) w Capital boasting 637 ha v'yds within city limits. Ancient tradition, reignited quality focus. Local field-blend tradition enshrined as DAC GEMISCHTER SATZ (2013). *Heurigen among vines; visit a must.* Best: CHRIST, LENIKUS, WIENINGER, Zahel.

Vinea Wachau Wach Pioneering quality WACH growers' association founded 1983. Strict charter with three-tier ripeness scale for dry wine: FEDERSPIEL, SMARAGD and STEINFEDER.

Vulkanland Steiermark (Southeast Styria) (r) Formerly Süd-Oststeiermark, DAC (2018), famous for GEWURZ. Best: NEUMEISTER, Winkler-Hermaden.

Wachau Nied Danube region of world repute for age-worthy RIES, GRÜNER V. Top: ALZINGER, DOMÄNE WACHAU, F PICHLER, HIRTZBERGER, JAMEK, KNOLL, NIKOLAIHOF, PICHLER-KRUTZLER, PRAGER, R PICHLER, Tegernseerhof, VEYDER-MALBERG.

Wachter-Wiesler, Weingut Burgen r ★★★ Young star of EISENBERG for muscular but graceful BLAUFRÄNKISCH, also WELSCHRIESLING.

Wagentristl Burgen r (w) ★★→★★★ Inspiring, intuitive youngster with wondrously elegant reds, esp PINOT N, BLAUFRÄNKISCH.

Wagram Nied (r) w Region just w of VIENNA, incl KLOSTERNEUBURG. Deep loess soils ideal for GRÜNER V and increasingly also PINOT N. Best: BAUER, Leth, OTT, STROBL.

Weingut Stadt Krems Krems r w ★★→★★★ KREMS municipal wine estate with 31 ha v'yds within city limits. Real focus on quality and sites.

Weinviertel (r) w ("Wine Quarter") Austria's largest wine region, with 13,858 ha between Danube and Czech border, eponymous DAC for GRÜNER V. Region once slaked VIENNA's thirst, now quality counts. Try: EBNER-EBENAUER, GROISS, GRUBER-RÖSCHITZ, PFAFFL.

Weninger, Franz Burgen r (w) ★★★★ 10 13 15 16 17 Long-lived BLAUFRÄNKISCH from single-v'yds Hochäcker, Kirchholz.

Werlitsch S Stei w ★★★ Ewald Tscheppe is one of S STEI's natural wine stars. Try SAUV BL/CHARD blend Ex Vero.

Weststeiermark (West Styria) Small wine region specializing in SCHILCHER. New DAC (2018).

Weszeli Kamp w Expressive, spicy GRÜNER V, esp Schenkenbichl. Fab RIES.

Wieninger, Fritz Vienna r w sp ★★→★★★ 13 14 VIENNA leading light; bio benchmark GEMISCHTER SATZ from Nussberg, Rosengartl. Great PINOT N. Viennese HEURIGE among Nussberg vines an institution.

Wohlmuth S Stei w ★★★ World-class STEI producer of dazzling CHARD, RIES, SAUV BL, esp single-v'yds Edelschuh, Gola, Hochsteinriegl.

England

We focus on sparkling here because that's where most of the fun is, but more and more still wines are making the grade, even reds in a year like 2018. The fizz style is super-fresh, often with notes of apples and flowers; and longer lees-ageing is giving balancing roundness. Quality varies greatly, but best are world-class. There's a new level of de luxe wines too: impressive, but prices at this level are, as everywhere, about market-making. Abbreviations: Berkshire (Berks), Buckinghamshire (Bucks), Cornwall (Corn), East/West Sussex (E/W S'x), Hampshire (Hants), Herefordshire (Heref).

Black Chalk Hants ★★★ Precise, elegant; flavoursome complex wines, delicate and tense; esp gd Wild Rose rosé, firm, v. pale.

Bluebell Vineyard Estates E S'x ★★ Blanc de Blancs is best bet here: subtle, balanced. Others can be a bit heavy; or richer, depending on your palate.

Bolney Estate W S'x ★ Classic Cuvée is best bet at this long-est v'yd.

Breaky Bottom E S'x ★★★ Wonderful elegance and precision from a tiny, long-est v'yd. Everything is lovely. Cuvée Oliver Minkley named after a comedian who told jokes to the BB sheep.

Bride Valley Dorset Steven and Bella Spurrier's v'yd nr sea. Super-freshness is aim, not time on lees. Appley Brut, gd Bella rosé and Crémant. Gd still CHARD.

Busi Jacobsohn E S'x New; vegan but not organic; unusual flavours. Wait and see.

Camel Valley Corn ★★ Juicy, pretty rosé, and flavoursome, weighty vintage based on SEYVAL BL.

Chapel Down Kent Big producer, underperforming wines. ★★ Kit's Coty, single-v'yd Blanc de Blancs and Coeur de Cuvée more interesting but super-pricey. Kit's Coty BACCHUS is among best.

Coates & Seely Hants ★★★ Lovely ripe, brisk Brut Res NV and savoury, spicy Rosé NV. Rich, creamy Blanc de Blancs. Lots of expertise, v. assured. Wines age well: 09 Rosé in fine fettle.

Cottonworth Hants ★★ Gd length and depth here. Classic Cuvée is best.

Kent grapes can be riper in a cool yr, S'x fresher in a hot one: clay vs chalk.

Court Garden E S'x ★★ Gd rich style with bit of power. Subtle, taut Blanc de Blancs; elegant, biscuity Classic Cuvée; rich Blanc de Noirs. Rosé is red-fruited, crunchy.

Denbies Surrey Large and commercial, marked toastiness. Well-organized tourism: new v'yd hotel.

Digby Hants, Kent, W S'x ★★★ Well-aged, assured wines (from bought-in grapes) made by Dermot SUGRUE under contract at WISTON, so gd pedigree. Res Brut crisp and taut; vintage Rosé ripe, red-fruit flavours. Leander Pink NV by appt to Henley Regatta.

Exton Park Hants ★★★ Gets better every year. Elegance balanced at quite low dosage. Lovely tense Rosé NV; rich, energetic Blanc des Noirs. *Half-bottles* too.

Grange, The Hants Newcomer to watch from the opera house. Rich, pretty Classic, structured pink. Made at HATTINGLEY VALLEY.

Greyfriars Surrey ★★ Best are balanced Classic Cuvée 13 and Blanc de Blancs 13. NV from CHARD/PINOTS N/M also worth a look.

Gusbourne Kent, W S'x ★★★ Gets better and better: ripeness, poise, precision, finesse. New de luxe cuvée v. elegant, tight; for release 2020ish. Still PINOT N 18 is remarkable.

Hambledon Vineyard Hants ★★★ Accomplished wines, lovely balance and richness. Sleek and complete Classic Cuvée; subtle and powerful Rosé. Mconhill label cheaper.

Harrow & Hope Bucks ★★ Subtle, flavoursome Brut Res, sleek, poised Blanc de Blancs, pale, spicy, raspberry Brut Rosé.

Hart of Gold Her ★★ Complex, layered, serious wine, and distinguished. With a wacky label.

Hattingley Valley Hants ★★★ Classic Res v.gd, pure and deep; Rosé 14 has lovely poise. Blanc de Blancs needs time, becomes savoury.

Henners E S'x ★★ Classic ripeness, balanced Brut, Brut Res, cherry-spice Rosé.

Herbert Hall Kent ★★ Promising wines, often a bit young, Rosé is pretty.

Hoffmann & Rathbone ★★ S'x-based, using bought-in fruit for gd toasty, creamy Blanc de Blancs; try also Classic Cuvée.

Hush Heath Estate Kent ★★★ Flagship Balfour Brut Rosé as gd as ever. Elegant Skye Blanc de Blancs, superb Winemaker's wines (cellar door only), v.gd still incl crunchy PINOT M.

Jenkyn Place Hants ★★ Fairly austere wines but balanced, made by Dermot SUGRUE. Classic Cuvée is precise; Rosé has a nice bitter herbs edge.

Langham Wine Estate Dorset Promising, punchy Blanc de Blancs, slightly green Classic Cuvée. To watch.

Leckford Estate Hants ★ From Waitrose's own estate, vinified by RIDGEVIEW. Fair balance, a bit short on oomph.

Nyetimber W S'x ★★★★ Best-known of top wines. NV incl 5 yrs res wines gives depth. Structured, fine Rosé, resonant Classic Cuvée, pure and deep Blanc de Blancs. Single-v'yd Tillington a treat and prestige cuvée 1086 for when you win the lottery.

Plumpton College E S'x ★★ UK's only wine college; own wines to gd standard incl excellent Rosé NV.

Pommery England Hants Champagne's 1st venture into England being made at HATTINGLEY VALLEY until its own vines come on stream; so far promisingly fresh, savoury, elegant.

Raimes Hants ★★ Gd wines, made at HATTINGLEY VALLEY. Classic is taut, ripe, concentrated; Blanc de Noirs rich, silky.

Rathfinny E S'x ★★★ Vast investment paying off with precise, assured wines, esp Blanc de Blancs. Sinuous Blanc de Noirs. Gd still whites. Cradle Valley still is sappy, v. pretty.

Ridgeview E S'x ★★ Well-made and enjoyable, reliable and gd-value. Various cuvées. tops are Blanc de Blancs, Blanc de Noirs, Rosé de Noirs. Contract maker of several brands

Simpsons Kent Appealing, smoky Classic Cuvée and well-judged Blanc de Noirs, plus gd still wines from this Canterbury newcomer.

Stopham Estate W S'x ★ Gd still PINOT BL – all sappy nuts and spice. PINOT GR is richly exotic.

Sugrue E S'x ★★★★ Dermot S is winemaker at WISTON, where he also makes Black Dog Hill, DIGBY, JENKYN PLACE under contract. He makes one (Chardonnay-dominant) wine a yr from his own v'yds, and it's glorious. Beg, borrow or steal.

Trotton W S'x Newcomer with toasty, citrus flavours in Spectacular Sparkling and gd still BACCHUS/PINOT GR.

Westwell Kent Elegant, precise fizz; gd, interesting still Ortega.

Wiston W S'x ★★★★ Brilliant winemaker Dermot SUGRUE makes tense, chalky Brut and steely vintage Blanc de Blancs that needs ageing. Uses oak cleverly for extra complexity. Beautiful, fine Blanc de Noirs.

Wyfold Oxon ★★ Tiny (1-ha) Champagne-variety v'yd at 120m (394ft) above sea level in Chilterns. Part-owned by hands-on Laithwaite family.

Central & Southeast Europe

More heavily shaded areas are the wine-growing regions.

Prague ○
CZECHIA
SLOVAKIA
Bratislava ○ *Danube*
Budapest ○
MOLDOVA
Ljubljana ○
Zagreb
Danube
HUNGARY
ROMANIA
Chișinău ○
Prut
SLOVENIA
Drava
CROATIA
Timișoara ○
Olt
BOSNIA-HERZEGOVINA
Sava
Belgrade ○
Bucharest ○
Danube
Split ○
SERBIA
Adriatic Sea
Sarajevo ○
MONTENEGRO
Dubrovnik ○
Podgorica ○
BULGARIA
Varna ○
Sofia ○
Black Sea
Plovdiv ○
Skopje ○
NORTH MACEDONIA
Tirana ○
ALBANIA

Abbeviations used in the text:

Bal	Balaton	N Hun	North Hungary
Cri & Mar	Crișana & Maramures	N/S Pann	North/South Pannonia
Cro Up	Croatian Uplands	Pod	Podravje
Dalm	Dalmatia	Pos	Posavje
Dan P	Danubian Plain	Prim	Primorje
Dob	Dobrogea	Sl & CD	Slavonia & Croatian Danube
Is & Kv	Istria & Kvarner	Thr L	Thracian Lowlands
Mold	Moldovan Hills	Tok	Tokaj
Mun	Muntenia & Oltenia Hills	Trnsyl	Transylvania

HUNGARY

Furmint, Hungary's great white grape, is flying the flag for a new generation of dry wines. From its Tokaj homeland it's now appearing on wine lists the world over – showing that Hungary's volcanic hills are capable of world-class dry whites. Indeed, dry whites are becoming more important than the region's legendary sweet Aszú. Elegant reds from local Kékfrankos (aka Blaufränkisch) and Kadarka, plus Cabernet Franc suit today's drinking – go with, not dominate, food. Most are still not exported, but a trip to Budapest is recommended, not least because it is now central Europe's culinary capital, with seven Michelin stars.

Aszú Tok Botrytis-shrivelled grapes and the resulting sweet wine from TOK. From 2014 rules were changed. Legal min for an Aszu is now 120g/l residual sugar, equivalent to 5 PUTTONYOS (similar to Château d'Yquem). 6 Puttonyos wine is even richer. Gd Aszú in 99' 05 06 07 08 09 13' 16 17. Not much botrytis in 11 12 15 18. 19 looks superb for Aszú and v.gd for dry.

Aszú Essencia / Eszencia Tok Term for 2nd-sweetest TOK level (7 PUTTONYOS+), not permitted since 2010. Do not confuse with ESSENCIA/ESZENCIA.

Badacsony Bal ★★→★★★ Volcanic slopes n of Lake Balaton; full, rich whites, esp rare and age-worthy KÉKNYELŰ. Look for Gilvesy, Laposa, Sabar, Szászi, *Szeremley*, Villa Sandahl, Villa Tolnay.

Balassa Tok w dr sw ★★→★★★ Excellent Mézes-Mály FURMINT, Villő ASZÚ and Bomboly SZAMORODNI.

Balatonboglár Bal r w dr ★★→★★★ Wine district s of Lake Balaton, also major winery of TÖRLEY. Gd: Budjosó, GARAMVÁRI, IKON, KONYÁRI, Kristinus, Légli Géza, Légli Otto, Pócz.

Barta Tok w dr sw sp ★★→★★★ Highest v'yd in TOK, HQ in MÁD's old Rákóczi mansion. impressive dry whites, esp Öreg Király FURMINT, HÁRSLEVELŰ, v.gd sweet SZAMORODNI, ASZÚ.

Béres Tok w dr sw ★★→★★★ Handsome winery at Erdőbénye producing v.gd ASZÚ and dry wines, esp Lőcse FURMINT, Diókút HÁRSLEVELŰ.

Bikavér r ★→★★★ Means "Bull's Blood". PDO only for EGER and SZEKSZÁRD. Always a blend, min four varieties. In Szekszárd, at least 5% KADARKA is compulsory, with min 45% KÉKFRANKOS, specified oak ageing. Look for: Eszterbauer (esp Tüke), HEIMANN, Meszáros, Sebestyén, TAKLER, Vestergombi, VIDA. Egri Bikavér is majority Kékfrankos and no grape more than 50%, oak-aged for min 6 mths. Superior and Grand Superior restricted yield, 12 mths in barrel. Best for Egri Bikavér: BOLYKI, Csutorás, DEMETER, GÁL TIBOR, Grof Buttler, ST ANDREA, Thummerer.

Bock, József S Pann r ★★→★★★ In VILLÁNY, making rich, full-bodied, oaked reds. Try: Bock CAB FR Fekete-Hegy, Bock & Roll, SYRAH, Capella Cuvée.

Bolyki N Hun r p w ★★ Dramatic winery in a quarry in EGER, great labels. V.gd EGRI CSILLAG, Meta Tema, rosé, Indián Nyár ("Indian Summer") and excellent BIKAVÉR, esp Bolyki & Bolyki.

Csányi S Pann r p ★→★★ Largest winery in VILLÁNY. Premium Ch Teleki, Kővilla labels best.

Hungarian alphabet has 44 letters. Magyar is one of hardest languages to learn.

Csopak Bal N of Lake BAL. Protected status for top v'yd OLASZRIZLING. Look for: FIGULA (esp Sáfránkert, Köves), Béla És Bandi, Dobosi, Homola, Jasdi (esp single-v'yd selections), St Donát.

Degenfeld, Gróf Tok w dr sw ★★→★★★ Improved Tarcal estate with luxury hotel. Sweet best Andante botrytis FURMINT, 6 PUTTONYOS, SZÁMORODNI.

Demeter, Zoltán Tok w dr sw sp ★★★★ Benchmark cellar in TOKAJ for elegant, intense dry wines, esp Boda, Veres FURMINTS; excellent Szerelmi HÁRSLEVELŰ, lovely Őszhegy MUSCAT. V.gd PEZSGŐ (sp). Eszter late-harvest cuvée, superb ASZÚ.

Dereszla, Ch Tok w dr sw ★★→★★★ Excellent ASZÚ. Gd dry FURMINT, Kabar. Also reliable PEZSGŐ. Rare flor-aged dry SZAMORODNI Experience.

DHC (Districtus Hungaricus Controllatus) Term for a more limited sub-category of Protected Designation of Origin (PDO): Oltalom alatt álló Eredetmegjelölés (OEM) in Hungary.

Disznókő Tok w dr sw ★★★→★★★★ Prominent "First Growth" estate; restaurant and winery tours. Fine expressive ASZÚ, superb *Kapi* cru in top yrs. Also gd-value late-harvest and Édes (sw) SZAMORODNI, new Inspiration dry white.

Dobogó Tok (r) w dr sw ★★★ Named for horses' "clip-clop" sound. Benchmark ASZÚ and late-harvest Mylitta, excellent long-lived dry FURMINT and *pioneering Pinot N* Izabella Utca.

Dúló Named single v'yd or cru. Top *dűlő* incl Betsek, Király, Mézes-Mály, Nyúlászó, SZENT TAMÁS, Úrágya.

Duna Duna Great Plain. Districts: Hajós-Baja (try Sümegi, Koch – also owns VinArt in VILLÁNY), Csongrád, Kunság (Frittmann, Font).

Eger N Hun ★→★★★ Burgundian-style reds and elegant fresh whites, esp noted for Egri BIKAVÉR. Try: BOLYKI, Csutorás, Gróf Buttler, DEMETER, *Gál Tibor*, Kaló Imre (natural), KOVÁCS NIMRÓD, Pók Tamás, ST ANDREA, Thummerer.

Egri Csillag N Hun "Star of Eger". Dry white blend of Carpathian grapes.

Essencia / Eszencia Tok ★★★★ Legendary, luscious free-run juice from ASZÚ grapes, occasionally bottled, alc usually well below 5%, sugar (and price) off the charts. Reputed to have medicinal/aphrodisiac properties.

Etyek-Buda N Pann Dynamic region, expressive, crisp whites, gd sparklers and promising PINOT N. Leading producers: ETYEKI KÚRIA, György-Villa, HARASZTHY, Nyakas, Kertész, Rókusfalvy.

Etyeki Kúria N Pann r p w ★★ Leading winery in ETYEK-BUDA, producing v.gd SAUV BL, elegant PINOT N and KÉKFRANKOS.

Figula Bal w ★★→★★★ Family winery nr BAL, notable v'yd selections of OLASZRIZLING, esp Öreghegy, Sáfránkert, Szákas. Excellent Köves (w blend).

Gál Tibor N Hun r w ★★ Try appealing EGRI CSILLAG, fine KADARKA and vibrant, modern TiTi BIKAVÉR. Hugely improved with new cellar.

Garamvári Bal r p w dr sp ★→★★★ Leading producer of bottle-fermented fizz (previously Ch Vincent). Try Optimum Brut, FURMINT Brut Natur, PINOT N Evolution Rosé. Gd Garamvári range, esp SAUV BL, IRSAI OLIVÉR. Lellei label is consistent, great-value varietals.

Gere, Attila S Pann r p ★★★→★★★★ Standard-setting family winery in VILLÁNY making some of country's best reds, esp elegant Solus MERLOT, intense Kopar Cuvée, top Attila barrel selection. V.gd Fekete-Járdovány from rare historic grape.

Gizella Tok w ★★★ Superb small winery. Impeccable v'yd selection dry whites, delicious SZAMORODNI.

Grand Tokaj Tok w dr sw ★→★★ Biggest producer in TOK. Relaunched 2013 with new winery and winemaker (Karoly Áts, ex-ROYAL TOKAJI). Appealing Arany Késői Late-Harvest, Dry FURMINT Kővágó DŰLŐ, v.gd Szarvaz ASZÚ 6 PUTTONYOS.

Haraszthy N Hun r w ★★ Beautiful estate at ETYEK-B, v.gd SAUV BL, zesty Sir Irsai (w).

Heimann S Pann r w ★★ Impressive family winery in SZEKSZÁRD, esp intense Barbár and BIKAVÉR. Fine KADARKA, Alte Reben KÉKFRANKOS.

Hétszőlő Tok w dr sw ★★★ Historic cellar and stunning v'yd, owned by Michel Reybier of Cos d'Estournel (B'x). Noted for delicacy. 6 PUTTONYOS ASZÚ impressive.

Heumann S Pann r p w ★★→★★★ German/Swiss-owned estate in Siklós making great KÉKFRANKOS Res, CAB FR Trinitás, delicious rosé, appealing SYRAH.

Hilltop Winery N Pann r p w ★★ In Neszmély. Meticulous, gd-value DYA varietals, Hilltop, Moonriver labels for export. V.gd Kamocsay Premium range (esp CHARD, Ihlet Cuvée).

Holdvölgy Tok w dr sw ★★→★★★ Super-modern winery in MÁD, noted for complex dry wines (esp Vision, Expression), plus v.gd sweet.

Ikon Bal r w ★★ Well-made wines from KONYÁRI and former Tihany abbey v'yds. Try Evanglista CAB FR.

Juliet Victor Tok w sw ★★→★★★ Ambitious new investment by founder of Wizzair. Already impressing with estate and v'yd selection dry FURMINTS (notable Betsek) and superb rich SZAMORODNI.

Hearts of oak

Hungarian oak is joining the roster of French, US, Slavonian and the rest in the world's cellars. Forests here are sustainably managed according to laws laid down by King Zsigmond in 1456. In n Hungary, the shallow volcanic soils are almost exclusively home to the superior tight-grained *Quercus petraea*. Big-name barrel-makers have invested in Hungarian coopers and the oak now appears in top wines as far afield as California, Burgundy, Italy as well as Hungary.

Kikelet w dr sw ★★★ Beautifully balanced wines from a small family estate in Tarcal owned by a French winemaker and her Hungarian husband.

Királyudvar Tok w dr sw sp ★★★ Bio producer in old royal cellars at Tarcal. Highly regarded for FURMINT Sec, Henye PEZSGŐ, Cuvée Ilona (late-harvest), flagship 6 PUTTONYOS Lapis ASZÚ.

Konyári Bal r p w ★★→★★★ Gd family estate nr BALATON. Try DYA rosé; Loliense (r w), lovely Szarhegy. Top reds: Jánoshegy KÉKFRANKOS, Páva.

Kovács Nimród Winery N Hun r p w ★★→★★★ EGER producer, jazz-inspired labels. Try Battonage CHARD, Blues KÉKFRANKOS, Monopole Rhapsody, 777 PINOT N, NJK.

World's 1st classified wine region: Tok, 1737. Pre-dates B'x or Burgundy.

Kreinbacher Somló w dr sp ★★→★★★ Hungary's best PEZSGŐ (fizz); Prestige Brut (pure FURMINT) is world class, also v.gd Classic Brut (based on Furmint). V.gd dry *Juhfark*, Öreg Tőkék (old vines).

Mád Tok Historic wine-trading town with superb v'yds, cellars and Mád Circle of leading producers: Árvay, Áts, BARTA, Budaházy, Demetervin (gd Mád FURMINT, Úrágya dűlő), HOLDVÖLGY, Lenkey, Orosz Gabor, ROYAL TOKAJI, SZENT TAMÁS WINERY, SZEPSY, TOK Classic, Úri Borok.

Malatinszky S Pann r p w ★★★ Certified organic VILLÁNY cellar. Top long-lived Kúria *Cab Fr*, Kövesföld (r). Gd: Noblesse labels.

Mátra N Hun ★→★★ Region for decent-value, fresh whites, rosé and lighter reds. Better producers: Balint, Benedek, Gábor Karner, NAG, NAGYRÉDE, Szőke Mátyás, Nagygombos (rosé).

Mór N Pann w ★→★★ Small region, famous for fiery local *Ezerjó*. Try Czetvei Winery.

Nagyréde N Hun (r) p w ★ Gd-value, commercial DYA varietals under Nagyréde and MÁTRA Hill labels.

Oremus Tok w dr sw ★★★→★★★★ Perfectionist Tolcsva winery owned by Spain's Vega Sicilia: top ASZÚ; v.gd late-harvest and dry FURMINT *Mandolás*.

Pajzos-Megyer Tok w dr sw ★★→★★★ Back-on-form, in Sárospatak. Megyer label for modern dry and late-harvest (sw) varietals. Pajzos for premium, esp age-worthy ASZÚ, lovely late-harvest HARSLEVELŰ.

Pannonhalma N Pann r p w ★★→★★★ 800-yr-old Abbey, focused, aromatic whites: RIES (esp Prior), SAUV BL, TRAMINI. Lovely *Hemina* (w), full-bodied PINOT N.

Patricius Tok w dr sw sp ★★→★★★ Beautiful estate. Consistent dry FURMINT, esp Selection, gd late harvest Katinka, ASZÚ. Appealing PEZSGŐ.

Pendits Winery Tok w dr sw ★★ Demeter-certified bio estate. Luscious long-ageing ASZÚ, pretty, dry DYA MUSCAT.

Pezsgő Hungarian for sparkling – a growing trend. Since 2017 must be bottle-fermented if PDO TOK.

Puttonyos (putts) Traditional indication of sweetness in TOK ASZÚ. Optional since 2013 (*see* ASZÚ). Historically a *puttony* was a 25kg bucket or hod of Aszú grapes, sweetness determined by number of Puttonyos added to a 136-litre barrel (*gönci*) of base must or wine.

Royal Tokaji Wine Co Tok dr sw ★★★→★★★★ MÁD winery that led renaissance of TOK in 1990 (I am a co-founder). Excellent 6-PUTTONYOS single-v'yd bottlings: esp Betsek, *Mézes-Mály*, Nyulászó, *Szent Tamás*. Blue and Red Labels are benchmark 5-puttonyos blends. Try appealing gd-value Late Harvest plus v.gd dry FURMINT, The Oddity.

St Andrea N Hun r p w ★★★ Leading name in EGER for modern, high-quality BIKAVÉR (Áldás, Hangács, Merengő). Gd white blends: Napbor, Örökké and delicious Szeretettel rosé. Flagships: Mária (w) and Nagy-Eged-Hegy (r). New Axios Bikavér made by son.

Sauska S Pann, Tok r p w ★★→★★★★ Immaculate wineries in VILLÁNY, TOK. V.gd

KADARKA, KÉKFRANKOS, CAB FR and impressive red blends, esp Cuvée 7 and Cuvée 5. Also Sauska-Tok with focus on excellent dry whites, esp Medve and Birsalmás FURMINTS. V.gd PEZSGŐ Extra Brut rosé and white.

Somló Bal Dramatic extinct volcano famous for long-lived, austere white *Juhfark* ("sheep's tail"), FURMINT, HÁRSLEVELŰ, OLASZRIZLING. Region of small producers, esp Fekete, Györgykovács, Kolonics, Royal Somló, Somlói Apátsági, Somlói Vándor, Spiegelberg. Bigger TORNAI, *Kreinbacher* also v.gd.

Sopron N Pann On Austrian border overlooking Lake Fertő. KÉKFRANKOS most important. Bio *Weninger* is excellent, maverick Ráspi for natural wines. Also Luka, Pfneiszl, Taschner.

Szamorodni Name of Polish origin for TOK made from whole bunches, with botrytis or not. Gaining popularity since 3 and 4 PUTTONYOS ASZÚ stopped – seen as more authentically Tok than late-harvest. *Édes* or sweet style is min 45g/l sugar (usually sweeter), 6 mths oak-ageing. Try BALASSA, BARTA, Bott, Gizella, HOLDVÖLGY, JULIET VICTOR, KIKELET, OREMUS, Pelle, SZENT TAMÁS, SZEPSY. Best dry (*száraz*) versions are flor-aged like Sherry; try CH DERESZLA, Karádi-Berger, *Tinon*.

Szekszárd S Pann Famous for ripe, rich reds. Increasing focus on BIKAVÉR, KÉKFRANKOS, reviving lighter KADARKA. Dúzsi (rosé), Eszterbauer (Nagyapám Kadarka, Tüke Bikavér), HEIMANN, Mészáros, Remete-Bor (Kadarka), Sebestyén (Ivan-Volgyi Bikavér), TAKLER, Vesztergombi (Csaba's Cuvée, Turul), Szent Gaál, VIDA.

Szent Tamás Winery Tok w sw ★★★ Significant winery (with handy café) named after top DŰLŐ in village of MÁD. Mád brand gd entry-point wines from bought-in grapes. Szent Tamás now focusing on own v'yds with excellent DŰLŐ dry wines, esp Dongó, Percze. Excellent SZAMORODNI; Dongó, Nyulászó. Top quality ASZÚ.

Szepsy, István Tok w dr sw ★★★★ Brilliant, soil-obsessed, no-compromise 17th and 18th-generation TOK producer in MÁD. Now focusing on dry FURMINT (esp Urágya, Betsek, Nyúlászó DŰLŐ) and reviving sweet SZAMORODNI (v.gd 13). Superb ASZÚ, amazing rare ESZENCIA.

Szeremley Bal w dr sw ★★ Pioneer in BADACSONY. Intense, fine RIES, *Szürkebarát*, (aka PINOT GR), age-worthy rare KÉKNYELŰ.

All 1st names must be from government list, which is why so many winemakers called László, István, Zoltán.

Takler S Pann r p ★★ Super-ripe, supple SZEKSZÁRD reds. Decent, gd-value, entry-point red, rosé. Best: Res selections of CAB FR, KÉKFRANKOS.

Tinon, Samuel Tok w dr sw ★★★ Bordelais in TOK since 1991. Exceptional v'yd-selection dry FURMINTS. Distinctive complex ASZÚ, v. long maceration, barrel-ageing. Wonderful dry flor-aged *Szamorodni*.

Tokaj Nobilis Tok w dr sw ★★★ Fine small producer run by Sarolta Bárdos, one of TOK's inspirational women. Excellent dry Barakonyi HÁRSLEVELŰ, FURMINT, v.gd SZAMORODNI, rare Kövérszőlő Edes (sw).

Tokaj / Tokaji Tok ★★→★★★★ Tokaj is the town and wine region; Tokaji the wine. Recommended producers without individual entries: Alana, Árvay, Áts, Bardon, Basilicus, Bodrog Borműhely, Bott Pince, Budahazy, Carpinus, Demetervin, Erzsébet, Espák, Füleky, Hommona Attila, Karádi-Berger, Kvaszinger, Lenkey, Orosz Gábor, Pelle, Sanzon, Zombory, Zsadányi.

Törley r p w dr sp ★→★★ Chapel Hill is major export brand. Well-made, gd-value DYA international and local varieties. Major fizz producer (esp *Törley*, Gala, Hungaria labels), v.gd classic method, esp François President Rosé Brut, CHARD Brut. György-Villa for top selections.

Tornai Bal w ★★ 2nd-largest SOMLÓ estate. Gd-value entry-level varietals, excellent Top Selection range FURMINT, Juhfark.

Túzkó S Pann r w ★★ Antinori-owned estate. Gd CAB FR, KÉKFRANKOS, MERLOT, TRAMINI.

Vida S Pann r ★★ V.gd BIKAVÉR, Bonsai (old-vine) KADARKA, Hidaspetrc KÉKFRANKOS, La Vida.

Villány S Pann Most s wine region. Noted for serious ripe B'x varieties esp CAB FR which now has own designation: VILLÁNYI FRANC. Juicy examples of *Kékfrankos*, PORTUGIESER. High-quality: ATTILA GERE, Bock, CSÁNYI, Gere Tamás & Zsolt (Aureus Cuvée), HEUMANN, Hummel, Jackfall, Janus, Kiss Gabor, Lelovits (Cab Fr), *Malatinszky*, Polgar, Riczu (Symbol Cuvée), Stier (MERLOT, Villányi Cuvée), Ruppert, *Sauska*, Tiffán, *Vylyan*, WENINGER-GERE.

Villányi Franc S Pann New classification for CAB FR from VILLÁNY. Premium version has restricted yield, 1 yr in oak. Super-premium from 2015 is max 35 hl/ha.

Vylyan S Pann r p ★★→★★★ Red specialist making v.gd v'yd selections, esp Gombás PINOT N, Mandolás CAB FR, Montenuovo, Pillangó MERLOT. *Duennium Cuvée* is flagship red. Also delicious rare Csoka.

Weninger N Hun r w ★★★ Standard-setting bio winery in SOPRON run by Austrian Franz Weninger Jr. Single-v'yd *Steiner Kékfrankos* is superb. SYRAH, CAB FR and red Frettner blend also impressive. Intriguing Orange Zenit.

Weninger-Gere S Pann r p ★★→★★★ Joint venture between WENINGER Sr and ATTILA GERE. Excellent CAB FR, tasty Tinta (TEMPRANILLO), Cuvée Phoenix, DYA fresh rosé.

BULGARIA

Bulgaria's wine scene today is dynamic, with increasing interest in local grapes and creative blends, and several new winery projects. Use of barrels is improving too, helped by a local market learning to like wine, not just oak. Exports largely confined to bargain-basement Merlot and Cabernet, which is a shame as the country has much more to offer.

Alexandra Estate Thr L r p w ★★ 60-ha estate; gd VERMENTINO, rosé, impressive Res.

Angel's Estate Thr L r p w ★★ Ripe, polished reds and smooth whites under Stallion label, also impressive Deneb.

Bessa Valley Thr L r p ★★★ Pioneering estate nr Pazardjik. Smooth, rich reds; new white blend 18. Try Enira, v.gd SYRAH and Enira Res, excellent *Grande Cuvée*.

Better Half Thr L r p w ★★★ True garage winery using amphorae. V.gd red blends, CHARD, MARSANNE/ROUSSANNE, VERMENTINO.

Black Sea Gold Thr L r p w ★ Large Black Sea coast winery. Better labels: Golden Rhythm, Pentagram, Salty Hills, Vera Terra.

Bononia Dan P r p w ★★ New winery in historic brewery nr Vidin. V.gd GAMZA, SAUV BL, VIOGNIER.

Borovitsa Dan P r w ★★★ Handcrafted terroir wines in far nw. Dux is long-lived flagship. V.gd: MRV (Rhône w), Cuvée Bella Rada (RKATSITELI), GAMZA (Black Pack), Sensum, rare local grape Bouquet.

Boyar, Domaine Thr L r p w ★→★★★ Pioneering large winery. DYA entry-level labels like Deer Point, Bolgare, via mid-range Elements, Platinum, Quantum, to top single-v'yd Solitaire (MERLOT). Owns boutique Korten winery for v.gd Merlot, SYRAH, CAB FR, esp Grand Vintage.

Bratanov Thr L r w ★★★ Low-intervention family estate in Sakar. V.gd Tamianka, CHARD, SYRAH and red blends.

Cyrillic alphabet was invented in C9 by two Bulgarian monks, Cyril and Methodius.

Burgozone Dan P r w ★★ Family estate overlooking Danube. Gd whites, fine reds, esp VIOGNIER, SAUV BL, Eva, Esperanto (r).

Damianitza Thr L r p w ★★ Holistic producer in STRUMA VALLEY. Try Ormano (w), Volcano SYRAH, Uniqato and flagship Kometa.

Domain Menada Thr L r p w ★ 3rd-biggest producer; cheerful Tcherga blends.

Dragomir Thr L r p w ★★→★★★ Intense long-lived reds, esp Pitos, flagship RUBIN Res, plus reliable Sarva (r p w). CAB FR impressive.

Eolis Thr L r w ★★ Tiny estate with bio principles. V.gd VIOGNIER, SEM and SYRAH, Inspiration blend.

Katarzyna Thr L r p w ★→★★ Large modern winery in s nr Greece making polished supple reds. Try Encore SYRAH, MAVRUD, Res.

Logodaj Thr L r p w sp ★★★ In STRUMA VALLEY with winemaking from protégé of Riccardo Cotarella. V.gd bottle-fermented Satin, esp rosé.

Bulgaria produces most of world's rose oil: takes 1000 flowers to make 1g.

Maryan Dan P r w ★→★★ Family winery making v.gd Res (r), Ivan Alexander (r), orange DIMIAT.

Medi Valley Thr L r p w ★★ Highest commercial v'yd in Bulgaria, plus plot nr Vidin. Try Great Bulgarian, Incanto Black, MELNIK 55, VIOGNIER.

Midalidare Estate Thr L r w ★★→★★★ Immaculate boutique winery. Precise whites and v.gd reds. New elegant Brut sparkling.

Minkov Brothers Thr L r p w ★→★★ Boutique arm of large Bulgarian producer. Try value CAB SAUV, Res Cab Sauv, Le Photografie PINOT N, flagship Oak Tree.

Miroglio, Edoardo Thr L r p w sp ★★→★★★ Italian-owned estate at Elenovo. Gd bottle-fermented sparkling. Exciting PINOT N in all styles, incl age-worthy Heritage, v.gd flagship Soli Invicto, also Elenovo CAB FR, MAVRUD.

Neragora Thr L r w ★★ Organic estate producing gd MAVRUD and Ares blend.

Orbelia Thr L r p w ★★ Family winery in STRUMA VALLEY. Bright Sandanski MISKET, gd Via Aristotelis (w), fine CAB FR.

Orbelus Thr L r p w ★★ Organic STRUMA winery. Gd unoaked whites, MELNIK 55.

Power Brands Thr L r p w ★ Formerly Vinprom Peshtera. Owns Villa Yambol (try gd-value Kabile range) and New Bloom for easy-drinking Pixels, Verano Azur labels, plus characterful F2F reds.

Rossidi Thr L r p w ★★→★★★ Boutique winery nr Sliven. Fine, part concrete-egg fermented CHARD. V.gd RUBIN, excellent SYRAH. Intriguing orange GEWURZ.

Rumelia Thr L r w ★★ V.gd MAVRUD specialist. Erelia, unoaked Merul, Merul Res.

Salla Estate Dan P r w ★★ Lively, precise whites, esp Vrachanski MISKET, RIES. Plus elegant CAB FR.

Santa Sarah Thr L r w ★★★ Quality pioneer, now with own estate. Bin reds v.gd; long-lived Privat is flagship. Smooth Petite Sarah (r), appealing No Saints rosé.

Slavyantsi, Vinex Thr L r p w ★→★★ "Fair for Life" certified for work with local Roma. Reliable budget varietals and blends, esp Leva brand.

Struma Valley Thr L Lovely scenery in Bulgaria's warmest region, dynamic wineries and well organized for tourism. Focus on local grapes: MELNIK 55, Sandanski

---

Rising stars

Wine estates barely existed 15 yrs ago in Bulgaria, but now they're everywhere. Also new projects with v'yds and rented space. Look for: Augeo (Ruen), Bendida (unoaked MAVRUD, RUBIN), Ch Copsa (Axl, Zeyla MISKET), Four Friends (Black Shadow, CHARD, MOURVÈDRE), Glushnik (CAB FR, Caladoc rosé), Haralambievi, Ivo Varbanov (Chard, VIOGNIER), Pink Pelican (RKATSITELI/FETEASCǍ ALBǍ), Rousse Wine House (Chard, GRENACHE rosé, Vrachanski Misket), Seewines (Colorito w, Ayano r), Staro Oryahovo (Varnenski Misket, Vrachanski Misket), Stefan Pirev Wines (Chard Kosara, Eager r blend), Stratsin (MERLOT, rosé, SAUV BL), Uva Nestum, Varna Winery (fresh DYA w, fruity PINOT N), Via Vinera (DIMIAT, Mavrud, Red Misket), Villa Yustina (4 Seasons range, Special Res, Monogram Rubin/Mavrud), Yalovo (Misket blend, Rubin, sp).

MISKET, Shiroka Melnik. Names (without own entries) to watch: Abdyika, Augeo, Kapatovo, Libera, Rupel, Seewines, Via Verde, Zlaten Rozhen.

Svishtov Dan P r p w ★→★★ Much-improved large producer close to Danube with Italian consultancy. Try Gorchivka, Imperium.

Terra Tangra Thr L r p w ★★ Large estate in Sakar, certified organic red v'yds. Gd MAVRUD (r p), MALBEC, serious Roto.

Tohun Dan P r p w ★→★★ Bright refreshing whites and rosé, esp Greus CHARD, Tohun rosé, promising Tohun CAB SAUV/MERLOT.

Tsarev Brod Dan P r w ★★ New dynamic estate. Try pét-nat RIES, rare local Gergana, Amber CHARD, Amber Ries, complex SAUV BL Res. V.gd Ries Icewine.

Villa Melnik Thr L r p w ★★ Family winery, focus on local grapes esp MELNIK, MAVRUD. Gd orange SAUV BL, and serious Res and Hailstorm labels.

Yamantiev's Thr L r w ★→★★ Sound commercial wines, plus excellent top Marble Land selections and Yamantiev's Grand Res CAB SAUV.

Zagreus Thr L r p w ★★ MAVRUD in all styles from acacia-fermented rosé to complex Amarone-style Vinica from semi-dried grapes.

Zelanos Dan P r w ★★ Pristine new winery. Try fresh Red MISKET, PINOT GR, elegant Z series PINOT N and CAB FR.

SLOVENIA

S lovenia's reputation for quality continues to rise, even as its vineyards shrink in area, to just 15,630 ha of often dramatically steep slopes in a green landscape of forests overlooked by the Julian Alps. It has picked up fame as a hotspot for orange and skin-contact whites, but these are only a small niche; most are classic whites, complex sparklers, fine reds and luscious sweet wines, produced by a dynamic group of winemakers.

Albiana Pos r p w ★★ Family with beautiful v'yds in Dolenjska. Try: MODRA FRANKINJA, Zeleni Silvanec.

Batič Prim r p w sw ★★ Bio/natural wines in VIPAVA. Top-selling rosé, also PINELA, REBULA, Angel blends, Valentino (sw).

Bjana Prim sp ★★→★★★ V.gd traditional-method PENINA from BRDA, esp fine Brut Rosé, NV, Brut Zero.

Blažič Prim w ★★→★★★ From BRDA. Long-ageing, complex REBULA esp Selekcija. Blaž in top yrs.

Brda Prim Top-quality district. Many leading wineries: BJANA, BLAŽIČ, Dobuje, DOLFO, EDI SIMČIČ, Erzetič, FERDINAND, JAKONČIČ, KLET BRDA, KRISTANČIČ, Medot, Mulit, MOVIA, Reya, ŠČUREK, Žanut. Orange wines from KABAJ and Klinec.

Burja Prim r w ★★★★ Exciting organic VIPAVA estate. Excellent Burja Bela, Burja Noir (PINOT N), Burja Reddo based on SCHIOPPETTINO.

Čotar Prim r w ★★ Intriguing organic/natural wines from KRAS, esp Vitovska (w), MALVAZIJA, SAUV BL, TERAN, Terra Rossa (r).

Cviček Pos Increasingly unfashionable, sharp, light red blend, based on Žametovka.

Slovenia is properly green. Over 60% forested and 53% protected landscape.

Dolfo Prim r w ★★→★★★ V.gd Spirito PENINA, Gredic (r w), CAB SAUV Res.

Dveri-Pax Pod r w sw ★★→★★★ Historic Benedictine-owned estate nr Maribor (name means Gate of Peace). Crisp, bright, gd-value whites. V.gd old-vine selections, also v.gd FURMINT PENINA.

Edi Simčič Prim r w ★★★★ Standard-setter in BRDA. Superb reds: Duet Lex, barrel-selection Kolos. Excellent whites: REBULA, SAUV BL, Triton Lex; lovely Kozana CHARD.

Erzetič Prim r p w ★★ Young winemaker making v.gd amphora wines, esp Amfora Belo (w), PINOT GR.

Ferdinand Prim r w ★★→★★★ Small estate making fine Sinefinis sparkling with Prinčič (Italy) and v.gd Brutus (r), Epoca REBULA.

Gašper Prim r w sp ★★ Brand of Slovenia's top sommelier with KLET BRDA. V.gd MALVAZIJA, PENINA, PINOT GR, REBULA. Refined CAB FR.

Gross Pod w ★★★→★★★★ Amazing terroir wines. Superb Gorca and Iglič FURMINTS, Colles SAUV BL, RIES.

Guerila Prim r w ★★ Bio producer in VIPAVA. V.gd PINELA, Retro (w) blend.

Slovenians love their honey – over 90,000 beekeepers.

Istenič Pos sp ★★ Gd fizz specialist: Prestige Extra Brut, Gourmet Rosé, N°1 Brut, Barbara Sec.

Istria Coastal zone partly in Croatia; main grapes: MALVAZIJA, REFOŠK. Best: Bordon, Korenika & Moškon, MonteMoro, Pucer z Vrha, Rodica (organic), Rojac, SANTOMAS, Steras, VINAKOPER.

Jakončič Prim r w sp ★★★ V.gd BRDA producer, esp Bela (w), Carolina REBULA, Grand Selection, Rdeča (r).

Joannes Pod r w ★★ RIES specialist nr Maribor. Also fresh light PINOT N.

Kabaj Prim r w ★★★ French-directed. Noted for long-aged Amfora, also skin-contact REBULA, Ravan (FRIULANO), Corpus, serious MERLOT.

Klet Brda Prim r w sp ★★→★★★ Slovenia's largest co-op, surprisingly gd and forward-thinking. Try Bagueri v'yd selections. Also v.gd Quercus varietal whites, unoaked Krasno, Colliano for US. Excellent flagship A+ (r w).

Kobal Pod w ★★ Personal vision of Bojan K. V.gd FURMINT, Black label SAUV BL.

Kogl Pod r p w ★★ Historic estate nr Ormož, from 1542. Vibrant, precise whites esp Mea Culpa AUXERROIS, Ranina (aka BOUVIER).

Kras Prim District on Terra Rossa soil in PRIM. Best-known for controversial TERAN denomination. Try Vinakras (esp Prestige).

Kristančič Prim r w ★★ Family producer in BRDA. Try Pavó wines from old vines.

Kupljen Pod r w ★★ Dry pioneer nr Jeruzalem. Try Aldebaran RIES, Loona.

Marof Pod r w ★★→★★★ Pioneering estate in Prekmurje. All wines now oak-fermented. Try Breg SAUV BL, Kramarovci CHARD, Mačkovci BLAUFRÄNKISCH.

Movia Prim r w sp ★★★ High-profile bio winery led by charismatic Aleš Kristančič. Excellent v. long-lived Veliko Belo (w), Rdeče (r), showstopping Puro Rosé (sp). V.gd MODRI PINOT. Orange Lunar spends eight full moons on skins.

Pasji Rep Prim r w ★★ Organic VIPAVA estate, now run by son. Much-improved, refined wines, esp MALVAZIJA, Jebatschin blends, PINOT N.

Penina Name for quality sparkling wine (charmat or traditional method). Trendy.

Podravje Largest wine region covering Štajerska and Prekmurje in e. Best for crisp dry whites, gd sweet, reds typically lighter styles from Modra Frankinja (aka BLAUFRÄNKISCH), PINOT N.

Posavje Region in se. V.gd sweet, esp Mavretič, Prus, Šturm. Improving PENINA and dry, esp in Dolenjska, Bizeljsko-Sremič regions.

PRA-VinO Pod sw 70s pioneer of private production. Best for sweet, incl Icewine (*ledeno vino*), botrytis wines from LAŠKI RIZLING, RIES, ŠIPON.

> **Orange segment**
> Slovenia is still an orange wine centre, esp close to border with Collio, where skin-contact whites were 1st revived by Gravner, Prinčič, Radikon. Time on skins varies from just a few days to wks or mths for a full orange style, often with min sulphites, sometimes in amphora. Long-lived, complex, astringent wines perfect with food are the result. BATIČ, ČOTAR, ERZETIČ, GROSS, JNK, KABAJ, Klinec, Mlecnik, MOVIA, Ražman, plus CLAI, KABOLA, Kozlović, ROXANICH, TOMAC over the Croatian border.

> ### Dolenjska
> Improving region now focusing on better quality as sales of traditional light red CVIČEK decline. Promising sparkling from local Žametovka (Domaine Slapšak), Rumeni Plavec. Watch: MODRA FRANKINJA from KOBAL (esp superb Luna), ALBIANA, Dular Selekcija, Frelih, Klet Krško, Kozinc.

Primorje Region in w covering Slovenian Is & Kv, BRDA, VIPAVA, KRAS. Aka Primorska.

Puklavec Family Wines Pod w sp ★★→★★★ Large family winery offering *consistent crisp aromatic whites* in Puklavec & Friends and Jeruzalem Ormož ranges. V.gd Seven Numbers label and new Ena Dva Tri FURMINT.

Pullus Pod p w ★★ Fresh whites: RIES, SAUV BL. Excellent "G" wines; LAŠKI RIZLING (SW), Rumeni MUSCAT.

Radgonske Gorice Pod sp ★→★★ Producer of bestselling Slovenian sparkler Srebrna (silver) PENINA, classic-method Zlata (golden) Penina. Now making new fizz in dark to avoid light-strike.

Santomas Prim r p w ★★→★★★ V.gd for *Refošk* and REFOŠK/CAB SAUV blends, esp Antonius from 60-yr-old vines. Also gd MALVAZIJA, SYRAH.

Ščurek Prim r p w sw ★★→★★★ Family estate in BRDA, five sons. Gd CAB FR, Jakot, PINOT BL, REBULA. Best wines from local grapes: Kontra, Pikolit, Stara Brajda (r w).

Simčič, Marjan Prim r w sw ★★★★ Single-v'yd Opoka range is world-class. Also gd Selekcija range, Teodor blends always v.gd. Leonardo (sw) consistently great.

Štajerska Pod Large e region incl important districts of Ljutomer-Ormož, Maribor, Haloze. Crisp, refined whites and top sw. Best (without individual entries): Doppler, Frešer, Gaube, Heaps Good Wine, Krainz, Miro, M-vina (esp ExtremM SAUV BL), Oskar, Šumenjak, Valdhuber, Zlati Grič.

Steyer Pod w sw ★★ TRAMINER specialist in ŠTAJERSKA.

Sutor Prim r w ★★★ Excellent small producer from VIPAVA. Try Sutor White from REBULA/MALVAZIJA, also v.gd Malvazija, fine CHARD, elegant red.

Tilia Prim r w ★★→★★★ "House of Pinots" in VIPAVA since co-owner gained PhD studying PINOT N. V.gd PINOT GR, appetizing SAUV BL. Juicy Vipava Pinot N.

Verus Pod w ★★★ Fine, focused, vibrant whites, esp v.gd FURMINT, crisp SAUV BL, flavoursome PINOT GR, refined RIES.

Vinakoper Prim r w ★★ Large producer nr coast. Look for Capo D'Istria and young MALVAZIJA, REFOŠK under Rex Fuscus, Capris labels.

Vipava Prim Valley noted for cool breezes in PRIM. Recommended without own entry: Denčina (PINOT N), Fedora (Goli Breg, Zelen), Jangus (MALVAZIJA, SAUV BL), JNK (orange), Lepa Vida (Malvazija, oOo orange), Miška (PINELA), Mlečnik (orange/natural), Štokelj (Pinela),Vina Krapež (excellent Lapor Belo).

Slovenia, Croatia still fighting: is Teran a denomination or a grape? Gd for lawyers.

CROATIA

Tourists flock to Croatia – nearly 20 million in 2018 – attracted by the climate, the beautiful coastline and islands, and perhaps the *Game of Thrones* factor (filmed in Dubrovnik). As a result Croatia imports five times more wine than it exports; you have to visit to taste the best of it. Don't expect it to be cheap; there are plenty of customers and a huge sense of local pride. With over 100 local grape varieties, several confined to specific islands or vineyards, and around 500 commercial winemakers, it will take several visits.

Ahearne Dalm r p w ★★ British Master of Wine Jo Ahearne makes elegant PLAVAC MALI, deep Rosina Darnekuša rosé, Wild Skins (w) on HVAR.

Arman, Franc Is & Kv r w ★★ 6th-generation family winery. V.gd TERAN, DYA MALVAZIJA, skin-contact Malvazija Classic. Gd MERLOT, CAB FR.

Arman, Marijan Is & Kv r w ★★ Excellent MALVAZIJA, esp Grand Cru, Res. Gd TERAN.

Badel 1862 r w ★→★★ Major drinks group. Best: smooth reds under Korlat label. Also sound PLAVAC MALI, DINGAČ 50°.

Benvenuti Is & Kv r w ★★★ Standard-setting family winery at Motovun. V.gd reds, esp Caldierosso, TERAN esp Santa Elisabetta. Benchmark fresh MALVAZIJA, complex Anno Domini (w), gorgeous San Salvatore MUŠKAT (sw).

BIBICh Dalm r w ★★→★★★ Focus on local grapes, esp Debit (w), plus SYRAH. Try Lučica single-v'yd and sweet Ambra. Also local rarity Lasin.

Bire Dalm r w ★★ Impressing with revival of local Grk on Korcula.

Bolfan Cro Up r p w ★→★★ Bio/natural wine. Gd Primus RIES, SAUV BL, PINOT N.

Bura-Mrgudić Dalm r p ★★ Rich, weighty wines from Pelješac; renowned for Bura, DINGAČ and Mare POSTUP.

Cattunar Is & Kv r w ★★ Hilltop estate. Gd MALVAZIJA from four soils. Excellent late-harvest Collina.

Clai Is & Kv r w ★★ Admired for skin-contact orange, esp Sveti Jakov MALVAZIJA, Ottocento blends.

Coronica Is & Kv r w ★★ Standard-setting producer, esp barrel-aged Gran MALVAZIJA and benchmark Gran TERAN.

Dalmatia Rocky coastline and lovely islands s of Zadar, many exciting wineries. Tourism hotspot, Dubrovnik is capital.

Damjanić Is & Kv r w ★★→★★★ Increasingly impressive family winery. V.gd Borgonja (aka BLAUFRÄNKISCH), MALVAZIJA, Clemente (r w).

Dingač Dalm Full-bodied reds from PLAVAC MALI. PDO, on Pelješac peninsula. Try: Benmosche, BURA-MRGUDIĆ, Kiridžija, Lučić, Madirazza, Matuško, Miličić, SAINTS HILLS, Skaramuča, Vinarija Dingač.

Enjingi, Ivan Sl & CD w sw ★★ Influential natural winemaker in SLAVONIJA. Noted for GRAŠEVINA and long-lived Venje.

Fakin Is & Kv r w ★★★ Exciting young *garagiste* winemaker impressing with MALVAZIJA, esp La Prima, Il Primo TERAN.

Feravino Sl & CD r p w ★→★★ Sound entry-level wines. Much-improved Miraz.

Galić Sl & CD r w ★★→★★★ High-tech winery: v.gd GRAŠEVINA, CHARD, Bijelo 9, Crno 9.

Geržinić Is & Kv r p w ★★ Brothers making v.gd TERAN (r p), MALVAZIJA, SYRAH, MUŠKAT Zuti (aka Yellow Muscat).

Gracin Dalm r p ★★→★★★ Dramatic rocky coastal v'yds nr Primošten making country's **best Babić**, Opol (p), Prošek (sw) guided by Prof. Leo Gracin.

Grgić Dalm r w ★★→★★★ Napa Valley legend Mike Grgich, ex-Ch Montelena (*see* California), returned to Croatian roots to make PLAVAC MALI, rich POŠIP on Pelješac peninsula with daughter and nephew.

Hvar Dalm Beautiful island, UNESCO protection for world's oldest continuously-cultivated v'yd. Noted for PLAVAC MALI. Gd: AHEARNE, Carič, Dubokovič, PZ Svirče, TOMIČ, ZLATAN OTOK.

Istria & Kvarner

N Adriatic peninsula and nearby islands. MALVAZIJA main grape. Gd CAB SAUV, MERLOT, TERAN. Banko Mario, BENVENUTI, Capo, CATTUNAR, CLAI, CORONICA, Cossetto, DAMJANIĆ, Degrassi (VIOGNIER, CAB FR), Dubrovac, Deklić, Domaine Koquelicot (Belaigra Grand Cru), FAKIN, FRANC ARMAN, Franković, GERŽINIĆ, KABOLA, KOZLOVIĆ, MARIJAN ARMAN, MATOŠEVIĆ, Medea, MENEGHETTI, Misal Peršurić (sparkling), Novacco, PILATO, Piquentum, Radovan (REFOŠK, Merlot), Rossi (Malvazija, Templara), ROXANICH, SAINTS HILLS, Sirotić, Tomaz, Trapan, Zigante. On islands: Boškinac (Cuvee), KATUNAR.

Iločki Podrumi Sl & CD r p w ★★ Ambitious renovated historic winery with C15 cellar. Try: Premium GRAŠEVINA, TRAMINAC, Principovac range.

Kabola Is & Kv r p w ★★→★★★ Immaculate organic estate. V.gd MALVAZIJA as fizz, young wine, cask-aged Unica, Amfora. Tasty DYA rosé; v.gd TERAN.

Katunar Is & Kv r w ★★ Leading producer of Žlahtina found only on island of Krk. Try Sv. Lucija. Also gd PLAVAC MALI.

Korta Katarina Dalm r p w ★★ →★★★ Modern winemaking from Korcula. Excellent POŠIP, PLAVAC MALI, esp Reuben's Res.

Istria is as gd for olives as vines: some (Istrians?) say Istrian olive oil is world's best.

Kozlović Is & Kv r w ★★★ *Benchmark Malvazija* in all its forms, esp exciting, complex Santa Lucia. Superb Santa Lucia Crna; MUŠKAT Momjanski.

Krajančić Dalm w ★★ Specialist in exciting white grape POŠIP: try Sur Lie, Intrada.

Krauthaker, Vlado Sl & CD r w sw ★★★ Pioneering producer from KUTJEVO, esp GRAŠEVINA Mitrovac, Vidim, Izborna Berba. Also v.gd PINOT N, Zelenac.

Kutjevo Cellars Sl & CD w dr sw ★★ 800-yr-old cellar in Kutjevo town; noted for gd GRAŠEVINA, esp De Gotho, Turković, lovely Icewine.

Matošević Is & Kv r w ★★→★★★ Benchmark MALVAZIJA, esp Alba, Alba Robinia (aged in acacia), Antiqua. V.gd Grimalda (r w).

Meneghetti Is & Kv r w ★★ Sleek, well-made blends (r w), fine precise MALVAZIJA.

Miloš, Frano Dalm r p ★★★ Much admired for legendary, long-lived Stagnum, also easier PLAVAC and rosé.

Pilato Is & Kv r w ★★ Consistent family winery; v.gd MALVAZIJA, PINOT BL, TERAN, MERLOT.

Postup Dalm 2nd-oldest v'yd designation nw of DINGAČ. Full-bodied, rich red from PLAVAC MALI. Try: Madirazza, Marija Mrgudić, Miličić, Vinarija Dingač.

Prošek Dalm Historic sweet made from sun-dried local grapes in DALM; 1st mention 1556. Gd versions: GRACIN, STINA, TOMIČ Prošek Hectorovich.

Roxanich Is & Kv r w ★★→★★★ Natural producer, intriguing orange wines (MALVAZIJA Antica, Ines U Bijelom). complex reds: TERAN Ré, Superistrian Cuvée, MERLOT.

Saints Hills Dalm, Is & Kv r p w ★★→★★★ Two wineries, three locations, consultant Michel Rolland. V.gd Nevina (w); richly fruity PLAVAC MALI St Roko, serious DINGAČ.

Skaramuča Dalm r w ★★ Modern family winery with v'yds in DINGAČ. Focus on gd PLAVAC MALI, also appealing POŠIP.

Slavonija Region in ne, famous for oak, and for whites, esp from GRAŠEVINA. Also gd reds now. Look for Adzić, Antunović, Bartolović, Belje, ENJINGI, FERAVINO, GALIĆ, KRAUTHAKER, KUTJEVO, Orahovica, Zdjelarević.

Stina Dalm r p w ★★→★★★ Dramatic steep v'yds on Brač island; v.gd POŠIP, Vugava, PLAVAC MALI, esp Majstor label. Gd Tribidrag (AKA ZIN), Opol rosé, PROŠEK.

Tomac Cro Up r w sp ★★ Estate nr Zagreb with 200-yr history, famous for sparkling and pioneering amphora wines.

Tomaz Is & Kv r w ★★ Family winery, serious Barbarossa TERAN, textured Avangarde MALVAZIJA, Sesto Senso aged in acacia.

Vrhunsko vino: premium-quality wine; *Kvalitetno vino:* quality wine; *Stolno vino:* table wine. *Suho:* dry; *Polsuho:* semi-dry.

Tomić Dalm r p w ★★ Outspoken personality, bold wines on HVAR; organic PLAVAC MALI. Gd reds, esp Plavac Barrique, PROŠEK Hectorovich.

Veralda Is & Kv r p w ★★ Smooth polished reds are hallmark, plus bright whites and orange amphora.

Vina Laguna Is & Kv r p w ★★ V.gd Festigia and Riserva ranges (notable MALVAZIJA, esp Vižinada). Try Riserva LV (r), TERAN. From red soils close to Adriatic coast.

Zlatan Otok Dalm r p w ★★ Family winery from HVAR with v'yds also at Makarska, Šibenik. Famous for ripe reds, esp BABIĆ, Crljenak, PLAVAC MALI. Gd DYA POŠIP.

BOSNIA & HERZEGOVINA, KOSOVO, NORTH MACEDONIA, SERBIA, MONTENEGRO

The rise of small family wineries, rediscovery of local grapes and a revolution in quality are common stories here. And welcome signs of cooperation through new groups such as Balkan Wine Network and Blaž.

Bosnia & Herzegovina exciting white Žilavka is this complex country's calling card, esp from sunny rocky v'yds of Herzegovina. Juicy supple red Blatina and some v.gd Vranac in s. Andrija, Begič, Brkič, Carska Vina, Hercegovina Produkt, Keža, Nuič, Rubis, Škegro, Tvrdos Monastery, Vilinka, Vino Milas, Vukoje.

Kosovo is recognized as a nation by 114 countries though it still has two sets of wine laws – its own and Serbia's. Now around 15 wineries; Stonecastle and Old Cellar (Bodrum i Vjeter) as biggest, also smaller producers like Kosova, Sefa getting better. Smederevka, WELSCHRIESLING, Vranac, Prokupac most planted grapes.

North Macedonia bottled wines and better quality are industry priorities, moving away from cheap bulk. Vranec (local spelling) is country's flagship, covering nearly 11,000 ha from 25,000 ha in total. Largest wineries are driving quality improvements. Giant Tikveš has a French-trained winemaker, an intensive research programme and impresses with Barovo, Bela Voda single-v'yd wines; gd Special Selection and rich, oaked Domaine Lepovo. Dalvina much improved, esp Armageddon, Dionis, Hermes, Synthesis labels. Stobi gd for Vranec Veritas, Vranec classic, Aminta (r), RKATSITELI, Žilavka. Ch Kamnik is leading boutique winery, with gd 10 Barrels, Cuvée Prestige, Vranec Terroir. Others: Bovin (try A'gupka, Imperator, Dissan), Ezimit (gd-value varietals), Imako (Black Diamond, Montovo River), Lazar (Kratošija, Erigon r), Puklavec Family V'yds (also Slovenia).

Montenegro Vranac dominates here (over 90% plantings) as does 13 Jul Plantaže with 2310 ha in one single site, one of Europe's largest. Wines are pretty gd (esp Vranac in many forms). Small wineries getting better. Lipovac (gd Amfora, Vranac Concept), Sjekloča, Bogojevic, Sefa, Rupice, Vukicevic. Recent research indicates Montenegro may be origin of Kratošija/ZIN.

Serbia has 25,000 ha of registered v'yds and c.400 wineries, many new and tiny. Focus on reinventing former workhorse grapes like Prokupac, obscurities like Bagrina, Seduša and trying newer local grapes like Probus, Morava, Neoplanta. Try: Aleksandrović (Trijumf range, Regent Res, Vizija), Aleksič (Amanet), Bikicki (Orange TRAMINER), Botunjac (Sveti Grai), Budimir (Triada, Svb Rosa, Boje Lila), Chichateau (CHARD, Fabula Mala, Fabula Lagum), Cilić, Čokot (Experiment), Despotika (Morava, Dokaz), Deurić (Probus, Talas, PINOT N), Doja (Prokupac), Dukay-Sagmeister (KADARKA, esp cru selections), Erdevik, Ivanović (Prokupac, No.1/2), Janko (Vrtlog, Zavet Stari, Zapis), Kovačević (Chard, Orange, Aurelius), Lastar (Pinot N, Chard, Tamjanika), Matalj (Kremen Chard, Kremen Kamen), Maurer (Kadarka, Fodor), Pusula (CAB FR), Radovanović (Cab Res, Saga), Šijački (Seduša), Temet (Ergo, Tri Morave), Tonković (Kadarka), Trisic (Dimasid), Virtus (Prokupac, MARSELAN), Zvonko Bogdan (Cuvée No.1, PINOT BL).

CZECHIA

Prague (1.3 million people) welcomes over eight million tourists annually, and a good many are only here for the beer. Until the fall of communism wine was treated as low quality and fit only for mass production, but over the past 30 years wine has become fashionable. The market leader remains the giant Bohemia Sekt Group, maker of the nation's favourite bubbly, while huge investments from the newly wealthy have led to a range of super-modern wineries springing up, mostly in the larger of the two wine regions, Moravia (Mor), which accounts for 96% of the total, although 25-times smaller Bohemia (Boh) near Prague is discovering its potential for Germanic-style Riesling as well as Pinot Noir, the latter having been brought here from Burgundy in the 14th century by Emperor Charles IV. Worth Czeching out...

Baloun, Radomil Mor ★★→★★★ Wide range of highly quaffable wines, all dry. White PINOT N, Blaufränkisch Blanc curiosities.

Dobrá Vinice Mor ★★★ SAUV BL, CHARD, PINOT N, WELSCHRIESLING in *qvevri* from Georgia. Top blends: V-D-B, V-D-Č, also Crème Brut Nature.

Dva Duby Mor ★★★ Dedicated terroirist, bio principles. Granodiorite subsoil of Dolní Kounice is esp suited to BLAUFRÄNKISCH/ST LAURENT grapes. Flagship labels: Vox In Excelso, Rosa Inferni, Ex Opere Operato.

Hartman, Jiří Mor ★★ Small producer of v.gd burgundy-style whites and reds in picturesque wine-cellar settlement.

Lobkowicz, Bettina Boh ★★→★★★ Outstanding PINOT N, classic-method sparklers: RIES, Pinot N Blanc de Noirs, CHARD. V.gd-value entry-level Lady Lobkowicz (r p w).

Mádl, František Mor ★★ Nicknamed *Malý vinař* (small vintner), family-run, top reds Mlask, Cuvée 1+1, also v.gd SAUV BL.

Mikrosvín Mikulov (Nikolsburg) Mor w ★★★ Based at foot of Pálava Hills where best WELSCHRIESLING is traditionally grown.

Stapleton & Springer Mor ★★★ *Remarkable* PINOT N, single-v'yd Čtvrtě, Záhřebenské, also Roučí blend under Jaroslav Springer label.

Stávek, Richard Mor ★★→★★★ Specialist in "raw" and pét-nat. Best labels: Kolberg, Špigle-Bočky, Veselý. Orange wines hit in trendy restaurants/bars (NY, London).

Vinselekt Michlovský Mor ★→★★★ Innovator, technologist, breeder. Extensive range, VOC Pálava a must.

Znovín-Znojmo r w ★→★★ Important wine centre in s, nr Austrian border, emphasis on aromatic whites, esp SAUV BL.

SLOVAKIA

Central European vines dominate, alongside international favourites. 12,000 ha of vineyards divided into six regions: Lesser Carpathians (L Car), Nitra (Nit), Central Slovakia (C Slo), Southern Slovakia (S Slo) and Tokaj (Tok). Château Topoľčianky and Sekt JE Hubert Sereď (both Nit) largest wine and Sekt producers, respectively

Château Belá S Slo ★★★ Fine RIES by Egon Müller (*see* Germany) and Miroslav Petrech joint venture.

Elesko L Car ★★ Large super-modern facility unrivalled in Central Europe.

Fedor Malík Jr L Car ★★ Son continues with still and Modragne classic sparkling.

J&J Ostrožovič Tok ★★★ V.gd Slovak Tok, traditional and modern products.

Karpatská Perla L Car ★★ V.gd wines from immaculately maintained v'yds.

Movino C Slo ★★ Most important Central Slovakia producer, est 1973 in Veľký Krtíš.

Pivnica Brhlovce Nit r p w ★★★ Est 2011 by photographer Ján Záborský. Artisanal production of "volcanic" wines in troglodyte dwellings.

Villa Víno Rača L Car ★★ Emphasis on reds, esp BLAUFRÄNKISCH. Recent relaunch of label Ch Palugyay (r w sp) by great-granddaughter of Jakob von Palugyay.

Víno Matyšák L Car ★→★★★ Large modern winery in Pezinok.

ROMANIA

No newcomer. Romania's Latin culture gives it a long wine history. It still makes more wine than NZ, yet its thirsty home market drinks most of it. Lots of foreign investment has helped raise standards enormously. There are significant areas of good quality local grapes like Fetească Albă, Neagră and Regală, Crâmpoşie and rarities like Negru de Drăgăşani and Novac being rescued from obscurity. Watch this space.

Avincis Mun r w p ★★ Dramatic state-of-art winery in DRĂGĂŞANI. Specialist in Negru de Drăgăşani (also in v.gd Cuvée Grandiflora). Crâmpoşie Selecţionată impresses and fine FETEASCĂ REGALĂ.

Balla Géza Cri & Mar r p w ★★ Estate in Miniş. Best is Stone Wine range from v'yd at 500m (1640ft). V.gd Cadarca (r p), FETEASCĂ NEAGRĂ, FURMINT.

Banat Wine region in w. New: Agape, Crama Aramic, Pivniţele Birăuaş, Thesaurus.

Bauer Winery Mun r p w ★★ Family winery of Oliver, Raluca Bauer (also at PRINCE ŞTIRBEY). V.gd Crâmpoşie, FETEASCĂ NEAGRĂ, PETIT VERDOT. Orange wine pioneer.

Budureasca Mun r w ★→★★ DEALU MARE estate with British winemaker. Consistent Budureasca (esp FUMÉ, TĂMÂIOASĂ, Noble 5), top Origini range.

Catleya Mun r p w ★★ Personal project of CORCOVA's French winemaker; gd Freamăt, excellent Epopée selection.

Corcova Mun r p w ★★ Exceptional v'yds and renovated C19 royal cellar. Try FETEASCĂ NEAGRĂ, SYRAH, appealing SAUV BL, rosé.

Cotnari Mold DOC region, only grows local varieties, esp FETEASCĂ ALBĂ, Frâncuşă, GRASĂ, TĂMÂIOASĂ.

Cotnari Wine House Mold p w ★→★★ Next-generation in COTNARI, 350 ha. Best is Colocviu label (esp GRASĂ de Cotnari, Busuioacă de Bohotin), Grasă Dulce (sw).

Cotnari Winery Mold w sw ★ Former collectivized winery with 1360 ha. Mostly dry, semi-dry whites from local grapes. Aged Collection (sw) *can be impressive.*

Crişana & Maramureş Region in nw. Look for Carastelec (Carassia bottle-fermented sp and gd PINOT N) and organic producer Nachbil (BLAUFRÄNKISCH, RIES, Grandpa).

Dagon Clan Mun r p w ★★ Small quality-focused v'yd; Aussie guidance.

Davino Winery Mun r p w ★★★ →★★★★ Excellent, consistent producer in DEALU MARE. Focus on blends for v.gd, age-worthy Dom Ceptura, Flamboyant, Revelatio (w). Local varieties feature in Monogram label.

Dealu Mare / Dealul Mare Mun Means "The Big Hill". Plans for Romania's 1st DOCG (DOC Garantat) by 2020 with Rezervă and Rezervă Specială categories too. Location of several leading producers.

Dobrogea Nr Black Sea. Incl DOC regions of Murfatlar, Badabag and Sarica Niculiţel.

DOC Romanian term for PDO. Sub-categories incl DOC-CMD: harvest at full maturity, DOC-CT: late-harvest and DOC-CIB: noble-harvest. PGI is Vin cu indicatie geografică or simply IG.

Domeniul Coroanei Segarcea Mun r p w ★→★★★ Historic royal estate. Famous for TĂMÂIOASĂ Roze. Also try Minima Moralia CAB SAUV, Principesa Margareta MARSELAN, Simfonia red blend.

Drăgăşani Mun Dynamic region on River Olt for aromatic whites and intriguing reds. Leading producers: AVINCIS, BAUER, PRINCE ŞTIRBEY.

Gîrboiu, Crama Mold r w ★→★★ 200 ha in earthquake-prone Vrancea, hence Tectonic label (try Şarba) and Epicentrum. Gd Bacanta whites, Cuartz (sp).

Iconic Estate Mun r p w sp ★→★★ Consistent gd-value commercial range, esp La Umbra, Colina Pietra, v'yd selections. Hyperion is top label: try FETEASCĂ NEAGRĂ, CAB SAUV. Also Rhein (sp).

Jidvei Trnsyl w ★→★★ Romania's largest single v'yd with 2460 ha. Best for crisp dry whites, esp Owner's Choice (with Marc Dworkin of Bulgaria's Bessa Valley).

LacertA Mun r w ★★ Quality estate in DEALU MARE. Cuvée IX (r), Cuvée X (w), SHIRAZ.

Licorna Wine House Mun r w ★★ In DEALUL MARE, opened 2013; impressing with Serafim for local grapes and Bon Viveur for international blends.

Liliac Trnsyl r p w sp ★★★ Impeccable Austrian-owned estate. Crisp fine whites, delicious sweet Nectar, Icewine with Kracher (Austria). V.gd super-premium Titan.

Metamorfosis, Viile Mun r w ★★★ Part-Antinori-owned (*see* Italy) estate in DEALU MARE. Top: Cantvs Primvs. V.gd Coltul Pietrei, (esp Negru de DRĂGĂȘANI, PINOT N, SAUV BL), fruit-driven Metamorfosis range.

Moldovan Hills Largest wine region ne of Carpathians. Crisp fresh whites and rosé incl Gramma, Hermeziu. Gd FETEASCĂ NEAGRĂ, Zghihara de Huși under Nativus label from improving Crama Avereşti.

Muntenia & Oltenia Hills Major wine region in s covering DOC areas of DEALU MARE, Dealurile Olteniei, DRĂGĂȘANI, Pietroasa, Sâmbureşti, Stefaneşti, Vanju Mare.

Francesco Illy, coffee-machine inventor, born in Romanian city of Timişoara.

Oprișor, Crama Mun r p w ★★→★★★ La Cetate range consistently gd. Also try Jiana Rosé, vibrant Rusalca Alba, Crama Oprișor CAB SAUV, excellent Smerenie red.

Petro Vaselo Ban r p w ★★ Italian investment in BANAT with organic v'yds. Gd Bendis (sp), Melgris FETEASCĂ NEAGRĂ, Ovas (r). V.gd entry-level Alb, Roşu and rosé.

Prince Ştirbey Mun r p w sp ★★→★★★ Pioneering estate in DRĂGĂȘANI. V.gd dry whites, esp Crâmpoşie Selectionată (still and sparkling), SAUV BL, FETEASCĂ REGALĂ, TĂMÂIOASĂ Sec and local reds (Novac, Negru de Drăgăşani).

Recaş, Cramele Ban r p w ★★→★★★ Romania's most successful exporter. Progressive, consistent wines with longstanding Australian and Spanish winemakers. V.gd-value, bright varietal wines sold under multiple labels incl Calusari, Castel Huniade, Paparuda, Schwaben Wein. Mid-range: Regno Recas, Solo. Excellent premium wines, esp Cuvée Uberland, Selene reds, Solo Quinta.

Sarica Niculiţel Dob r p w ·★★ Raising standards in overlooked region. Gd FETEASCĂ NEAGRĂ, rosé.

S.E.R.V.E. Mun r p w ★★→★★★ 1st private winery in Romania, founded by late Corsican Count Guy de Poix. The de Poix family were rivals of their Ajaccio neighbours, the Bonapartes. Reliable entry-point Vinul Cavalerului. V.gd Terra Romana, esp PINOT N, rosé, Cuvée Amaury (w) and impressive Guy de Poix FETEASCĂ NEAGRĂ. *Cuvée Charlotte* quality red benchmark.

Transylvania Cool mtn plateau encircled by Carpathians. Noted for crisp whites. New producers in Lechinţa zone incl organic Lechberg, Jelna.

Valahorum Mun r w ★★ New premium project in DEALU MARE from owners of Tohani and Mennini from DRĂGĂȘANI.

Villa Vinèa Trnsyl r w ★★ Italian owned Târnave estate. Gd whites, esp Diamant, GEWURZ, KERNER and red blend Rubin.

Vinarte, Domenu Mun r w ★★ V'yds in Sâmbureşti (Castel Bolovanu) and Starmina (Mehedinţi). Best: Nedeea (r), Soare, Sirena Dunarii (sw).

Vişinescu, Aurelia Mun r w ★★ DEALU MARE estate. Try Artizan (w) using local grapes, TĂMÂIOASĂ Dulce. Anima is top label, esp CHARD, Fete Negre.

MALTA

A hot, humid climate makes for robust wines, with reds from Cabernet Sauvignon, Merlot and more Med varieties like Syrah, Grenache, with whites from Sauvignon Blanc, Chardonnay, Chenin Blanc, Moscato. Most growers sell their crop to large wineries, including Delicata, Marsovin, Antinori-owned Meridiana. Wines are also produced from imported Italian grapes. For typically Maltese, look for local grapes Gellewza (light r, also acceptable fizz), Girgentina (w, often blended with Chardonnay). Most celebrated is Marsovin's Grand Maître (Cabernets Sauvignon/ Franc), costly but hardly good value. A handful of boutique wineries, some rustic, but others, such as San Niklaw (Vermentino, Sangiovese, Syrah), make high quality but in very limited quantities.

Greece

G reek wine is, at the moment, among the most exciting in the world. Millennia-old grape varieties are saved from extinction every year. A young generation of winemakers is embracing both new knowledge and old techniques. Extreme vineyards and terroirs, often on remote islands, used to be anathema but are now pushing Greek wine forward. Greek wines are sophisticated, vibrant and utterly delicious. Abbreviations: Aegean Islands (Aeg), Central Greece (C Gr), Ionian Islands (Ion), Macedonia (Mac), Peloponnese (Pelop), Thessaloniki (Thess).

Alpha Estate Mac ★★★ Acclaimed KTIMA in AMYNTEO with largest v'yd investment in Greece. Ktima Alpha (r w) modern classics. Ecosyste range a terroir study, Barba Yiannis XINOMAVRO from century-old vines.

Amynteo Mac (POP) Captivating XINOMAVRO reds, excellent rosés (and sp) from coolest Greek POP – and cold it is.

Argyros Aeg ★★★★ Top SANTORINI producer; old magnificent VINSANTOS (older the better). Groundbreaking Monsigniori and Evdemon ASSYRTIKOS: age for a decade.

Avantis C Gr ★★★ Boutique winery in Evia and SANTORINI. Exquisite Aghios Chronos SYRAH/VIOGNIER and Afoura Santorini.

Biblia Chora Mac ★★★ Classic SAUV BL/ASSYRTIKO. Ovilos range (r w) could rival B'x at triple the price. Increasing focus on Greek varieties: Vidiano, AGIORGITIKO. Sister estate of GEROVASSILIOU. New wineries in Pelop (Dyo Ipsi), GOUMENISSA to watch.

Boutari, J & Son ★ →★★★ Historic brand, initially from NAOUSSA. Excellent value, esp **Grande Res Naoussa** to age 40 yrs+. Top: 1879 Legacy Naoussa, from v. old v'yd.

Carras, Domaine Mac ★★ Historic estate at Halkidiki. Ch Carras is a classic, SYRAH not far off.

Cephalonia Ion Important Ionian island with three POPs: mineral ROBOLA (w), rare MUSCAT (w sw) and excellent MAVRODAPHNE (r sw). Dry Mavrodaphne is a trend and a must-try but cannot be POP.

Dalamaras ★★★ →★★★★ Young yet prodigious producer in NAOUSSA. XINOMAVROS of great purity. Delectable range; Palaiokalias single-v'yd is world class.

Dougos C Gr ★★★ Rich reds, an ambassador for RAPSANI, esp Old Vines. Try Mavrotragano (r): pricey but worth it.

Economou Ktima Crete ★★★★ One of great artisans of Greece; brilliant, burgundy-like Sitia (r).

Gaia Aeg, Pelop ★★★ Top NEMEA and SANTORINI producer. Great Thalassitis Santorini (rare Submerged is aged underwater), thought-provoking **wild-ferment Assyrtiko**. Top from Nemea: ever-evolving **Gaia Estate**. Juicy "S" red (AGIORGITIKO with a touch of SYRAH).

Gentilini Ion ★★→★★★ Leading CEPHALONIA name, incl **steely Robola**. Wild Paths redefines ROBOLA. Marvellous dry MAVRODAPHNE Eclipse (r) is benchmark.

Gerovassiliou Mac ★★★ Quality and trend leader. Original ASSYRTIKO/MALAGOUSIA and top Malagousia (he's the specialist). Top reds: Avaton and Evangelo. Linked with BIBLIA CHORA.

Island-hopping
New frontiers of Greek wine are the islands: dozens of them, in both Aegean and Ionian Seas, and they're perfect settings for unique wines – isolated ecosystems, special terroirs, rare varieties and crazy people. Tinos already chasing SANTORINI; Chios, Ikaria, Syros, Thassos, Zakynthos and others will follow.

Goumenissa Mac (POP) ★★→★★★ Elegant XINOMAVRO/Negoska (r), for those who find pure Xinomavro too much. Chatzyvaritis, natural Tatsis, Aidarinis (single v'yd).

Hatzidakis Aeg ★★★ Top-class producer of SANTORINI; children now in charge, less adventurous but v. fine. Try Skytali ASSYRTIKO.

Karydas Mac ★★★ Tiny family estate and amazing v'yd in NAOUSSA, crafting classic, refined, age-worthy XINOMAVRO.

Kotsifali (Cretan red grape) named after blackbird. So is Merlot, but no relation.

Katogi Averoff Pelop, Epir ★★→★★★ Original cult Greek wine now decent large-volume brand. Top: Rossiu di Munte range from plots at 1000m (3281ft)+. Vlahiko is impressive.

Katsaros Thess ★★★ Tiny winery on Mt Olympus. KTIMA (CAB SAUV/MERLOT) is a Greek classic. XINOMAVRO Valos getting there.

Kechris ★★→★★★ Makes The Tear of the Pine, possibly *world's best Retsina*: fantastic wine. No kidding. Even Kehrimbari Retsina, volume brand, is great summer wine with Greek food.

Kir-Yianni Mac ★★→★★★ V'yds across Mac. Age-worthy, XINOMAVRO-based reds incl Ramnista, Diaporos, Blue Fox. Tarsanas ASSYRTIKO is fab. Range is expanding, going from strength to strength.

Ktima Estate. Should be used on export labels instead of "estate". Not that difficult.

Lazaridi, Ktima Costa Att, Mac ★★★ Wineries in Drama and Att (under Oenotria Land label). Refreshing, aromatic Amethystos redefined Greek whites. Top: Cava Amethystos CAB FR, then Oenotria Land CAB SAUV/AGIORGITIKO. New plantings in upper Drama v. promising; try MALAGOUSIA.

Lazaridi, Nico Mac ★ →★★★ Originally from Drama. Several large-volume, gd-value ranges. Top: Magiko Vouno (CAB SAUV), Perpetuus (Cab Sauv/SANGIOVESE).

Ligas Mac ★★ Full-blown natural producer in Pella. For hard-core fans of style.

Lyrarakis Crete ★★ →★★★ Heraklio-based, reviving old, almost extinct Cretan varieties like Plyto, Dafni, Melissaki, as well as Liatiko. *Single-v'yd versions* extraordinary.

Malvasia Group of POPs recreating famous Medieval "Malmsey". Not from MALVASIA grapes, but a reflection of local varieties. Four POPs: Monemvassia-Malvasia in Laconia from Monemvassia/ASSYRTIKO/Kydonitsa), Malvasia of Paros (from Monemvassia/Assyrtiko), Malvasia Chandakas-Candia (from Assyrtiko/Vidiano/MUSCAT) and Malvasia of Sitia (ditto plus Thrapsathiri), both from Crete. Surprisingly, not many producers capitalize on those.

Manousakis (Nostos) Crete ★★★ Great KTIMA, initially making Rhône-inspired blends, but Greek varieties here to stay: ASSYRTIKO, revealing MUSCAT of Spinas, Vidiano and even Romeiko (r).

Mantinia Pelop (POP) w sp High-altitude, cool region. Fresh, crisp, low-alc, almost Germanic styles from MUSCAT-like *Moschofilero*. Excellent sparklers from TSELEPOS. Troupis and Bosinakis rising stars.

Mercouri Pelop ★★★ Beautiful estate on w coast. V.gd KTIMA (r), delicious Foloi RODITIS (w), complex dry MAVRODAPHNE (r), REFOSCO (r).

Naoussa Mac ★★★ (POP) Top-quality region for sophisticated, structured XINOMAVRO. Best on par in quality, style (but not price) with Barolo. Top: DALAMARAS, KARYDAS, KIR-YIANNI, THIMIOPOULOS, but hardly a bad producer on this appellation.

Nemea Pelop ★★→★★★ (POP) AGIORGITIKO reds. Can be stunning; styles from

---

Naturally Greek

Natural wines on rise here, in visibility and popularity, even if they're still tiny overall. Common problems persist: some are excuse for incompetent winemaking, while others are magnificent. As usual, top natural producers don't even bother with term.

fresh to classic to exotic. Try Driopi from TSELEPOS, GAIA, Ieropoulos, Mitravelas, PAPAÏOANNOU, SKOURAS. Single-v'yd bottlings on rise.

Palyvos Pelop ★★→★★★ Excellent producer in NEMEA making modern, big-framed, almost New World-style reds. Fine single-v'yd selections.

Papaïoannou Ktima Pelop ★★★ If NEMEA were Burgundy, Thanassis Papaïoannou would have been Jayer. Sadly died 2019; now in able hands of son Giorgos. Excellent value KTIMA, Palea Klimata (old vines), Microklima (micro-v'yd), top-end Terroir. Age everything.

To make 500ml of Santorini Vinsanto needs about 16m² of v'yd land. Rare stuff.

Pavlidis Mac ★★★ Based in Drama. Trendy Thema (w) ASSYRTIKO/SAUV BL. Emphasis: expressive varietals incl AGIORGITIKO, Assyrtiko. High quality all round.

POP Greek equivalent of Appellation Controlée. More or less.

Rapsani Thess POP on Mt Olympus. Made famous in 90s by TSANTALIS (try Grande Res); now DOUGOS, THIMIOPOULOS add excitement. XINOMAVRO, Stavroto, Krasato.

Retsina New Retsinas (eg. GAIA, KECHRIS, natural-style Kamara), packed with freshness and a great alternative to Fino Sherry. Revolution in making, and one you mustn't miss.

Samos Aeg ★★→★★★★ (POP) Island famed for sweet MUSCAT Blanc. Esp fortified Anthemis, sun-dried Nectar. Rare old bottlings are steals, eg. hard-to-find Nectar 75 or 80.

Santo Aeg ★★→★★★ Most successful SANTORINI co-op. Solid portfolio with rich Grande Res, dry Irini (aged in VINSANTO barrels), complex Vinsantos. Great value by Santorinian standards.

Santorini Aeg ★★★→★★★★ Dramatic volcanic island with POP white (d, sw) wines to match. Luscious VINSANTO, salty, ***bone-dry Assyrtiko***. Top: ARGYROS, GAIA, HATZIDAKIS, SANTO, SIGALAS. Although getting expensive, still cheapest ★★★★ whites around, age 20 yrs. MAVROTRAGANO reds (not incl in POP) can be sublime. Image leader for Greek wine.

Semeli Ktima C Gr, Pelop ★★ Umbrella name for Semeli, Nassiakos and Orinos Helios, gd value across rapidly expanding range. Top Thea MANTINIA.

Sigalas Aeg ★★★★ Leading light of SANTORINI. Trail-blazing MAVROTRAGANO. Kavalieros, Nyhteri, Roptro out of this world; Seven Villages micro-cuvée a thesis on local terroir. New bottlings incorporate fruit from other parts of Greece.

Skouras Pelop ★★★ Solid range. Lean, wild-yeast Salto MOSCHOFILERO. Top reds: high-altitude Grande Cuvée NEMEA, Megas Oenos. Solera-aged Labyrinth is weird but beautiful, Peplo a thought-provoking premium rosé.

Tatsis Mac ★★★ Natural producer in GOUMENISSA with pure, clean, beautiful style.

Thimiopoulos Mac ★★★★ Top producer in NAOUSSA, RAPSANI (Terra Petra). Aftorizo Naoussa (own-rooted vines) best.

Tsantalis Mac ★→★★★ Long-est producer. Huge range. Gd RAPSANI Res, Grande Res, gd value from Thrace. Monastery wines from ***Mount Athos*** noteworthy, eg. excellent Avaton.

Tselepos Pelop ★★★ Leader in MANTINIA, NEMEA (as Driopi) and SANTORINI (Canava Chrysou). Greece's best MERLOT (★★★★ Kokkinomylos). Avlotopi CAB SAUV not far behind. Great Driopi Res and clay amphora-aged Laoudia Santorini.

Vinsanto Aeg ★★★★ Sun-dried, cask-aged luscious ASSYRTIKO and Aidani from SANTORINI that can age forever. Insanely low yields.

Greek appellations
Terms are changing in line with other EU countries. Quality appellations of OPAP and OPE now fused together into POP (or PDO) category. Regional wines, known as TO, will now be PGE (or PGI).

Eastern Mediterranean & North Africa

EASTERN MEDITERRANEAN

The Eastern Med gave wine culture to the world. Islamic prohibition hasn't helped. But the past 20 years have seen a great revival, using French (Lebanon, Turkey) and Californian (Israel) expertise. Now there's a revaluation of grapes like Cinsault (Lebanon), Carignan (Israel), Öküzgözü (Turkey) and Mavro, Xynisteri (Cyprus) and local vines like Obeideh, Merwah (Lebanon), Argaman, Marawi (Israel), Maratheftiko, Yiannoudi, Promara (Cyprus) and Boğaskere, Kalecik Karasi and Narince (Turkey). Let's hear it for local flavours.

Cyprus

Aiming higher and focusing on local varieties are the strategies changing the image of Cyprus wine, known before only for low prices, apart from its historic sweet Commandaria. Better producers are reporting growing sales; customers are paying a premium for rediscovered, still-rare grapes like Promara and Yiannoudi, and even previously unregarded Mavro from centenarian vines is showing its quality potential. New open-cellar initiatives are attracting visitors into the foothills of Mt Olympus, and Cypriot restaurants are increasingly proud of the island's own wines, rather than touting imports.

Cyprus has ungrafted vines and no phylloxera: not unique in Europe, but rare.

Aes Ambelis ★→★★ V.gd modern COMMANDARIA. DYA Morokanella, XYNISTERI, rosé.

Anama Concept br ★★ Husband-and-wife handcrafting amazing rich COMMANDARIA from old-vine MAVRO only.

Argyrides Vineyards r w ★★ Immaculate pioneering estate winery with new visitor facility. Excellent MARATHEFTIKO, MOURVÈDRE. V.gd VIOGNIER.

Commandaria Rich, sweet PDO wine from sun-dried XYNISTERI and MAVRO grapes. Probably most ancient named wine still in production – since 800BC. Survived Ottomans by exporting to Christian churches. New-generation producers: AES AMBELIS, ANAMA CONCEPT, Gerolemo, KYPEROUNDA, TSIAKKAS. Traditional styles: Alasia (Loel), Centurion (ETKO), St Barnabas (KAMANTERENA), St John (KEO).

Constantinou r w ★→★★ Lemesos region. Gd CAB SAUV, SHIRAZ.

ETKO & Olympus r w br ★→★★ Better wines since move to Olympus winery. Best for COMMANDARIA, Haggipavlou label for dry wines.

Gerolemo r p w br ★ Modern family winery at 900m (2953ft), noted for expressive aromatic whites and improving reds

Kamanterena (SODAP) r p w ★→★★ Co-op in Pafos hills. Gd-value DYA whites and rosé, young unoaked MARATHEFTIKO.

KEO r p w br ★ Winemaking now at Mallia Estate in hills. Ktima Keo range is best. St John ★★ COMMANDARIA.

Kyperounda r p w br ★★→★★★ Some of Europe's highest v'yds at 1450m (4757ft). *Petritis* remains standard-setting XYNISTERI. Flagship Epos CHARD and red from own v'yd. V.gd: Skopos SHIRAZ, Andessitis blend. Excellent modern COMMANDARIA.

Makkas r p w ★→★★ Pafos region. Garage winery. Gd MARATHEFTIKO, SYRAH, XYNISTERI.

Tsiakkas r p w br ★★→★★★ Expressive whites: SAUV BL, Promara, XYNISTERI. Also v gd COMMANDARIA, Vamvakada (aka MARATHEFTIKO), Yiannoudi, organic Rodinos rosé.

Vasilikon, K&K Winery r p w ★★ Only female winemaker on Cyprus. V.gd whites: XYNISTERI, Morokanella. Appealing reds: Ayios Onoufrios, MARATHEFTIKO, Methy.

Vlassides r p w ★★→★★★ UC Davis-trained Vlassides makes superb SHIRAZ, gd DYA Grifos, promising Yiannoudi, excellent long-ageing Opus Artis from best sites.

All Cypriot v'yds registered with date of planting, even if a century old.

Vouni Panayia r w ★★ Dynamic family winery with local grape focus. Try Alina XYNISTERI, MARATHEFTIKO, Promara, Spourtiko, Yiannoudi.

Zambartas r p w ★★→★★★ Australia-trained winemaker; v.gd single-v'yd range incl Margelina from centenarian vines. Zambartas label, new Koukouvagia range.

Israel

Many wines still have jammy overripe fruit, high alc and use bags of oak. There is, however, a new Israel striving for elegance, lower alc for better balance and more v'yd identity. More blends use appropriate Med varieties, but the biggest progress is in quality whites. Best regions tend to be high-altitude Upper Galilee (Up Gal), Golan Heights (Gol) and Judean Hills (Jud). Other abbreviations: Gailiee (Gal), Negev (Neg), Samson (Sam), Shomron (Shom).

Large Crusader winery found in village called Mi'ilya. Puts the cru in crusader.

Abaya Gal ★★ A terroirist. Natural, always experimenting and learning.

Amphorae ★→★★ Beautiful winery. Crisp, fresh whites, dry fruity rosés.

Ashkar Gal ★→★★ Connects a people, land and their heritage. Unique SAUV BL.

Barkan-Segal Gal, Sam ★★ Israel's largest winery. Fruity Beta Argaman, crisp COLOMBARD, flowery Marawi. Excellent whole-cluster Segal SYRAH, PINOT N.

Bar-Maor Shom ★★ Min-intervention winemaking. Crisp whites, fresh rosé.

Carmel Up Gal ★→★★ Historic winery, focus on basic wines. Private Collection reds gd value. Med blend is agreeably complex.

Château Golan Gol ★★★ Innovative winemaker. V.gd Geshem (r w) Med blends. Excellent SYRAH, bold Eliad. Quality small-batch ROUSSANNE.

Clos de Gat Jud ★★★ Estate exuding quality, style, individuality. Powerful Sycra SYRAH. Rare, rich MERLOT. Gd CHARD. Great-value Harel Syrah, entry-level Chanson (w).

Covenant Israel Gal ★→★★ Fresh, flavourful Blue C (r). Bold SYRAH. Gd VIOGNIER.

Cremisan Jud ★→★★ Palestinian wines made in a monastery from indigenous grapes: Baladi, Dabouki, Hamdani, Jandali. Hamdani/Jandali (w blend) best.

Dalton Up Gal ★★→★★★ Family winery, creative winemaker. Refreshing PINOT GR, mineral CHENIN BL and spicy Alma (r). Fun pét-nat.

Domaine du Castel Jud ★★★★ Pioneer of Jud, setting standards. Grand Vin is rich, deep, complex. Plush Petit Castel. Exquisite CHARD. La Vie entry-level. New subsidiary Raziel: lean, elegant SYRAH/CARIGNAN.

Feldstein Jud, Gal ★★→★★★ Artisan. Great rosés. Concentrated, rich ROUSSANNE, sharp Dabouki and dried-grape Argaman. Expensive.

Flam Jud, Gal ★★★→★★★★ Elegant B'x blend Noble, fruit-forward SYRAH, deep MERLOT. Classico always great value. Fresh, fragrant white (SAUV BL/CHARD), crisp rosé. New prestige Camellia (w) silky, elegant.

Galil Mountain Up Gal ★ Prestige blend Yiron always gd value.

Gush Etzion Jud r w ★★★ Central mtn v'yds. Gd GRENACHE/SYRAH/MOURVÈDRE.

Jezreel Valley Shom ★→★★ Big and oaky CARIGNAN, Argaman. Pét-nat Dabouki. Prestige Icon.

Kosher Necessary for religious Jews; irrelevant to quality. Wines can be v.gd; 90%+ Israeli wine is kosher. Largest wineries only make kosher wine.

Lahat Gol, Jud ★★ Rhône specialist making precise wines; white has ageing ability.

Lewinsohn Gal ★★★ Quality *garagiste*. Gd CHARD. Chunky, spicy SYRAH/PETITE SIRAH. Whole-cluster Petite Sirah.

Lotem Gal ★→★★ Spiritual, rare organic winery. Surprising NEBBIOLO.

Maia Shom ★→★★ Med-style. Greek consultants. Refreshing, drinkable wines.

Margalit Gal, Shom ★★★→★★★★ Israel's 1st cult wine. Father and son. B'x blend Enigma. Complex CAB FR, fine CAB SAUV. Gd cellaring potential. Perfumed Paradigma, Optima (w), RIES from Zichron v'yd.

Mia Luce Gal ★★→★★★ SYRAH: N Rhône feel. Superb MARSELAN. Fine-tuned COLOMBARD.

Nana Neg ★→★★ Desert pioneer making crisp CHENIN BL, big reds.

Ortal Gol ★★ Juicy CAB FR; SAUV BL represents great value.

Pelter-Matar Gol ★★ V. popular brand. Gd, clean whites. New dry RIES.

Psagot Jud ★★ Central mtn v'yds. Peak succulent, refreshing CHARD, gd VIOGNIER.

Recanati Gal ★★→★★★ Complex, wild CARIGNAN. Special Res deep, velvety prestige red. Local Bittuni, Marawi vines. Palestinian grower, Israeli winemaker.

Sea Horse Jud ★★ →★★★ Idiosyncratic winemaker. Crunchy, aromatic GRENACHE.

Shvo Up Gal ★★★ Non-interventionist winemaker. True vigneron. Super-rustic chewy red. Rare Gershon SAUV BL. Great value Cheninchik, characterful rosé.

Sphera Jud ★★★→★★★★ White only; sparkling on its way. Racy, crisp White Concept varietals (RIES, CHARD, SAUV BL). Harmonious First Page (SEM, ROUSSANNE, CHENIN BL). Complex, rare White Signature (Sem/Chard). Cool-climate style.

Tabor Gal ★★ V.gd whites, esp fantastic-value SAUV BL. Complex Malkiya.

Teperberg Jud, Sam ★ →★★ Israel's largest family winery; 5th generation. Flavourful full-bodied CAB FR; crisp PINOT GR.

Tulip Gal ★★→★★★ Innovative, progressive. Opulent Black Tulip, deep SHIRAZ, complex CAB FR/MERLOT. Works with adults with special needs.

Tzora Jud ★★★★ Terroir-led, precision winemaking. Talented winemaker (Israel's only MW). Wines show intensity, balance, elegance. Crisp, minerally Shoresh SAUV BL; Jud red and white always superb value. Complex, elegant prestige Misty Hills (CAB SAUV/SYRAH). Rare luscious Or dessert. Now Fair'n Green certified.

Vitkin Jud ★★→★★★ Quality CARIGNAN pioneer. Complex GRENACHE BL, new MACABEO.

Vortman Shom ★ →★★ Watch. COLOMBARD, FUMÉ BL, GRENACHE/CARIGNAN blend all gd.

Yaacov Oryah Sam ★★→★★★ Creative artisan. Bold. Orange wines, bottle-aged SEM.

Yarden Gol ★★ →★★★★ Pioneering winery farming sustainably. Advanced v'yd technology. Rare, prestige Katzrin. Bold Bar'on CAB/SYRAH. Big-selling Mt Hermon (r). Cab Sauv shows quality, consistency, value. Superb Blanc de Blancs. Delicious sweet HeightsWine. Second label: Gamla. SANGIOVESE of interest.

Yatir Jud ★★ Desert winery. Velvety Yatir Forest. Deep PETIT VERDOT.

Lebanon

Small-volume, maverick producers are moving out of the squeaky-clean B'x/ Rhône comfort zone, with min-intervention wines from indigenous Merwah, Obeideh, Meksessi and red "heritage" varieties CINSAULT, GRENACHE, CARIGNAN, with CAB FR also showing well. However, mainstream France is still the dominant influence. There are high-altitude (1000m/3281ft+) CHARD, SAUV BL, VIOGNIER; and rosé is being taken more seriously for a discerning local market.

Aniseed used to flavour arak sourced in Hina, on Syrian slopes of Mt Hermon.

Atibaia r w ★★★ *Garagiste*. Elegant red B'x blend, soft tannins, brooding black fruit.

Chateau Belle-Vue r (w) ★★★ Le Chateau and Le Renaissance, plush blends of B'x grapes/SYRAH. Also Petit Geste (SAUV BL/VIOGNIER).

Château Ka r w ★ →★★ Great-value, cherry-berry Cadet de Ka. Source Rouge, Blanche, more scrubbed-up. Fleur de Ka is top dog.

Château Kefraya r w ★★→★★★★ Fine, ripe, rich, complex *Comte de M*. Full, fragrant, oaky Comtesse de M (CHARD/VIOGNIER). Les Breteches is fruity, CINSAULT-based red.

Château Ksara r w ★★★ Founded 1857. Res du Couvent is fruity, easy-drinking, full of flavour. Blanc de Blancs (CHARD/SAUV BL/SEM) and Chard outstanding whites.

Château Marsyas r (w) ★★ Powerful, fruity CAB/SYRAH; B-Qa de Marsyas more Rhône-y. Owner of complex ★★★ Bargylus (Syria), miracle wine made in impossible conditions.

Chateau Musar r w ★★★→★★★★ Icon wine of the e Med, CAB SAUV/CINSAULT/CARIGNAN 02 03 05' 07' 08 09 10 11 12. *Unique recognizable style*. Best after 15–20 yrs in bottle. Indigenous Obaideh, Merweh age indefinitely. Second label: Hochar (r) now higher profile. Musar Jeune is softer, easy-drinking.

Clos St. Thomas r w ★→★★★ The Toumas are a famous Bekaa wine family. Fruity, elegant CINSAULT-based Les Gourmets. Aromatic Obaidy (sic).

Domaine de Baal r (w) ★★ Crisp CHARD/SAUV BL, organic heady estate red.

Domaine des Tourelles r w ★★→★★★ Blockbuster SYRAH, gd Marquis des Beys. Outstanding CINSAULT from 70-yr-old vines; equally gd classic red. Oaky CHARD.

Domaine Wardy r (w) ★★★ Les Terroirs (CAB SAUV/MERLOT/CINSAULT) and Clos Blanc (Obeideh/CHARD/SAUV BL/VIOGNIER/MUSCAT) excellent value.

Now at least four Lebanese small-batch gins and one malt whiskey.

IXSIR r w ★★→★★★ Stony SYRAH-based blends, floral whites and prestige El. Altitudes range and Grande Res Rosé excellent.

Massaya r w ★★ Terraces de Baalbeck: refined, elegant GSM. Entry-level Les Colombiers v.gd value. Also Cap Est (r) from E Bekaa v'yds on Anti-Lebanon Mtns. Punchy rosé.

Vertical 33 r w ★→★★ Organic CINSAULT, CARIGNAN, Obeideh varietals. Neo-MUSAR!

Turkey

Everything is stacked against Turkish winemakers, but they have such variety of climates, unpronounceable indigenous vines and a few heroes investing in quality that they are worth investigating.

Boğaskere means "throat scratcher". Lots of tannin.

Büyülübağ ★★ One of new small, quality wineries. Gd CAB SAUV.

Corvus ★★→★★★ Bozcaada island. Intense Corpus, luscious Passito.

Doluca ★→★★ Theodora label showcases local ÖKÜZGÖZÜ/BOĞASKERE (r), NARINCE (w).

Kavaklidere ★→★★★ Well-made modern wines. Pendore estate is best, esp ÖKÜZGÖZÜ, SYRAH. Cherry-berry Yakut. Stéphane Derenoncourt of B'x consults.

Kayra ★→★★ Spicy SHIRAZ, fresh NARINCE. Ripe ÖKÜZGÖZÜ from E Anatolia.

Pasaeli ★→★★ Fresh, vibrant B'x-style blends from single v'yd.

Sevilen ★→★★ International variety specialist. Spicy SYRAH, aromatic SAUV BL.

Suvla ★→★★ Full-bodied B'x blend Sur, and fruity SYRAH backed by oak.

Urla ★★ Tempus (r) has complexity, depth. Bold, spicy BOĞASKERE.

Vinkara ★ Charming NARINCE and cherry-berry KALECIK KARASI.

North Africa

Baccari ★★ Perhaps Mor's best red; Premiere de Baccari from Meknes.

Bernard Magrez Mor ★★ Investment by B'x tycoon. Tannic, spicy SYRAH/GRENACHE.

Castel Frères Mor ★ Gd-value brands. Vin Gris – delicate rosé best.

Celliers de Meknès, Les Mor ★→★★ Seen everywhere in Mor. Ch Roslane SYRAH gd.

Domaine Neferis Tun ★→★★ 90-yr-old CARIGNAN v'yds. Selian Carignan best.

Ouled Thaleb Mor ★★ Interesting Ait Souala (r) incl new vine-crossing Arinarnoa. Lively SYRAH Tandem (Syrocco in US): Thalvin and Graillot (Rhône) joint venture.

Val d'Argan Mor ★→★★ At Essaouira. Gd value: Mogador. Best: Orients.

Vignerons de Carthage Tun r p w ★ Best from UCCV co-op: Magon Magnus (r).

Vin Gris ★ Pale-pink resort of the thirsty. Castel Boulaouane brand best-known.

Volubilia Mor r p w ★→★★ Best delicate pink *vin gris* in Morocco.

Asia & The Old Russian Empire

ASIA

China Of c.870,000 ha v'yd, only 10% is used for wine, but things are changing fast. Far-flung Xinjiang and Ningxia each have a quarter of plantings; Hebei and coastal Shandong, combined, share another 25%, with the remaining quarter in other provinces incl Yunnan. CAB SAUV leads with 60%, followed by MERLOT, CHARD, Cabernet Gernischt (aka CARMENÈRE), MARSELAN, SYRAH, CAB FR and WELSCHRIESLING. Others incl RIES, UGNI BL, SEM, PETIT MANSENG, PINOT N, GAMAY and PETIT VERDOT. The biggest challenge remains having to bury vines in winter in Xinjiang, Ningxia, Shanxi, Gansu, Hebei and Jilin where temperatures can plummet to -20°C (-4°F). Such buried vines seldom live beyond 25 yrs, as the trunk ultimately snaps. Investors have piled in. At the top, Lafite-Rothschild's Domaine de Long Dai's B'x blend joins Moët-Hennessy's Ao Yun ("Sacred Cloud"). Gd B'x blends incl Ch Rongzi, Ch Zhongfei, Grace V'yd, Jia Bei Lan, Leirenshou, Li's Family, Silver Heights, Tiansai, Shangri-La Winery. Newcomer Ch Chanson does well with pure Cab Fr. Gd, inexpensive, everyday reds come from Changyu, China's largest producer. The most interesting red remains Marselan; best incl Xinjiang's Ch Zhongfei and Shanxi's Grace V'yd. Tiansai is fast on their heels with cracking inaugural Marselan Grand Res 15. Pinot N is more ticklish. Domaine du 1er Juin is gd; Jade Valley in Xi'an commendable. Domaine Chandon in Ningxia wears the crown for sparkling wine. Whites are less popular but Ch Guofei has gd dry and off-dry Ries. Legacy Peak, Domaine Helan Mtn (Pernod Ricard), Tiansai and Mystic Island show the way for Chard. China's best rosé is Ch Changyu-Moser XV, its best sweet, Taila Winery of Shandong's Petit Manseng.

Indians prefer imported wines (and whisky) to their own, but India has c.115,000 ha of vines of which c.2000 ha make wine, mostly in Maharashtra, Karnataka, Andhra Pradesh. Try Moët Hennessy's Domaine Chandon and York for bubbles; Rajeev Samant's Sula, Charosa VIOGNIER, Fratelli SANGIOVESE/CAB SAUV Sette, Grover Zampa La Reserva (Cab Sauv/SHIRAZ). Varietal Shiraz from Sula Dindori and Myra Winery. Experiments continue: Krsma Sangiovese, Vallonne MALBEC, Charosa TEMPRANILLO. Fratelli Sangiovese Bianco lights way for rosé. Considering that world's best rosés are produced from relatively warm regions, India could well blush more.

Japan is now making more Koshu than all its reds put together.

Japan's growers are torn between turning their native KOSHU grapes into wine or selling them at eye-watering prices as postcard-perfect bunches in supermarkets. Yamanashi is birthplace of Japanese winemaking since the 2nd half of the C19. In 2002, Grace Wine planted a revolutionary v'yd in Akeno, training the vines along vertical shoot positioning. Its Cuvée Misawa Akeno has kudos, but gd Koshu is also made by Ch Lumière, Ch Mercian, Domaine Hide, Haramo, L'Orient, Marquis, Soryu, Suntory Tomi No Oka. Best have no oak influence. Grace Blanc de Blanc is Japan's best sparkling, candyfloss-scented Muscat Bailey A the most planted red grape. Domaine Hide is best. Chitose Winery makes gd PINOT N. There are c.280 wineries, dominated by brewers Kirin, Sapporo and Suntory. Best, though, are small family-owned wineries.

THE OLD RUSSIAN EMPIRE

Georgia continues to be the focus of wine geeks and those open to try rare, unusual and quirky wines made with first-class grapes. Its tradition of making wine in buried clay *qvevris* (amphorae), long considered archaic, is now super-fashionable worldwide. Armenia, Georgia's neighbour and another possible birthplace of wine, is not short of real wine heritage either. In Russia, locally produced quality wines are becoming an increasing consumer choice. Moldova bets on tourism to get its rather splendid wines – and some unique wine cellars – better known. Off the beaten track are large-scale plantings, above 1000m (3281ft), for quality wines in Kazakhstan.

Armenia like Georgia, has millenia of winemaking history (the most ancient winery dates back 6100 yrs). Its remote mountainous v'yds are phylloxera-free. Indigenous white Voskeat and Garandamak, red Areni, Hindogny and Kakhet can give gd quality. Private investment and international consultants drive standards at ArmAs, Armenia Wines, Karas Wines, v.gd Zorah.

Georgia's prehistoric winemaking methods, using buried *qvevris*, its indigenous grapes (around 500, though only a handful dominate) and 8000 yrs of unbroken viticultural history set Georgia apart from the rest of the world. Like reds, whites are often skin-macerated, known as Kakheti-method in Georgia and orange wines elsewhere. Signature varieties are red SAPERAVI, made in many styles from light semi-sweet to powerful, dry, tannic and rewarding age, and white RKATSITELI (vibrant). White Mtsvane and Kisi gain recognition. While every family in Georgia appears to make wine, most production comes from Kakheti in se. Leading producers incl Badagoni, Ch Mukhrani, Dakishvili, GWS, Jakeli Khashmi, Kindzmarauli Marani, Marani (TWC), *Pheasant's Tears*, Schuchmann, Tbilvino, *Tsinandali* (a noble domaine restored).

Moldova is 6th in Europe by v'yd area, yet remains little known. With the Crimea, source of the tsar's best wines in the past, now Moldova offers excellent value. Production centred around international grapes, but worth seeking local: (w) FETEASCĂ ALBĂ, FETEASCĂ REGALĂ, Viorica, (r) Rară Neagră, FETEASCĂ NEAGRĂ. Don't miss red blends Roşu de Purcari (Cab Sauv/MERLOT/MALBEC) and *Negru de Purcari* (CAB SAUV/SAPERAVI/Rară Neagră), plus Icewine. State-owned Cricova (sp) and Milestii Mici are largest; Vinăria Purcari is most acclaimed. Some gd producers incl Asconi, Castel Mimi, Ch Vartely, Et Cetera, Lion Gri, Vinăria Bostavan.

Russian wines are trendy, but not necessarily affordable, in the home market; they have yet to make an impression elsewhere. "Champanski" is hugely popular (and not always to be sneezed at). Best conditions for production are by the Black Sea and the River Kuban, but some v'yds are planted as far e as Astrakhan. International grapes (incl CAB FR, RIES) lead. Harsh climate in the Don Valley, known for indigenous grapes (red Krasnostop, Tsimliansky), requires vines to be buried in winter. Ch le Grand Vostock and Lefkadia have consistent gd quality. Est large producers: Ch Tamagne, Fanagoria, Myskhako, Yubileinaya, Abrau Durso (sp); small: Burnier, Gai-Kodzor, Usadba Markoth.

Ukraine's temperate climate by the Black Sea allows for quality wine production. The Crimea is buzzing thanks to small estates (Uppa Winery, Oleg Repin) that make wines with a sense of place. Grapes are mainly international, also some est local varieties (w Kokur, r Ekim Kara) and hybrids. Massandra, Solnechnaya Dolina, Koktebel continue strong Tsarist tradition of fortified styles. Wines modelled on Champagne are an important/popular heritage: try Artyomovsk Winery, Novy Svet, Zolotaya Balka. Quality dry wines by Guliev Wines, Inkerman (Special Res), Prince Trubetskoy Winery, Satera (Esse, Kacha Valley), Veles.

United States

WASHINGTON
OREGON
IDAHO
NEVADA
CALIFORNIA
COLORADO
ARIZONA
NEW MEXICO
OKLAHOMA
TEXAS
WISCONSIN
MICHIGAN
OHIO
MISSOURI
PENNSYLVANIA
NEW YORK/
NEW JERSEY
MARYLAND
VIRGINIA
NORTH CAROLINA
GEORGIA

NORTH COAST
Anderson Valley
Mendocino
Redwood Valley
Sierra Foothills
Clear Lake
Clear Lake
Sonoma Coast
Northern Sonoma
Sacramento
El Dorado
Shenandoah Valley
Amador
Carneros
Sonoma Valley
Coombsville/Oak Knoll
Lodi
Clarksburg
Calaveras
CENTRAL VALLEY
Napa Valley
Lake Tahoe
NEVADA
San Francisco
Livermore Valley
Santa Clara Valley
Santa Cruz Mountains
Monterey
Salinas
Carmel Valley/
Santa Lucia Highlands
Arroyo Seco
San Lucas
CENTRAL COAST
San Joaquin
Fresno
Pacific Ocean
Paso Robles
CALIFORNIA
San Luis Obispo
Edna Valley/Arroyo GV
Santa Maria Valley
Santa Barbara
Sta Rita Hills
Santa Ynez Valley
Santa Barbara
Los Angeles

UNITED STATES

Abbreviations used in the text
(*see also* Principal Vineyard/
Viticultural Areas pp.245, 262, 268):

Clark	Clarksburg, CA
Coomb	Coombsville, CA
Mad	Madera, CA
Mend	Mendocino, CA
Oak Knoll	Oak K, CA
PNW	Pacific Northwest
San LO	San Luis Obispo, CA
Santa B	Santa Barbera, CA
Santa Cz Mts	Santa Cruz Mountains, CA
Son	Sonoma, CA

None of us, Americans included, has really taken in the range of wines the US now has to offer. Yes, it's tempting to open a wine list at the pages chock-full of Napa Cabernets and Meritage blends; but if you're in Sonoma, why would you do that? If you're in Virginia, or New York, why would you do that? It would be like going to Burgundy and insisting on drinking Bordeaux. Virginia, to take a state full of promise, makes wines of tension and purity, of subtlety and freshness, and its Viogniers and Petit Verdots are becoming a signature style. The Sonoma Coast has a justified reputation for edgy Pinot Noir and Chardonnay. Texan strengths are in Rhône varieties. Oregon is one of the world's best places for Pinot Noir; Washington State makes supple Merlot. And New York? Most New Yorkers have little idea of what's growing around them.

American Viticultural Areas

I'm not sure AVAs will ever catch on with the public as USPs. They're not exactly (or even approximately) like appellations contrôlées. One thing they do is stimulate local feelings and claims for specialness, which is, overall, a gd thing. Federal regulations on appellation of origin in the US were approved in 1977. Do you need to know? There are two categories: 1st is a straightforward political AVA, which can incl an entire state, ie. CA, WA, OR and so on. Individual counties can also be used, ie. Santa B or Son. When the county designation is used, all grapes must come from that county. The 2nd category is a geographical designation, such as Napa V or Will V, within the state. These AVAs are supposed to be based on similarity of soils, weather, etc. In practice, they tend to be inclusive rather than exclusive. Within these AVAs there can be further sub-appellations, eg. the Napa V AVA contains Ruth, Stags L and others. When these geographical designations are used, all grapes must come from that region. A producer who has met the regulatory standards can choose a purely political listing, such as Napa, or a geographical listing, such as Napa V. It will probably be many yrs before the public recognizes the differences, but there is no doubt that some AVAs already fetch hefty premiums.

Arizona

A hip vibe drives the local scene, but wines are getting better known outside the state. Soils and climate zones mimic Burgundy (!). High-desert terroir of volcanic rock and limestone soils, gd ripening weather. Wineries incl **Arizona Stronghold** ★→★★ flagship red Rhône blend Nachise and excellent white blend Tazi and VIDAL Blanc dessert wine. **Alcantara V'yds** elegant and earthy reds, esp red blends Confluence IV and Grand Rouge. **Burning Tree Cellars** artisanal, small-batch, intense red blends. **Bodega Pierce** estate-grown grapes from family-run winery; gd SAUV BL. **Caduceus Cellars** ★★★ ownership by Alt-rocker Maynard James Keenan, once a disruptor, now est; excellent white blend Dos Ladrones, top reds Sancha, Nagual del Marzo. **Callaghan V'yds** ★★ awarded AGLIANICO and TANNAT, quality red blends; top-rated Caitlin made by vintner's daughter. **Dos Cabezas WineWorks** best-in-show traditional-method sparklers, rosé. **Javelina Leap V'yd & Winery** awarded ZIN. **Page Springs Cellars** GSM, other Rhône single-varietal white, esp Dragoon MARSANNE and blends. **Pillsbury Wine Company** ★★ Filmmaker Sam Pillsbury producing 100% estate-grown wine. PETITE SIRAH Special Res, v.gd CHENIN BL; lauded WildChild aromatic blend (w), MALVASIA, VIOGNIER and Guns & Kisses SHIRAZ.

California (CA)

The diversity of CA wine has never been greater, in terms of branding, grape varieties and winemaking styles. But at the same time there is constant consolidation, with medium and large brands sucked up by big players. Once a dream business for retiring romantics, the wine biz is no longer for the faint of heart. For 1st time in decades, wine consumption is flat. There are grumblings that young Americans may not be thrilled by wine. Beer and cocktails are big, while some eschew alc in favour of cannabis. But CA wine is increasingly sophisticated, elegant and precise. Winemakers are using less new oak and striving for "freshness", a word that was seldom uttered in the "unctuous", "hedonistic" era of the 90s and 2000s. Wildfires are becoming a regular fall threat; 18 was a harrowing experience, while Sonoma was hit again in 19, it was largely post-harvest, without casualties.

Recent vintages

CA is too diverse for simple summaries. There can certainly be differences between the N, Central and S thirds of the state, but no "bad" vintages in over a decade. Wildfires and smoke have proved challenging in recent yrs, but only for latest-picked grapes.

2019 Mellow, cool summer, dry autumn, solid harvest. Minor late losses in Alex V to fires, smoke taint.

2018 Bumper crop of great quality, but smoke ruined some grapes in Lake County.

2017 Catastrophic wildfires in Napa, Son after most grapes picked; quality mostly v.gd.

2016 Gd quality: reds/whites show great freshness, charm.

2015 Dry yr, low yields, but quality surprisingly gd, concentrated.

2014 Despite 3rd year of drought, quality high.

2013 Another large harvest with excellent quality prospects.

2012 Cab Sauv oustanding. V. promising for most varieties.

2011 Difficult yr. Those who picked later reported gd Cab Sauv, Pinot N. Zin suffered.

Principal vineyard areas

There are well over 100 AVAs in CA. Below are the key players.

Alexander Valley (Alex V) Son. Warm region in upper Son. Best-known for gd Zin, Cab Sauv on hillsides.

Amador County (Am Co) Warm Sierra County with wealth of old-vine Zin; Rhône grapes also flourish.

Anderson Valley (And V) Mend. Pacific fog and winds follow Navarro River inland. Superb Pinot N, Chard, sparkling, v.gd Ries, Gewurz, some stellar Syrah. Watch out.

Atlas Peak E Napa. Exceptional Cab Sauv, Merlot.

Calistoga (Cal) Warmer n end of Napa V. Red wine territory esp Cab Sauv.

Carneros (Car) Napa, Son. Cool AVA at n tip of SF Bay. Gd Pinot N, Chard; Merlot, Syrah, Cab Sauv on warmer sites. V.gd sparkling.

Coombsville (Coomb) Napa. Cool region nr SF Bay; top Cab Sauv in B'x pattern.

Diamond Mtn Napa. High-elevation vines, outstanding Cab Sauv.

Dry Creek Valley (Dry CV) Son. Outstanding Zin, gd Sauv Bl; gd hillside Cab Sauv and Zin.

Edna Valley (Edna V) San LO. Cool Pacific winds; v.gd Chard.

El Dorado County (El Dor Co) High-altitude inland area surrounding Placerville. Some real talent emerging with Rhône grapes, Zin, Cab and more.

Howell Mountain Napa. Briary Napa Cab Sauv from steep, volcanic hillsides.
Livermore Valley (Liv V) Suburban, gravelly, warm region e of San Francisco, with gd potential.
Mendocino Ridge (Mend Rdg). Emerging region in Mendocino County, dictated by elevation over 365m (1198ft). Cool, above fog, lean soils.
Monterey County (Mont) Big ranches in Salinas V provide affordable Chard, Pinot N in cool, windy conditions. Carmel V bit warmer, Arroyo Seco moderate.
Mt Veeder Napa. High mtn v'yds for gd Chard, Cab Sauv.
Napa Valley (Napa V) Cab Sauv, Merlot, Cab Fr. Look to sub-AVAs for meaningful terroir-based wines, and mtn areas for most complex, age-worthy.
Oakville (Oak) Napa. Prime Cab Sauv territory on gravelly bench.
Paso Robles (P Rob) San LO. Popular with visitors; Rhône, B'x varieties, reds prominent.
Pritchard Hill (P Hill) E Napa. Elevated, woodsy, rugged, prime terrritory for Cab Sauv.
Red Hills of Lake County (R Hills) N extension of Mayacama range, great Cab Sauv country.
Redwood Valley (Red V) Mend. Warmer inland region; gd Zin, Cab Sauv, Sauv Bl.
Russian River Valley (RRV) Son. Pacific fog lingers; Pinot N, Chard, gd Zin on benchland.
Rutherford (Ruth) Napa. Outstanding Cab Sauv, esp hillside v'yds.
Saint Helena (St H) Napa. Lovely balanced Cab Sauv.
Santa Lucia Highlands (Santa LH) Mont. Higher elevation, s-facing hillsides, great Pinot N, Syrah, Rhônes.
Santa Maria Valley (Santa MV) Santa B. Coastal cool; gd Pinot N, Chard and Viognier.
Sta Rita Hills (Sta RH) Santa B. Excellent Pinot N.
Santa Ynez (Santa Ynz) Santa B. Rhônes (r w), Chard, Sauv Bl best bet.
Sierra Foothills El Dor Co, Am Co, Calaveras County. All improving.
Sonoma Coast (Son Coast) V. cool climate; edgy Pinot N, Chard.
Sonoma Valley (Son V) Gd Chard, v.gd Zin; excellent Cab Sauv from Sonoma Mountain (Son Mtn) sub-AVA. Note Son V is area within Sonoma County.
Spring Mtn Napa. Elevated Cab Sauv, complex soil mixes and exposures.
Stags Leap (Stags L) Napa. Classic red, black fruited Cab Sauv; v.gd Merlot.

A Tribute to Grace N Coast ★★★ Kiwi Angela Osborne's homage to GRENACHE. Fruit from exceptional v'yds, diverse terroirs all over state, none more exciting than 975m (3199ft), mtn-ringed Santa B Highlands.
Abreu Vineyards Napa V ★★★→★★★★ Supple CAB SAUV-based wines from selected v'yds. Madrona v'yd leads way with powerful, balanced opening, long, layered finish. V.gd cellar choice for 10–12 yrs.
Alban Vineyards Edna V ★★★→★★★★ John Alban, a SYRAH frontiersman, specialist and original Rhône Ranger, still making great wine in EDNA V sweet spot. Top VIOGNIER, GRENACHE too.
Albatross Ridge Mont ★★★ Bowlus family rules high-elevation roost 11km (7 miles) from Pacific nr Carmel. Early CHARDS, PINOT NS fresh and lively, warrant watching.
Alma Rosa Sta RH ★★★→★★★★ Pioneer Dick Sanford's 2nd act after selling namesake winery. Continuing tradition of refined PINOT N, CHARD, also v.gd rosé.
Andrew Murray Santa B ★★★ Rhônes around the clock, hits keep coming. SYRAH leads pack, but VIOGNIER, ROUSSANNE, fresh GRENACHE BLANC hits too.
Antica Napa Valley Napa V ★★★ Piero Antinori's ATLAS PEAK project initially flopped, subsequent lessees improved v'yds, proving potential for fine CAB SAUV, CHARD. Antinori wisely reclaimed property.

Au Bon Climat Santa B ★★★ Jim Clendenen made PINOT N, crisp CHARD before it was hip, and advocated balanced style now trending. Relevant as ever.

Banshee Wines Son Coast ★★★ Growing, scrappy PINOT N-driven label with no v'yds, but gd connections. Well-made single-v'yd wines.

Baxter And V ★★★→★★★★ 2nd-generation Phil B's subtle, burgundian PINOT N (Oppenlander or Valenti v'yds), exudes passion, confidence, competence. Small, but influential.

Beaulieu Vineyard ("BV") Napa V ★→★★★ Iconic Georges de Latour Private Res CAB SAUV is back on track, as are other reds. Cheap Coastal Estate brand is fine in a pinch.

Beckmen Vineyards Santa B ★★★ Steve Beckmen's bio Purisma Mtn estate produces formidable SYRAH, GRENACHE, GRENACHE BLANC. Rhône blend Cuvée le Bec rightly popular nationwide.

Bedrock Wine Co. Son V ★★★ Morgan Peterson's label is a paean to historic ZIN v'yds, techniques. Wisdom of ages seen through clear young eyes.

Beringer Napa ★★→★★★★ (Private Res) Big producer of average to high-level wines. Private Res CAB SAUV, single-v'yd Cabs serious, age-worthy. HOWELL MTN Cab Sauv strong, CHARDS now fresher, better. Grand historic estate well worth a visit.

Berryessa Gap Vineyards Central V ★★★ Upstart Yolo Co project nr Sacramento making fresh, lightly oaked, Iberian-inspired wines. TEMP dazzling, VERDEJO and DURIF also delicious. Popular in community, beyond.

Bevan Cellars Napa V, Son ★★★→★★★★ Outsized personality with great taste, Russell Bevan sources from prime single v'yds making superb B'x varieties, a bit of PINOT N and luscious Dry Stack v'yd SAUV BL from Bennett V. Cultish, pricey.

Bogle Central V, Lodi ★★ Dependable under-$15 grocery-store family-owned brand delivers ever-reliable varietal wines from LODI, Clarksburg and now more coastal zones, all aged in real barrels. Respect.

Bokisch Lodi ★★→★★★ Markus B is leading pack when it comes to Spanish varieties. V.gd TEMPRANILLO heads list backed by superb GRACIANO, ALBARIÑO, flirty rosado.

Bonny Doon Mont ★★★ Randall Grahm's marketing is whimsical, but his wines are serious, more terroir-driven than ever. *Vin gris* is superb, juicy Clos de Gilroy GRENACHE, *Le Cigare Volant* Rhône blend revised, younger, price down, 19; many tricks up his sleeve. Sold 2020; Grahm stays on as chief winemaker.

Brewer Clifton Santa B ★★★ Estate STA RH PINOT N, CHARD producer now owned by JACKSON FAMILY WINES, OG PINOT N brand still in fine form, zesty Chards matured in neutral oak.

Bronco Wine Company ★→★★ Provocateur, populist Fred Franzia's company, famous for Two-Buck Chuck and scores of other commercial labels.

CADE Napa V ★★★ Superb wines, CAB SAUV, SAUV BL, from swanky, ultra-modern winery atop HOWELL MTN. Partnership between Getty family, CA Gov. Newsom, GM John Conover.

Cakebread Napa V ★★★ CAB SAUV still has massive cachet with baby boomers. SAUV BL popular, CHARD v.gd. Diverse direct-to-consumer offerings.

Calera ★★★→★★★★ Central Coast pioneer Josh Jensen sought limestone and altitude for PINOT N, CHARD and struck gold. Sold to DUCKHORN (2017), brand in gd hands. Selleck and Jensen v'yds always stylish.

Caymus Napa V ★★★ One of Napa's foremost international status brands. Special Selection CAB SAUVS, esp iconic, also on sweet, sappy end of style spectrum.

Cedarville Sierra F'hills ★★★ Bootstrappers Jonathan Lachs and Susan Marks built a powerhouse in the granite-rich Fairplay District of EL DOR CO. Superb wines across board, mostly red. GRENACHE and SYRAH, fine CAB SAUV, ZIN.

Chalone Mont ★★★ Historic property, been kicked around a bit, seeking rejuvenation under Foley Family umbrella. Known for subtle CHARD, PINOT N.

Chappellet Napa V ★★★★ Pritchard Hill original, great since 60s. Rugged terrain gives v. durable reds. Signature series CAB SAUV superb, dry CHENIN BL a rare treat. Still family-owned, also owns PINOT N, CHARD themed SONOMA-LOEB.

Charles Krug Napa V ★★→★★★ Historically important winery made recent comeback, demanding recognition for role in modern NAPA V. Late owner Peter Mondavi was Robert's estranged brother. Supple CAB SAUV, crisp, pure SAUV BL.

Chateau Montelena Napa V ★★★ Tons of history, great continuity of ownership and style. Serious, if slightly funky CAB SAUV is cellar-worthy; CHARD holds up well too. Castle-like winery, impressive grounds.

Chateau St Jean Son V ★★★ Rock of SON V, solid wines on all fronts, but consensus flagship wine for decades has been Cinq Cépages blend of five B'x varieties, reliable and age-worthy CA classic.

Chimney Rock Stags L ★★★→★★★★ Underrated Terlato-owned brand making best wines ever under steady hand of WM Elizabeth Vianna. Tomahawk V'yd CAB SAUV top-notch.

Cliff Lede Stags L ★★★ Excellent CAB SAUVS, big but balanced with tannin, gd acid. Small production Cabs from NAPA V hillsides, leesy SAUV BL notable. Owns Fel brand in AND V, SON V.

Clos Pegase Napa V ★★★ Look-at-me winery, v.gd MERLOT from CAR v'yd, gd CAB SAUV. Wine shares stage with paintings, sculpture.

Clos du Val Napa V ★★★ Stags L classic. New owners slashed production, moved to upscale, estate-based model. Can estate v'yds make cut? CAB SAUV can improve, CAR PINOT N solid, jury still out.

Cobb Wines Son Coast ★★★ Ross Cobb makes restrained, natural SON COAST PINOT N, CHARD from select v'yds. Pinots improve with few yrs. Emaline Ann, Coastlands top sites.

Constellation ★→★★★ Publicly traded major wine/beer/spirits company owns famed ROBERT MONDAVI brand, Meiomi, The Prisoner, Woodbridge. Lately re-focusing on beer and cannabis products.

Continuum St H, Napa V ★★★★ Scion Tim Mondavi's P HILL estate spares no expense making one of NAPA V's greatest, most complex B'x blends. Second label Novicium from younger vines.

Copain Cellars And V ★★★ Old World-influenced, classically proportioned wines; recently sold to JACKSON FAMILY. PINOT N is strong suit, esp bright, spicy Kiser v'yd versions. Tous Ensemble line easy-going, friendly.

Corison Napa V ★★★★ While many in NAPA V follow \$iren call of bloated wines for big scores, diminishing pleasure, Kathy Corison consistently makes elegant, fresh Cabs, esp focused age-worthy Kronos v'yd CAB SAUV.

Cuvaison Car ★★★ Quiet historic property, making great wine yr after yr. Top marks to PINOT N, CHARD from CAR estate; gd SYRAH, CAB SAUV from MT VEEDER. Single Block bottlings incl lovely rosé, SAUV BL.

Dalla Valle Oak ★★★★ 1st-rate hillside estate transitioning to 2nd generation. Maya CAB SAUV is legendary, eponymous Cab Sauv a cult wine, Collina label best *affordable introduction* to luxury NAPA Cab.

Daou P Rob ★★★ Elevated estate in Adelaida Dist is driving CAB SAUV in P ROB to new heights in altitude and price.

Oh, Garnacha!
Heat- and drought-tolerant Spanish and Portuguese grapes settling in next to ZIN in Lodi, EL DOR CO and Central Valley. With rising temperatures, expect to see more GARNACHA, GARNACHA BLANCA, VERDEJO, VERDELHO, TEMPRANILLO and TOURIGA N. Wouldn't "Garnacha!" make a gd swear word?

Dashe Cellars Dry CV, N Coast ★★★ RIDGE veteran Mike Dashe makes tasteful, balanced DRY CV and ALEX V ZIN from urban winery in Oakland. Also terrific old-vine CARIGNANE, zesty GRENACHE rosé.

Dehlinger RRV ★★★ PINOT N specialist still on par after more than four decades. Also v.gd CHARD, SYRAH and balanced CAB SAUV.

DeLoach Winery Son ★★★ Flamboyant maestro JC Boisset saw gd value in this progressive organic, bio-oriented winery making great PINOT N, CHARD, even great dry GEWURZ. Solid investment, if not his sexiest.

Most dazzling Pinot N, Chard? From wild, cool Son Coast, And V, Mend Rdg.

Diamond Creek Napa V ★★★★ 09 11 12 Napa Mtn jewel. Prices v. high for age-worthy, minerally CAB SAUV from famous hillside v'yds on DIAMOND MTN. Patience always rewarded.

Domaine Carneros Car ★★★→★★★★ Taittinger outpost in CAR offering consistently gd bubbly, esp Vintage Blanc de Blancs Le Rêve. V.gd NV Rosé. Vintage Brut impressive. The Famous Gate PINOT N formidable.

Domaine Chandon Napa V ★★→★★★ Moët outpost in Yountville, top bubbly is v.gd. NV Res Étoile Blanc and Rosé. Pairings and great nibbles on outdoor patio.

Domaine de la Côte Sta RH ★★★ Exacting, burgundy-style estate PINOT N from coastal reaches of STA RH by former sommelier Rajat Parr and winemaker Sashi Moorman. *See also* SANDHI.

Dominus Estate Napa V ★★★★ Moueix-owned (*see* France) Brilliant winery is dazzling but not open to public. Wines from gravelly bench soils consistently elegant, impressive. Second label: Napanook, v.gd. Important voice of E NAPA V.

Donum Estate N Coast ★★★→★★★★ Anne Moller-Racke has passionately worked CAR soils since 1981. PINOT from four sites is focused, generous, complex. Adding v'yds in AND V, SON COAST.

Drew Family And V ★★★→★★★★ MEND RDG visionary making minimalist, savage PINOT N from AND V and higher up hills. Look for estate Field Selections Pinot N from MEND RDG, SYRAH from coastal Valenti V'yd. Seek out.

Dry Creek Vineyard Dry CV ★★★ DRY CV standard-bearer back on A-game. Trustworthy, Loire-inspired, grassy FUMÉ BLANC and other SAUV BL always delicious, reds like CAB SAUV, MERLOT, ZIN all improved.

Duckhorn Vineyards Napa V ★★★ ★★★★ Crowd-pleasing, super-consistent CAB SAUV, MERLOT, esp Three Palms v'yd, gd SAUV BL. Second label Decoy wines better than ever. Parent company owns Migration brand, Goldeneye in AND V, CALERA and Kosta Browne.

Dunn Vineyards Howell Mtn ★★★→★★★★ Mtn man Randy Dunn stubbornly resisted stampede to jammy, lush CAB SAUV styles, favouring restraint, age-ability. Wines aren't always spotless, but when great can last decades.

Dutton-Goldfield RRV ★★★ Steady-handed, classical cool-climate CA PINOT N, CHARD from RRV-based powerhouse grower, not super-edgy or risky, maybe a gd thing.

Edna Valley Vineyard Edna V ★★★ Easy-drinking varietals from gentle Central Coast. Lovely, lilting SAUV BL, crisp but tropical CHARD. Impressive SYRAH, gd CAB SAUV from top v'yd.

Emeritus RRV ★★★→★★★★ Emergent estate, three home dry-farmed (!) v'yds making focused, structured PINOT N under supervision of talented Dave Lattin. Wesley's Res from Hallberg v'yd reigns supreme.

Ernest Vineyards Son ★★★ Newcomer brings verve, acidity, style to a hip roster of excellent regional, single-v'yd with intriguing labels. Complex, racy, drinkable.

Etude Car ★★★→★★★★ Ever-trustworthy brand that always succeeded at making great CAB SAUV, PINOT N under same roof, using same attentive techniques. Now owned by Treasury Wine Estates, but legacy stays true. PINOT rosé to die for.

Failla Son Coast ★★★ One of savviest, most talented winemakers in CA, Ehren Jordan effortlessly tempers CA fruit to make savoury, compelling, complex PINOT N, SYRAH, CHARD from cool coastal sites.

Far Niente Napa V ★★★→★★★★ Pioneer of single-v'yd CAB SAUV, CHARD in big, generous, NAPA style. Hedonism with soul. Dolce: celebrated dessert wine.

Fetzer Vineyards N Coast ★★→★★★ Early champion of organic/bio viticulture in MEND CO, still gd, best under Bonterra brand. Owned by Concha y Toro (Chile).

Field Recordings P Rob ★★★ Impressively subtle, perceptive wines from P ROB'S Andrew Jones. Best are blends Neverland and Barter & Trade, but don't miss Alloy and Fiction, delicious in 500ml *cans.*

Flowers Vineyard & Winery Son Coast ★★★→★★★★ Extreme SON COAST pioneer 3 km (2 miles) from Pacific. PINOT N, CHARD remain great illustrations of that climate and elevation.

Foppiano Son ★★→★★★ Honest RRV wines loaded with sunny fruit and little pretence. PETITE SIRAH, SAUV BL notable.

Forman Vineyard Napa V ★★★ Ric F is dedicated veteran terroirist making elegant, age-worthy CAB SAUV-based wines from hillside v'yds. Also v.gd CHARD with a nod to Chablis.

Fort Ross Vineyard Son Coast ★★★ Dazzling high-elevation estate a stone's-throw from Pacific; terrific, savoury PINOT N, zesty CHARD, surprisingly gd PINOTAGE (!).

Freeman RRV, Son Coast ★★★→★★★★ Restrained terroir-driven PINOT N, CHARD from cool-climate SON COAST and RRV, with nod to Burgundy. The Ryo-fu Chard ("cool breeze" in Japanese) is amazing, as is Akiko's Cuvée Pinot N.

Freemark Abbey Napa V ★★★ Classic name claimed by JACKSON FAMILY WINES in 2006, improved. Great values: classic single-v'yd Sycamore, Bosché CAB SAUV bottlings.

Indie film *Sideways* didn't spoil idyllic Santa B County. Still worth going.

Freestone Son Coast ★★★ Fine expression of SON COAST. Intense, racy CHARD, PINOT N from vines only few miles from Pacific show gd structure, long finish, esp Chard. Investment by NAPA V'S JOSEPH PHELPS.

Frog's Leap Ruth ★★★ John Williams, pioneer champion of organic and bio viticulture, coaxes best out of valley floor estate. Supple CAB SAUV, MERLOT, elegant CHARD, popular SAUV BL, great ZIN.

Gallo of Sonoma Son ★★★ Formidable wines from great Son v'yd sources and further, unfussy as founders would have wanted. Fruit quality speaks loudly.

Gallo Winery, E&J ★→★★★ Privately held, secretive company, titan in under-$20 sector. Major CA brands incl Apothic, Barefoot, Louis Martini. Recent buys: Black Box, Clos du Bois, RAVENSWOOD, Jayson. *See also* GALLO OF SONOMA.

Gary Farrell RRV, Son Coast ★★★ Namesake vintner sold it yrs ago, but high performance continues. Excellent PINOT N, CHARD from cool-climate v'yds. Rocholi v'yd and basic RRV Chards both superb.

Gloria Ferrer Car ★★★→★★★★ Exceptional CA bubbly. Toast to decades-long team of owners, growers, winemakers that made this Freixenet-owned venture extraordinary. All wines v.gd, Vintage Royal Cuvée best of all.

Grace Family Vineyard Napa V ★★★★ 05 06 07 09 10 11 12 Stunning CAB SAUV shaped for long ageing. One of few cult wines that might actually be worth price.

Graziano Family Mend ★★★ Best-known for Italian varietals made under Enotria and Monte Volpe labels, namely BARBERA, MONTEPULCIANO, PINOT GRIGIO, SANGIOVESE. Savvy veteran, Champion of MEND CO.

Grgich Hills Cellars Napa V ★★★ Beret-wearing hall-of-famer Mike Grgich built one of NAPA v's great early achievers, esp with CHARD, but CAB, delicious ZIN also blossomed. In latter days he adopted bio growing.

Gundlach Bundschu Son V ★★★ Terrific wines, welcoming vibe, popular tasting destination with adventurous cool Huichica Fest music concerts for hipster set. Best bets MERLOT, CAB SAUV, GEWURZ.

Hahn Santa LH ★★★ Always overdelivers for $. B'x varieties combine MONT/P ROB fruit to great effect, MERITAGE often killer. Lucienne PINOT N releases fantastic.

Hall Napa V ★★★→★★★★ Glitzy ST H winery makes great NAPA CAB SAUV, but bewildering variety of selections. Signature offering best, velvety SAUV BL v.gd, MERLOT among best in CA. Also owns WALT coastal PINOT N, CHARD brand.

Hanzell Son ★★★ PINOT pioneer of 50s still making CHARD, Pinot N from estate vines. Both reward cellar time. Arguably among best of CA. Sebella Chard, from young vines, all bright, crisp fruit.

Harlan Estate Napa V ★★★★ Concentrated, robust CAB SAUV – one of original cult wines only available via mailing list at luxury prices. Still all those things today. Son Will makes The Mascot from younger vines and now Promontory, at OAK.

HdV Wines Car ★★★ Underrated Son gem makes fine complex *Chard* with a honed edge and v.gd PINOT N, from grower Larry Hyde in conjunction with Aubert de Villaine of DRC (*see* France). V.gd CAB SAUV, SYRAH.

Heitz Cellar Napa V ★★★ Once iconic, now steady source of gd CAB SAUV at fair price, sold in 2018. Gd *Sauv Bl*, even GRIGNOLINO.

Hendry Oak K ★★★ Classic, pure, minimalist wines. Est 1939, brambly, distinctive CAB SAUV, ZIN (try Block 28) from cool pocket of valley nr Napa town.

Hess Collection, The Napa V ★★★ Great mtn-top visit with world-class art gallery, also makes gd wine. CAB SAUV from MT VEEDER speciality, esp exceptional 19 Block Cuvée, blockbuster with gd manners.

Hirsch Son Coast ★★★ Pioneer of SON COAST, David H's v'yd won acclaim growing premium grapes; now family label gets cream of crop from towering Pacific ridge. Lithe PINOT N, breathtaking CHARD.

Honig Napa V ★★★→★★★★ Sustainably grown NAPA V CAB SAUV and SAUV BL are nationwide benchmarks thanks to consistent quality, hard-working family and team. Top Cab from Bartolucci v'yd in ST H.

Inglenook Oak ★★★★ FF Coppola's Rubicon reclaims original brand with classic central NAPA V CAB SAUV: balanced, elegant and historic. Also v.gd CHARD, MERLOT. Victorian showpiece.

Iron Horse Vineyards Son ★★★ Amazing selection of 12 vintage bubblies, all wonderfully made. Ocean Res Blanc de Blancs is v.gd, Wedding Cuvée a winner. V.gd CHARD, PINOT N.

Jackson Family Wines ★★→★★★★ Visionary, massive v'yd owner in CA with prime elevated sites, owns popular Kendall-Jackson brand, and high achievers like COPAIN, FREEMARK ABBEY, Hartford Family. Lokoya, MAIANZAS CREEK, Verité. Jackson Estate series great for mtn CAB SAUV.

Jordan Alex V ★★★ Adjustments in grape sourcing led to brilliant revival of balanced, elegant wines from showcase ALEX V estate. CAB SAUV homage to B'x: and it lasts. Zesty, delicious CHARD.

Joseph Phelps Napa V ★★★→★★★★ Expensive NAPA "First Growth" Insignia, one of CA's 1st ambitious B'x blends, still dependably great, as is Napa CAB SAUV. Most offerings excellent quality, esp SYRAH. *See also* Son brand FREESTONE.

Joseph Swan Son ★★★ Long-time RRV producer of intense old-vine ZIN and single-v'yd PINOT N. Often overlooked Rhône varieties also v.gd, esp SYRAH, ROUSSANNE/MARSANNE blend.

Josh N Coast ★★ Shooting-star brand from Joseph Carr. Solid varietal bulk brand successfully competing with GALLO and CONSTELLATION offerings.

Keller Estate Son Coast ★★★ Lump it in with rest of new wave, but one more example of balanced, elegant CA wine coming off the cool coastal regions. PINOT N, CHARD thrilling.

CALIFORNIA

> **Natty wines**
> Natural wine bars and retail shops are everywhere. Trend that once was
> limited to hipster cities like Brooklyn and Oakland is gaining traction
> nationwide. Young consumers, squeezed economically, are exploring
> eccentric wine, but is faulty wine driving them to hard seltzer? My
> advice: ask for a taste before buying.

Kenwood Vineyards Son V ★★→★★★ Landmark SON V producer steadily cranks out
palatable reds and whites. CAB SAUV leads way, esp Jack London v'yd bottling.

Kistler Vineyards RRV ★★★★ Style of PINOT N and CHARD has adapted over yrs,
wines have only improved. Still from a dozen designated v'yds in any given yr.
Highly sought.

Kongsgaard Napa V ★★★★ 5th-generation Napanista John K is valley fixture,
influencer. *Remarkable Chard* from Judge v'yd in SON, excellent CAB SAUV, SYRAH.

Korbel ★★ Cheap fizz sold in grocery stores, but all traditional-method and
remarkably decent for price. And a fun visit by Russian River.

Ladera Napa V ★★★→★★★★ Thesp Stotesbery clan sold HOWELL MTN winery and
set up shop in ST H. Hillside CAB SAUVS, MALBEC great; don't miss superb SAUV BL
from NZ winemaker.

Lagier-Meredith Mt Veeder ★★★ Wine from renowned UC Davis vine researcher
Carole Meredith and oenologist husband Steve Lagier. Handmade, tiny
production of beautiful, pure SYRAH, MONDEUSE, MALBEC and ZIN (whose genetic
ancestry in Croatia Meredith famously decoded).

Lang & Reed Mend, Napa V ★★★ No one in CA has flown CAB FR banner more
passionately than L&R's John Skupny. Wines capture perfume, litheness with
NAPA generosity. Also delicious MEND CHENIN BL.

Larkmead Napa V ★★★★ Historic gravel-laced NAPA V estate revived; *outstanding Cab
Sauv*, supple, balanced; bright, delicious SAUV BL. Rare Tocai FRIULANO a delight.

Laurel Glen Son ★★★ Brand has shrunk, but hillside CAB SAUV Counterpoint from
SON MTN is high-quality and has international clout.

Lewis Cellars Napa V ★★★★ CAB SAUV, CHARDS, SYRAH from former racing driver
Randy Lewis can only be described as full-throttle. Unapologetic NAPA hedonism
at its best.

Lioco N Coast ★★★ Influential minimalist brand champions elegant, subtle PINOT N,
CHARD, CARIGNAN. Wines are dependable, restrained, satisfying.

Littorai Son Coast ★★★★ Burgundy-trained Ted Lemon's N Coast PINOT N, CHARD are
pure, inspiring wines with sense of place. Breathtaking, modern, worth seeking.

Lohr, J ★★→★★★ Prolific producer of Central Coast makes CAB SAUV, PINOT N, CHARD
for balance and gd value. Cuvée Pau and Cuvée St E pay homage to B'x. Don't
miss floral red Wildflower Valdiguié.

Long Meadow Ranch Napa V ★★★→★★★★ Smart, holistic vision incl destination
winery with restaurant, cattle on organic farm. Supple, age-worthy, fresh CAB
SAUV has reached ★★★★ status; lively Graves-style SAUV BL.

Louis M Martini Napa V ★★→★★★ Since buying the Martini brand and epic Monte
Rosso v'yd, GALLO has restored latter to greatness. Martini brand is solid for
workaday CABS, ZINS.

MacPhail Son Coast ★★★ Now owned by HESS COLLECTION, making mostly PINOT N
from cool sites in SON and MEND, highlights are Gap's Crown, Sundawg Ridge
and Toulouse v'yd bottlings.

MacRostie Son Coast ★★★→★★★★ New tasting room is a modern beauty;
screwcapped wines steadily improving. Lovely PINOT N, SYRAH; SON COAST CHARD
is absolute delight.

Marimar Torres Estate RRV ★★★ Great Catalan family's CA outpost issues several bottlings of CHARD, PINOT N. V'yds now all bio. Chards are excellent, long-lived, esp Acero (unoaked, *fresh, expressive*). Pinot N from Doña Margarita v'yd nr ocean is intense, rich.

Masút Mend ★★★ Newish elevated Eagle Peak property run by Ben and Jake FETZER shines brightly. Estate PINOT NS lithe, ethereal. Will inspire others to explore area.

Matanzas Creek Son ★★★ Exceptional JACKSON FAMILY WINES property in cool Bennett V focuses on excellent MERLOT, SAUV BL from lavender-perfumed estate.

Matthiasson Napa V ★★★ Experimental wines have become cult hits. Racy CHARD, elegant CAB SAUV, epic white blend, plus esoterica like RIBOLLA GIALLA, SCHIOPPETINO.

Mayacamas Vineyards Mt Veeder ★★★★ Now owned by Charles Banks, former partner in SCREAMING EAGLE. CA classic has not changed classic big-boned style, only improved. Age-worthy CAB SAUV, CHARD recall great bottles of 70s, 80s.

Meritage Basically a B'x blend (r or w). Term invented for CA but has spread. It's a trademark, and users have to belong to The Meritage Alliance. Insider tip: not a French term – it rhymes with "heritage".

Merry Edwards RRV ★★★ One of great CA winemakers, and PINOT N pioneer. Single-v'yds from SON always wildly popular. Ripe, rounded, layered, tad sweet by today's standards. Slightly sweet musqué SAUV BL also popular.

Miraflores ★★★ Sierra F'hills Marco Cappelli left NAPA V to set up in Sierra Mtns, vinifies subtle, sublime, broad array of wines from an estate, region he rightly believes in.

Mount Eden Vineyards Santa Cz Mts ★★★ V'yd high up in Santa Cz Mts with gorgeous vistas, one of CA's 1st boutique wineries. Taut, mineral CAB SAUV, PINOT N, stunning CHARD since 1945.

Mount Veeder Winery Mt Veeder ★★★ Classic CA mtn CAB SAUV and CAB FR grown at 500m (1640ft) on rugged, steep hillsides. Big and dense wines with ripe, integrated tannins.

Mumm Napa Valley ★★★ At RUTH since 1970. Quality bubbly, notably Blanc de Noirs and pricier, complex DVX single-v'yd left on lees for a few yrs.

Nalle Dry CV ★★★ Family winery crafts elegant, lower alc, claret-style ZIN. Once quite fashionable, still excellent. Great stop nr Healdsburg.

Newton Vineyards Spring Mtn ★★→★★★★ Beautiful estate at base of SPRING MTN, now LMVH-owned; wines have improved recently. Look for CAB SAUV and opulent unfiltered CHARD.

Niner Edna V, P Rob Young, ambitious family estate with excellent CAB SAUV from P ROB, great CHARD and ALBARIÑO from EDNA V. CA cuisine restaurant gd for lunch in P Rob countryside.

Obsidian Ridge Lake ★★★ Star of Lake County extension of Mayacamas mtn range. Super CAB SAUV, SYRAH from hillside v'yds at 780m (2559ft), volcanic soils scattered with glassy obsidian. Half Mile Cab 1st rate. Also owns Poseidon brand from CAR.

Ojai Santa B ★★★ In a change of style from big, super-ripe to leaner, finer, former AU BON CLIMAT partner Adam Tolmach making best wines of his career. V.gd PINOT N, CHARD, Rhône styles. SYRAH-based rosé is delicious.

Next hotbed of innovation? Mont: oversupply of grapes. Just watch.

Opus One Oak ★★★★ Mouton-Rothschild family-controlled standard-bearer for fine NAPA CAB SAUV in gd form; popular luxury export. Wines designed to cellar for 10 yrs+.

Pahlmeyer Napa V ★★★ Jammy, pricey, well-made NAPA V wines, B'x blend, MERLOT, lavish CHARD most notable. Popular volume label Jayson. Now in GALLO portfolio.

Palmina Santa B ★★★ Accurate Italian styles with CA flair, unusual vines. Sumptuous, meant for food. FRIULANO, VERMENTINO, LAGREIN, NEBBIOLO.

Patz & Hall N Coast ★★★★ James Hall is one of CA's most thoughtful winemakers; culls fruit from top v'yds from Central Coast to MEND. Style is generous, tasteful, super-reliable. Zio Tony *Chard* v. special, lemony, electric, opulent.

Paul Hobbs N Coast ★★★→★★★★ Globe-trotting winemaker Paul Hobbs still a local hotshot. Bottlings of single-v'yd CAB SAUV, CHARD, PINOT N, SYRAH are top. V.gd-value second label: Crossbarn.

Peay Vineyards Son Coast ★★★→★★★★ Standout brand from one of coast's coldest zones. Finesse-driven CHARD, PINOT N, SYRAH superb. Second label, Cep, also v.gd, esp rosé. Weightless, impeccably made wines.

Be warned. Napa V has ★★★★ traffic problems.

Pedroncelli Son ★★ Old-school DRY CV winery updated v'yds, winery; still makes bright, elbow-bending CAB SAUV, ZIN, solid CHARD. Refreshingly unpretentious.

Peter Michael Winery Mont, Son ★★★★ *Sir* Peter Michael to you. Brit in Knight's Valley, NAPA V, SON COAST sells mostly to restaurants, mailing list. Quality outstanding: rich, dense CHARD, B'x blend Les Pavots, hedonist's PINOT N.

Philip Togni Vineyards Spring Mtn ★★★★ 09 10 12 14, Legendarily age-worthy SPRING MTN CAB SAUV. Stiff mtn terroir generally needs time, rewards patience. All class.

Pine Ridge Napa V ★★★ Outstanding CAB SAUV from several NAPA V v'yds. Estate STAGS L bottling is silky, graceful. Lively CHENIN BL/VIOGNIER blend innovative classic.

Pisoni Vineyards Santa LH ★★★ Family winery in SANTA LH became synonymous with PINOT N explosion and big, jammy wines. Still, Pinot N is and always was well-made and remains popular.

Presqu'ile Santa MV ★★★ New Central Coast winery, elegantly styled PINOT N, SYRAH. On watch list.

Pride Mountain Napa V, Spring Mtn ★★★→★★★★ Epic Mayacamas mtn-top estate straddles NAPA V and SON border; superb, bold CAB SAUV and amazing MERLOT.

Quintessa Ruth ★★★★ Magnificent estate at heart of NAPA V owned by Chilean international player Augustin Huneeus makes single wine: superb, refined B'x blend justifies triple-digit price.

Qupé Santa B ★★★→★★★★ One of original SYRAH champions, brilliant Rhône range, esp X Block, from one of CA's oldest v'yds. Hillside Estate also epic; don't miss impeccable MARSANNE, ROUSSANNE. Central Coast SYRAH unbeatable for $.

Ravenswood ★★★ Owned by CONSTELLATION; single-v'yd ZINS still from remarkable sites like Bedrock, Old Hill, Teldeschi. "No wimpy wines" motto still applies.

Red Car Son Coast ★★★ Hip, artsy estate-based label making precise CHARD, lacy, fruit-forward PINOT N and killer rosé.

Ridge N Coast, Santa Cz Mts ★★★★ Saintly founder Paul Draper has retired, but his spirit lives on. Majestic, legendary, age-worthy estate Montebello CAB SAUV is always superb. Keep 10 yrs. Outstanding single-v'yd field-blend ZINS are special. Don't overlook top-rank, minerally CHARD.

Robert Mondavi ★★→★★★ Owned by CONSTELLATION since 2004, many wines could be better; changing of winemaking guard appears at hand. Home To Kalon v'yd, still a great site; potential there.

Robert Sinskey Vineyards Car ★★★ Great, idiosyncratic NAPA estate favouring balance, restraint. Impressive CAB SAUV and CAR PINOT N. Racy Abraxas white blend and Pinot rosé excellent.

Rodney Strong Son ★★★ Strong indeed, across the board, from 14 significant v'yds. Sinewy coastal PINOT N, CHARD, super ALEX V CAB SAUV from Alexander's Crown, Rockaway V'yds. Also owns David Bynum.

Roederer Estate And V ★★★★ Adventurous Champagne Roederer venture brought glamour to AND V. Finesse, class off the charts, esp luxury cuvée L'Ermitage. Also makes Scharffenberger fizz. Domaine Anderson PINOT NS also excellent.

Saintsbury Car ★★★ Regional pioneer and benchmark still making v.gd, highly relevant PINOT N, CHARD, yummy Vincent Van Gris rosé.

St-Supéry Napa V ★★★ Bought by owners of Chanel (and Ch Rausan-Ségla, B'x), but some continuity of talent. Tasteful, balanced Virtú (w) and Élu (r) B'x blends and SAUV BL, esp *Dollarhide Ranch*, thrilling.

Sandhi Sta RH ★★★ *See* DOMAINE DE LA CÔTE. Same winemaking team, grapes bought from top local v'yds. Must for lovers of white burgundy: racy, intense CHARD. Gd PINOT N.

Sanford Santa B, Sta RH ★★★→★★★★ Now owned by Terlato family, wines still exceptional: La Rinconada, Sanford & Benedict PINOT N, Santa B County CHARD.

Schramsberg Napa V ★★★★ Best bubbles in CA? Exacting quality in every cuvée, esp luxurious J Schram and Blanc de Noirs. Great tours of historic caves.

Screaming Eagle Napa V, Oak ★★★★ Original "cult" CAB SAUV, famously ripe, rare, and four-figures *cher*. Highly collectible, and now edging towards freshness. Also limited-production SAUV BL. Sister winery is Jonata.

Scribe Son ★★★ Hipster gentleman-farmer aesthetic a hit with younger set. Tasting room pours well-made esoterica like SYLVANER, ST-LAURENT and PINOT N rosé all day.

Sea Smoke Sta RH ★★★ Cultish, high-end, opulent PINOT N, CHARD, sparkling estate-driven model. Great wines, but drink while young (them, and you).

Seghesio Son ★★★ Classic SON ZINS. Rich, strong, but graceful. Old Vine bottling benchmark for price range. Rockpile Zin dynamite.

Shafer Vineyards Napa V, Stags L ★★★→★★★★ Artful, widely respected brand. Hillside Select CAB SAUV a lavish CA classic; Relentless SYRAH/PETITE SIRAH blend powerful, clever. One Point Five a beauty. Fine MERLOT, CHARD from nearby CAR.

Shannon Ridge Lake ★★★ Ambitious large estate in elevated High Valley. Honest wines that overdeliver for $. Fast-growing brand, incl second label Vigilance.

Silverado Vineyards Stags L ★★★ V'yds owned by Walt Disney descendants since 1976, has kept up with times. Single-v'yd Solo CAB SAUV is powerful, smooth; new release of Geo B'x blend from COOMB AVA is dark, dense. CAB FR excellent.

Silver Oak Alex V, Napa V ★★★ Juicy, plush NAPA V and ALEX V CABS still made in consistent, fruit-driven style.

Smith-Madrone Spring Mtn ★★★ Serious, purist producer, daring to dry-farm in lean mtn soils. Superb RIES with brilliant floral briskness. Also powerful CAB SAUV from high-elevation v'yd.

Sonoma-Cutrer Vineyards Son ★★★ Flagship CHARD, classic, by-the-glass pour at restaurants all over country, rich but zesty, bright. Owsley PINOT N from RRV lush.

Sonoma-Loeb Car, RRV ★★★ CHAPPELLET'S Son PINOT N, CHARD label. Sangiacomo v'yd Chard stands out.

Spottswoode St H ★★★★ Crown jewel of ST H, sublime estate always chasing perfection. CAB SAUV pricey, not bulky; worth it. Value Lyndenhurst Cab Sauv 2nd wine, Spottswoode SAUV BL delightful.

Spring Mountain Vineyard Spring Mtn ★★★★ Top-notch estate delivers site-driven, age-worthy mtn wines. Signature Elivette B'x blend layered and sturdy, Estate CAB SAUV v.gd, estate SAUV BL is Rubenesque treat.

CALIFORNIA

Napa Valley: structure from terroir

Is all NAPA V terroir interesting? Some valley-floor v'yds are loamy and undistinguished. Better wines come from benchlands and mtn v'yds where rockier soils produce lower fruit yields, more minerality, concentration, structure. Forested mtn sites infuse wines with signature pine, sage, "madrone scrub" notes in CAB SAUVS from eg. CHAPPELLET on P HILL, DIAMOND CREEK, DUNN, SPRING MOUNTAIN VINEYARD. (If "Garnacha!" is a swear word, Madrone Scrub is something you'd have in the bath.)

Staglin Family Vineyard Ruth ★★★★ Perennial 1st-class, potent CAB SAUV from family-owned estate. Also powerful, complex Salus CHARD.

Stags' Leap Winery Napa V, Stags L ★★★→★★★★ Important, beautiful estate recently restored by corporate owners, exceptional spot to visit (by appt only), wines great now. CAB SAUV leads, but PETITE SIRAH and field-blend Ne Cede Malis have always been special.

Stag's Leap Wine Cellars Stags L ★★★→★★★★ Gd to see quality maintained since founder Winiarski sold to large corp. Flagships still silky, seductive CABS (top-of-line Cask 23, Fay, SLV).

Sterling Napa V ★★ A great spot to visit; take aerial tram to tasting room with 90m (295ft)-high view of valley.

Stony Hill Vineyard Spring Mtn ★★★★ Revered NAPA V estate mostly famous for whites, esp minerally, ageable CHARD, plus RIES, GEWURZ. Sold to LONG MEADOW RANCH (2018). Expect to hold steady.

Sutter Home *See* TRINCHERO FAMILY ESTATES.

Tablas Creek P Rob ★★★ Joint venture between Ch Beaucastel (*see* France) and Hass family with vine cuttings from Châteauneuf. Red and white blends the way to go: Patelin, Côtes de Tablas and Esprit wines all 1st rate, killer VERMENTINO.

Terra Valentine Napa V, Spring Mtn ★★★ Wurtele family lovingly rehabbed this property in early 2000s, then handed it to winemaker Sam Baxter. Mtn CAB SAUV is focus, but always romanticizing SANGIOVESE. Try Amore Cab/Sangiovese.

Terre Rouge / Easton Sierra F'hills ★★★ Single company with two sides: traditional (old vine) ZIN (Easton) and Rhône variety (Terre Rouge). Mostly reds. Affordable Tête-à-tête blend exemplary.

Wine tourism goes luxe: young tourists linger longer in comfy "experiential" tasting rooms.

Trefethen Family Vineyards Oak K ★★★ 2nd-generation winery in cool Oak Knoll Dist known for tasteful, elegant CAB SAUV, MERLOT, CHARD and delicious dry RIES. Value.

Trinchero Family Estates ★→★★★ Major producer; bewildering slew of labels incl mass-market Sutter Home. Esp pleasing Napa Wine Company label CAB SAUV.

Truchard Car ★★★ Estate est 1974, planted to pastiche of B'x, Rhône, Burg; lovely MERLOT, CHARD, ROUSSANNE.

Turley Wine Cellars P Rob ★★★★ Selling mostly to mailing list, so rare in market. Brambly old-vine ZINS from old v'yds scattered across state. True CA treasures.

Unti Dry CV ★★★ Wines start in DRY CV v'yds; grower always refining range of luscious, tasty BARBERA, GRENACHE, SYRAH, ZIN.

Viader Estate Howell Mtn ★★★★ Ripe, powerful expression of HOWELL MTN still turns heads. "V" is marvellous B'x blend based on PETIT VERDOT, CAB FR.

Vineyard 29 Napa V ★★★→★★★★ Top winemaker Philippe Melka's fingerprints all over gorgeous CAB SAUVS at maturing estate venture. Gd but oaky SAUV BL.

Vino Noceto Sierra F'hills ★★ Down-to-earth, classic Cal-Ital winery in Plymouth loved for BARBERA, SANGIOVESE, reasonable prices.

Volker Eisele Family Estate Napa V ★★★ Special site tucked way back in NAPA's Chiles V continues to overdeliver with CAB SAUV and more. Looking for an adventure? Try a twisty Chiles V road trip.

Wente Vineyards ★★→★★★ Oldest continuing family winery in CA makes better whites than reds. Outstanding gravel-grown SAUV BL leads way.

Wind Gap Son Coast ★★★→★★★★ Pax Mahle one of CA's most talented winemakers, esp in cool climates. PINOT N, CHARD v.gd, in best vintages SON COAST SYRAH is genius.

Wine Group, The Central V ★ By volume, world's 2nd-largest wine producer; budget brands like Almaden, Big House, Concannon, Cupcake, Glen Ellen.

Colorado

High desert, and some of highest-altitude v'yds in nation. 136+ wineries. Climate similar to Rhône, Central Coast and Mendoza. CAB FR rising star. AVAs incl Grand Valley and West Elks. Wineries on radar: **Bookcliff** ★★ excellent MALBEC, SYRAH, Res Cab Fr, CAB SAUV, VIOGNIER. **Carlson** ★ family-run winery working with GEWURZ, RIES, LEMBERGER (r) and other fruit wines. **Colterris** premium 100% locally grown grapes; B'x style with callout to Coloradeaux blend (r). **Creekside** v.gd Robusto blend aged in Appalachian oak, gd Cab Fr. **Grande River** focus on traditional B'x, Rhône styles; noted SAUV BL, Lavande Vin Blanc, Viognier blend infused with lavender. **Infinite Monkey Theorem** ★★ counter-culture winery marching to its own beat with fruit sourced from Palisade and Texas, now in a can. V.gd red blend, 100th Monkey, 1st traditional-method sparkling in CO with Grand Valley ALBARIÑO. **Jack Rabbit Hill Farm** ★ only certified bio winery in state, notable Ries, trendy grower ciders. **Snowy Peaks (Grande Valley)** v. high-altitude vines, 100% Colorado grapes Rhône varieties; Oso (r) blend uses hybrid grapes. **Sutcliffe V'yds** v.gd Cab Fr, CINSAULT, CHARD, MERLOT from Welsh winemaker. **Turquoise Mesa** ★ award-winning Vintner's Res, Cabs Sauv and Fr. **Two Rivers** excellent Cab Sauv, v.gd Chard, Ries and Port-style. **Whitewater Hill V'yds** exceptional red blend Ethereal. **Winery at Holy Cross** ★ red-driven historic winery; award-winning Res Merlot, Sangre de Cristo Nouveau; surprising Norton.

Georgia (GA)

Shares Upper Hiwassee Highlands AVA with NC; 1st all-GA AVA Dahlonega Plateau in 2018, rocky hills. CHARD, B'x, PINOT N, PETIT MANSENG. Best: **Crane Creek**, **Engelheim** (PINOT GR), **Frogtown** (SANGIOVESE), **Habersham**, **Sharp Mtn** (Sangiovese, GEWURZ), **Stonewall Creek** (NORTON), **Three Sisters**, **Tiger Mtn**, **Wolf Mtn** (traditional-method sparkling), **Yonah Mtn**.

Idaho (ID)

It's early days here, with only 52 wineries and 1300 acres of v'yds. Growers and winemakers still finding their way, but SYRAH looking to be a star.

Cinder Wines Snake RV ★★ Former Ste Michelle (WA) assistant winemaker Melanie Krause shown knack for high-quality SYRAH, RIES, VIOGNIER also v.gd. **Coiled** Snake RV ★★ One of state's top producers, making tasty dry RIES, SYRAH **Ste Chapelle** Snake RV ★ ID's 1st and largest winery, owned by WA-based Precept Wines. Dry and off-dry style reds and whites, incl quaffable RIES and SAUV BL.

Maryland (MD)

East Shore sandy soils, hills of Garrett and Allegheny mtns, blue-crab-rich Chesapeake Bay checks freezing winters, stifling summers. B'x incl PETIT VERDOT, SAUV BL; MERLOT, ALBARIÑO. For MD-terroirs tastes: **Black Ankle**, 1st post-Prohibition winery **Boordy** (*Sun*-journalist founded), **Bordeleau**, **Crow** (BARBERA r p), **Dodon**, **Elk Run** (PINOT N), **Linganore**, **Links Bridge**, **Old Westminster** (bio, MOSCATO, SYRAH; GAMAY; 90 new hybrids on unfarmed hillside).

Michigan (MI)

Dubbed the "Third Coast" due to huge Lake Michigan. Five AVAs, two downstate, three up n; 3050 acres. Production steadily increasing; 148 wineries, many distracted by cider, fruit wine. Three PINOTS do well, also RIES, CHARD, CAB FR, MERLOT. Terroir, quality: **2 Lads**, **Bel Lago** (incl AUXERROIS); **Big Little** (playful: still Blanc de Noirs, sparkling GEWURZ blends; traditional-method

Pinots N/M); **Black Star Farms** (GAMAY); **Left Foot Charley**, **Mari**, **Mawby** (all-sparkling, incl traditional-method Chard, spontaneous-fermentation blends), **Nathaniel Rose** (single-v'yd focus), **Rove Estate** (fresh whites); **Shady Cellars** (MUSCAT); **Wyncroft** (single-v'yd Pinot N, Chard, BLAUFRÄNKISCH).

5-yr-old Michigan Wine Collaborative promotes sustainability: 16 wineries on board so far.

Missouri
The University of Missouri has a new experimental winery to test techniques and grape varieties in local (warm and humid) conditions. Best so far: Chambourcin, SEYVAL BL, VIDAL, Vignoles (sweet and dry). **Stone Hill** in Hermann produces v.gd Chardonel (frost-hardy hybrid, Seyval Bl x CHARD), Norton and gd Seyval Bl, Vidal. **Hermannhof** is notable for Vignoles, Chardonel, Norton. Also: **Adam Puchta** for fortifieds and Norton, Vignoles, Vidal; **Augusta Winery** for Chambourcin, Chardonel, Icewine; **Les Bourgeois** for SYRAH, Norton, Chardonel, Montelle, v.gd Cynthiana, Chambourcin; **Mount Pleasant** in Augusta for rich fortified and Norton; **St James** for Vignoles, Seyval, Norton.

Nevada
Few commercial wineries. **Churchill V'yds** in high-desert region producing gd SEM/CHARD; all Nevada-grown grapes. **Pahrump Valley** oldest winery here, v.gd PRIMITIVO, ZIN; Nevada Ridge signature label.

New Jersey (NJ)
More than 50 wineries, some among best in e. 150 growers, 1600 acres vines, increasing. Three AVAs: B'x varieties in s Jersey's flat gravelly Outer Coast Plain (incl Cape May); limestone, granite hills in n's Warren Hills for elegant PINOT N, BLAUFRÄNKISCH, SYRAH, RIES, GRÜNER V, GEWURZ; Central Delaware Valley shared with PA. A few passionate explorations, incl Austrian-grapes-focused Mount Salem.

Alba ★★★ Limestone, granite in Warren Hills AVA. One of largest PINOT N plantings on e coast; Burgundy aspiration, incl earthy 30-mths Grand Res. Excellent CHARD, gd RIES, GEWURZ, 30-days-macerated CAB FR.

Beneduce Vineyards ★★★ Family estate; serious PINOT N, CHARD, BLAUFRÄNKISCH, Alto-Adige-style GEWURZ, also made as orange wine.

Mount Salem ★★★ Austrian varieties to match terroir. Wild ferment, min sulphur: no-temp-control RIES; barrel-ferm whites. Outside-fermented reds, incl BLAUFRÄNKISCH, ST LAURENT, ZWEIGELT, CAB FR.

Unionville ★★★ Nutty concentrated RHÔNE W, v.gd SYRAH, aromatic CAB SAUV. Counoise rosé; PINOT N, CAB FR, PICPOUL, MOURVÈDRE. PINOT GR, GEWURZ planted.

Ice age glaciers reached mid-New Jersey, left v'yd-perfect Pattenberg gravelly loam.

William Heritage ★★★ Cape May. Top sparkling (CHARD/PINOT N co-fermented). Meaty MERLOT-led rosè. Attention-getting SYRAH/VIOGNIER. Own-rooted Chard. Experimental side incl pét-nats, piquette, carbonic Chambourcin.

Working Dog ★★★ Barrel-fermented VIOGNER, CHARD have ripeness, lift; SYRAH, MERLOT. Oak-aged CAB FR Retrieves is flagship: creamy, rustic, lasting.

New Mexico
High-altitude v'yds and diurnal variations give wines crisp character and lower alc. French-hybrid grape varieties; sparkling wines excellent. **Black Mesa ★★** red-driven winery with local grapes and award-winning MERLOT, PETITE SIRAH.

Casa Abril family-owned, Spanish and Argentine varieties. **Gruet ★★★** long-time regional benchmark for sparkling, blending WA and Lodi, CA fruit; esp Blanc de Noirs, Brut rosé, Sauvage; v.gd CHARD, PINOT N. **La Chirapada ★** oldest winery in state, 20+ varieties; top-notch Res CAB SAUV; new DOLCETTO and late-harvest RIES. **Noisy Water ★★** ambitious winery breathing life in old region with gd Res CAB, Absolución Cab/MERLOT blend; unfiltered Dirty brand. **Vivác ★★** excellent red blends Divino (Italian grapes), Diavolo (French), v.gd Port-style Amante.

New York (NY)

US's 3rd-largest producer. Ten AVAs, winter freeze and hybrids (increasingly making serious dry wine) throughout; lakes, rivers, ocean influence crucial for vinifera relief. Climate like n Europe's. 113km (70 miles) from NYC: maritime Long Island (Long I) and colder Hudson Valley (Hudson V; most complex soils). Largest is remote Finger Lakes (Finger L; sunlight hours equal to Napa's, over fewer days). Farther n and w: Champlain Valley (sp, Icewine), Niagara Escarpment (NY-rare limestone), Lake Erie (CHARD, sea of grape-juice-bound Concord).

21 Brix ★★ Estate on Lake Erie with 1st-rate RIES, CHARD, GEWURZ, GRÜNER V; aromatic BLAUFRÄNKISCH, CAB SAUV. Serious Noiret. VIDAL Blanc Icewine.

Anthony Nappa Long I **★★** Interesting blends incl traditional-method sparkling PINOT N/VIOGNIER, SAUV BL/skin-contact SEM. From Long I's only bio, organic v'yd: Bordo Antico (CAB FR), La Strega (MALBEC), MERLOT-led rosé. All wild ferment.

Life's a beach: Channing Daughters grows own-rooted Sauv Bl on sand. No phylloxera. Brave, though.

Anthony Road Finger L **★★★** Dry and semi-dry RIES some of US's best. Outstanding GRÜNER V, PINOT GR, CAB FR, LEMBERGER, MERLOT, PINOT N. New styles, same quality: skin-contact CHARD, Ries, barrel-fermented Pinot Gr, GEWURZ. Fresh sweet *federweisser* in early autumn.

Arrowhead Spring Vineyards ★★★ Niagara Escarpment estate est 2006. Noteworthy B'x blends, SYRAH, CAB FR, PINOT N manage richness, spice, grippy tannins via 13% alc and cool-climate acidity. Focused CHARD, 12% alc.

Bedell Long I **★★★** Leading, stalwart Long I estate since 1980. Native-yeasts *pied de cuve*, maritime climate shows in powerful, saline wines. Musée (MERLOT/PETIT VERDOT/MALBEC) is top label; varietal bottlings of same, also SYRAH. VIOGNIER, SAUV BL, CHARD blended and varietal. Artist labels, eg, April Gornik. Chuck Close.

Benmarl Winery Hudson V Pioneer overlooking Hudson River: NY Farm Winery license no.1. Serious, dry, estate Baco Noir, SEYVAL BL. Gd CAB FR, MERLOT, SAUV BL, Blanc de Blancs. BLAUFRÄNKISCH, MUSCAT Ottonel, SAPERAVI coming soon.

Bloomer Creek Finger L **★★★** Min-intervention winemaking; grapes of two sites, separate bottlings. Tanzen Dame RIES in vintage-variance, late-harvest, or EDELZWICKER styles; also GEWURZ, GRÜNER V. CHENIN BL coming soon. White Horse is clever CAB FR/MERLOT blend.

Boundary Breaks Finger L **★★★** Top-notch dry to dessert RIES, all lush, acid-driven. Serious GEWURZ. Expanding: planted CAB FR 2019.

Channing Daughters Long I **★★★** Deliciously experimental wines from S Fork: BLAUFRÄNKISCH, DORNFELDER, LAGREIN, MALVASIA, RIBOLLA GIALLA, single-v'yd FRIULANO and SAUV BL, a range of *pétillants*, plus CAB FR, MERLOT, SYRAH. CHARD made masterfully and playfully, in styles from strong oak influence to hands-off, skin-macerated. Seasonal vermouths from foraged leaves, herbs, petals.

Element Winery Finger L **★★** Christopher Bates MS explores terroir-driven CHARD, RIES, CAB FR, LEMBERGER, PINOT N, SYRAH. Small production, cult status. Latest focus is MERLOT, ripe and herbal in Finger L.

Let's hear it for hybrids

Hybrid grapes like Catawba, Baco Noir, SEYVAL BL have been planted in e US for so long that some are practically autochthonous, a match for freezing winters and casual drinkers' taste for sweet simple wines, but ignored by more serious customers. Now, throughout NY, the search for sustainability (and craft) is setting a new course: hybrids farmed with vinifera-quality attention for concentrated, site-specific, ageable wines, dry or with balanced sweetness. At the forefront is grape-historian Steve Casscles, and there's drinkable proof he's on to something. In Hudson V, HUDSON-CHATHAM's Chelois, Noiret and multi-plot Baco Noirs, BENMARL's old-vine Baco Noir. In Finger L, Chëpìka's Catawba; KEUKA LAKE V'YDS' Vignoles, and even a forgotten barrel of "sherry": RED NEWT's 06 Legacy.

Fjord Hudson V ★★★ Owner/winemaker also 2nd-generation of BENMARL WINERY. V.gd, floral ALBARIÑO. Excellent, vintage-reflecting CAB FR (some spontaneous fermentation) is setting Hudson V standards. CHARD (still and Icewine).

Frank, Dr. Konstantin Finger L ★★★★ Vinifera pioneer, still leading RIES producer. Also v.gd GRÜNER V, PINOT GR, RKATSITELI, SAPERAVI; gd GEWURZ; old-vine PINOT N; impeccable Blancs de Blancs/de Noirs, Ries nature.

Heart & Hands Finger L ★★★ Small production, excellent, and just three grapes: RIES, PINOT N, CHARD for exploring limestone (Devonian, and rare in Finger L) v'yd on shores of Cayuga Lake. Classic cool-climate white, rosé; delicate red.

Hermann J Wiemer Finger L ★★★ One of best RIES producers in US from three main v'yds, incl bio original. Also fine CHARD (now farmed bio too), GEWURZ, CAB FR, PINOT N and superlative fizz. Owns fantastic Standing Stone (SAPERAVI).

Hudson-Chatham Winery Hudson V ★★★ As serious about (dry) hybrids as vinifera: Baco Noir in old-vines and plot-specific bottlings; rare maker of Leon Millot; Chelois; hybrid field blends; CAB FR, PINOT N, CHARD. Res can age 5 yrs+.

Keuka Lake Vineyards Finger L ★★★ Vivacious RIES incl Falling Man from steep slopes on Keuka Lake. V.gd CAB FR. Hybrids incl Vignoles and old-vine Alsatian Leon Millot (cult bottles).

Keuka Spring Vineyards Finger L ★★★ On scenic Keuka Lake. World-class GEWURZ line-up incl site blends, single-site expressions; Alto-Adige-style; 19 esp age-worthy. LEMBERGER, MERLOT, CAB FR.

Lakewood Vineyards Finger L ★★★ 3rd-generation estate. High-quality Res CAB FR. Impressive GEWURZ, PINOT N, multiple RIES. Rare Leon Millot vines.

Lamoreaux Landing Finger L ★★★ Excellent RIES, *Chard*, GEWURZ, Icewine, plus CAB FR, MERLOT, PINOT N. Sparkling incl special 07 Blanc de Blancs release: 11 yrs on lees. Library wines available. Greek Revival building, views of Lake Seneca.

Liten Buffel ★★★ Niagara Escarpment estate PINOT N (c.12% abv), PINOT GR, RIES (vinified whole-cluster) planted on long slope from escarpment to ancient lake ridge. Wild yeasts in neutral oak, no filtering, no sulphur. Noble rot some yrs.

Macari Long I, North F ★★★ Clifftop estate on Long I Sound: bio-minded: cows, pigs, compost. Top-notch SAUV BL. Premium reds: TexSom-winning CAB FR. MERLOT, B'x blends, incl Alexandra in best vintages. Popular Horses is bottle-fermented Cab Fr rosé.

McCall Long I, North F ★★★ Known for top PINOT N, incl single-v'yd, Res, rosé. Gd CAB FR, SAUV BL. Red B'x blends. Also home to French-origin Charolais cattle.

Millbrook Hudson V ★★★ 1st to grow vinifera in Hudson V; estate RIES, CHARD, PINOT N. Single-v'yd Tocai (FRIULANO), CAB FR. Acidity lets reds age a few yrs.

Paumanok Long I ★★★ Racy CHENIN BL, excellent B'x blends, RIES, CAB SAUV. Fine CHARD in steel- or barrel-fermented or Blanc de Blancs, single-v'yd MERLOT, CAB FR; occasional Grand Vintage (all three). Spontaneous ferments, little to no sulphur.

Ravines Finger L ★★★ *Inspired Ries*, GEWURZ, CAB FR, PINOT N, plus sparkling CHARD/ Pinot N. Since 15 Le Petit Capora is Res Cab Fr/CAB SAUV/MERLOT blend, from limestone Argetsinger v'yd on Seneca Lake. Sophisticated bistro.

Red Newt Finger L ★★★ RIES-focused, incl Seneca Lake cru bottlings, top US quality. Elegant GEWURZ, PINOT GR; gd CAB FR, MERLOT, PINOT N. Tierce Ries is collaboration with ANTHONY ROAD, ARROWHEAD SPRING V'YDS. Winery bistro.

Red Tail Ridge Finger L ★★★ Seneca. Super CHARD, RIES (incl one-block, wild ferment), PINOT N, BLAUFRÄNKISCH. Lean, fruity TEROLDEGO, sells out. Sparkling incl Blanc de Noirs, pét-nats, Sekt.

Shaw Vineyard Finger L ★★★ On Seneca Lake, quieter w side. Once trellised, vines grow "wild" with min interference, multiple trunks allowed. Output is full-body, acidity-lifted reds like PINOT N, CAB SAUV, MERLOT; focused RIES, GEWURZ. Orange wine PINOT GR, SAUV BL blend.

Reclaim the suburbs: Floral Terranes harvesting from Long I's abandoned apple trees, cider and wine cellar in two-car garage.

Sheldrake Point Finger L ★★ Cayuga Lake since 1997. Exuberant cool-climate GAMAY, fresh earthy B'x blends, multiple RIES, single-plot PINOT GR, MUSCAT Ottonel.

Sparkling Pointe Long I ★★★ Convincing *fizz*; French winemaker, traditional Champagne grapes, loam soil. Traditional-method. Cuvée Carnaval line (r p w) lets MERLOT into mix.

Whitecliff Hudson V ★★ Site-, soils-driven, incl ex-cherry orchard, quartz-rich historic Olana slope for v.gd barrel-aged GAMAY, CAB FR, PINOT N on limestone ridge. Peachy, stony CHARD; robust Res RIES.

Wölffer Estate Long I ★★★ Premier S Fork estate and destination; classical approach yields quality MERLOT, CAB SAUV, PINOT N. Premium whites incl CHARD made burgundy-style; new maritime-minded SAUV BL. Gd rosé set off Hamptons vacationers' craze for the stuff.

North Carolina (NC)

Long hot summers; dry or rainy yrs. Winters can turn frigid. Four AVAs incl Yadkin Valley. Blue Ridge mtns offer High Country elevation or Piedmont hills for best quality; unremarkable v'yds clustered on old tobacco fields. Plenty of native Scuppernong grape, traditional se US wine. Gd CAB FR, MERLOT, CHARD, VIOGNIER. Best: **Jones Von Drehle** (TEMPRANILLO; Res MALBEC), **Junius Lindsay** (SYRAH), **McRitchie** (dry MUSCAT), **Raffaldini** (SANGIOVESE, MONTEPULCIANO), **RayLen**, **Sanctuary V'yds** (Outer Banks ALBARIÑO), **Shelton**.

Ohio (OH)

Lake Erie moderates winters. Five AVAs. CHARD, RIES, PINOT N, PINOT GR, B'x varieties, MÜLLER-T, DOLCETTO. **Debonne**, family-run **Ferrante** (GRÜNER V, GEWURZ), **Firelands**, **Harpersfield** (KERNER/RIES/MUSCAT Ottonel blend), **Laurentia** (concrete-tank whites), **Markko** (several Chards), **M Cellars** (RKATSITELI); **St Joseph V'yd** (experiments with CORVINA, SANGIOVESE).

The Skeleton Root (Cincinnati) has re-created once-famous Longworth Sparkling Catawba, 150 yrs later.

Oklahoma (OK)

Two AVAs: Ozark Mtn and Texoma, with c.40 wineries. Mostly reds, esp CAB SAUV. **Clauren Ridge** gd Meritage, PETITE SIRAH, VIOGNIER. For better or worse, now on canned wine trend: **Stable Ridge** v.gd Bedlam CHARD; **The Range Winery** 13 varieties of OK grapes; gd white blend Jackwagon.

Oregon (OR)

State made name with Will V PINOT N, but warmer areas make powerful Rhône and Iberian styles; cooler Columbia Gorge makes mtn-inflected whites and Rocks District makes long-lived SYRAH. A string of warm, dry vintages extended 2014–16, resuming 18, though unusually cool, wet weather returned for 19. Climate change and cheap real estate are encouraging both big companies and small, sommelier-led projects to invest: bit of a hotspot for experimentation.

Principal viticultural areas

Southern Oregon (S OR) encompasses much of w OR, s of Will V, incl sub-AVAs Rogue (Rog V), Applegate (App V), Umpqua (Um V) Valleys and Elkton OR. Amid expansive experimentation, Albariño, Gewurz, Grüner V, Viognier, Cab Fr, Grenache, Syrah and Tempranillo are standouts.
Willamette Valley (Will V) has sub-AVAs incl Dundee Hills (Dun H), Chehalem Mts, Ribbon Ridge (Rib R), Yamhill-Carlton (Y-Car), Eola-Amity Hills (E-A Hills), McMinnville (McM) and now Van Duzer Corridor. Chard, Pinots Bl/Gr and Ries excel here, but Pinot N is clearly the star. More new soil-based sub-AVAs are expected: Laurelwood District covers n face of existing Chehalem Mtns AVA; neighbouring Tualatin Hills is w of Portland, while Lower Long Tom is nw of Eugene and will be most s AVA in valley. Mount Pisgah, Polk County will incl Freedom Hill v'yd, a prestigious site with dozens of clients.
Rocks District of Milton-Freewater (Walla Walla Valley [Walla]) entirely in OR, cult wines from Cayuse and dense, rich Syrah from others. Important new projects are Willamette Valley V'yds (Pambrun, Maison Bleue), Force Majeure.

Abacela Um V ★★★ 15′ 16′ 17 (18) Planted 1st TEMPRANILLO in US; Barrel Select v.gd; Fiesta, NV Vintner's Blend for value. Deep Res MALBEC, SYRAH, snappy ALBARIÑO.
Adelsheim Will V ★★★ 15′ 16′ 17 Founders retired, new owners still making reliable PINOT N, CHARD, esp single-v'yd Pinot N (with snazzy new labels). New Staking Claim CHARD and Breaking Ground PINOT N both v.gd value.
Alloro Chehalem Mts ★★★ 15′ 16′ 17′ Beautiful site with elegant PINOT N, CHARD, RIES. Riservata, Justina age v. well.
Andrew Rich Will V ★★→★★★ 15′ 16′ 17 Fine, value Volcanic, Marine Sedimentary PINOT N and v.gd *Sauv Bl*.
Archery Summit Dun H ★★★ 15′ 16 17 Impact of recent winemaker change still uncertain. Try CHARD, and Arcus, Looney and Summit for PINOT N; exotic Ab Ovo PINOT GR fermented in concrete egg.
Argyle Will V ★★ 14 15 16 17 Decent Vintage *bubbly*, esp Brut Rosé; Nuthouse, Spirithouse CHARD, RIES better than spotty PINOT NS.
A to Z Wineworks S OR ★★ 16 17 18′ Value-priced, soundly made, widely available. RIES, CHARD, PINOT GR, rosé best.
Ayoub Dun H ★★★→★★★★ 15′ 16′ (17) Brilliant handling of oak; superb v'yd sources for well-defined, cult-quality CHARD, PINOT N.
Beaux Frères Rib R Robert Parker co-founded with brother-in-law Mike Etzel. Now owned by Maisons & Domaines Henriot. Etzel doing Sequitur label while son Mikey runs Beaux Frères.
Bergström Rib R ★★★★ 15′ 16′ 17′ Elegant, powerful PINOT N, CHARD. Sigrid Chard ethereal, Old Stones Chard v.gd value. Bergström, Shea, Temperance Hill, Winery Block best Pinot N.
Bethel Heights E-A Hills ★★★★ 15′ 16′ 17′ Brilliant old-vine High Wire, Justice, Casteel CHARD; Res Casteel PINOT N dark, muscular. Mid-priced Aeolian v.gd value.
Big Table Farm Will V ★★★ 15′ 16′ 17′ (18) Hand-drawn, letterpress labels; quirky, complex wines: Elusive Queen CHARD, Laughing Pig rosé, all single-v'yd PINOT N.

Brick House Rib R ★★★→★★★★ 15′ 16′ 17 All bio; owned by ex-newsman Doug Tunnell. Fine Cascadia CHARD; Evelyn's, Les Dijonnais and Cuvée du Tonnelier PINOT N show gamey, earthy, textural strengths. Rare GAMAY Noir v.gd.

Brittan Vineyards McM ★★★★ 14′ 15′ 16′ (17) Veteran Robert Brittan makes superb, estate-driven portfolio of austere, age-worthy PINOT N; full-bodied CHARD.

Broadley Will V ★★★ 15′ 16′ 17 Estate (esp Jessica) and purchased (Zenith and Shea) grapes yield spicy, polished PINOT N.

Brooks E-A Hills ★★★ 15′ 16′ 17′ 18 Exceptional RIES (up to 20 cuvées, incl fizz). Also v.gd PINOT BL, PINOT N, Amycas (w blend).

Cowhorn App V ★★★ 15′ 16′ 17 18 Bio, best for dense, age-worthy VIOGNIER, GRENACHE, SYRAH; v.gd Rhône blends (r w).

Cristom Will V ★★★ 15′ 16′ Highly regarded producer of lightly earthy, herbal, long-lived PINOT N from expanding estate v'yds. V.gd VIOGNIER, PINOT GR, rare SYRAH.

DanCin S OR ★★★ 15′ 16 17′ Fine range, v.gd value CHARDS (esp Chassé), PINOT N. Exciting, powerful BARBERA, SYRAH.

Domaine Divio Dun H ★★★★ 15 16′ 17′ 18 French-born/-trained Bruno Corneaux's estate already among OR best. Aromatic, sleek, polished PINOT N, CHARD; lovely Pinot-based rosé and rare OR Passetoutgrain.

Domaine Drouhin Oregon Dun H ★★★ →★★★★ 15 16′ 17′ 1st Burgundy-owned OR winery (1987). Édition Limitée, Louise PINOT N, CHARD best. Newer Drouhin Oregon Roserock Eola-Amity Hills v'yd making equally fine wines, esp Zéphirine Pinot N.

Domaine Serene Dun H ★★★★ 15′ 16′ 17′ Superb single-v'yd CHARD (Récolte, Clos de Lune), PINOT N (Grace and Mark Bradford). Cœur Blanc (w Pinot N); Grand Cheval Pinot N/SYRAH blend. 1st sparklings v.gd.

Dozens OR wineries make racy *méthode champenoise* bubbly (Chard/Pinots N/M).

Elk Cove Will V ★★★ 16′ 17′ 18 V. fine single-v'yd PINOT N, esp Clay Court, Mt Richmond, also v.gd PINOT BL, RIES. New Pike Road wines well made from purchased grapes.

Evening Land E-A Hills ★★★ 15′ 16′ 17 Re-righting the ship after turbulent changes. Estate-grown CHARD, PINOT N. Summum, La Source tops; Seven Springs value.

Evesham Wood ★★★ →★★★★ 15 16 17′ Brilliant, restrained, evocative PINOT N from multiple v'yds. Cuvée J best of all-star line-up.

Eyrie Vineyards, The Dun H ★★★★ 14′ 15′ 16′ 17′ 2nd generation Jason Lett continues pioneering tradition with elegant, age-worthy, low alc wines. Original Vines CHARD, PINOT GR, PINOT N textural wonders; rare Trousseau, PINOT M.

Failla E-A Hills ★★★ 15 16′ 17 CA superstar Ehren Jordan builds OR portfolio with stunning GAMAY and PINOT N, esp Björnson, Eola Springs, Seven Springs v'yds.

Foris S OR ★★ 15′ 16′ 17 18′ Estate-grown RIES, GEWURZ, PINOT BL mirror Alsace with acid-driven, low alc, refreshing dry style. Well-balanced Cedar Ranch PINOT N.

Hyland ★★→★★★ 15′ 16 17 Legacy v'yd with numerous old-vine offerings, incl tart, dry RIES, GEWURZ and medium-bodied PINOT N.

Kelley Fox McM ★★★ 15′ 16 17 One-time Scott Paul winemaker steps out on her own: seductive, elegant old-vine Maresh-v'yd PINOT NS; wonderful rare PINOT BL.

Ken Wright Will V ★★★★ 14 15′ 16′ Superb v'yd knowledge informs deeply fruited PINOT N. Formerly high alc levels have come way down, enhancing aromatics. Overall quality v. high, esp Shea, Bryce, Freedom Hill and Carter cuvées.

King Estate Will V ★★★ 16′ 17 18 Estate now 100% bio. PINOT GR (esp Domaine and Backbone) remains core strength in rapidly expanding portfolio, incl a dozen PINOT NS plus SAUV BL, GEWURZ, CHARD and more.

Lange Estate Will V ★★★ 15′ 16′ 17′ Fine PINOT GR and estate PINOT N; CHARD still tops here. Classique line v.gd value.

Lavinea Will V ★★★→★★★★ 14' 15 16' 17' EVENING LAND alum Isabelle Meunier makes vivid, AVA-defined CHARD, PINOT N from top v'yd sources.

Lingua Franca E-A Hills ★★★ 15' 16' 17' Veteran sommelier Larry Stone co-founded this project, focus on dense, stylish CHARD (Avni, Sisters), PINOT N (Mimi's Mind). French-trained Thomas Savre oversees winemaking.

Ovum ★★★ 16' 17' 18 Artisanal, quirky, ethereal RIES, GEWURZ from both n and s OR. Big Salt (w) blend fine value.

Panther Creek Will V ★★★ 15' 16' 17' Revitalized single-v'yd PINOT N line-up. Lazy River, Carter, Kalita selects best.

Patricia Green Will V ★★★ 15' 16' 17' Cult-calibre line-up of over two dozen single-v'yd PINOT NS. Founder deceased; commitment to value and quality remains. Etzel Block, Bonshaw Block, Mysterious, Notorious superb. Rare OR SAUV BL.

Ponzi Will V ★★★→★★★★★ 15' 16 17 2nd-generation Luisa P making outstanding wines across all price points. Aurora, Abetina PINOT N knockout; Avellana, Aurora CHARD also. Entry Tavola, mid-priced Classico Pinot N v.gd value.

Purple Hands Will V ★★★ 15' 16' 17' Cody Wright (son of Ken) offers juicy, well-balanced portfolio of single-v'yd PINOT N. Holstein, Shea standouts.

Quady North App V, S OR, Rog V ★★→★★★ 16' 17 18' Exploring numerous s OR Rhône and Loire varieties, Herb Quady excels with VIOGNIER and fascinating line-up of rosé. GRENACHE, MALBEC, SYRAH, Pistoleta (w blend) noteworthy.

Résonance Will V ★★★ 14' 15' 16 Jadot's OR project now in dedicated winery; age-worthy PINOT N, CHARD from winemaker Guillaume Large. Estate wines tops; Will V cuvée best value.

Rex Hill Will V ★★★ 15' 16 17 Well-chosen v'yd selects; Jacob-Hart PINOT N tops. Fine, toasty barrel-fermented Seven Soils CHARD.

Roco ★★★ 14 15 16' (17) Former ARGYLE wizard Rollin Soles crafts superb vintage RMS Brut bubbly and layered, age-worthy PINOT N, CHARD. Private Stash is Res, Gravel Rd for value.

Shea Wine Cellars Will V ★★★★ 15' 16' (17) Top-tier winemakers clamour for Shea fruit; in-house wines just as gd. Block selections and Homer PINOT N top the list. Some excellent CHARD also.

Sineann Will V ★★★ 16 17' 18' Brightly fruity PINOT N from WILL V and Columbia Gorge v'yds; TFL and Yates-Conwill tops. Dense, lush GEWURZ.

Sokol Blosser Will V ★★→★★★ 15' 16' 17 Value Evolution series v.gd. PINOT NS Peach Tree and Orchard tops; rather ordinary CHARD.

Soter Will V ★★★ 15' 16 17 CA legend Tony Soter moved to OR to make PINOT N, but his world-class OR bubbly (all styles) is just as gd. Three labels: estate-grown Mineral Springs Ranch; mid-priced North Valley; Planet Oregon for value.

Stoller Family Estate Dun H ★★ 15 16' 17 Gd, rarely great PINOT N, CHARD from expansive estate. Growing Stoller Wine Group now owns Canned Oregon, Chehalem, History and Chemistry brands.

Styring Rib R ★★★ 14' 15' 16' Small estate with superb RIES, black-fruited, toasty estate PINOT N, all value-priced.

Trisaetum Will V, Rib R ★★★★ 15' 16' 17' Meticulous estate wines; all styles of RIES, v.gd PINOT N, CHARD, impressive sparkling under Pashey label.

Walter Scott ★★★ 16' 17' (18') PATRICIA GREEN/EVENING LAND alums craft tense, detailed, buzz-worthy CHARD, GAMAY, PINOT N.

Willamette Valley Vineyards Will V ★★→★★★ 15 16 17 Hundreds of shareholder/owners; extensive v'yd holdings, diverse line-up, principally PINOT N, CHARD. Rocks District AVA now home to Pambrun v'yd and winery. Purchased Maison Bleue.

Winderlea Will V ★★★→★★★★★ 15' 16' 17' Top bio producer with vibrant single-v'yd PINOT N, notably Shea, Weber, Winderlea Legacy. V.gd, age-worthy CHARD. Robert Brittan is consulting winemaker.

Pennsylvania (PA)

200+ wineries. 1723 v'yds (!) Three AVAs, continental climate. Lake Erie-softened nw; gentler temperatures in se, rocky soils, thick cover crops help in rainy growing seasons. RIES, GRÜNER V, CHARD, PINOT N, CAB FR, PETIT VERDOT, MERLOT can fare well. **Allegro**, reliable since 70s; **Fero V'yds** (v.gd SAPERAVI); **Galen Glen**; v.gd ALBARIÑO at **Galer, Maple Springs**; **Karamoor**; **Presque Isle**; **Va La** (cult Avondale field blends incl CORVINA, nine NEBBIOLO clones, six BARBERA clones); **Vox Vineti** (*garagiste* n-Piedmont-style Nebbiolo); **Vynecrest** (luscious lifted LEMBERGER blend, limestone); **Waltz** (SAUV BL); **Wayvine** (gd barrique-aged Carmine),

Rhode Island

Smallest US state, B'x reds, PINOT N, CHARD. **Diamond Hill** (chemical-free farming, min sulphur, barrel-aged Pinot N), **Mulberry V'yds** (PINOT GR, SYRAH), **Newport** (GEWURZ), **Sakonnet** (Gewurz, red blends), **Verde** (biology prof turned small farmer; BLAUFRÄNKISCH; St Croix, other hybrids).

Texas (TX)

A growing number of serious wineries, led by Texas Fine Wine, association promoting state-appellation wines (requiring 75% or more of grapes to be grown in state). Main AVAS of Hill County and High Plains enormous – respectively, nine and eight million square miles – but there's increased focus on regionality to distinguish styles. Med varieties star here: MARSANNE, ROUSSANNE, VIOGNIER, GARNACHA/GRENACHE, MONTEPULCIANO, MOURVÈDRE, TEMPRANILLO. Quality rosé of varying styles on rise.

Becker Vineyards ★★★ Wide range B'x-, Burgundian- and Rhône-styled wines. V.gd TEMPRANILLO Res, Prairie Rotie and MALBEC/PETIT VERDOT blend Raven, Res Newsom V'yd CAB SAUV, Res MALBEC, silky VIOGNIER Res, rosé.

Bending Branch ★★★ Pioneering sustainable winery; Med varieties. TANNAT signature grape (still, sparkling rosé). V.gd Souzão, ROUSSANNE, Newsom V'yds CAB SAUV, PETITE SIRAH.

Brennan Vineyards ★★★ Known for dry VIOGNIER, white Rhône blend Lily; Res TEMPRANILLO; v.gd NERO D'AVOLA Super Nero. MOURVÈDRE dry rosé signature; drink all yr.

Burklee Hill High Plains growers turned winemakers. Worth trying SANGIOVESE- and MONTEPULCIANO-driven blends, fresh rosé

Crowson Wines New up-and-comer with low-intervention natural wines. Fresh, fruity MALVASIA Bianco; rich, intense rosé.

Duchman Family Winery ★★★★ Specializes in Italian varieties. AGLIANICO a consistent award-winner along with DOLCETTO, VERMENTINO. Gd TEMPRANILLO, refreshing Grape Growers (w) Blend; v.gd Salt Lick Cellars GSM and BBQ White.

With rosé on the rise, the Yellow Rose of Texas is now all shades of pink.

Fall Creek Vineyards ★★★ Pioneering Hill Country winery. Excellent B'x blend Meritus, old-vine whites. New pricey ExTerra focused on MOURVÈDRE, TEMPRANILLO, SYRAH from Salt Lick v'yds.

Haak Winery ★★ Gd dry and aromatic Blanc du Bois from coastal area; exceptional "Madeira" copies from a Spanish winemaker.

Kuhlman Cellars Young winery with proprietary red blends driven by PETITE SIRAH. Signature is Kankar (r). Estate CARIGNAN rosé and MARSANNE/ROUSSANNE blend.

Lewis Wines ★★→★★★★ Focus on single-v'yds; rosé specialist, esp Tinta Cão (also red) and MOURVÈDRE. Impressive red TEMPRANILLO.

Llano Estacado ★★★ Historic winery. Excellent MALBEC, 1836 (r w). V.gd Viviana (w),

Viviano (r) that mimics Super Tuscan. Outstanding THP TEMPRANILLO from all-TX fruit. Affordable signature rosé is powerful, exotic.

Lost Draw Cellars ★★★ Small-batch wines, Med varieties: CARIGNAN, PICPOUL Blanc, VIOGNIER. Signature is Gemütlich white blend GRENACHE BL/VIOGNIER/ROUSSANNE. New "Grower Project" experiments.

McPherson Cellars **★★★** Delicious Les Copains, excellent Res ROUSSANNE. Serviceable ALBARIÑO, MARSANNE. 40 yrs+ of winemaking; 1st to plant SANGIOVESE in TX.

Messina Hof Wine Cellars ★→★★★ Big range. Excellent RIES, esp late-harvest. V.gd Papa Paolo Port-style, Res CAB FR, unoaked CHARD.

Pedernales Cellars ★★★★ Top VIOGNIER Res, excellent TEMPRANILLO, GSM. Launched Signature Series of single varietal, single v'yds; classically styled wines.

Perissos Vineyard and Winery ★★ Family-run winery with compelling reds, esp excellent AGLIANICO, PETITE SIRAH, TEMPRANILLO. Strong in Italian blends.

Ron Yates ★★ High achieving, same owner as SPICEWOOD. Focus on Rhône, Spanish, Italian styles. Stellar GSM blend and VIOGNIER, award-winning CAB SAUV from Friesen v'yd.

Southold Farm and Cellar ★★★ Attention-getting Hill Country winery. "Natural" winemakers Regan and Carey Meador, moved to TX after successful run on Long I. Proprietary names like Foregone Conclusion change yearly. Plantings of ALICANTE BOUSCHET, MOURVÈDRE, Terret Noir. Putting pét-nat on radar.

Spicewood Vineyards ★★★ Estate-grown; outstanding Sancerre-like SAUV BL. TEMPRANILLO-driven Good Guy red blend is top, award-winning CAB Claret; v.gd estate SYRAH, dry MOURVÈDRE rosé.

William Chris Vineyards ★★ Pét-nat pioneer. V'yd focused wines; look for MALBEC rosé, MOURVÈDRE, flagship blend Enchante.

Vermont

Mtns, harsh winters, brief, sunny summers, frost, hail, humidity: a few hardy souls plant v'yds anyway, rely on own-rooted hybrids like Frontenac Noir, La Crescent plus RIES and BLAUFRÄNKISCH in extreme n terroirs. **Lincoln Peak** (incl nouveau Marquette), bio-farmed natural-thinking **La Garagista**, same-minded, fizz-focused **Shelburne V'yds**, **ZAFA Wines** pioneers to look for. **Iapetus** is experimental bio side of Shelburne winemaker, 72-day-maceration Marquette.

Virginia (VA)

Continental climate, weather tied to East Coast. Ten AVAs; challenge is to beat humidity, winter freeze, uneven autumns; heights of Shenandoah Valley offer mtn quality for still and sparkling. Some elegant outcomes in classical (lots of rain on well-drained clay soils for CAB FR, PETIT VERDOT) and experimental (hardy, high-acid, Jurançon-native PETIT MANSENG sings in sweet and still; noteworthy high-altitude PINOT N). VIOGNIER, TANNAT favoured too. Plenty of CAB SAUV, MERLOT. V'yd is 79% vinifera (20% of that is CHARD, 16% Cab Fr), 15% hybrids, <1% American incl Norton, US's oldest wine grape.

Ankida Ridge ★★★ Top, low-alc, age-able PINOT N widely considered best in VA, for a lucky few: production is <1000 cases. Steep granite slopes, 518m (1700ft) up in Blue Ridge Mts. CHARD, gd Blanc de Blancs. GAMAY promising.

Barboursville **★★★★** Founded by Italy's Zonin family in Monticello where phylloxera-thwarted Thomas Jefferson tried a century earlier. Elegant B'x-style reds, led by Octagon, plus fine CAB SAUV, NEBBIOLO, PETIT VERDOT. VIOGNIER Res is steady award-winner. Paxxito is luscious VIDAL/MUSCAT Ottonel blend. On-site inn and restaurant set tone for VA country elegance.

Boxwood ★★ Founder of Middleburg AVA. Blends based on CAB FR, MERLOT, plus CAB

SAUV, PETIT VERDOT in both classic B'x style and drink-now; rosé of same grapes; SAUV BL. Short drive from Washington DC.

Chrysalis ★★ In protected agricultural district, producer is advocate for Norton, VA's native grape. Among 1st to grow VIOGNIER; also ALBARIÑO, PETIT VERDOT, TEMPRANILLO.

Early Mountain ★★★ Quality B'x blends (heft, e-coast acidity): luxury-bottling Rise and flagship Eluvium; poised PETIT MANSENG; terroir-driven line-up of four CAB FR; pét-nats of SYRAH, MERLOT. In 2019 released 1st all-TANNAT: vintage 17, single-v'yd. Tasting room pours other VA producers too; winemaker is founding member of state's Winemaker Research Exchange.

VA's experimental roots reach back to 1835, when Dr. Norton created state's native grape.

Gabriele Rausse ★★★ Small, quality estate nr Monticello, owned by VA's 1st commercial grape-grower: Italian viticulturalist who planted BARBOURSVILLLE with Gianni Zonin. Varietal NEBBIOLO, MALBEC, CHARD, PINOT GR. PINOT N as *vin gris*.

Glen Manor Vineyards ★★ 5th-generation historic farm. Vines on steep rocky slopes in Blue Ridge Mts 305m (1000ft)+ up. Began with planting SAUV BL in 1995, now joined by rich CAB FR from 20–30-yr-old vines, off-dry PETIT MANSENG, PETIT VERDOT.

King Family Vineyards ★★★ French winemaker producing dignified, age-worthy Meritage; outstanding, tiny-production *vin de paille*-style PETIT MANSENG; and experimental Small Batch Series line-up (he's a member of Winemaker Research Exchange): recent yrs incl lively no-sulphur CHARD, whole-cluster CAB FR, skin-contact VIOGNER.

Lightwell Survey ★★★ FAIRY MOUNTAIN winemaker Ben Jordan's Shenandoah-Valley-based project for less-known mtn v'yds, fruit from small, top-quality, pioneer growers. Curiosity and skill then guide wild wines: co-fermented SYRAH and RIES; red (CAB FR, Syrah)/white (Ries, PETIT MANSENG) blends; various single-site blends of Ries plus Petit Manseng.

Linden ★★★ Estate founded in n VA in 80s by early believer in site over fruit; VA-wine mentor ever since. Notable high-altitude wines from three sites: rich CHARD, vivacious SAUV BL, savoury PETIT VERDOT, elegant, complex B'x-style red blends often require ageing. Tasting room offers vertical pours.

Michael Shaps Wineworks ★★ Longtime VA producer with solid examples of VIOGNIER, CHARD, luscious PETIT MANSENG; tasty TANNAT and PETIT VERDOT, Raisin d'Être in white (Petit Manseng), red (blend) from grapes dried in old tobacco barns. Custom crusher too.

Pollak ★★ Estate since 2003, wines on VA's international side: heftier CABS SAUV/FR, MERLOT, Meritage; lush, spicy VIOGNIER, creamy PINOT GR.

Ramiiisol ★★★ New, no-expense-spared CAB FR from Blue Ridge Mtns, v. elegant.

RdV Vineyards ★★★★ B'x-inspired red blends, from granite soil hillside. Elegance, complexity, power. MERLOT-led Rendevous; CAB SAUV–driven Lost Mountain label is VA's 1st $100 wine.

Veritas ★★★ Solid estate, founded 1995, incl steep 20-yr-old forest v'yds. Concentrated, floral CAB FR can age 10 yrs+. Paul Schaffer PETIT VERDOT is flagship wine, for tannin lovers. Restrained SAUV BL, richer VIOGNIER. Gd CHARD, MERLOT. Soon traditional-method sparkling too, at producer's new Virginia Sparkling Company: VA-sourced grapes.

Washington (WA)

With a modern history just 50 yrs old, WA is a young industry in a period of grand discovery. It has 1000-odd wineries, but only one-tenth the acreage of CA (though more than twice that of OR). CAB SAUV is dominant, making up over one-quarter of production and rising. Other key varieties are CHARD, MERLOT, RIES,

TEXAS-WA

SYRAH. However, 85 varieties are planted, with some niche ones making the state's best wines. WA is dominated by small producers, with the best wines seldom travelling far outside the state's borders but worth seeking out. The rising-star status of the area means many wines offer v.gd value – for now.

Principal viticultural areas

Columbia Valley (Col V) Huge AVA in central and e WA with a touch in OR. High-quality Cab Sauv, Merlot, Ries, Chard, Syrah. Key sub-divisions incl Yakima Valley (Yak V), Red Mtn, Walla AVAs.

Red Mountain (Red Mtn) Sub-AVA of Col V and Yak V. Hot region known for Cabs and B'x blends.

Walla Walla Valley (Walla) Sub-AVA of Col V with own identity and vines in WA and OR. Home of important boutique brands and prestige labels. Syrah, Cab Sauv and Merlot.

Yakima Valley (Yak V) Sub-AVA of Col V. Focus on Merlot, Syrah, Ries.

WA bonded its 1000th winery in 2019. In 2000 there were 74.

Abeja Col V, Walla ★★★ High-quality COL V CAB SAUV, CHARD. VIOGNIER also v.gd.

Andrew Will Col V, Red Mtn ★★★→★★★★ 10' 12' 14' Long-time producer of some of state's best and most age-worthy B'x blends. Sorella flagship. Champoux V'yd also top notch but no misses in line-up. Involuntary Commitment second, value label.

Avennia Yak V, Col V ★★★ 10 12' 14' 16' Rising star focusing on old vines, top v'yds. Cellar-worthy B'x, Rhône styles. Sestina B'x blend and Arnaut SYRAH tops. SAUV BL v.gd. Lydian value label.

Betz Family Winery Col V ★★★→★★★★ 10 12' 14' 16' Long-time producer of high-quality Rhône, B'x styles. New winemaker Louis Skinner has reserved style. Pére de Famille CAB SAUV and La Côte Patriarche SYRAH consistent standouts. Untold Story gd value.

B Leighton Yak V ★★★ Side project for Brennon Leighton, former white winemaker at CHATEAU STE MICHELLE and current winemaker for K VINTNERS. Rhône blend Gratitude, SYRAH, PETIT VERDOT v.gd.

Cadence Red Mtn ★★★ 10' 12' 14 16' Producer of some of state's best, most age-worthy wines. Dedicated to B'x-style blends from RED MTN fruit. Coda from declassified barrels exceptional value.

Cayuse Walla ★★★★ 10 11 12' 14 16' Mailing list only, and yrs-long wait. All estate-v'yd wines from WALLA; stratospheric scores. Earthy, savoury SYRAH (esp Cailloux, Bionic Frog) and GRENACHE worth seeking out. Sister wineries Hors Categorie, Horsepower, No Girls also top quality. Beg, borrow or steal.

Charles Smith Wines Col V ★★ Focus on value wines. Look for Kung Fu Girl RIES, Chateau Smith CAB SAUV.

Chateau Ste Michelle Col V ★★ →★★★ Largest single producer of RIES in world; all prices/styles: v.gd quaffers (excellent COL V Ries) to TBA-style rarities (Eroica Single Berry Select). Gd-value reds and whites from Col V as well as estate offerings from Cold Creek, Canoe Ridge.

Col Solare Red Mtn ★★★→★★★★ 10 12' Partnership between CH STE MICHELLE and Tuscany's Antinori. Focus on CAB SAUV from RED MTN. Complex, long-lasting.

Columbia Crest Col V ★★ →★★★ WA's top-value producer and by far largest winery makes oodles of v.gd, affordable wines under Grand Estates, H3, Res labels. Res wines offer v.gd value esp CAB SAUV and Walter Clore (r).

Corliss Estates Col V 08' 10 12' WALLA producer of cult B'x blend and CAB SAUV. Extended time in barrel and bottle before release. Sister winery Tranche v.gd value.

Côte Bonneville Yak V ★★★ 09' 10 12' Estate winery for highly regarded DuBrul v'yd making age-worthy wines in sophisticated style, additional bottle-age before release. *Carriage House* v.gd value. Don't overlook CHARD.

DeLille Cellars Col V, Red Mtn ★★★ 10' 12' 14' 16 Long-time Woodinville producer of B'x and Rhône styles. Chaleur Blanc one of state's best whites. Chaleur Estate red age-worthy. D2 red v.gd value. No misses in line-up.

Doubleback Walla ★★★ 10' 12 Former footballer Drew Bledsoe makes one wine, CAB SAUV, feminine expression of WALLA fruit, cellaring potential. Bledsoe Family sister winery making v.gd SYRAH. Bledsoe-McDaniels new OR PINOT project.

Dusted Valley Vintners Walla ★★★ B'x and Rhône styles. Stoney Vine SYRAH worth seeking out. Stained Tooth Syrah gd value. Boomtown value label.

Efeste Yak V, Col V, Red Mtn ★★★ Woodinville maker of zesty RIES, racy SAUV BL from cool Evergreen V'yd, also v.gd SYRAH, old-vine CAB SAUV. Final Final gd value.

Fielding Hills Col V ★★★ Small producer, estate v'yd on Wahluke Slope. CAB SAUV, MERLOT outstanding plus v.gd value.

Figgins Walla ★★★ 10 12 14 16 2nd-generation winemaker Chris F also makes wine at famed LEONETTI. Single-estate B'x blend, RIES from WALLA fruit. Toil OR PINOT N project.

Force Majeure Red Mtn ★★★ Estate B'x and Rhône styles. VIOGNIER a standout. Recently moved to WALLA.

Gorman Red Mtn ★★★ Rich, ripe wines. Evil Twin standout CAB SAUV/SYRAH blend. Ashan CHARD project.

Gramercy Cellars Walla ★★★ 10 12' 13 Master sommelier Greg Harrington produces lower-alc/oak, higher-acid wines made to go with food. Speciality earthy SYRAHS (esp Lagniappe), herby CAB SAUV. MOURVÈDRE sneaky gd. Lower East value label.

Hedges Family Estate Red Mtn ★★★ Long-time family winery, bio wines. Check out estate red blend and CAB SAUV. CMS blend gd value.

Januik Col V ★★★ 10 12' Former CH STE MICHELLE winemaker Mike Januik makes v.gd-value B'x styles and some of state's best CHARD. Single v'yds a step above. Novelty Hill sister winery. Son Andrew assists and has eponymous label.

Kevin White Winery Yak V ★★★ Woodinville producer of Rhône styles focusing on purity. Production small, value outrageous.

K Vintners Col V, Walla ★★★ Owner and former rock band manager CHARLES SMITH focuses on single-v'yd SYRAH, plus Syrah/CAB SAUV blends. Sixto CHARD-focused sister winery. Other brands CasaSmith, Substance, ViNo.

Latta Wines Col V ★★★ Former K VINTNERS winemaker Andrew Latta makes stunning GRENACHE, SYRAH, MOURVÈDRE. ROUSSANNE one of state's best whites. Latta Latta gd value. Disruption side project, value.

L'Ecole No 41 Walla ★★★ 10 12' 14 Wide range; Ferguson top B'x blend. CHENIN BL, SEM v.gd value.

Col V a desert, just 12–20cm (5–8in) rain/yr. No irrigation = no grapes.

Leonetti Cellar Walla ★★★★ 08 10' 12' 14 Celebrated 40th harvest in 18. Well-deserved cult status, steep prices for cellar-worthy CAB SAUV, MERLOT, SANGIOVESE. AGLIANICO recently added. Res B'x blend flagship.

Long Shadows Walla ★★★→★★★★ Former Ste Michelle Wine Estates CEO Allen Shoup brings a group of globally famous winemakers to WA to make one wine each. Poet's Leap RIES one of best in state, but all high-quality, worth seeking.

Mark Ryan Yak V, Red Mtn ★★★ Big, bold but still refined B'x and Rhône styles. MERLOT-based Long Haul and Dead Horse CAB SAUV stand out. Board Track Racer second label gd value.

Milbrandt Vineyards Col V ★★ Focus on value. The Estates are higher-tier, single-v'yd wines from Wahluke Slope and Ancient Lakes. Look for PINOT GR, RIES.

Northstar Walla ★★★ 10 MERLOT-focused winery. Premier (made to age) is gorgeous. Also v.gd CAB.

Owen Roe Yak V ★★★ SYRAH, CAB SAUV and B'x blends emphasizing elegance and restraint. CHARD also v.gd.

Pacific Rim Col V ★★ RIES specialist, with oceans of tasty, inexpensive yet eloquent Dry to Sweet and Organic. Chinese-inspired labels. For more depth try *single-v'yd releases*.

Passing Time Col V ★★★ Project from former NFL quarterbacks Dan Marino and Damon Huard focusing on CAB SAUV. Wines made by Chris Peterson (AVENNIA). Horse Heaven Hills wine tops. Quickly earning cult status.

Pepper Bridge Walla ★★★ B'x styles: Pepper Bridge and Seven Hills v'yd blends among WA's best. Give time in cellar.

Quilceda Creek Col V ★★★★ 04' 07 10 12' 14' 16' Flagship producer of cult CAB SAUV known for ageing potential. One of most-lauded producers in US. Sold by allocation. Find it if you can.

Most WA wineries are w of Cascade Mtns, nr Seattle. Most v'yds to e.

Reynvaan Family Vineyards Walla ★★★ →★★★★ 10' 11 12' 14 16 Estate v'yds in Rocks District and foothills of Blue Mtn. Dedicated to SYRAH, CAB SAUV, Rhône-style whites. Wait-list winery but worth it.

Rôtie Cellars Walla ★★★ Rhône-style specialist. Northern Blend consistent standout. Also v.gd GRENACHE BL.

Saviah Cellars Walla ★★★ Under-the-radar producer of exquisite SYRAH, B'x blends. Funk V'yd Syrah is knee-buckler. Une Vallée top B'x blend. The Jack label gd value.

Seven Hills Winery Walla ★★★ 10 12' 14 V. respected for age-worthy reds made in reserved style. Recently bought by Crimson Wine Group but same winemaker.

Sleight of Hand Walla ★★★ Trey Busch makes dazzling B'x blends and Rhône styles. Funkadelic SYRAH from Rocks District worth seeking out. Also v.gd CHARD, RIES. Renegade value label.

Sparkman Cellars Yak V, Red Mtn ★★★ Woodinville producer focuses on power. Stella Mae and Ruby Leigh B'x blends offer excellent quality, value. Evermore Old Vines CAB SAUV cellar-worthy.

Spring Valley Vineyard Walla ★★★ Estate reds. Uriah MERLOT B'x blend consistent standout. Katherine Corkrum CAB FR also superb.

Syncline Cellars Col V ★★★ Rhône-dedicated producer in Columbia Gorge; distinct, fresh style. Subduction Red is v.gd value. NB MOURVÈDRE. Sparkling GRÜNER V delicious. PICPOUL consistent standout.

Tamarack Cellars Col V ★★★ CAB SAUV, MERLOT and CAB FR. Firehouse Red gd value.

Waterbrook Walla ★→★★ Large gd-value producer: multiple styles and many price points.

Woodward Canyon Walla ★★★★ 07 10 12' 14 Founding WALLA producer, B'x styles. *Old Vines Cab Sauv* is complex, age-worthy. CHARD consistently best in state. Nelms Road value label.

WT Vintners Yak V ★★★ Sommelier-winemaker Jeff Lindsay-Thorsen makes single-v'yd SYRAH, GRENACHE and GRÜNER V with restraint, purity.

WA is 2nd biggest producer of Ries after Germany.

Wisconsin

Wollersheim Winery (est 1840s) is one of best estates in midwest, with hybrid and Wisconsin-native American hybrid grapes. Look for Prairie Fumé (SEYVAL BL), Prairie Blush (Marechal Foch).

Mexico

Wine tourism is on the rise in the Guadalupe Valley in Baja California, with new wine resort projects; a far cry from its somewhat rustic past. It might still lack Napa's polish or Europe's patina, but that's its charm. There are over 100 wineries on the Ruta del Vino, and the valley's "godfather", Hugo d'Acosta, has trained some 300 winemakers in his wineries and oenology school, with alumni moving on to start their own endeavours; the valley is now known for its food. Most wineries are small-scale, most of the grape varieties are Mediterranean, and benefit from high-altitude, cool-climate sites. The better wines are produced on hillsides where the water comes from mountain springs. Flavours are ripe, rich and often with a touch of rusticity to keep it real.

Adobe Guadalupe ★★★★ Hugo d'Acosta helped est this showcase winery. Wines incl B'x blends, named after angels. Serafiel blend of CAB SAUV/SYRAH is top. Fresh Jardín Romántico CHARD is the only varietal.

Alximia Founded by mathematician-turned-winemaker, featuring Italian, Spanish, French varieties grown organically. Vertical Libis Special Res (r) blend v.gd.

Bichi New family-run winery in Tecate specializing in natural wines from heritage vines. Look for trendy pét-nat and No Sapiens (r).

Bruma Valle de Guadalupe Chic winery attached to luxury hotel has young B'x-trained Lulu Martinez Ojeda as winemaker plus top chef in kitchen.

Carrodilla, Finca La 1st certified organic winery in Valle de Guadalupe. Focus on monovarietals; CAB SAUV, SHIRAZ among best; gd Canto de Luna blend.

Casa de Piedra ★★★ Modern winery in historic stone house, 1st project of Hugo d'Acosta. V.gd blend of CAB SAUV/TEMPRANILLO; San Antonio de las Minas CHARD.

Château Camou ★★★ Pioneering, French-inspired winery; serious B'x credentials. Award-winning Gran Vinos: complex, elegant CAB SAUV-driven, worthy of ageing.

Pioneering wineries like Bodegas Magoni considered the OG of Guadalupe Valley.

Cielo Winery, El Another wine-resort project with a range of wines named after heavenly elements. Stars in the making.

Mogor Badan ★★ Small-production passion project; B'x informed.

Monte Xanic ★★→★★★ 1st modern premium winery here with excellent CAB SAUV, v.gd MERLOT. Awarded whites; look for SAUV BL, unoaked CHARD, fresh CHENIN BL. Interesting CAB FR Res Limitada. Now trying high-elevation PINOT N.

Nubes, Las 11-yr-old sustainable winery with French focus. Gd Res red blends, NEBBIOLO and Kuiiy, fresh SAUV BL-driven blend.

Paralelo ★★ Early pioneering eco-constructed winery by Hugo d'Acosta, ultra-modern. Small production. Emblema SAUV BL, two versions red B'x-style Ensamble.

Pijoan, Viñas Honest *garagista*-style wines named for the winemaker's family. Mostly French varieties. B'x blend Leonora is flagship wine. Approachable Coordinates line.

Tres Mujeres ★★★ Rustic co-op owned by women who also cook on site. Top TEMPRANILLO; v.gd GRENACHE/CAB SAUV, La Mezcla del Rancho; Isme MERLOT.

Tres Valles ★★ Est 1999, producing powerful reds from Guadalupe, San Antonio, San Vicente Valleys. V.gd Kuwal blend driven by TEMPRANILLO. Top-rated single varieties: Maat (GRENACHE), Kojaa (PETITE SIRAH) cult quality.

Vena Cava ★★★ Hip winery, founded by expats and constructed from reclaimed fishing boats. Well-priced, modern, organic wines, bottles signed by winemaker. Focus on CAB SAUV, SAUV BL. V.gd, complex, oak-aged Cab Sauv, TEMPRANILLO.

Canada

Canada's rather unexpected vineyards range from British Columbia to Ontario, Quebec and Nova Scotia. Their wines started well in the 90s and have reached standards that surprise everyone. There is an ever-expanding list of grapes: white favourites are Chardonnay and Riesling, while Pinot Noir, Syrah, Cabernet Franc and red blends lead a style shift to reds with more finesse. There are new wild cards too: Albariño, Gamay, Grüner Veltliner, L'Acadie Blanc, Malbec, Marsanne, Viognier and more. Young, enterprising and growing, the story here is about fresh, cool-climate wines well worth adding to your must-taste list, even if the vogue for Icewine has faded. Canada can do better things.

Ontario

Prime appellations of origin: Niagara Peninsula (Niag), Lake Erie North Shore (LENS) and Prince Edward County (P Ed). Within the Niagara Peninsula: two regional appellations – Niagara Escarpment (Niag E) and Niagara-on-the-Lake (Niag L) – and ten sub-appellations. New LENS sub-appellation South Islands.

Bacchelder Niag r w ★★★ 15′ 16′ 17 18′ Thomas B is the passionate conscience of Niag; pure, precise, elegant, age-worthy CHARD, PINOT N. Book to visit.

Cave Spring Niag r w sw (sp) ★★★ 16′ 17 18′ (19) Regional pioneer with impressive old-vine v'yds. Elegant CSV wines age effortlessly: RIES, CAB FR, CHARD, late-harvest, Icewine.

Henry of Pelham Niag r w sw sp ★★★ 16′ 17 18′ (19) 6th-generation, clay-limestone soils, CHARD, RIES, Baco Noir. Top label Speck Family Res, Cuvée Catherine Brut, Ries Icewine.

Hidden Bench Niag r w ★★★★ 16′ 17 18′ 19 Reference producer of RIES, PINOT N, CHARD; certified organic v'yds: Felseck, Locust Lane, Rosomel; top picks Nuit Blanche, Tête de Cuvée Chard, Felseck Ries.

Inniskillin Niag r w sw ★★★ 16′ 17 18′ (19) Icon Icewine pioneer, plus juicy Res RIES, PINOT GR, PINOT N and fine CAB FR; Single-v'yd series CHARD, Pinot N in best yrs.

Malivoire Niag r p w (sp) ★★★ 16′ 17 18′ (19) Promotes economic, social and environmental success. V.gd CAB FR, CHARD, GAMAY, GEWURZ, PINOT N, trio of rosés.

Norman Hardie P Ed r w ★★★★ 15 16′ 17 (18) Iconic P ED pioneer hand-crafting CAB FR, CHARD, PINOT N, RIES on limestone-clay. Cuvée L Chard, Pinot N in best yrs.

Organized Crime Niag r w ★★★ 17 18′ (19) Cool-climate, craftsman-styled, hands-on winemaking, moderately priced, quality CAB FR, CHARD, RIES.

Prince Edward County Home to some 40 producers at e end of Lake Ontario on limestone-rich soils. High-quality CHARD, PINOT N. Casa Dea, Closson Chase, Hinterland, Huff, Rosehall Run, The Grange.

Ravine Vineyard Niag r p w sp ★★★ 17 18′ (19) Organic 14-ha St David's Bench v'yd spans ancient riverbed. Top Res CAB FR, CHARD; drink-now: Sand and Gravel.

Stratus Niag r w ★★★★ 15 16′ 17 (18) Winemakers JL Groux, v.gd CAB FR/GAMAY blends and Charles Baker age-worthy RIES both speak to origin and terroir.

Canada's 49er fizz

Canadian fizz is all grown up, and each region brings a distinctive style to the table. In the Ok V and on Vancouver Island it is pure, fruit-forward and bright; in Niag and P ED the bubble is all about lean fruit and toast, while Nova Scotia's Annapolis Valley gives super-bright acidity and high-toned minerality. Did we mention we share the 49th parallel with Champagne?

Tawse Niag r w (sp) p ★★★★ 16' 17' 18' (19) Steady, award-winning certified organic/bio producer; v.gd CHARD, RIES; gd CAB FR, MERLOT, PINOT N.

Thirty Bench Niag r w ★★★ 17 18' (19) Limited Small Lot RIES, from original single-block, old-vine Post, Triangle and Wood Post estate v'yds on Beamsville Bench.

Two Sisters Niag r p w ★★★ 14 15 16' 17 18' (19) New to Niag scene, less is more, aged, high-end estate reds CAB FR, CAB SAUV, MERLOT; flagship Stone Eagle blend.

British Columbia

Geographical Indications for BC wines of distinction and BCVQA are BC, Fraser Valley, Gulf Islands, Kootenays, Lillooet, Okanagan Valley (Ok V), Golden Mile Bench, Okanagan Falls, Naramata Bench, Skaha Bench (subdivisions of Ok V), Shuswap, Similkameen Valley, Thompson Valley, Vancouver Island.

Blue Mountain Ok V r w sp ★★★ 17 18 (19) 2nd generation renews fizz legacy incl smart RD versions; reliable age-worthy PINOT N, CHARD, new single-block Pinot N series.

CedarCreek Ok V r w ★★★★ 16' 17 18 (19) Electric aromatic RIES, GEWURZ; Ehrenfelser, Platinum single-v'yd blocks PINOT N, CHARD. New restaurant, visitor centre.

Checkmate Ok V r w ★★★ 15' 16' 17 18 Sophisticated CHARD (six) and MERLOT (five) from micro-blocks. Top w: Attack, Queen Taken. Top r: End Game, Silent Bishop. Modern visitor centre.

Cowichan Valley Vancouver Island region: exciting PINOT N, PINOT GR, ancient volcanic soils. Top: Averill Creek, Blue Grouse, Emandare, Rathjen, Unsworth.

Haywire Ok V r p w sp ★★★ 16' 17 18 (19) Summerland growers use natural, pét-nat, organics, concrete ferments, amphorae, for wines of precision and grace. CHARD, PINOT GR, PINOT N, GAMAY, sparkling.

Ok V is a true desert: water comes from lake, 245m (800ft) deep, and snowmelt.

Martin's Lane Ok V r w ★★★★ 15 16' 17 18 PINOT N and equally compelling RIES from single v'yds in Naramata and Kelowna.

Mission Hill Ok V r p w (sp) ★★★★ 15 16' 17 18 (19) Large, high-quality: Legacy, Terroir, Rcs and Five V'yds series. Experiential visitor centre, restaurant.

Nk'Mip Ok V r w ★★★ 16' 17 18 (19) Bright RIES, top-end Qwam Qwmt CHARD, SYRAH, Mer'r'iym White. Part of $25 million aboriginal resort, Desert Cultural Centre.

Osoyoos Larose Ok V r ★★★ 16' 17 18 (19) B x-based Groupe Taillan owns this 33-ha single v'yd. Track record of age-worthy Le Grand Vin echoes B'x.

Osoyoos Indian Band elders call Ok V *garrigue Sín i tmx ulax*, "sin eet tim whoo lough".

Painted Rock Ok V r w ★★★ 16' 17 18 (19) Skaha Bench, steep, 24-ha, 14-yr-old estate v'yd below 500-yr-old native pictographs. SYRAH, CAB FR, CHARD, signature red Icon.

Phantom Creek r w ★★★ 16' 17 18 (19) New, with v'yds on Black Sage Bench, Golden Mile Bench, Similkameen. Impressive SYRAH, CAB SAUV, PINOT GR, red blends. Consultants Phillipe Melka, Olivier Humbrecht.

Quails' Gate Ok V r w ★★★ 16' 17 18 (19) Inspired style. Fresh, aromatic RIES, CHENIN BL; continued refinement of core PINOT N, CHARD and limited Collector Series.

Road 13 Ok V r w (sp) ★★★★ 16' 17 (18) Excellent Golden Mile Bench producer of Rhône style VIOGNIER, SYRAH and treasured old-vine CHENIN BL (w sp), plus premium Jackpot series, PINOT N, CAB FR.

Tantalus Ok V r w sp ★★★ Electric producer of lauded old-vine RIES, silky PINOT N and quality CHARD farmed sustainably, produced in Leeds facility.

Nova Scotia

Benjamin Bridge sp ★★★ 04' 12' 14 15 Traditional-method fizz. Excellent, age-worthy Vintage and NV Brut from CHARD/PINOTS M/N BLENDS.

South America

CHILE

Chile's diversity isn't new. Its wine got serious in the 19th century, then lagged behind with primitive methods until the 90s, when stainless steel replaced old wooden vats. It all started round Santiago, the capital, with wines modelled on, and vines coming from, Bordeaux. Now new (and some old) planting ranges from the Atacama Desert to Patagonia, from classy Bordeaux blends of the Andes and signature Carmenère from the valley floor, to the racy Sauvignon Blanc and hard-to-beat Chardonnay of the coast, down to the distinctive old-vine Carignan, Cinsault, Muscat and País of Maule, Itata and Bío-Bío, the Riesling of Malleco and crunchy Pinot Noir of Osorno. Every variety finds a niche somewhere in this great long, lanky country.

Recent vintages

While Chile's extremes will often experience different vintage conditions, the Central Valley is reasonably consistent. Both 18 and 19 were excellent, with rich, concentrated reds and balanced whites. But a mega-drought over the past several years has brought increasing complications to a country that relies heavily on irrigation.

Abolengo r ★★ Promising 1st CARMENÈRE/SYRAH from Peumo, Cach.

Aconcagua Region stretching e-w, all n of Santiago. Mtn reds rich and fruity, while coastal CHARD, SYRAH, PINOT N give fresh face to traditional region.

Almaviva Mai ★★★★ Grande dame of MAI since mid-90s. Opulent B'x blend from top Puente Alto terroir is collaboration between Rothschild family (Mouton) and CONCHA Y TORO.

Altaïr Wines Rap ★★★ Finest wine of SAN PEDRO, named after a star, leading light in Cach. Complex B'x blend from Andes.

Antiyal Mai ★★★ Family winery of top bio consultant in Chile, Alvaro Espinoza. Heartfelt, earth-conscious, elegant reds from MAI.

Apaltagua ★★ Based in Col, but with 390 ha all over major valleys. Gd-value, wide portfolio incl Pacifico Sur brand.

Aquitania, Viña Mai ★★★ Lazuli is one of MAI's most polished CAB SAUVS, but this classy producer excels in many reds incl v.gd CHARD, PINOT N, SAUV BL from Mal.

Arboleda, Viña Aco ★★→★★★ Fresh, modern wines from ACO Costa by ERRÁZURIZ clan. Try SYRAH, SAUV BL, CHARD, PINOT N.

Aristos Cach ★★★→★★★★ Burgundy's L-M Liger-Belair's boutique brand with Pedro Parra and François Massoc. Complex, layered CAB SAUV, CHARD from Cach Andes.

Bío-Bío Milder climate in s region offers fresh styles of RIES, SAUV BL, PINOT N. Revived interest in old vines (PAÍS, MUSCAT) also shining new light on region.

Bouchon Mau ★★→★★★ Top producer with renewed vision, energy. Vibrant reds, stellar SEM. Putting PAÍS on map.

Caliboro Mau ★★→★★★ Small but smart collection of reds and old-vine, botrytized Torontel. Organic, boutique winery of Count Cinzano (Italy).

Calyptra Cach ★★★ High-altitude v'yds and winery making intense but fresh mtn reds. Barrel-aged SAUV BL a modern classic.

Carmen, Viña Casa, Col, Mai ★★→★★★ Historic 1850 MAI winery, now in SANTA RITA stable. Modern wines often v.gd value. Exciting DO range, v.gd Gold Res CAB SAUV.

Casablanca Chile's 1st cool coastal region, already one of New World's best-known. Best for CHARD, SAUV BL, PINOT N, SYRAH.

Casa Marín San A ★★★ Top-notch; run by mother-and-son winemakers. Excellent RIES, SAUV BL, Sauvignon Gris. Racy, intense wines, 4km (2.5 miles) from Pacific.

Casas del Bosque Casa, Mai ★★→★★★ Cool hillside v'yds. Top pours: SAUV BL, SYRAH, CHARD, csp Pequeñas Producciones.

Casa Silva Col, S Regions ★★→★★★★ Wine dynasty, still family-run, with wines from Andes to coast. V.gd CARMENÈRE, exciting Lago Ranco PINOT N, SAUV BL in Pat.

Clos des Fous Cach, Casa, S Regions ★★→★★★ Soil-seeking micro-terroir wines by Pedro Parra and François Massoc. Fruit-driven reds, lean RIES, big CHARD.

Concha y Toro Cent V ★→★★★★ A titan not only of Chile, but of one of largest worldwide. Hard to find wine that isn't in CyT's portfolio, with almost every valley and vine in its stable of brands (Casillero del Diablo, Marquis, Terruño). Top picks: vibrant coastal whites in MAYCAS DE LIMARÍ range, Gravas SYRAH from MAI, and top CAB SAUV Don Melchor. *See also* ALMAVIVA, TRIVENTO (Argentina).

Cono Sur Casa, Col, Bío ★★→★★★ Sister winery of CONCHA Y TORO, now biggest PINOT N producer in S America. Bicicleta range v.gd value, Silencio top CAB SAUV (MAI).

Cousiño Macul Mai ★★→★★★ Founded 1856, still in same family hands. Diverse portfolio. Lota top CAB SAUV.

De Martino Cach, Casa, Elq, Mai, Mau, Ita ★★→★★★★ Traditional MAI winery with modern outlook. Low intervention, lean, fresh wines from all over. V.gd old-vine MALBEC, ITA CINSAULT.

Elqui Transverse valley in n with vines from coast to high-altitude Andes (over 2000m/6562ft). PISCO area but also distinctive SYRAH, SAUV BL.

Emiliana Casa, Rap, Bío ★→★★★ One of world's largest organic producers, overseen

> Andes to Costa
> Getting to grips with Chile's wine regions isn't just a vertical
> conundrum; you have to think horizontally too: from Andes in e to coast
> (Costa) in w. Appellations are split into Andes (altitude, mtn influence),
> Entre Valles (valley floor) and Costa (cooler, coastal influence).

by bio expert Álvaro Espinoza (*see* ANTIYAL). Bright, refreshing whites and varietals, complex blends incl top G and Coyam.

Errázuriz Aco, Casa ★★→★★★★ Leader in ACO with v'yds from coast to Andes. Rich mtn reds (incl top Don Maximiliano CAB SAUV blend), fresh coastal SYRAH, SAUV BL; excellent Pizzarras CHARD and PINOT N grown on schist. *See also* SEÑA, VIÑA ARBOLEDA, VIÑEDO CHADWICK.

Falernia, Viña Elq ★★→★★★ Pioneer of fine wine. Several sites in valley ranging from racy SAUV BL towards coast to meaty SYRAH in mtns. Mayu is sister winery.

Garcés Silva, Viña San A ★★→★★★ Grower-turned-producer with 175 ha in Ley with two labels: Amayna (voluptuous, complex), Boya (fresh, fruity).

Haras de Pirque Mai ★★→★★★ Classy Alto MAI wines with B'x focus. Now owned by Italy's Antinori.

Itata Treasure trove of old vines with CINSAULT, PAÍS, MUSCAT supreme. Haven of small producers and artisanal wines.

Koyle, Ita ★★→★★★ Bio, organic Col estate in foothills. Juicy reds (Cerro Basalto v.gd), coastal whites with small ITA line.

Lapostolle Cach, Casa, Col ★★→★★★★ Handsome bio estate in Apalta, famous for complex Clos Apalta B'x blend. Rich estate wines are complemented by fresh Collection range from around Chile.

Leyda, Viña Col, Mai, San A ★★→★★★ Large estate that put Ley on map. Cool coastal whites (CHARD, SAUV BL, Sauvignon Gris) and peppery SYRAH, PINOT N.

Limarí The rare phenomenon of a cool, coastal desert with limestone soils has pushed Lim into limelight for distinctive CHARD, SAUV BL, SYRAH, PINOT N.

Louis Antoine Luyt (Pipeño) Mau ★★ Burgundian Louis Antoine Luyt moved to MAU and kick-started natural wine and old-vine revival. Champion of PAÍS, and white field blends.

Luis Felipe Edwards ★★★ Huge family business based in Col, more v'yds in Ley and MAU. Wide range of spot-on varietals, incl fine single v'yds. Generally value.

Maipo Chile's premier wine region; finest CAB SAUV in country. Best sites on river gravels close to Andes: Pirque, Puente Alto, Alto Jahuel.

Matetic Casa, San A ★★★ Superb bio estate on CASA/SAN A border. Consistent, classy SYRAH, **Sauv Bl** stand out.

Maule Biggest wine region in Chile but still overlooked. Old-vine CARIGNAN (*see* VIGNO) and PAÍS are now garnering deserved attention.

Maycas del Limarí Lim ★★→★★★ Cool coastal wines from limestone-rich LIM region in n. Racy SAUV BL, lean CHARD top. Part of CONCHA Y TORO.

Montes Casa, Col, Cur, Ley ★★→★★★★ One of leading COL producers but v'yds extend from ACO coast to Chiloe. Best known for rich, opulent reds and **Folly Syrah**.

MontGras Col, Ley, Mai ★★ COL-based producer with v'yds in MAI (rich reds under Intriga label) and Ley (zippy SAUV BL under Amaral).

Montsecano Casa ★★★ Bio, natural wines from boutique producer. Excellent PINOT N, and now MALBEC blend.

Morandé Casa, Mai, Mau ★★→★★★ Large widely spread producer still takes its name from founder Pablo Morandé (1996). Gd-value varietals, but best is Brut Nature NV bubbly.

Neyen Col ★★★ Old-vine CAB SAUV/CARMENÈRE from bio estate in Apalta, planted 1889. Complex, collectors' wines from VERAMONTE group.

Odfjell Cur, Mai, Mau ★→★★★ Norwegian shipping merchant breeds horses and rich reds in MAI. V.gd old-vine MALBEC, CARIGNAN further s.

Pérez Cruz, Viña Mai ★★→★★★★ Estate with stylish red B'x varieties and increasingly Med portfolio with top GRENACHE, SYRAH too.

Pisco Chile's local brandy, often whipped with egg whites, lime, sugar for a "sour".

Polkura Col ★★→★★★ Full-bodied reds from Sven Bruchfeld. SYRAH v.gd.

Quebrada de Macul, Viña Mai ★★→★★★ Top-notch CAB SAUV and B'x blends. Domus Aurea highly sought-after.

Rapel Large, slightly useless denomination covering COL and CACH in Central Valley. Appears on big brand blends and table wines.

RE, Bodegas Casa ★★★ Family enterprise: Pablo Morandé Sr, Jr and daughter. Weird and wonderful world of orange wines, cloudy bubbles and bold reds from CASA, MAU.

San Antonio Cool coastal region with Ley subregion as star. Top for SAUV BL, CHARD, SYRAH, PINOT N. Cooler than neighbouring CASA.

San Pedro Cur ★→★★★ 2nd-biggest group in Chile with host of brands all over country, and base in CUR. Top picks: 1865 line (esp SAUV BL, SYRAH from ELQ), Cabo de Hornos CAB SAUV. Everyday supermarket brands are 35 Sur, Castillo de Molina. (Also owns ALTAÍR, Missiones de Rengo, Santa Helena, TARAPACÁ, Viña Mar).

Santa Carolina, Viña Mai ★★→★★★★ Historic producer with v'yds in major valleys with large, diverse portfolio covering all bases, incl specialities like field blends. Top CARMENÈRE (Herencia) and Luis Pereira CAB SAUV two of Chile's best.

Santa Rita Mai ★★→★★★★ Historic hacienda was hideout for freedom-fighters in C19; today home to one of Chile's biggest wine players. Diverse range with MAI focus. V.gd, consistent CAB SAUV, CARMENÈRE, expressive reds in all ranges esp Floresta and iconic *Casa Real Cab Sauv*. Tres Medallas, 120, Medalla Real everyday labels.

Seña Aco ★★★★ One of finest B'x blends; shows ACO potential. Complex stuff from Chadwick/ERRÁZURIZ stable.

Tabalí Lim ★★→★★★ Top producer of cool coastal LIM with searing CHARD, SAUV BL; fresh PINOT N, SYRAH.

Tarapacá, Viña Casa, Ley, Mai ★★ VSPT-owned mega-winery with 600-ha estate in Isla de Mai. V.gd CAB SAUV, esp Etiqueta Negra.

Torres, Miguel Cur ★★→★★★★ One of Chile's most influential wineries ever since Miguel Torres (*see* Spain) arrived in 1979. Champions of sparkling PAÍS, fairtrade and youthful, gd-value wines, as well as complex CAB SAUV. Escaleras de Empedrado is exciting PINOT N project on MAU schist.

Undurraga Casa, Ley, Lim, Mai ★→★★★ Large producer of still and sparkling, often v.gd value. Superb TH range is polished, single-v'yd series from around Chile.

Valdivieso Cur, San A ★→★★★ Producing fizz since 1879, and sparkling reigns. Plenty of still wines too, incl v.gd single-v'yd CAB FR, MALBEC, *Ley Chard*. Caballo Loco top of range blends and bubbles.

Is Pisco Peruvian, Chilean or both? Don't get too sour over it.

Vascos, Los Rap ★→★★★ B'x varieties in COL. A Lafite-Rothschild venture in Chile, but more "fifth growth" than "first". La Dix is top, from 70-yr-old vines.

Ventisquero, Viña Casa, Col, Mai ★→★★★ Large, widespread producer in Chile: vines reaching up to Atacama Desert (exciting Tara range). Rich, fruity reds (top Enclave CAB SAUV, Pangea SYRAH), fresh whites (Kalfu).

Veramonte Casa, Col ★★→★★★ Large CASA producer. Fresh, fruity whites (v.gd Ritual SAUV BL), juicy reds. Organic vines. Owned by González Byass (*see* Spain).

Vigno Mau Groundbreaking association in MAU championing old-vine, dry-farmed CARIGNAN. 20 producers, each with own style, but all bottled as Vigno.

VIK ★★→★★★ Extraordinary investment in CACH by Norwegian billionaire. 400 ha vines in 4000 ha estate and less than 10% of grapes make cut for top VIK wine: a rich, B'x blend.

Villard Casa, Mai ★★ Early adopters in CASA, est 1989. Family winery with fresh whites, perfumed reds.

Viñedo Chadwick Mai ★★★→★★★★ World-class CAB SAUV from Puente Alto, owned by Chadwick/ERRÁZURIZ. One wine, on allocation only.

Viu Manent Casa, Col ★★ Family winery in COL with focus on MALBEC and hearty reds. Coastal SAUV BL also gd.

Von Siebenthal, Viña Aco ★★→★★★ Full-bodied reds from Mauro von Siebenthal and family in ACO. Parcela 7 blend v.gd value; try Toknar PETIT VERDOT.

ARGENTINA

A titan of New World wine, Argentina's economy may have been shaky this past decade but its wine production has stood firm and its producers unwavering. The country used to drink over 90l of Italian-style wine a year. Today consumption is much less, primitive wines have almost gone and quality has never been higher – nor have vineyards. Altitude is still Argentina's greatest weapon against climate change: winemakers are exploring the Andes in Mendoza, Salta, San Juan and Jujuy. Cooler climates at altitude, down south in Patagonia and by the coast have brought a flurry of new whites with great Chardonnay, Semillon and even Albariño, though spicy Torrontés is still mainstream. Malbec is, however, still king and Argentina is a kingdom of red – with Cabernet Franc, Criolla and Petit Verdot as rising stars.

Achaval Ferrer Men ★★→★★★ MALBEC-focused winery with plush, polished single v'yd range. Luján-based, owned by Stolichnaya spirits group.

Aleanna Men ★★→★★★★ Aka El Enemigo wines, quickly achieved cult status in Latin America. Uco V-focused MALBEC, CAB FR, CHARD, BONARDA by one of Argentina's top winemakers, Alejandro Vigil.

Alicia, Viña Men ★★★ Personal project of Arizu dynasty (LUIGI BOSCA family) using old vines and unusual varieties incl ALBARIÑO, SAVAGNIN, NEBBIOLO.

Alta Vista Men ★→★★★ 1st winery in MEN to release single-v'yd range, back in early 2000s. French vision of Luján terroir.

Altocedro Men ★★→★★★ Sustainable v'yd, low-intervention wines in La Consulta by Karim Mussi. Top TEMPRANILLO, old-vine MALBEC.

Altos las Hormigas Men ★★★→★★★★ MALBEC specialist, founded 1995, top-notch single-v'yds in Luján and Uco V. Dream team of Italian winemakers Alberto Antonini, Attilio Pagli and Chilean Leo Erazo with soil specialist Pedro Parra. Bio BONARDA under Colonia Las Liebres label.

Anita, Finca La Men ★★→★★★ Classy reds with Old World philosophy and patience. V.gd CAB SAUV, SYRAH, PETIT VERDOT.

Atamisque Men ★→★★★ Tupungato estate, modern Uco V wines. V.gd CHARD, MERLOT in Catalpa range. Serbal top value.

Benegas Men ★★→★★★ Founding family of TRAPICHE; complex B'x blends, v.gd CAB FR.

Bianchi, Bodegas Men ★→★★ Historic San Rafael producer with new sites in Uco V (Enzo Bianchi). Gd bubbles, v.gd Enzo Bianchi MALBEC.

Bressia Men ★★→★★★ Walter Bressia is master of complex, age-worthy red blends (Ultima Hoja is top), but family winery also has younger Sylvestra line. Monteagrelo in middle is v.gd value.

Caelum Men ★★ Boutique family production: attractive MALBEC, interesting FIANO.

Callia San J ★→★★ Leading producer making everyday, gd value. V.gd SYRAH.

Canale, Bodegas Humberto Rio N ★→★★★ Leading producer with 110-yr history. V.gd old-vine RIES, MALBEC, PINOT N.

Caro Men ★★★→★★★★ Opulent, velvety B'x blends from CATENA ZAPATA and (Lafite) Rothschild collaboration.

Casarena Men ★★→★★★ Luján producer with v.gd single-v'yd MALBEC, CAB FR, CAB SAUV. Also younger lines Ramanegra, 505.

Catena Zapata, Bodega Men ★★→★★★★ Renowned wine family in Argentina, wide influence, many wineries. Top are from Adrianna v'yd in Gualtallary (MALBEC, CHARD, CAB FR), but range is huge (*see also* CARO). Alamos is entry-level label.

Chacra Rio N ★★★→★★★★ Classy old-vine, bio production of Piero Incisa della Rocchetta of Sassicaia (*see* Italy). Sophisticated *Pinot N*, and now CHARD.

Clos de los Siete Men ★★ 850-ha clos in Uco V with handful of B'x producers (*see* BODEGA ROLLAND, CUVELIER LOS ANDES, DIAMANDES, MONTEVIEJO) working together to make one red, blended by Michel Rolland.

Cobos, Viña Men ★★★→★★★★ Premium producer in Luján but with wines from Uco V too. V.gd MALBEC, CAB SAUV, opulent CHARD. Paul Hobbs' winery also owned by Grupo Molinos.

Colomé, Bodega Sal ★★→★★★ Intense, vibrant wines from high altitude (up to 3100m/10,171ft!) Dense reds, perfumed TORRONTÉS.

Cuvelier Los Andes Men ★★→★★★ Ripe, concentrated reds from B'x family (Léoville-Poyferré) in Uco V in CLOS DE LOS 7.

Decero, Finca Men ★★→★★★ Precise, modern wines from Luján. V.gd PETIT VERDOT and Owl and the Dust Devil blend.

DiamAndes Men ★★★ B'x varieties plus VIOGNIER, CHARD from Bonnie family (B'x's Malartic-Lagravière) in CLOS DE LOS SIETE stamping ground.

Doña Paula Men ★★→★★★★ Large producer, innovative team, quality at all levels. MALBEC Parcel series is excellent, Los Cardos is v.gd value entry range. Owned by Chile's SANTA RITA.

Durigutti Men ★★→★★★ Durigutti brothers make classic portfolio but Las Compuertas line is step up; 5 Suelos MALBEC one of Luján's best.

Esteco, El Sal ★★→★★★ Distinctive altitude wines from Cafayate, best is El Esteco CAB SAUV, Don David, Ciclos gd value.

Etchart Sal ★→★★ One of Cafayate's most important wineries, founded 1850. Smart reds, top for TORRONTÉS.

Fabre Montmayou Men, Rio N ★★→★★★ Classy reds from old vines in MEN and RIO N. V.gd CAB SAUV, MALBEC, MERLOT.

Fin del Mundo, Bodega Del Neu ★→★★ Ripe, concentrated reds (v.gd CAB FR) at top end. Postales, Ventus, Newen are everyday gd value. Pioneer in NEU and biggest today.

Flichman, Finca Men ★★→★★★ Maipú-based but v'yds split with Uco V. Dedicado is top line, Caballero de la Cepa is easier-drinking range. Owned by Sogrape (*see* Portugal).

Kaikén Men ★★→★★★ Chilean MONTES family adventure in Argentina with classic B'x varieties. V.gd Ultra CAB SAUV.

Luca / Tikal / Tahuan / Alma Negra / Animal Men ★★→★★★ Stable of wines by CATENA kids, who aren't kids anymore. Laura makes luxury Luca line, Ernest makes natural wines (T/T/A/A).

GIs picking up pace

First it was Paraje Altamira GI (Geographical Indication), now Pampa El Cepillo GI, Los Chacayes GI, San Pablo GI. What's next? Several Uco V wine regions lining up to have own GIs, emerging as Argentina's modern appellation system based on soil, climate and altitude.

Luigi Bosca Men ★★→★★★ Diverse portfolio from this historic Luján producer. Richer, old-vine Las Nobles reds are age-worthy, whites are classy (v.gd RIES). La Linda is entry-level.

Manos Negras / Tinto Negro / TeHo / ZaHa Men ★★→★★★ Terroir-focused series made by soil specialist Alejandro Sejanovich. Excellent MALBEC, CAB FR.

Marcelo Pelleriti Men ★★→★★★ One of Argentina's top winemakers (also MONTEVIEJO) makes eponymous line of classy reds. V.gd MALBEC, CAB FR.

Heard of Criolla Number 1? Malbec/Criolla Grande cross, made in Argentina.

Masi Tupungato Men ★★→★★★ Ripasso-style wines, CORVINA and PINOT GRIGIO. Italians in Uco V show MEN how they do theirs. Intense reds, light white.

Matias Riccitelli Men ★★→★★★ One of best in his generation, Riccitelli Jr makes exciting, sophisticated wines. Excellent MALBEC, SEM, CHARD.

Mendel Men ★★★ Roberto de la Motta, a living legend with gentle hand in winery, inherited from his father. Excellent Finca Remota, Lunta superb value, solid SEM.

Mendoza Argentine wine mecca, 70% national production. Maipú is historic, best for old vines; Luján is Malbec territory; higher altitude Uco V is leading fresher and leaner style.

Michel Torino Sal ★★ Torino TORRONTÉS is bread-and-butter wine of Cafayate. Everyday value, large production.

Moët-Hennessy Argentina Men ★→★★ One of original foreign investors in MEN, since 50s, making simple sparkling under Chandon label. Baron B is top, traditional method. *See* TERRAZAS DE LOS ANDES.

Monteviejo Men ★★→★★★★ V.gd Vista Flores producer: top-value B'x varieties from everyday Festivo line, mid-range Petite Fleur and top-end Lindaflor. La Violeta is superb. B'x family behind Ch Le Gay.

Moras, Finca Las San J ★→★★ Gd-value wines. SYRAH is perfect barbecue bottle. Owned by TRAPICHE.

Neuquén Pat Warm, continental s region, 1700 ha vines planted where dinosaurs used to roam. Rich reds (MALBEC, CAB FR), fruity PINOT N.

Nieto Senetiner, Bodegas Men ★→★★★ Founded 1888 and hasn't looked back. One of MEN's biggest players, ranging from value lines (Benjamin, Emilia) to top Don Nicanor. Gd MALBEC, fresh SEM. Big in bubbles.

Noemia Pat ★★★→★★★★ Sophisticated, complex wines from old vines in RIO N. One of Argentina's best MALBECS.

Norton, Bodega Men ★→★★★ English rail engineer got Norton on its tracks in 1895; today it's one of MEN's biggest wineries, now owned by Swarovski. Gd-value wines across board, esp bubbles. Lot MALBEC is excellent.

Passionate Wine Men ★★→★★★ Innovative, bio and natural wines by wild child of Uco V, Matias Michelini. Mouthwatering Agua de Roca SAUV BL.

Peñaflor Men ★→★★★ Mammoth group with EL ESTECO, FINCA LAS MORAS, Mascota, Navarro Correas, Santa Ana, Suter, TRAPICHE in its fold.

Piatelli Sal ★★ Two terroirs, Cafayate and Luján, give different expressions but same polished style, esp in CAB SAUV, MALBEC.

Piedra Negra Men ★→★★★ François Lurton's Argentine outpost in Los Chacayes, Uco V. Excellent whites, polished MALBEC.

Porvenir de Cafayate, El Sal ★★→★★★ Modern portfolio from historical Cafayate family producer. Floral TORRONTÉS complements intense reds. Amauta blends v.gd.

Pulenta Estate Men ★★→★★★ Pulenta brothers focus on fast cars, slow wines. Well-aged, complex reds (v.gd CABS FR/SAUV), intense whites. La Flor is everyday line.

Renacer Men ★★→★★★ Youthful team and 2nd gen in charge of modern winery. Rich reds (incl Amarone-style Enamore), fresh SAUV BL from CASA in Chile.

Riglos Men ★★→★★★ Gualtallary estate making modern reds (v.gd CAB FR); CHARD.

Riojana, La R ★→★★ Notable producer making gd-value wines from 500 growers. Fairtrade, organic, one of Argentina's only co-ops.

Río Negro Milder temperatures and old vines in this s region produce distinguished wines. Top for MALBEC, PINOT N, SEM.

Rolland, Bodega Men ★★★ Winery made by a winemaker, and you can tell. Functional unit in middle of Vista Flores v'yds, where Michel Rolland's team make complex B'x reds, plush PINOT N, intense SAUV BL.

Salentein, Bodegas Men ★★→★★★ One of 1st modern investors in Uco V, striking showcase winery. Fruit-forward, fresh wines: gd-value El Portillo and Res range.

Salta Historic n province with highest altitude vines. Home of TORRONTÉS, intense reds (TANNAT, MALBEC). Cafayate is main wine region in Calchaquí valleys.

San Juan Powerhouse of BONARDA, SYRAH and gd-value rcds, just n of MEN. Argentina's 2nd-biggest region.

San Pedro de Yacochuya Sal ★★★ Boutique, premium wines by ETCHART family with Michel Rolland (*see* France). Intense MALBEC and fragrant TORRONTÉS.

Schroeder, Familia Neu ★★ Dinosaur fossils found in cellar now on display between rows of PINOT N bottles. Gd-value NEU wines.

Sophenia, Finca Men ★★→★★★ Gualtallary estate making bold, fresh wines. V.gd-value Altosur range (esp BONARDA). Synthesis is top-line.

Susana Balbo Wines Men ★★→★★★★ Family estate of Argentina's 1st female winemaker, now working with her son. Complex reds (try Ben Marco), vibrant whites (White Blend is excellent).

Tapiz Men ★★→★★★ Now with v'yds in Uco V and coastal RÍO N, has moved beyond its Luján base. Diverse portfolio, incl top MALBEC Black Tears.

Terrazas de los Andes Men ★★→★★★★ B'x influence in MEN with classy reds, intense whites. Superb *Cheval des Andes* blend, in collab with Cheval Blanc (*see* B'x).

Toso, Pascual Men ★★→★★★ Old vines and experience at MALBEC-focused winery. Magdalena Toso top pour.

Trapiche Men ★★→★★★ Biggest brand in Argentina? Peñaflor flagship reaches far and wide with v'yds across Andean corridor and on coast where exciting Costa & Pampa range comes from. Highlights incl *Medulla* CAB SAUV; Iscay MALBEC/CAB FR blend; single-v'yd Finca range.

Trivento Men ★→★★ Gd-value, mainly everyday reds from Chile's CONCHA Y TORO in MEN. Eolo is winery's icon.

Vines of Mendoza / Winemaker's Village Men ★★ Block of micro-wineries and micro-plots in Uco V owned by hundreds of Vines of Mendoza associates. Also Winemaker's Village with Abremundos (*see* MARCELO PELLERITI), Corazon del Sol, Gimenez Riili, Recuerdo, Super Uco.

Zorzal Men ★★★ Uco V producer with wide range, fresh wines, from Gualtallary. Min oak, max texture in Eggo line (v.gd CAB FR, SAUV BL).

Zuccardi Men ★★→★★★★ One of leaders in MEN with modern wines and single-v'yd focus. Excellent Alluvional MALBEC, Concreto, Emma BONARDA, Fossil CHARD. Santa Julia in Maipú is weekday brand, easy-drinking.

BRAZIL

It took 400 years for Brazil's wine industry to work out how to make quality wines in spite of the humid climate, but now it's making headway. Wine regions run from the border of Uruguay to the highlands of Bahia, and there's an emerging wine zone just outside São Paulo. Top vines planted include hybrids, but winemakers are mainly getting serious about Syrah, Merlot, Chardonnay, Cabernet Sauvignon and Malbec. Bubbles are also big business, with top sparklers coming from Pinto Bandeira. Brazil is one to keep your eye on.

Aurora ★→★★ Big name. Brazil's largest co-op with 1000+ growers in Serra Gaucha. Diverse portfolio.

Casa Valduga ★→★★★ Serious about sparkling, but this family winery also has top MERLOT and red blends.

Cave Geisse ★★★ Chilean winemaker Mario Geisse makes top-notch traditional-method bubbly in Pinto Bandeira.

Lidio Carraro ★→★★ Easy-drinking, enjoyable wines that respect fruit from Vale dos Vinhedos and Serra do Sudeste. Quorum blend more complex.

Miolo ★→★★★ Big player in Brazil with premium focus, diverse portfolio. V.gd TOURIGA N blend, top bubbles too.

Pizzato ★→★★★ Flavio Pizzato puts his heart into his wines, and the result is some of Bento's best. V.gd CHARD, MERLOT. Fausto is young range.

Salton ★→★★ Founded 1878, Salton is Brazil's oldest winery and one of biggest. Gd party fizz, v.gd Salton Gerações blend.

URUGUAY

One of S America's hidden gems, but word is getting out. The warm Atlantic influence makes Uruguay more similar to Bordeaux than Chile or Argentina, and its wines are distinctive from the rest of the New World. Tannat is king, tannic fleshy red, formerly rustic, now being made in complex, age-worthy styles as well as fresher, energetic styles with limited or no oak. Albariño is becoming the queen of Uruguay, at least by reputation if not in number. But in Uruguay's court, you'll find Nebbiolo, Cabernet Franc, Syrah and even Zinfandel.

Alto de la Ballena ★→★★ Pioneer of cooler, coastal Maldonado. Excellent SYRAH, CAB FR and juicy TANNAT/VIOGNIER from boutique family producer.

Bouza ★★→★★★ Consistent in quality and innovation. Superb ALBARIÑO and top RIES, MERLOT, TANNAT, from Canelones and Pan de Azucar.

Garzón, Bodega ★→★★★ Impressive Argentine investment in world-class winery and estate nr Punta del Este. Alberto Antonini is consultant and wines are energetic, lively, characterful. Balasto blend superb.

Juanico Establecimiento ★→★★★ Historic family estate with innovation aplenty. Single-v'yd TANNAT series in Deicas is star, but entry-level Pueblo del Sol and Don Pascual are what locals drink everyday.

Marichal ★→★★ Family of growers with 3rd generation now making wines. Gd TANNAT, PINOT N (and blend of both).

Pisano ★→★★★ Three Pisano brothers work together to make some of Uruguay's most consistent, exciting wines. V.gd PINOT N, TANNAT, TORRONTÉS, VIOGNIER.

Viñedo de los Vientos ★★→★★★ Boutique production, mainly Italian varieties (ARNEIS, NEBBIOLO) and TANNAT, of course. Natural wines on the coast.

OTHER SOUTH AMERICAN WINES

Bolivia Ancient vines, over 300 yrs old, a highlight of this high-altitude wine country. With wines made from v'yds up to 3000m (9843ft) altitude, expect reds of intense colour, vibrant acidity and full body (esp CAB SAUV, SYRAH, MALBEC and TANNAT). Top white is fragrant dry MUSCAT, also distilled to make Bolivia's national liqueur, Singani. Biggest wineries: Campos de Solana, Kohlberg, Kuhlmann and La Concepción.

Peru 1st wine country in S America (since 1500s) but slowest to develop serious wine industry. There's a grassroots revival happening with Pisco producers now returning to wine, and big wineries are moving to more exports. Look for Intipalka, Mimo, Tacama, Vista Alegre.

Australia

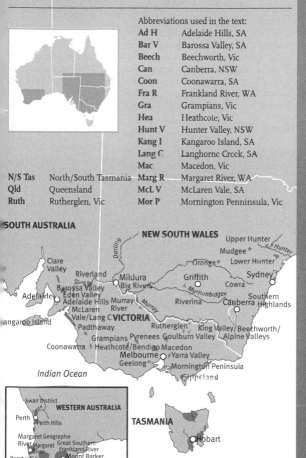

Abbreviations used in the text:

Ad H	Adelaide Hills, SA
Bar V	Barossa Valley, SA
Beech	Beechworth, Vic
Can	Canberra, NSW
Coon	Coonawarra, SA
Fra R	Frankland River, WA
Gra	Grampians, Vic
Hea	Heathcote, Vic
Hunt V	Hunter Valley, NSW
Kang I	Kangaroo Island, SA
Lang C	Langhorne Creek, SA
Mac	Macedon, Vic
N/S Tas North/South Tasmania	**Marg R** Margaret River, WA
Qld Queensland	**McL V** McLaren Vale, SA
Ruth Rutherglen, Vic	**Mor P** Mornington Penninsula, Vic

A ustralia used to be all about brands and blending. The following entries show that the reinvention has been deep, wide and long-running: now it's regions that are having their shout. The massive growth of Australian wine in the 90s gradually became the chronic oversupply of the mid-noughties, but Australia is never short of ideas. Necessity has been the mother of wholesale reinvention. And heatwaves are going to call for even more. Reinvention is not just seen in the trendy abundance of natural wines, pét-nat and skin-contact whites. Every category seems to be having a rethink. It's in the way Grenache has been completely reimagined, now resembling Pinot Noir more than Shiraz; it's in the way whole-bunch fermentation in reds (*see* A little learning) has gone from quirky to the norm; it's in the way Chardonnay is now made in a full gamut of styles rather than one orthodoxy; it's in the way mature vines are now standard rather than exceptional (those 90s' plantings are nearly 25 years old now); it's in the overall thirst for all that is new. The supermarket was once the be-all and end-all of almost all Australian wine. The wine bar is no longer a curio; it's a driver of change, and a significant avenue to market. In short, you'd almost call the Australian wine scene sophisticated now; a thought that once might have seemed laughable. All the energy and maturity in these changes will no doubt be called on in spades as the nation works through the ramifications of the devastatingly widespread 2020 vintage bushfires.

Recent vintages

New South Wales (NSW)

2019 Hot, with just enough rain for generous whites/reds.
2018 Big ripe reds for long haul. Hot/tough yr for whites.
2017 Wet spring, hot summer: whites lapped it up; reds too, in general.
2016 Be wary of Hun V reds (drink early), gd mid-term ageing (r w) elsewhere.
2015 Difficult in most parts, but Orange and Can excellent; Hilltops v.gd.
2014 Hun V Shiraz will be exceptional. Can Ries, Shiraz right up there.

Victoria (Vic)

2019 Compressed season but wines (r w) look vibrant.
2018 Overshadowed by yr before. Slow-evolving.
2017 Excellent yr. Reds/whites looked gd young and will stay.
2016 Warm, dry season produced many overripe reds. Tread carefully.
2015 Strong yr across board.
2014 Frost damage galore but v.gd red vintage for most.

South Australia (SA)

2019 Exceptionally low yields should produce concentrated wines.
2018 Expect gutsy reds with yrs up their sleeve. Whites gd, not in same class.
2017 High yield, high quality, highly drinkable young.
2016 Hopes are high for a special vintage for red/white.
2015 Warm regions coped well with summer of wild temperature swings.
2014 Hot, low-yield yr produced generous white/red.

Western Australia (WA)

2019 Cool vintage, the kind to sort the wheat from the chaff.
2018 Reds will outlive most of us; whites will do medium-term in a canter.
2017 Tricky vintage. Medium-term wines.
2016 Humid, sultry vintage. Nothing wrong with wines; mid-termers.
2015 Challenging, mixed results; be selective.
2014 Luck continued; almost getting monotonous. Another tiptop yr.

A. Rodda Beech, Vic r w ★★ Bright CHARD from est v'yds; whole-bunch-fermented *Tempranillo* grown at high altitude is routinely a beauty.

Accolade Wines r w Name of once-mighty Constellation, HARDYS groups. BAY OF FIRES, Hardys, HOUSE OF ARRAS, PETALUMA, ST HALLETT chief quality brands.

Adelaide Hills SA Cool 450m (1476ft) sites in Mt Lofty ranges. CHARD, SAUV BL, SHIRAZ outgun PINOT N. ASHTON HILLS, HAHNDORF HILL, HENSCHKE, JERICHO, MIKE PRESS, SHAW & SMITH, TAPANAPPA all in excellent form.

Adelina Clare V, SA r w ★★ Reds (SHIRAZ, GRENACHE, MATARO, NEBBIOLO) the stars here. Imposing, intense, but polished. Label designs quite something too.

Alkoomi Mt Barker, WA r w (RIES) 05′ 10′ 17′ 18 (CAB SAUV) 10′ 12′ 16 Veteran maker of fine Ries; rustic reds; more accessible young than they were.

All Saints Estate Ruth, Vic br ★★ Great fortifieds. Hearty table wines, led by B'x blend.

Alpine Valleys Vic In valleys of Victorian Alps. BILLY BUTTON, MAYFORD, Ringer Reef. TEMPRANILLO the star, closely followed by SHIRAZ and aromatic whites.

Andrew Thomas Hun V, NSW r w ★★ Old-vine SEM; silken SHIRAZ. Reds particularly gutsy in HUN V context.

Angove's SA r w (br) ★ MURRAY V family business. Cheapies (r w) often standouts of a broad range. Mainstream face of organic grape-growing, both value and premium ends.

Arenberg, d' McL V, SA r w (br) (sw) (sp) ★★ Sumptuous SHIRAZ, GRENACHE. Many varieties, wacky labels (incl The Cenosilicaphobic Cat SAGRANTINO). Fancy shirts, fancy names, no-nonsense flavours/styles.

Ashton Hills Ad H, SA r w (sp) ★★★ (PINOT N) 10′ 15′ 17′ 18 Totemic AD HILLS producer. Compelling Pinot N from 30-yr-old+ v'yds. Bought in 2015 by WIRRA WIRRA.

Bailey's NE Vic, Glenrowan, Vic r w br ★★ Rich SHIRAZ, magnificent dessert MUSCAT (★★★★) and TOPAQUE. V'yds all organic. Sold by TWE to CASELLA in 2017.

Balgownie Estate Bendigo, Vic, Yarra V, Vic r ★★ Capable of medium-bodied, well-balanced, minty CAB of elegance, finesse, character from its BENDIGO heartland. Separate YARRA V arm.

Balnaves of Coonawarra SA r w ★★★ Family-owned COON champion. Lusty CHARD; v.gd spicy, SHIRAZ, full-bodied Tally CAB SAUV flagship. "Joven-style" Cab gd.

Bannockburn Vic r w ★ (CHARD) 14′ 15′ 17′ 18 (PINOT N) 10′ 12′ 17 Intense, complex Chard, spice-shot Pinot N. Put GEELONG region on map, but star not as bright as once was.

Barossa Valley SA Ground zero of Aussie red. V.-old-vine SHIRAZ, MOURVÈDRE, CAB SAUV, GRENACHE. Can produce bold, black, beautiful reds with its eyes closed, and has done for just about ever. ELDERTON, GLAETZER, GRANT BURGE, HENTLEY FARM, JOHN DUVAL, LANGMEIL, OCHOTA BARRELS, PETER LEHMANN, ROCKFORD, RUGGABELLUS, ST HALLETT, SALTRAM, SEPPELTSFIELD, SPINIFEX, TEUSNER, WOLF BLASS, YALUMBA.

Bass Phillip Gippsland, Vic r ★★★★ (PINOT N) 10′ 14′ 15′ 16 17 Tiny amounts of variable but mostly exceptional Pinot N. In top form.

Bay of Fires N Tas r w sp ★★★ PINOT N shouldn't be passed over but HOUSE OF ARRAS *super-cuvee sparklings* rightly dominate.

Beechworth Vic The rock-strewn highlands of ne Vic. Tough country. CHARD, SHIRAZ best-performing varieties, but NEBBIOLO fast rising from (winter) fog. A. RODDA, CASTAGNA, DOMENICA, FIGHTING GULLY ROAD, GIACONDA, SAVATERRE, SCHMÖLZER & BROWN, SORRENBERG essential producers.

Bendigo Vic Hot central Vic region. BALGOWNIE ESTATE, PASSING CLOUDS, SUTTON GRANGE. Home of rich CAB SAUV, SHIRAZ.

Best's Great Western Gra, Vic r w ★★★ (SHIRAZ) 05′ 10′ 15′ 17 Shiraz master; *v.gd mid-weight reds*. Thomson Family Shiraz from 120-yr-old vines superb. Wines generally on plush side of elegant. Old-vine PINOT M insider's tip.

Billy Button Vic r w ★ So many wines, such small quantities. Everything from SHIRAZ and CHARD to Verduzzo, VERMENTINO, SCHIOPPETTINO, SAPERAVI and more.

Bindi Mac, Vic r w ★★★ (PINOT N) 04′ 10′ 15′ 17′ 18 Ultra-fastidious maker of long-lived Pinot N (esp), CHARD. Tiny production.

Bortoli, De Griffith, NSW, Yarra V, Vic r w (br) dr sw ★★★ (Noble SEM) Both irrigation-area winery and leading producer. Excellent cool-climate PINOT N, SHIRAZ, CHARD; gd sweet, botrytized, Sauternes-style Noble Sem. YARRA V arm where quality is.

Brash Higgins McL V, SA r ★ Brad Hickey is a smart cookie. He has degrees in

AUSTRALIA

English and botany, but has also worked as a brewer, baker, sommelier and now makes radical expressions of MCL V (r w). Modern face of McL V.

Bremerton Lang C, SA r w ★★ Silken CAB, SHIRAZ with mounds of flavour. Never mean, always generous.

Brokenwood Hun V, NSW r w ★★ (ILR Res SEM) 07' 11' (Graveyard SHIRAZ) 00' 06' 09' 13' 14. HUN V classic. Outside of *Cricket Pitch* Sem/SAUV BL can be hard to find value, but quality generally gd.

Brown Brothers King V, Vic r w br dr sw sp ★ Wide range of crowd-pleasing styles, varieties. General emphasis on sweetness. Innocent Bystander (YARRA V) and Devil's Corner/Tamar Ridge (TAS) savvy acquisitions.

By Farr / Farr Rising Vic r w ★★★★ 10' 14 15' 17' 18 (PINOT N) Superb producer, at Bannockburn. CHARD, Pinot N can be minor masterpieces.

Campbells Ruth, Vic r (w) br ★ Smooth ripe reds (esp Bobbie Burns SHIRAZ); extraordinary Merchant Prince Rare *Muscat*, Isabella Rare TOPAQUE (★★★★).

Canberra District NSW One of the most significant cool-climate regions in Oz, for quality at least. CLONAKILLA the leader but GUNDOG ESTATE, MOUNT MAJURA and RAVENSWORTH all v.gd.

Cape Mentelle Marg R, WA r w ★★★ (CAB SAUV) 01' 10' 14' 15' 16 MR pioneer on great form. Robust Cab has become more elegant (with lower alc), CHARD v.gd; old-vine ZIN has been tamed; v. popular SAUV BL/SEM. LVMH Veuve Clicquot-owned.

Casella Riverina, NSW r w ★ Casella's Yellow Tail range of budget reds/whites has helped build an Australian wine empire. Now owner of BAILEY'S, Brand's of Coonawarra, MORRIS, PETER LEHMANN.

Castagna Beech, Vic r w ★★ (SYRAH) 06' 10' 12' 14' 15 Julian C leads Oz bio brigade. Estate-grown SHIRAZ/VIOGNIER, SANGIOVESE/Shiraz excellent.

Chambers Rosewood NE Vic (r) (w) br Viewed with MORRIS as greatest maker of sticky TOPAQUE (★★★★), *Muscat*.

Chapel Hill McL V, SA r (w) ★★ High-profile MCL V producer. SHIRAZ, CAB the bread and butter, but TEMPRANILLO and esp GRENACHE on rise. Changed hands in 2019.

Chatto Tas r ★★★★ Young v'yd, but already among Australia's best PINOT N producers. Fruit, spice and all things nice. Smoky savouriness abounds. Must taste.

Clarendon Hills McL V, SA r ★★ Full-Monty reds (high alc, intense fruit) from grapes grown on hills above MCL V. Cigar wines.

Clare Valley SA Small, pretty, high-quality area 160-km (100-miles) n of Adelaide. Best toured by bike, some say. Australia's most prominent RIES region. Gumleaf-scented SHIRAZ; earthen, tannic CAB SAUV. ADELINA, GROSSET, KILIKANOON, KIRRIHILL, MOUNT HORROCKS, TIM ADAMS, WENDOUREE (esp) lead way.

Clonakilla Can, NSW r w ★★★★ (SHIRAZ) 05' 07' 09' 10' 14' 15' 17' 18 CAN region superstar. RIES, VIOGNIER excellent, Shiraz/Viognier famous, SYRAH a rare treat.

Clos du Tertre Fra R, WA ★★ Stunning RIES. Textural, intense, long. Made by and for Ries fanatics.

Clyde Park Vic r w ★★ Broody single-v'yd CHARD, PINOT N v.gd. SHIRAZ turning heads.

Coldstream Hills Yarra V, Vic r w (sp) ★★★★ (CHARD) 10' 12' 15' 17' 18 (PINOT N) 10' 15' 17' Est 1985 by critic James Halliday. Delicious Pinot N to drink young, *Res to age*. Excellent Chard (esp Res). Head-turning *single-v'yd releases*. Part of TWE.

Lusatia lassoed by De Bortoli

The 18-ha, 1985-planted Lusatia Park v'yd has been one of the most popular among quality YARRA VALLEY producers over the past decade. GIANT STEPS, OAKRIDGE, OUT OF STEP, PHI, Riorret and SENTIO have all sourced CHARD, PINOT N & SHIRAZ from this prime Upper Yarra v'yd, but their party stops with its sale to the DE BORTOLI family. 1st wines under new ownership were released in late 2019 under the name Lusatia Park.

Coonawarra SA Home to some of Australia's best (value, quality) CAB SAUV; land of richest red soil (on limestone). WYNNS is senior, and champion, resident. BALNAVES, KATNOOK, LINDEMANS, MAJELLA, PARKER COONAWARRA ESTATE, WYNNS COONAWARRA ESTATE, YALUMBA all key.

Coriole McL V, SA r w ★★ (Lloyd Res SHIRAZ) 04' 10' 14' 16 Renowned producer of SANGIOVESE, old-vine SHIRAZ Lloyd Res. Interesting Italians, esp FIANO, NERO D'AVOLA.

Corymbia Marg R, WA, Swan V, WA r w ★★★ Hit ground running with beautifully pitched CAB SAUV, CHENIN BL, among others.

Craiglee Mac, Vic r w ★★★ (SHIRAZ) 10' 12' 14' 15' **16** Salt-of-the-earth producer. N Rhône inspired. Fragrant, peppery Shiraz, age-worthy CHARD.

Crawford River Henty, Vic w ★★★ Outstanding RIES producer. Cool, cold, scintillatingly (dry) style, great for seafood, highly age-worthy.

Cullen Wines Marg R, WA r w ★★★★ (CHARD) 10' 13' 14' 15' 16' 17 (CAB SAUV/MERLOT) 09' 12' 13' 14' 15' 16' 17 2nd-generation star Vanya Cullen makes substantial but subtle SEM/SAUV BL, outstanding Chard, elegant, sinewy Cab/Merlot. Bio in all she does. The best of Oz.

Curly Flat Mac, Vic r w ★★★ (PINOT N) 10' 12' 13' 14' 15' 16' 17 Robust but perfumed Pinot N on two price/quality levels. Full-flavoured CHARD. Both age-worthy. Gradual turn towards elegance.

Dalwhinnie Pyrenees, Vic r ★★ In order of priority it produces SHIRAZ 1st, 2nd and 3rd, usually a smidgen above medium weight, and smooth-skinned.

Dappled Yarra V, Vic ★★★ Top-flight, top-value PINOT N, CHARD in elegant mode.

Deep Woods Estate Marg R, WA r w ★★★ Compelling CHARD, CAB SAUV. Powerhouse wines, built to impress/last.

Devil's Lair Marg R, WA r w ★ Opulent CHARD, 16 CAB SAUV/MERLOT is this estate at its best. At cool s end of MARG R. Owned by TWE.

Domaine A S Tas r w ★★ V.gd oak-matured SAUV BL. Polarizing cool-climate CAB SAUV. Charismatic, let's call it. Owned by MOORILLA.

Domaine Chandon Yarra V, Vic r (w) sp ★★ Cool-climate sparkling and table wine. Owned by Moët & Chandon. Known in UK as Green Point. NV cuvées in best ever shape. Tony Jordan RIP 2019.

Domenica Beech, Vic ★★★ Exciting BEECH producer with est v'yds. Exuberant, spicy SHIRAZ. Textural MARSANNE. NEBBIOLO stealing show.

Dr Edge Tas ★★★ Peter Dredge makes wine for various TAS wineries, but his own brand CHARD, (esp) PINOT N have more personality than clients would likely allow.

Eden Valley SA BAR's closest neighbour. Hilly region to e, home to CLARE Ringland, HENSCHKE, PEWSEY VALE, Radford, TORZI MATTHEWS and others; racy RIES, (perfumed, bright) SHIRAZ, CAB SAUV of top quality.

Elderton Bar V, SA r w (br) (sp) ★★ Old vines; rich, oaked CAB SAUV, SHIRAZ. All bases covered. Some organics/bio. Rich reds in excellent form.

Eldorado Road Ruth, Vic r ★★ Pet project of winemaker Paul Dahlenburg. DURIF, SHIRAZ, NERO D'AVOLA all show elegance and power, are not mutually exclusive.

Eldridge Estate Mor P, Vic r w ★★★ Winemaker David Lloyd is a fastidious experimenter. PINOT N, CHARD worth the fuss. Varietal GAMAY really quite special.

Epis Mac, Vic r w ★ (PINOT N) Long-lived Pinot N; elegant CHARD. Cold climate. Powerful at release; complexity takes time.

Faber Vineyards Swan V, WA r ★★★ (Res SHIRAZ) 12' 14' 15' 17 John Griffiths is a guru of WA winemaking. Home estate redefines what's possible for SWAN V Shiraz. Polished power.

Fighting Gully Road Beech, Vic r w ★★ Touchstone producer of BEECH region. CHARD, AGLIANICO, TEMPRANILLO kicking goals SANGIOVESE is king. Quality continues to rise.

Flametree Marg R, WA r w ★★ Exceptional CAB SAUV; spicy, seductive SHIRAZ; CHARD often compelling.

AUSTRALIA

Fraser Gallop Estate Marg R, WA r w ★★ Concentrated CAB SAUV, CHARD, (wooded) SEM/SAUV BL. Cab has been particularly strong in recent yrs.

Freycinet Tas r w (sp) ★★★ (PINOT N) 10' 12' 13' 17 Pioneer family winery on TAS's e coast producing dense Pinot N, gd CHARD, excellent Radenti sparkling.

Garagiste Mor P, Vic r w ★★★ CHARD, PINOT N of intensity, finesse. Quality always seems to be high or higher. Multi-v'yd blends really take value cake.

Geelong Vic Region w of Melbourne. Cool, dry climate. Best names: BANNOCKBURN, BY FARR, CLYDE PARK, LETHBRIDGE, Provenance.

Gembrook Hill Yarra V, Vic r w ★★★ Cool site on upper reaches of Yarra River: fine-boned PINOT N, CHARD par excellence.

Gemtree Vineyards McL V, SA r (w) ★★ Warm-hearted SHIRAZ alongside TEMPRANILLO and other exotica, linked by quality. Largely bio.

Giaconda Beech, Vic r w ★★★★ (CHARD) 10' 11' 15' 16' 17 (SHIRAZ) 10' 14' 15' 17 In mid-80s Rick Kinzbrunner kickstarted BEECH region. Australian Chard royalty. Tiny production of powerhouse wines.

Giant Steps Yarra V, Vic r w ★★★ Top single-v'yd CHARD, PINOT N, SHIRAZ. Vintages 14' 15' 17' 18 all exciting for three main varieties.

Glaetzer-Dixon Tas r w ★★★ Nick Glaetzer turned his family history on its head by setting up camp in cool TAS. Euro-style RIES, Rhôney SHIRAZ, autumnal PINOT N. Strength to strength.

Glaetzer Wines Bar V, SA r ★ Big, polished reds with eye-catching packaging to match. V. ripe old-vine SHIRAZ led by iconic Amon-Ra.

Goulburn Valley Vic Temperate region in mid-Vic. Full-bodied, earthy table wines. MARSANNE, CAB SAUV, SHIRAZ the pick, MITCHELTON, TAHBILK perpetual flagbearers. Aka Nagambie Lakes.

Grampians Vic Temperate region in nw Vic previously known as Great Western. Spicy SHIRAZ, sparkling Shiraz, limey RIES. Home to SEPPELT (for now), BEST'S, MOUNT LANGI, Montara, The Story.

Granite Belt Qld High-altitude, (relatively) cool, improbable region just n of Qld/NSW border. Spicy SHIRAZ, rich SEM, eg. Boireann, Golden Grove.

Grant Burge Bar V, SA r w (br) (sw) (sp) ★★ Smooth red/white from best grapes of Burge's large v'yd holdings. Acquired by ACCOLADE (2015).

Great Southern WA Remote cool area at bottom left corner of Oz; Albany, Denmark, Frankland River, Mount Barker, Porongurup are official subregions. 1st-class RIES, SHIRAZ, CAB SAUV. *Style, quality, value here.*

Grosset Clare V, SA r w ★★★ (RIES) 10' 15' 17' 18 (Gaia) 05' 12' 13' 14' 15' 16 Fastidious winemaker. Foremost Oz Ries, elegant CHARD, v.gd *Gaia* CAB SAUV/MERLOT. Beetrooty PINOT N.

Gundog Estate Can, NSW r w ★★ Highly aspirational SEM SHIRAZ from CAN, HUN V.

Hahndorf Hill Ad H, SA r w ★★ Much experimentation across wide range, made GRÜNER V its own in Oz, consistently producing richly spiced, textured examples.

Hardys r w (sw) sp ★★ (Eileen CHARD) 12' 15' 16 (Eileen SHIRAZ) 06' 10' 12' 15 Historic company now part of ACCOLADE. Chard excellent. Shiraz not far off.

Heathcote Vic The region's 500-million-yr-old Cambrian soil has great potential for high-quality reds, esp SHIRAZ, with ample body and spice. JASPER HILL, PAUL OSICKA, Place of Changing Winds, TAR & ROSES, Whistling Eagle, Wild Duck Creek.

Henschke Eden V, SA r w ★★★★ (SHIRAZ) 04' 06' 12' 14 (CAB SAUV) 04' 06' 10' 15 Pre-eminent 150-yr-old family business: delectable Hill of Grace (Shiraz), v.gd Cab Sauv, red blends, gd whites, scary prices and more. Wonder of modern world.

Hentley Farm Bar V, SA r ★★★ Consistently produces SHIRAZ of immense power, concentration – wall-of-flavour territory – though importantly in a (generally) fresh, almost frisky, context.

Hewitson SE Aus r (w) ★★★ (*Old Garden Mourvèdre*) 10' 12' 14' Dean Hewitson sources parcels off the "oldest MOURVÈDRE vines on the planet". V.gd SHIRAZ at various price levels.

Hoddles Creek Estate Yarra V, Vic r w ★★★ Made name as value producer of CHARD, PINOT N, but it's more than just value; quality is outstanding, full stop.

Houghton Swan V, WA r w ★★ (Jack Mann) 08' 11' 12' 13' 14' 15' 16 Once-legendary winery of Swan Valley nr Perth. Part of ACCOLADE. Inexpensive white blend was long *a national classic*. V.gd CAB SAUV, SHIRAZ, etc. sourced from GREAT SOUTHERN, MARG R. Jack Mann Cab blend is seriously gd.

House of Arras Tas ★★★ Best-performing and most prestigious sparkling house in Oz. Part of ACCOLADE.

Howard Park WA r w ★★ (RIES) 12' 14' 16' 17 (CAB SAUV) 09' 10' 11' 12' 13' CHARD 14' 15' 18 Scented Ries, Chard; earthy Cab. Second label *MadFish* can be gd value. PINOT N improving.

Hunter Valley NSW Sub-tropical coal-mining area 160-km (100-miles) n of Sydney. Mid-weight, earthy SHIRAZ, gentle SEM can live for 30 yrs. Arguably most terroir-driven styles of Oz. ANDREW THOMAS, BROKENWOOD, MOUNT PLEASANT, *Tyrrell's* (esp).

Hutton Wines Marg R, WA ★★ V.gd CAB SAUV, SHIRAZ, but CHARD is where things tip into outstanding territory. Powerful palate, powerhouse finish.

Inkwell McL V, SA r (w) ★★ High polish, high opinion, high character. Full house of intriguing wines, mostly SHIRAZ-based.

Jacob's Creek Bar V, SA r w (br) (sw) sp ★ Owned by Pernod Ricard. Almost totally focused on various tiers of uninspiring-but-reliable Jacob's Creek wines, covering all varieties, prices. New red range, aged in whisky barrels, proves that bourbon, Coke and wine aren't always dissimilar.

Lloyd family celebrated 50 yrs of Coriole in 2019. Mark L at helm for 40 of them.

Jasper Hill Hea, Vic r ★★ (SHIRAZ) 09' 10' 17 Emily's Paddock Shiraz/CAB FR blend, Georgia's Paddock Shiraz from dry-land estate are intense, burly, long-lived. NEBBIOLO to watch. Bio.

Jericho Ad H, SA, McL V, SA r w ★★ Excellent fruit selection and skilled winemaking produce a suite of modern, tasty, well-presented wines, esp SHIRAZ, TEMPRANILLO.

Jim Barry Clare V, SA r w ★★★ Great v'yds provide v.gd RIES, McCrae Wood SHIRAZ and richly robed, pricey, oaked-to-the-devil The Armagh Shiraz.

John Duval Wines Bar V, SA r ▲▲▲ John Duval – former maker of PENFOLDS Grange – makes *delicious Rhôney reds* of great intensity, character.

Kalleske r ★★ Old family farm at Greenock, nw corner of BAR V, makes rather special single-v'yd SHIRAZ among many other intensely flavoured things. Bio/organic.

Katnook Estate Coon, SA r w (sw) (sp) ★★★ (Odyssey CAB SAUV) 10' 13' 14' 15 Pricey icons Odyssey, Prodigy SHIRAZ. Concentrated fruit slathered in oak.

Kilikanoon Clare V, SA r w ★★★ RIES, SHIRAZ excellent performers. Luscious and generous, beautifully made. Sold to Chinese investment group (2017).

King Valley Vic Altitude range 155–860m (509–2821ft) has massive impact on varieties, styles. Over 20 brands, headed quality-wise by BROWN BROTHERS, Chrismont, Dal Zotto, PIZZINI (esp).

Kirrihill Clare V, SA r w ★ V.gd CAB SAUV, SHIRAZ, RIES at, often, excellent prices.

Knappstein Wines Clare V, SA r w ★★ Reliable RIES, SHIRAZ, CAB SAUV. Sold by Lion Nathan to ACCOLADE (2016). Always has some gd-value offerings.

Kooyong Mor P, Vic r w ★★★ PINOT N, *superb Chard* of harmony, structure. PINOT GR of charm. High-quality single-v'yd wines.

Lake Breeze Lang C, SA r (w) ★★ Succulently smooth, gutsy, value SHIRAZ, CAB SAUV; has mid-level wines thoroughly licked.

Lake's Folly Hun V, NSW r w ★★★ (CHARD) 13' 14' 17' 18 (CAB SAUV) 13' 14' 17 Founded by surgeon Max Lake, pioneer of HUN V Cab Sauv. Chard often best. Idiosyncratic.

Langmeil Bar V, SA r w ★★ Holder of some of the world's oldest SHIRAZ vines (planted mid-1800s), plus other old v'yds, all employed to produce full-throttle Shiraz, GRENACHE, CAB SAUV.

Larry Cherubino Wines Fra R, WA r w ★★★ Intense SAUV BL, RIES, *spicy Shiraz*, polished CAB SAUV. Ambitious label now fulfilling early promise. Expansive range but pretty much every egg yields a bird.

Leasingham Clare V, SA r w Once-important brand now a husk of its former self, at best. Owned by ACCOLADE.

Mildara's 1963 Coon Cab Sauv nicknamed Peppermint Pattie: eucalypt character...

Leeuwin Estate Marg R, WA r w ★★★★ (CHARD) 10' 13' 14' 15' 16 Iconic producer. All about Chard. Full-bodied, age-worthy Art Series rendition. SAUV BL, RIES less brilliant. *Cab Sauv* occasionally v.gd.

Leo Buring Bar V, SA w ★★ 02' 13' 14' 18 Part of TWE. Exclusively RIES; Leonay top label, *ages superbly*. Doesn't get much love, but nothing wrong with wine quality.

Lethbridge Vic r w ★★★ Small, stylish producer of CHARD, SHIRAZ, PINOT N, RIES. Forever experimenting. Cool climate but wines are meaty, substantial.

Limestone Coast Zone SA Important zone, incl Bordertown, COON, Mt Benson, Mt Gambier, PADTHAWAY, Robe, WRATTONBULLY.

Lindemans r w ★ Owned by TWE. Low-price Bin range now main focus, far cry from glory days.

Luke Lambert Yarra V, Vic ★ Off-beat producer of variable but at times v.gd (cool-climate, mostly) SHIRAZ, PINOT N, NEBBIOLO.

Macedon and Sunbury Vic Adjacent regions: Macedon higher elevation, Sunbury nr Melbourne airport. Quality from BINDI, CRAIGLEE, CURLY FLAT, EPIS, Hanging Rock.

Mac Forbes Yarra V, Vic ★★★ Mover and shaker of YARRA V. Myriad (in both number, styles) single-v'yd releases, mainly PINOT N, CHARD, RIES. Wines more about structure than brightness; unusual in Oz.

McHenry Hohnen Marg R, WA ★★ Among best producers of MARG R CHARD. More sophisticated than most.

McLaren Vale SA Beloved maritime region on s outskirts of Adelaide. Big-flavoured reds in general but BRASH HIGGINS, CHAPEL HILL, CLARENDON HILLS, CORIOLE, D'ARENBERG, GEMTREE, INKWELL, JERICHO, MARIUS, PAXTON, SAMUEL'S GORGE, SC PANNELL, WIRRA WIRRA, YANGARRA can show elegance too. SHIRAZ the hero but old-vine, dry-grown GRENACHE often outshines it.

McWilliam's SE Aus r w (br) (sw) ★★★ Family-owned. Hanwood for value, *Mount Pleasant* for quality. Went into administration 2020.

Main Ridge Estate Mor P, Vic r w ★★ Rich, age-worthy CHARD, PINOT N. Founder Nat White is legend of MOR P wine; hard to imagine place without him. New owners (2015), quality still tiptop.

Majella Coon, SA r (w) ★★ As reliable as the day is long. Opulent SHIRAZ, CAB SAUV. Essence of modern COON.

Margaret River WA Temperate coastal area s of Perth. Powerful CHARD, structured CAB SAUV. CAPE MENTELLE, CULLEN, DEEP WOODS ESTATE, DEVIL'S LAIR, FLAMETREE, FRASER GALLOP, HUTTON WINES, LEEUWIN ESTATE, MCHENRY HOHNEN, MOSS WOOD, PIERRO, STELLA BELLA, TRIPE.ISCARIOT, VASSE FELIX, VOYAGER ESTATE, WOODLANDS and others. Great touring (and surfing) region.

Marius McL V, SA r ★★★ Varietal SHIRAZ and blends of dramatic concentration. Quality in inverse proportion to fuss; latter kept to a min.

Mayford NE Vic, Vic r w ★★★ Tiny v'yd in hidden valley. Put ALPINE VALLEYS region on map. SHIRAZ, CHARD, exciting spice-shot TEMPRANILLO.

Meerea Park Hun V, NSW r w ★ Brothers Garth and Rhys Eather create age-worthy SEM, SHIRAZ often as single-v'yd expressions.

Mike Press Wines Ad H, SA r (w) ★★ Tiny production, tiny pricing. CAB SAUV, SHIRAZ. Crowd favourite of bargain hunters.

Mitchelton Goulburn V, Vic r w (sw) ★★★ Stalwart producer of CAB SAUV, SHIRAZ, RIES, plus speciality of *Marsanne*, ROUSSANNE. Top spot to visit; fancy new hotel set among those fab river red gums.

Montalto Mor P, Vic r w ★★★ For some yrs was nice restaurant, gallery. Then wine quality skyrocketed. Firmly "must try" of MOR P. Single-v'yd releases top-notch.

Moorilla Estate Tas r w (sp) ★★★ Pioneer nr Hobart on Derwent River. Gd CHARD, RIES; PINOT N. V.gd restaurant, extraordinary art gallery. Now also owns nearby DOMAINE A.

Moorooduc Estate Mor P, Vic r w ★★★ Long-term producer of rich and complex CHARD and PINOT N.

Moppity Vineyards Hilltops, NSW r w ★★ Affable SHIRAZ, CAB SAUV (Hilltops). Elegant CHARD (TUMBARUMBA). Quality ambitions but best known as a value producer.

Mornington Peninsula Vic Coastal area 40-km (25-miles) se of Melbourne. Quality boutique wineries abound. Cool climate. PINOT N, CHARD, PINOT GR. Wine/surf/beach/food playground. ELDRIDGE ESTATE, GARAGISTE, KOOYONG, MAIN RIDGE ESTATE, MONTALTO, MOOROODUC ESTATE, PARINGA ESTATE, STONIER, TEN MINUTES BY TRACTOR, WILLOW CREEK, YABBY LAKE and more.

Morris NE Vic (r) (w) br ★★★ RUTH producer of Oz's (the world's?) greatest dessert *Muscats*, TOPAQUES. Owned by CASELLA.

Moss Wood Marg R, WA r w ★★★ (CAB SAUV) 05' 12' 14' 15' 16 MARG R's most opulent (red) wines. SEM, CHARD, super-smooth *Cab Sauv*. Oak- and fruit-rich.

Mount Horrocks Clare V, SA r w ★★★ Fine dry RIES, celebrated sweet Cordon Cut Ries. SHIRAZ, CAB SAUV in fine form.

Mount Langi Ghiran Gra, Vic r w ★★★ (SHIRAZ) 10' 12' 13' 14' 15 17 Rich, peppery, *Rhône-like Shiraz*. Excellent Cliff Edge Shiraz. Special patch of dirt. Special run of form.

Mount Majura Can, NSW r w ★★ Leading TEMPRANILLO producer. RIES, SHIRAZ, CHARD all gd. Reds sturdy, spicy.

Mount Mary Yarra V, Vic r w ★★★★ (PINOT N) 13' 14' 15' 16' 17 (Quintet) 10' 14' 15' 16' 17 Late Dr. Middleton made tiny amounts of suave CHARD, complex Pinot N, elegant CAB SAUV blend. All age impeccably. Post-Dr. era has brought improvement, if anything.

Mount Pleasant Hun V, NSW ★★★★ Old HUN V producer owned by MCWILLIAM'S, now re-invigorated. ND single-v'yd SEMS (esp *Lovedale*) and SHIRAZ. 18 reds quite incredible.

Mudgee NSW Region nw of Sydney. Earthy reds, fine SEM, full CHARD. Gd quality but needs a hero.

Ngeringa Ad H, SA r w ★ Perfumed PINOT N and NEBBIOLO. Rhôney SHIRAZ. Savoury rosé. Bio.

Nick Spencer Can, NSW r w ★★ Former Eden Road winemaker. CHARD and red blend (SHIRAZ/TEMPRANILLO/TOURIGA/CAB SAUV) of particular interest.

...today winemakers often cut down mature eucalypts to prevent same flavour.

Oakridge Yarra V, Vic r w ★★★★ A top-five producer of CHARD in Australia, and more recently a noteworthy producer of PINOT N. Multiple single-v'yd releases.

Ochota Barrels Bar V, SA r w ★★ Quixotic producer making hay with (mostly) old-vine GRENACHE, SHIRAZ from MCL V, BAR V.

O'Leary Walker Clare V, SA r w ★★ Low profile but excellent quality. CLARE V RIES, CAB SAUV standout. MCL V SHIRAZ oak-heavy but gd.

Orange NSW Cool-climate, high-elevation region. Lively SHIRAZ (when ripe), but best suited to (intense) aromatic whites and CHARD.

Out of Step Yarra V, Vic r w ★★★ Took on YARRA V SAUV BL and won. Now doing likewise with CHARD, PINOT N and NEBBIOLO from various v'yds. Only the brave.

Padthaway SA V.gd SHIRAZ, CAB SAUV. Rarely mentioned but important region. Soil salinity ongoing issue.

Penfolds now has Champagne (from Thienot). Cab Sauv from California is next.

Paringa Estate Mor P, Vic r (w) ★★★ Maker of irresistible PINOT N, SHIRAZ. Fleshy, fruity, flashy styles.

Parker Coonawarra Estate Coon, SA r ★★★ 1st vintage 1988. In gd yrs it produces full-bodied, age-worthy, tannic CAB SAUV of authority, distinction.

Passing Clouds Bendigo, Vic ★★★ Pioneer of modern Vic wine. Off radar for many yrs but burst back in 2016 with a gloriously elegant, textured signature CAB blend. Gd form since.

Paul Osicka Hea, Vic r ★★★ Vines dating back to 50s. Both character and flavour writ large. Small-scale, low-profile, high-impact and quality SHIRAZ, CAB SAUV.

Paxton McL V, SA r ★ Prominent organic/bio grower/producer. Ripe but elegant SHIRAZ, GRENACHE.

Pemberton WA Region between MARG R and GREAT SOUTHERN; initial enthusiasm for PINOT N replaced by RIES, CHARD, SHIRAZ.

Penfolds r w (br) ★★★★ (Grange) 90' 96' 04' 06' 08' 10' 12' 14 15 (CAB SAUV Bin 707) 02' 05' 06' 10' 12' 15' 16 and of course *St Henri*, "simple" SHIRAZ. Originally Adelaide, now SA. Oz's best warm-climate red wine company. Superb *Yattarna* CHARD, Bin Chard now right up there with reds.

Petaluma Ad H, SA r w sp ★★ (RIES) 11' 12' 17' 18 (CHARD) 12' 16' 17 (CAB SAUV COON) 05' 08' 12' 13 Seems to miss ex-owner/creator Brian Croser. Gd but low-key now.

Peter Lehmann Wines Bar V, SA r w (br) (sw) (sp) ★★ Well-priced wines incl easy RIES. Luxurious/sexy Stonewell SHIRAZ among many others (r w). Heroic Peter L died 2013; company sold 2014 to CASELLA (Yellow Tail).

Pewsey Vale Eden V, SA w ★ V.gd RIES, standard and (aged-release) The Contours, grown on lovely tiered v'yd.

Pierro Marg R, WA r w ★★★ (CHARD) 14' 15' 16' 17 Producer of expensive, tangy SEM/SAUV BL and full-throttle Chard.

Pipers Brook Tas r w sp ★★★ (RIES) 09' 13' 17 (CHARD) 13' 14' 16 Cool-area pioneer; gd Ries, *restrained Chard and sparkling* from Tamar Valley. Second label: Ninth Island. Owned by Belgian Kreglinger family.

Pizzini King V, Vic r ★★ (SANGIOVESE) 14' 15' 16' A leader of Italian varieties in Oz, esp NEBBIOLO, SANGIOVESE (recently stepped up a gear). Dominant KING VALLEY producer.

Pooley Tas r w ★★★ Est 1985 in Coal River Valley. Age-worthy PINOT N, CHARD. Family-run across generations. Among TAS's best.

Primo Estate SA r w dr (sw) ★★★ Joe Grilli's many successes incl rich MCL V SHIRAZ, tangy COLOMBARD, potent Joseph CAB SAUV/MERLOT and (exceptionally) complex sparkling Shiraz.

Punch Yarra V, Vic r w ★★★ Lance family ran Diamond Valley for decades. When they sold, they retained close-planted PINOT N v'yd. It can grow detailed, decisive, age-worthy wines.

Pyrenees Vic Central Vic region making rich, often minty reds. Blue Pyrenees, DALWHINNIE, Dog Rock, Summerfield, Mount Avoca, TALTARNI leading players, though it's also a happy hunting ground for assorted small producers.

Ravensworth Can, NSW r w ★★ Suddenly in hot demand for various wine experiments. SANGIOVESE best-known but there's a buzz over skin-contact whites and GAMAY Noir.

Riverina NSW Large-volume irrigated zone centred on Griffith.

Robert Oatley Wines Mudgee, NSW r w ★★ Ambitious venture of ROSEMOUNT ESTATE creator Robert Oatley. Quality/price ratio usually well aligned.

Rochford Yarra V, Vic r w ★★ Main outdoor entertainment venue in YARRA VALLEY now makes complex CHARD, PINOT N of note.

Rockford Bar V, SA r (w) sp ★★ Sourced from various old, low-yielding v'yds; reds best; iconic Basket Press SHIRAZ and noted *sparkling Black Shiraz*.

Rosemount Estate r w ★★ Once the pacesetter. Periodically loses way, reds can be gd.

Ruggabellus Bar V, SA r ★★ Causing a stir. Funkier, more savoury version of BAR V. Old oak, min sulphur, wild yeast, whole bunches/stems. Blends of CINSAULT, GRENACHE, MATARO, SHIRAZ.

Rutherglen & Glenrowan Vic Two of four regions in warm ne Vic zone, justly famous for sturdy reds, magnificent fortified dessert wines. ALL SAINTS, CAMPBELLS, ELDORADO ROAD, SCION, SIMAO & CO, STANTON & KILLEEN, TAMINICK CELLARS.

St Hallett Bar V, SA r w ★★ (Old Block) 12' 13' 14' 15 Old Block SHIRAZ the star, rest of range is smooth, sound, stylish. ACCOLADE-owned.

Saltram Bar V, SA r w ★★ Value Mamre Brook (SHIRAZ, CAB SAUV) and (rarely sighted) No.1 Shiraz are leaders. Main claim to fame is ubiquitous Pepperjack Shiraz.

Samuel's Gorge McL V, SA r ★★ Justin McNamee makes (at times) stunning GRENACHE, SHIRAZ, TEMPRANILLO of character and place.

Savaterre Beech, Vic r w ★★ (PINOT N) 12' 13' 16 Excellent producer of full-bodied CHARD, meaty Pinot N, close-planted SHIRAZ, SAGRANTINO.

Schmolzer & Brown Beech, Vic r w ★★ CHARD, PINOT N and rosé of intense, spice-drenched interest. Textural RIES of note.

Scion Ruth, Vic r w ★★ Fresh, vibrant approach to region's stalwart SHIRAZ, Durif.

SC Pannell McL V, SA r ★★★ Excellent (spicy, whole-bunch-fermented) SHIRAZ (often labelled SYRAH) and (esp) GRENACHE-based wines. NEBBIOLO to watch. Meticulous.

Sentio Beech, Vic r w ★★ Picks eyes out of various cool-climate regions to produce compelling CHARD, PINOT N. Shine.

Seppelt Gra, Vic r w br sp ★★★ (St Peter's SHIRAZ) 10' 12' 13' 14' 16' 17 Historic name owned by TWE. Impressive CHARD, RIES, (esp) peppery Shiraz.

Seppeltsfield Bar V, SA r br National Trust Heritage Winery bought by Warren Randall (2013). Fortified wine stocks back to 1878.

Serrat Yarra V, Vic r w ★★★ Micro-v'yd ot noted winemaker Tom Carson (YABBY LAKE) and wife Nadege. Complex, powerful, precise SHIRAZ/VIOGNIER, PINOT N, CHARD.

Seville Estate Yarra V, Vic r w ★★★ (SHIRAZ) 10' 14' 17 Excellent CHARD, spicy Shiraz, structured PINOT N. YARRA V pioneer still showing 'em how it's done.

Shaw & Smith Ad H, SA r w ★★★ Savvy outfit. Crisp *harmonious* SAUV BL, complex M3 CHARD and, surpassing them both, *Shiraz*. PINOT N slowly improving.

Shy Susan Tas r w ★★ New range by winemaker Glenn James, former maker of top-end HARDYS and PENFOLDS whites. CHARD, RIES, PINOT N particularly strong, age-worthy initial releases. Gorgeous packaging.

Simao & Co Ruth, Vic r w ★★ Young Simon Killeen, of STANTON & KILLEEN family, makes scrumptious TEMPRANILLO, UGNI BL, SHIRAZ and more. Personality+.

Sorrenberg Beech, Vic r w ★★★★ No fuss but highest quality. SAUV BL/SEM, CHARD, (Australia's best) GAMAY, B'x blend. Ultimate "in the know" winery of Oz.

Warren's world

Aussie wine daredevil Warren Randall is in the middle of amazing v'yd-buying spree. He now owns 3500 ha of premium v'yd, bulk of it in the BAR V and MCL V regions (now largest v'yd holder in both). In past yr alone he's snapped up 750 ha+. Historic wineries Quelltaler, Ryecroft, SEPPELTSFIELD all now part of his plan to conquer China.

Southern NSW Zone NSW Incl CAN, Gundagai, Hilltops, TUMBARUMBA. Savoury SHIRAZ; lengthy CHARD.

Spinifex Bar V, SA r w ★★★ Bespoke BAR V producer. Complex SHIRAZ, GRENACHE blends. Routinely turns out rich but polished reds.

Stanton & Killeen Ruth, Vic (r) br ★★ Fortified vintage is dominant attraction.

Stefano Lubiana S Tas r w sp ★★★ Beautiful v'yds on banks of Derwent River, 20 mins from Hobart. Excellent PINOT N, sparkling, MERLOT, CHARD. Homely but driven and ambitious. Bio.

Stella Bella Marg R, WA r w ★★★ Humdinger wines. CAB SAUV, SEM/SAUV BL, CHARD, SHIRAZ, SANGIOVESE/Cab Sauv. Sturdy, characterful.

Stoney Rise Tas r w ★★★ Joe Holyman used to be a world-class wicketkeeper; he's an even better winemaker/grower, particularly with PINOT N, CHARD.

Stonier Wines Mor P, Vic r w ★★★ (CHARD) 15' 17' 18 (PINOT N) 12' 15' 17' 18 Consistently gd; Res notable for elegance. Pinot N in particularly fine form, tense, resonant. Excellent single-v'yd releases now.

Sunbury Vic *See* MACEDON AND SUNBURY.

Sutton Grange Bendigo, Vic ★★★ w r (AGLIANICO) 17' 18 (SHIRAZ) 16' 17' 18 Organic producer of intense-but-savoury Shiraz and impeccably crafted whites, FIANO the leader.

Swan Valley WA Birthplace of wine in the w, 20 mins n of Perth. Hot climate makes strong, low-acid wines. FABER V'YDS leads way.

Swinney Fra R, WA ★★★ A grape-grower for decades; 1st release of wines under own steam is outstanding. GRENACHE, SHIRAZ, RIES all shine.

Tahbilk Goulburn V, Vic r w ★★★ (MARSANNE) 14' 16' 17' 18' 19 (SHIRAZ) 10' 12' 16 Historic Purbrick family estate: long-ageing reds, also some of Oz's best old-vine *Marsanne*. Res CAB SAUV can be v.gd. Rare 1860 Vines Shiraz. For lovers of rustic.

Taltarni Pyrenees, Vic r w sp ★★ SHIRAZ, CAB SAUV in gd shape. Long-haul wines but jackhammer no longer required to remove tannin from your gums.

Taminick Cellars Glenrowan, Vic r ★ The Booth family have been farming this tough, dry patch since 1914. A fair amount of character has been accumulated along the way.

Tapanappa SA r ★★★★ WRATTONBULLY collaboration between Brian Croser, Bollinger, J-M Cazes of Pauillac. Splendid CAB SAUV blend, SHIRAZ, MERLOT, CHARD. Surprising *Pinot N* from Fleurieu Peninsula.

Tar & Roses Hea, Vic r w ★★★ SHIRAZ, TEMPRANILLO, SANGIOVESE of impeccable polish, presentation. Modern success story. 2017 death of co-founder Don Lewis a great loss but quality remains strong.

Tarrawarra Estate Yarra V, Vic r w ★★ (Res CHARD) 12' 13' 17 (Res PINOT N) 12' 13' 17 Moved from hefty, idiosyncratic to elegant, long. Res generally a big step up on standard.

Tasmania Tas Cold island region with hot reputation. Outstanding sparkling, PINOT N, RIES. V.gd CHARD, PINOT GR, SAUV BL.

Taylors Wines Clare V, SA r w ★ Large-scale production led by RIES, SHIRAZ, CAB SAUV. Exports under Wakefield Wines brand.

Ten Minutes by Tractor Mor P, Vic r w ★★★ Wacky name, smart packaging, even better wines. *Chard, Pinot N both excellent* and will age. Style meets substance.

Adelaide, naturally

The AD H is a sleepy region on the doorstep of Adelaide known for such stalwarts as ASHTON HILLS, HENSCHKE, PETALUMA, SHAW & SMITH. Or at least it was. If there's any such thing as a centre for natural wine in Australia it's these gorgeous hills: Basket Range Wine, Commune of Buttons, Lucy Margaux, OCHOTA BARRELS, Unico Zelo all call these hills home.

Teusner Bar V, SA r ★★ Old vines, clever winemaking, pure fruit flavours. Leads a BAR V trend towards "more wood, no good".

Thousand Candles Yarra V, Vic r w ★★ Beautiful site producing beautiful wines. Delicate PINOT N, spicy SHIRAZ, lively field blend. Cool climate. Quality on steady march forward.

Tim Adams Clare V, SA r w ★ Ever-reliable (in gd way) RIES, CAB SAUV/MALBEC blend, SHIRAZ and (full-bodied) TEMPRANILLO.

2020 crop wiped out in Hunter V, Canberra, Hilltops, Beechworth, Alpine Valleys.

Tolpuddle Tas ★★★ SHAW & SMITH bought this outstanding 1988-planted v'yd in TAS's Coal River Valley in 2011. Scintillating PINOT N, CHARD in lean, lengthy style.

Topaque Vic Replacement name for iconic RUTH sticky "Tokay", thanks to EU; a decade later it's still hard to find anyone who likes the name.

Torbreck Bar V, SA r (w) ★★★ Dedicated to (often old-vine) Rhône varieties led by SHIRAZ, GRENACHE. Ultimate expression of rich, sweet, high-alc style.

Torzi Matthews Eden V, SA r ★★ Aromatic, stylish, big-hearted SHIRAZ. Value RIES, SANGIOVESE. Incredible consistency yr-on-yr.

tripe.Iscariot Marg R, WA r w ★★ Hard to spell, easy to drink. Complex whites/reds by its own design. Natural wine specialist.

Tumbarumba NSW Cool-climate NSW region tucked into Australian Alps. Sites 500–800m (1640–2625ft). CHARD the star. PINOT N not in same class.

Turkey Flat Bar V, SA r p ★★★ Top producer of bright-coloured rosé, GRENACHE, SHIRAZ from core of 150-yr-old v'yd. Controlled alc and oak. New single-v'yd wines. Old but modern.

TWE (Treasury Wine Estates) Aussie wine behemoth. COLDSTREAM HILLS, DEVIL'S LAIR, LINDEMANS, PENFOLDS, ROSEMOUNT, SALTRAM, WOLF BLASS, WYNNS COONAWARRA ESTATE among them.

Two Hands Bar V, SA r ★★★ Big reds and many of them. They've turned the volume down a fraction lately; glory of fruit seems all the clearer.

Tyrrell's Hun V, NSW r w ★★★★ (SEM) 14' 15' 16' 17' 18 (Vat 47 CHARD) 14' 15' 16' 17' 18 Oz's greatest maker of Sem, Vat 1 now joined with series of individual v'yd or subregional wines. *Vat 47*, Oz's 1st Chard, continues to defy climatic odds. Outstanding old-vine 4 Acres SHIRAZ, Vat 9 Shiraz. One of the true greats.

Vasse Felix Marg R, WA r w ★★★ (CHARD) 14' 15' 16' 17' 18 (CAB SAUV) 10' 12' 14' 15 With CULLEN, pioneer of MARG R. Elegant Cab Sauv for mid-weight balance. Complex/funkified Chard. Returned to estate-grown roots.

Voyager Estate Marg R, WA r w ★★★ Big volume of (mostly) estate-grown, rich, powerful SEM, SAUV BL, (esp) CHARD and CAB SAUV/MERLOT.

Wanderer, The Yarra V, Vic ★★★ Upper YARRA VALLEY producer of exceptionally fine-boned PINOT N.

Wantirna Estate Yarra V, Vic r w ★★★ Regional pioneer showing no sign of slowing down. CHARD, PINOT N, B'x blend all in excellent form. Small on quantity, big on quality.

Wendouree Clare V, SA r ★★★★ Treasured maker (tiny quantities) of powerful, tannic, concentrated reds, based on CAB SAUV, MALBEC, MATARO, SHIRAZ. Recently moved to screwcap; the word "longevity" best defined with a picture of a Wendouree red. Beg, borrow or steal.

West Cape Howe Denmark, WA r w ★ Affordable, flavoursome reds the speciality.

Westend Estate Riverina, NSW r w ★★ Thriving family producer of **tasty bargains**, esp Private Bin SHIRAZ/Durif. Recent cool-climate additions gd value.

Willow Creek Mor P, Vic r w ★★ Gd gear. Impressive producer of CHARD, PINOT N in particular. Power and poise.

Wirra Wirra McL V, SA r w (sw) (sp) ★★ (RSW SHIRAZ) 05' 10' 12' 15 (The Angelus

CAB SAUV) **10' 12' 13'** 15 High-quality, concentrated wines in flashy livery. The Angelus Cab Sauv named Dead Ringer outside Australia.

Wolf Blass Bar V, SA r w (br) (sw) (sp) ★★ (Black Label CAB SAUV blend) **06' 08' 12' 13'** 14 Owned by TWE. Not the shouty player it once was but still churns through an enormous volume of clean, inoffensive wines.

Woodlands Marg R, WA r (w) ★★ 7 ha of 40-yr-old+ CAB SAUV among top v'yds in region, plus younger but v.gd plantings of other B'x reds. Brooding impact.

Wrattonbully SA Important grape-growing region in LIMESTONE COAST ZONE; profile lifted by activity of TAPANAPPA, Terre à Terre, Peppertree.

Wynns Coonawarra Estate Coon, SA r w ★★★★ (SHIRAZ) **10' 12' 14' 16'** 17 (CAB SAUV) **06' 12' 13' 14'** 15 16' 17 TWE-owned COON classic. RIES, CHARD, Shiraz all v.gd, Cab Sauv outstanding, esp Black Label Cab Sauv, *John Riddoch Cab Sauv*. Recent single-v'yd releases are the icing.

Yabby Lake Mor P, Vic r w ★★★ Made its name with estate CHARD, PINOT N, boosted with single-site releases, now spice-shot SHIRAZ adds yet more to reputation.

Yalumba Bar V, SA, SA r w sp ★★★ 170 yrs young, family-owned. *Full spectrum of high-quality wines*, from budget to elite single-v'yd. Entry-level Y Series v.gd value.

Yangarra Estate McL V, SA r w ★★★★ Conventional in some ways, inventive in others. Whatever it takes to make great wine. Full box and dice here, across most price points. Varietal GRENACHE, SHIRAZ particularly strong.

Yarraloch Yarra V, Vic r w ★★ Complex CHARD can be terrific. Capable of exceptional PINOT N too.

Yarra Valley Vic Thriving area just ne of Melbourne. Emphasis on CHARD, PINOT N, SHIRAZ, sparkling. Understated, elegant CAB SAUV. COLDSTREAM HILLS, DE BORTOLI, DOMAINE CHANDON, GEMBROOK HILL, GIANT STEPS, HODDLES CREEK ESTATE, LUKE LAMBERT, MAC FORBES, OUT OF STEP, PUNCH, ROCHFORD, SERRAT, SEVILLE ESTATE, TARRAWARRA, THOUSAND CANDLES, WANTIRNA ESTATE, YARRALOCH, YARRA YERING, YERINGBERG, YERING STATION is a formidable line-up.

Yarra Yering Yarra V, Vic r w ★★★ (Dry Reds) **06' 15' 17'** 18 One-of-a-kind YARRA V pioneer. Powerful PINOT N; deep, herby CAB SAUV (Dry Red No.1); SHIRAZ (Dry Red No.2). Luscious, daring flavours (r w).

Yellow Tail NSW *See* CASELLA.

Yeringberg Yarra V, Vic r w ★★★★ (MARSANNE/ROUSSANNE) **13' 14' 15'** 16 (CAB SAUV) **05' 10' 12' 13' 14' 15'** 16 Historic estate still in hands of founding (1862) Swiss family, the de Purys. Extremely small quantities of v. high quality CHARD, Marsanne, Roussanne, Cab Sauv, PINOT N.

Yering Station / Yarrabank Yarra V, Vic r w sp ★★ On site of Vic's 1st v'yd; replanted after 80-yr gap. Snazzy table wines (Res CHARD, PINOT N, SHIRAZ, VIOGNIER); Yarrabank (sparkling wines in joint venture with Champagne Devaux).

Upper Yarra magic

The words "Upper Yarra" only started to appear on wine labels a decade ago though they're now hot property. It's often (wrongly) assumed that the phrase refers to higher-altitude sections of the YARRA VALLEY, but it actually refers to the upper reaches of the Yarra River. This area is generally cooler; until relatively recently it was considered too cool to ripen grapes. Upper Yarra PINOT N is about as clear an expression of terroir as you will find in Oz. Compared with the warmer, fruitier examples coming off the main Yarra Valley floor, the Pinot N of the Upper Yarra is finer (drastically), lighter in colour and seems to carry as much bone as flesh. GEMBROOK HILL is the standard flagbearer, though DAPPLED, HODDLES CREEK ESTATE, OAKRIDGE (single-v'yd releases), THE WANDERER are key exponents.

New Zealand

Abbreviations used
in the text:

Auck	Auckland
B of P	Bay of Plenty
Cant	Canterbury
Gis	Gisborne
Hawk	Hawke's Bay
Hend	Henderson
Marl	Marlborough
Mart	Martinborough
Nel	Nelson
N/C Ot	North/Central Otago
Waih	Waiheke Island
Waip	Waipara Valley
Wair	Wairarapa

N Z is Sauvignon Blanc Central: made in that familiar pungent, green,
often easy-drinking style, it is still many drinkers' first choice. As a
result, in the 19 vintage winemakers here harvested more than ten times
as much Sauv as Pinot Noir, but produced and labelled twice as many
Pinots as Sauvs. This snapshot of NZ wine says several things: most Sauv
goes into big brands that sell millions of cases, whereas Pinot Noir is often
made in tiny amounts and sold locally. Those big brands offer safety
and familiarity, but NZ offers far more interesting and better things to
explore; and if you want fine wines, explore you must. Check out
www.winesearcher.com for stockists near you. These wines (Pinot Noir,
Syrah, Bordeaux blends) are what to look for, because NZ's reds are some
of its best wines. Riesling, Chardonnay, Gewurztraminer, Pinot Gris are
on the rise too, and Albariño is stirring up interest; Tempranillo is looking
promising. It's northern Europe upside down, with more sunshine.

Recent vintages

2019 High quality expectations; slightly less aromatic, but weighty, textured
Marl Sauv Bl. Hawk Chard and reds esp promising.

2018 Hottest-ever summer, destructive storms in April. Ripe, less herbaceous
Marl Sauv Bl.

2017 Challenging vintage, with rain before harvest. C Ot more successful.

2016 Ripe, tropical fruit-flavoured Marl Sauv Bl. In Hawk, excellent Chard but
autumn rain hit Merlot

Akarua C Ot r (p) (w) (sp) ★★★ PINOT N: outstanding Bannockburn 17'; drink-young
Rua **17'**; esp powerful The Siren 16. Lively fizz, incl Vintage Brut. PINOT GR **18'**.

Allan Scott Marl (r) (p) w (sp) ★★ Family firm. Fresh, easy-drinking wines: SAUV BL 19'; CHARD 18'; off-dry PINOT GR 18'. Upper tier: Generations range.

Alpha Domus Hawk r w (sp) ★★ Family winery recently celebrated 30 yrs. B'x-style reds, esp MERLOT-based The Navigator 14 and v. classy AD CAB SAUV The Aviator 15'. CHARD-based bubbly, Cumulus 17'. AD is top range (savoury Chard 16').

Amisfield C Ot r (p) (w) (sp) ★★★ Intense sparkling 15'. Strong CHENIN BL 17. Classy RIES (dry 17, medium-sweet 18'). Fragrant PINOT N 17' (RKV Res is Rolls-Royce model 16'). Lake Hayes to drink young.

Astrolabe Marl (r) w ★★→★★★ Impressive range: herbal SAUV BL 19'; pure Awatere Valley Sauv Bl 19'. Crisp ALBARIÑO 18; slightly sweet CHENIN BL 19'. Gd Dry RIES 18. Creamy CHARD 18. Earthy PINOT N 17'. Durvillea is 2nd tier. v.gd value.

Ata Rangi Mart r (p) (w) ★★★ →★★★★ Much-respected family affair. PINOT N 14' 15' 16' 17 is a NZ classic; perfumed, savoury (1st vines 1980). Notable Craighall CHARD 15' 16 17 (planted 1983); complex Lismore PINOT GR 18'.

Auckland Largest city (n, warm, cloudy) in NZ; 1.0% v'yd area but 14% of producers (incl head offices of big firms.) Nearby wine districts: Kumeu/Huapai/ Waimauku (long est); newer (since 80s): Matakana, Clevedon, WAIH (island v'yds, popular with tourists). Savoury B'x blends in dry seasons 13' 14' 19', bold SYRAH 13' 14' 19' is fast-expanding and rivals HAWK for quality; underrated CHARD 13' 14' 15 19'; promising PINOT GR.

Auntsfield Marl r w ★★→★★★ Classy wines from site of region's 1st (1873) v'yd (replanted 1999). Weighty SAUV BL 18'; tight CHARD 17 (esp single-block Cob Cottage 16', dense PINOT N 17'.

Awatere Valley Marl Key subregion (pronounced "Awa-terry"), with few wineries, but huge v'yd area (more than HAWK), pioneered in 1986 by VAVASOUR. YEALANDS is key producer. Slightly cooler, drier, windier, less fertile than WAIRAU VALLEY, with racy, ("tomato stalk") SAUV BL (rated higher by UK than US critics); vibrant RIES, PINOT GR; slightly herbal PINOT N.

Babich Hend r w ★★→★★★ NZ's oldest family-owned winery (1916), Croatian origin. HAWK, MARL v'yds; wineries in AUCK, MARL. Age-worthy Irongate from GIMBLETT GRAVELS: weighty CHARD 18 and B'x-like Irongate CAB/MERLOT/CAB FR 13' 14' 15' 16. Biggest seller: tropical Marl SAUV BL (classy Res 17'). Top red: stylish The Patriarch (B'x-style, MALBEC-influenced) 13' 14' 15' 16.

Blackenbrook Nel r w ★★ Small winery, impressive aromatic whites, esp Alsace-style GEWURZ 19'; PINOT GR 19', PINOT BL 19'. Punchy SAUV BL 19'; generous CHARD 18; vivacious Rosé 19'; graceful PINOT N 17.

Black Estate Cant r w ★★ Small organic WAIP producer with mature (1994) vines. Tight Home V'yd CHARD 17; graceful CAB FR 17; savoury PINOT N 16'; outstanding Damsteep Pinot N 16'.

Blank Canvas Marl r w ★★ Owners Matt Thomson (ex-SAINT CLAIR) and Sophie Parker-Thomson. Tropical SAUV BL 18; light RIES 18; complex Upton Downs PINOT N 16'.

Borthwick Wair r w ★★ V'yd at Gladstone with Paddy Borthwick brand. Pungent SAUV BL 19'; citrus CHARD; scented PINOT GR 18; full Pinot Rosé 18. Rich PINOT N 16'.

Brancott Estate Marl r (p) w ★ →★★★ Major brand of PERNOD RICARD NZ that replaced Montana worldwide (except in NZ). Top wines: Letter Series eg. fleshy "B" Brancott SAUV BL 18'; rich "O" CHARD 17; savoury "T" PINOT N 17. Huge-selling, herbaceous Sauv Bl, gd value. Identity range: subregional focus. Living Land: organic. Flight: plain, low alc. Top-value, lively, bottle-fermented Brut Cuvée.

Brightwater Nel (r) w ★★ Impressive whites, esp intense SAUV BL 19'; medium-dry RIES 17; gently oaked CHARD; poised PINOT GR 18. Rich PINOT N 17. Top: Lord Rutherford (incl deep Sauv Bl 17; refined Chard 16').

Brookfields Hawk r w ★★→★★★ Smallish, long-est winery. Gd value CAB SAUV 17, Sun-Dried MALBEC 18. Top wines: Marshall Bank CHARD 18'; Hillside SYRAH 16.

Burn Cottage C Ot r w ★★ →★★★ Organic, v'yds at Pisa and Bannockburn, owned by Nevada-based Sauvage family. Moonlight Race PINOT N, delicious young. Refined Burn Cottage V'yd Pinot N 17'. Also RIES/GRÜNER V 17'.

Canterbury NZ's 4th-largest wine region (just ahead of GISBORNE); most v'yds in relatively warm n WAIP district (increasingly called North Cant). Greatest success with aromatic RIES and savoury PINOT N. Emerging strength in Alsace-style PINOT GR. SAUV BL heavily planted, but often minor component in other regions' wines.

Carrick C Ot r w ★★★ Bannockburn winery with organic focus. Classy RIES (dry, sweetish Josephine 17); elegant CHARD 17, esp EBM 16'; partly oak-aged PINOT GR. PINOT N, built to last 17'. Attractive Unravelled Pinot N 18.

Catalina Sounds Marl r w ★★ →★★★ Export-focused, Australian-owned producer, with large Sound of White v'yd in upper Waihopai Valley. SAUV BL, PINOT N.

Central Otago (r) 16 17' 19 (w) 16 17 19' High-altitude, dry inland region (now NZ's 3rd largest) in s of S Island, with many tiny, "weekend" producers. Sunny, hot days, v. cold nights. Most vines in Cromwell Basin. Crisp RIES, PINOT GR; well-justified interest in vibrant, tight-knit CHARD; famous PINOT N (78%+ v'yd area) has drink-young charm; older vines now more savoury, complex. Excellent Pinot N Rosé and traditional-method fizz.

Chard Farm C Ot r w ★★ Pioneer in striking gorge setting. Fleshy PINOT GR; lemony, slightly sweet RIES 17'; mid-weight PINOT N (fruity River Run 17'; single-v'yd The Tiger 17' and The Viper 17' more complex). Drink-young Rabbit Ranch Pinot N. Mata-Au Pinot N is sweet-fruited, signature red 17'.

25% of acreage of C Ot is bio or organic; 7% in NZ as a whole.

Church Road Hawk r (p) w ★★ →★★★ PERNOD RICARD NZ winery with historic HAWK roots. Buttery CHARD; partly oak-aged SAUV BL; Alsace-style PINOT GR; dark MERLOT/CAB SAUV; drink-young SYRAH. Impressive Grand Res wines (esp Merlot/Cab Sauv 15'.) McDonald Series, between standard and Grand Res, offers eye-catching quality, value (incl Chard, Syrah, Cab Sauv, Merlot). Delicate Gwen Rosé. Prestige (NZ$150-$220) TOM selection: lush Cab Sauv/Merlot 15', powerful Chard 15', refined Syrah 15'.

Churton Marl r w ★★ Elevated Waihopai Valley site with bone-dry SAUV BL 17'; intense Best End Sauv Bl 17'; sturdy VIOGNIER; savoury PINOT N 16 (esp The Abyss; oldest vines, greater depth 13). Honeyed PETIT MANSENG 17'.

Clearview Hawk r (p) w ★★ →★★★ Coastal v'yd at Te Awanga (also grapes from inland). Hedonistic Res CHARD 17 (Beachhead Chard 17 is junior version); rich Enigma (MERLOT-based) 16'; savoury Old Olive Block (CAB SAUV/MALBEC/CAB FR blend). Top-value Cape Kidnappers Merlot 16'.

Clos Henri Marl r w ★★ →★★★ Organic, founded 2001 by Henri Bourgeois of Sancerre. Weighty SAUV BL from stony soils, one of NZ's best 17'; sturdy PINOT N 15' (on clay). Second label: Bel Echo (reverses variety/soil match). Third label: Petit Clos, from young vines. Distinctive, satisfying wines, priced right.

Cloudy Bay Marl r w sp ★★★ Large-volume, still-classy SAUV BL is NZ's most famous wine. Also complex CHARD 17', supple PINOT N 17. Stylish Pelorus (sp): Rosé, NV; esp intense Vintage 10', 7 yrs on lees. Te Koko (oak-aged Sauv Bl) has personality 16'. More involvement in C OT for Te Wahi Pinot N (fleshy, rich 14' 15' 16 17'). Owned by LVMH.

Constellation New Zealand Auck r (p) w ★ →★★★ Largest producer, previously Nobilo Wine Group, now owned by Constellation Brands (New York-based). Strong in US market (KIM CRAWFORD MARL SAUV BL is no 1-selling NZ wine.) Strength mainly in solid, moderately priced wines (esp Sauv Bl) under Kim Crawford, Monkey Bay, NOBILO and SELAKS brands.

Cooper's Creek Auck r w ★★ →★★★ Innovative, gd value. Rich Swamp Res HAWK

CHARD 18'; gd SAUV BL, RIES; MERLOT; easy-drinking PINOT N; rich SYRAH (esp sturdy SV Chalk Ridge Hawk Syrah 18'). Res is top range; SV (Select V'yd) range is mid-tier. NZ's 1st: ALBARIÑO, ARNEIS, GRÜNER V, MARSANNE.

Craggy Range Hawk r (p) w ★★★→★★★★ High-profile, top restaurant, large v'yds in HAWK, MART. Stylish CHARD 18, PINOT N 16; excellent mid-range MERLOT, SYRAH, Te Kahu (B'x red blend 17) from GIMBLETT GRAVELS; dense Sophia (Merlot) 16; show-stopping Syrah Le Sol 14' 15' 16'; sturdy The Quarry (CAB SAUV) 16; refined Aroha (Pinot N) 14' 15' 16' 17'.

Delegat Auck r w p ★★ Large listed family company: 2000 ha v'yds in HAWK, MARL; three wineries. Hugely successful OYSTER BAY brand, esp SAUV BL 19', plummy MERLOT 18', excellent dry Rosé 19'. Top-value Delegat range: citrus CHARD; full Sauv Bl; vibrant Merlot; savoury PINOT N. Owns Barossa Valley Estates.

Delta Marl ★★→★★★ Owned by SAINT CLAIR. V.gd-value, full-flavoured PINOT N 17', aromatic SAUV BL 19'; creamy CHARD 18. Hatters Hill range a step up.

Destiny Bay Waih r ★★→★★★ Expat Americans make classy, high-priced (but cheaper to Patron Club members) brambly B'x-style reds. Flagship is savoury Magna Praemia (mostly CAB SAUV). Mystae is mid-tier: lush. Destinae: softly textured, for earlier drinking.

Deutz Auck sp ★★★ Champagne house gives name to refined, great-value fizz from MARL by PERNOD RICARD NZ. Popular Brut NV has min 2 yrs on lees. Much-awarded Blanc de Blancs 15' is vivacious. Crisp Rosé NV; outstanding Prestige (disgorged after 3 yrs), mostly CHARD 15'.

Dog Point Marl r w ★★★ Grower Ivan Sutherland and winemaker James Healy make incisive SAUV BL (Section 94) 16'; CHARD (elegant) 17; complex PINOT N 13' 14' 15 16 17', all among region's finest. Larger-volume, but v.gd unoaked Sauv Bl 19'.

Domaine-Thomson C Ot r ★★→★★★ Small, organic, v'yds in two hemispheres: Gevrey-Chambertin and Lowburn, overlooking Cromwell Basin. Fragrant and full-flavoured wines.

Dry River Mart r w ★★★ Small pioneer winery, now US-owned. Reputation for long-lived whites: savoury CHARD 18'; intense RIES 18'; oily PINOT GR (NZ's 1st outstanding example) 18'; heady GEWURZ 18'; late-harvest whites; graceful PINOT N 16' 17. Dense TEMPRANILLO 16'.

Elephant Hill Hawk r (p) w ★★→★★★ Stylish winery/restaurant on coast at Te Awanga, also grapes from inland. Rich CHARD 17; bold MERLOT/MALBEC; deep SYRAH. Outstanding Res range, incl Chard 16'. Top pair: Airavata Syrah (notably complex 15'); Hieronymus (powerful blended red 15').

Escarpment Mart r (w) ★★★ Australian-owned, winemaker Larry McKenna (ex-MARTINBOROUGH V'YD). Known for savoury PINOT N. Top label: Kupe. Single-v'yd, old-vine reds esp gd. MART Pinot N is regional blend. Lower tier: The Edge.

Esk Valley Hawk r p w ★★→★★★★ Owned by VILLA MARIA. Acclaimed MERLOT-based blends (esp Winemakers Res, but also excellent, lower-tier Merlot/CAB/MALBEC. Supple SYRAH (rich Res 16). Popular Merlot Rosé 19; barrel-fermented CHARD superb value 18; full-bodied VERDELHO 18; v.gd, dryish PINOT GR 19'. Striking flagship red Heipipi The Terraces: spicy single-v'yd blend, Malbec/Merlot/CAB FR 16 .

Price Trends

The average price of NZ wines has fallen by more than a 3rd in past 15 yrs. Chasing huge-volume sales of MARL SAUV BL is key reason for price drop. It often means shipping the stuff in bulk, rather than bottling it in NZ. Same statistics say that NZ wine remains highest or 2nd-highest priced in the US, Canada, UK and Hong Kong, which makes you wonder mostly about statistics.

Felton Road C Ot r w ★★★★ Celebrated winery at Bannockburn, best-known for PINOT N, but RIES, CHARD notably classy too. Bold yet graceful Pinot N Block 3 15' 17' 18', more powerful Block 5 15' 17' 18' from The Elms V'yd; intense Ries (dr s/sw); tight Chard (esp Block 2 18'); key label is Bannockburn Pinot N, four-v'yd blend 18'. Other fine single-v'yd Pinot N: Calvert, Cornish Point.

Forrest Marl r (p) w ★★ Big success with The Doctors' MARL SAUV BL, low alc (9.5%), delicate. Wide range of attractive, value Marl whites, esp Sauv Bl (lively 19'); floral Rosé 19'. Rich Botrytis RIES 18'. Tatty Bogler: complex C OT, WAITAKI VALLEY wines. Top range: John Forrest Collection (v. classy CHARD 13').

Müller-T was c.40% of NZ v'yd in 1980. Now, 2 ha. But not a single wine...

Framingham Marl (r) w ★★→★★★ Owned by Sogrape (*see* Portugal). Aromatic whites: intense RIES, esp zesty, organic Classic from mature vines. Perfumed, rich PINOT GR 18'. Vibrant CHARD. Subtle SAUV BL; lush Noble Ries 17'; silky PINOT N. Rare F Series wines (incl Old-Vine Ries and brilliant botrytized sweet whites), full of personality.

Fromm Marl r w ★★★ Swiss-owned. Distinguished PINOT N, esp hill-grown Clayvin V'yd 15' 16' 17'. Fromm V'yd, sturdier 17'. Cuvee H Pinot N: multi-site blend 17'. Powerful Fromm V'yd SYRAH 16'. Refined Clayvin CHARD 16'. Earlier-drinking La Strada range, incl tangy SAUV BL; excellent Rosé.

Gibbston Valley C Ot r (p) w ★★→★★★ Strong name for PINOT N, esp fragrant, bold GV Collection 17'. Silky Le Maitre (17' best ever), mostly from 1st vines planted in 80s. Racy GV RIES; intense Red Shed Ries 18'. Full-bodied GV PINOT GR (esp refined, organic School House 17'); classy CHARD (esp China Terrace 17'). Gold River Pinot N: drink-young charm.

Giesen Cant (r) w ★★ Large winery, family-owned (ex-Neustadt, in Germany's Rheinpfalz) making huge volume MARL SAUV BL. Popular RIES, intensity and value 19'; single-v'yd Gemstone Ries, partly fermented in granite tanks 19'. Multi-region PINOT GR also gd value. Fast-improving PINOT N.

Gimblett Gravels Hawk Defined area (800 ha planted, mostly since early 80s) of old riverbed, so arid and inhospitable that rabbits rarely venture onto it without taking a cut lunch. Noted for rich B'x-style reds (mostly MERLOT-led, but stony soils also suit CAB SAUV – recent renewed interest). Super SYRAH. Best reds world-class. Also age-worthy CHARD, gd MARSANNE/VIOGNIER.

Gisborne NZ's 5th-largest region, on e coast of N Island. Declining in planted area and producer numbers. Abundant sunshine but often rainy; fertile soils. Key is CHARD (ripe, soft). Excellent GEWURZ, CHENIN BL, VIOGNIER; MERLOT, PINOT GR more variable. Interest in ALBARIÑO (rain-resistant). Top wines from MILLTON.

Gladstone Vineyard Wair r w ★★ Largest producer in n WAIR, bought by Asian investment company (2018). Tropical SAUV BL 18; spicy PINOT GR 18; v.gd RIES; full VIOGNIER. Fruit-packed PINOT N 18' (18 single-v'yd reds finest yet). 12,000 Miles lower-priced, early-drinking range.

Grasshopper Rock C Ot r ★★→★★★ Estate-grown since 2003 at Alexandra by PINOT N specialist. Subregion's finest red 14' 17': cherry, spice, dried-herb flavours. Age-worthy, great value.

Greenhough Nel r w ★★→★★★ One of region's best; intense Apple Valley RIES 19', lively River Garden SAUV BL 19', consistently gd CHARD 17, PINOT N 16. Top label: Hope V'yd (organic Chard 16'; old-vine PINOT BL is NZ's finest 16; mushroomy Pinot N 17).

Greystone Waip (r) w ★★★ Star producer (also owns Muddy Water), partly organic, with aromatic whites (dry and medium RIES, Alsace-style PINOT GR 18'; GEWURZ 18'; lush CHARD 17'; oak-aged SAUV BL 18'; PINOT N (complex 17). Thomas Brothers is top, notably Pinot N 16'.

Greywacke Marl r w ★★★ Distinguished wines from Kevin Judd, ex-CLOUDY BAY. Fleshy SAUV BL 19'; weighty CHARD 16; rich PINOT GR 17'; gently sweet RIES 18; powerful PINOT N 17'. Barrel-fermented Wild Sauv 17' full of personality.

Grove Mill Marl r w ★★ Attractive, gd-value whites with WAIRAU VALLEY subregional focus: strong SAUV BL 19'; generous CHARD 18; oily-textured PINOT GR 18; slightly sweet RIES. Full PINOT N 18. Owned by Foley Family Wines.

Haha Hawk, Marl r w ★★ Fast-growing producer, v.gd value. Notable CHARD, MERLOT, PINOT GR, SAUV BL.

Hans Herzog Marl r w ★★★ Warm, stony, organic v'yd, many varieties planted. Classy MERLOT/CAB 14'; savoury PINOT N Duc 15'. Fleshy CHARD 17'; apricot-coloured PINOT GR 17'; oak-aged SAUV BL 16'; deep VIOGNIER 17'. Classy TEMPRANILLO 15', MONTEPULCIANO 15. Sold under Hans brand in the EU, US.

Hawke's Bay NZ's 2nd-largest region (12.4% v'yd area). Founded 1850s; sunny, dryish climate. Classy, B'x-like MERLOT and CAB SAUV-based reds in favourable yrs; SYRAH (perfumed) fast-rising star; weighty CHARD; rounded SAUV BL (suits oak); NZ's best VIOGNIER. Promising PINOT N from cooler, elevated, inland districts. *See also* GIMBLETT GRAVELS.

Hunter's Marl (r) (p) w (sp) ★★→★★★ Strength in whites and fizz. Crisp SAUV BL 19', oak-aged Kaho Roa Sauv Bl 17. Vibrant CHARD 18. Excellent fizz Miru Miru NV (esp late-disgorged Res 15'). RIES (off-dry) 18, GEWURZ 18, PINOT GR 18 (dry), GRÜNER V 18 all rewarding, value. Easy-drinking PINOT N.

Invivo Auck (r) w ★★ Young producer, aromatic MARL SAUV BL 19'. Focus on celebrity labels, esp "chief winemaker" Graham Norton's Own Sauv Bl (v. easy-drinking) 19'; weighty Sarah Jessica Parker Sauv Bl 19' aimed at US.

Johanneshof Marl (r) w sp ★★ Small winery; perfumed GEWURZ 18' (one of NZ's finest). Excellent fizz; v.gd RIES 16', dryish PINOT GR 18'.

Jules Taylor Gis, Marl (r) (p) w ★★ Stylish, gd value. Refined MARL CHARD 18, incisive Marl SAUV BL 19'; scented PINOT GR 18'; generous PINOT N 18. Complex top wines: OTQ ("On The Quiet").

Kim Crawford Hawk ★→★★ Brand owned by CONSTELLATION NEW ZEALAND. Easy-drinking, incl high-impact MARL SAUV BL (huge seller in US, where many assume wrongly that co-founder, Kim Crawford, no longer involved, is a woman); floral PINOT GR; fruity HAWK MERLOT; generous Marl PINOT N. Top range: Small Parcels.

Kumeu River Auck (r) w ★★★★ Complex Estate CHARD 18 is multi-site blend; value. Top, single-v'yd Mate's V'yd Chard (planted 1990) is more opulent 18; single-v'yd *Hunting Hill Chard* 18 rising star. Lower-tier Village Chard great value (18 incl HAWK fruit.) New range from recently acquired v'yd in Hawk: SAUV BL, tight 18'; Rays Road Chard, Chablis-like, 18'; Rays Road PINOT N, savoury 18.

Lawson's Dry Hills Marl (r) (p) w ★★→★★★ Best-known for incisive SAUV BL 19', exotic GEWURZ 18. Lightly oaked CHARD 18; dry PINOT GR 19. Top range: The Pioneer (gorgeous Gewurz). Mid-tier Res range.

Lindauer Auck ★→★★ Hugely popular (in NZ), low-priced fizz, esp bottle-fermented Lindauer Brut Cuvée NV. Latest batches easy-drinking. Special Res (2 yrs lees) offers complexity, value. Ever-expanding range: low-alc, single-variety, "strawberry-infused" and Rosé.

Mahi Marl r w ★★ Weighty SAUV BL (part oak-aged) 18'; fresh CHARD 17 (esp Twin Valleys V'yd); dry PINOT GR 18; floral Rosé; supple PINOT N 17.

Man O' War Auck r w ★★ Largest v'yd on WAIH. Penetrating Valhalla CHARD; full-flavoured PINOT GR from adjacent Ponui Island. Reds incl generous MERLOT/CAB/MALBEC/PETIT VERDOT, delicious Death Valley Malbec and spicy Dreadnought SYRAH.

Marisco Marl r w ★★ Large Waihopai Valley producer (owned by Brent Marris, ex-WITHER HILLS) with two brands, The Ned and Marisco The King's Series.

Marlborough NZ's dominant region (69% plantings) at top of S Island (land ideal for v'yd expansion now in short supply.) 1st vines in modern era planted 1973 (SAUV BL in 1975.) Hot, sunny days and cold nights give aromatic, crisp whites and PINOT N-based rosés. Intense Sauv Bl, from green capsicum to ripe tropical fruit (some top wines faintly or clearly oak-influenced). Fresh RIES (recent wave of sweet, low-alc wines); some of NZ's best PINOT GR, GEWURZ; CHARD is slightly leaner than HAWK but more vibrant and can mature well. High-quality, gd-value fizz and classy botrytized Ries. Pinot N underrated, top examples (from n-facing clay hillsides) among NZ's finest. Interest stirring in ALBARIÑO, GRÜNER V.

Martinborough Wair Small, prestigious district in s WAIR (foot of N Island). Cold s winds reduce yields, warm summers, usually dry autumns, free-draining soils (esp on Martinborough Terrace.) Success with several whites (SAUV BL, PINOT GR, CHARD, RIES, GEWURZ), but renowned for sturdy PINOT N (higher % of mature vines than other regions).

Martinborough Vineyard Mart r (p) (w) ★★★ Famous PINOT N (perfumed Home Block 17'). Classy Home Block CHARD 18'; intense Manu RIES 18'. Gd-value Te Tera range. Owned by American Bill Foley (2014).

Matawhero Gis r (p) w ★★ Former star GEWURZ producer of 80s, now different ownership. Peachy CHARD 18; bold Church House Chard 18. Perfumed Gewurz; scented PINOT GR; promising ALBARIÑO; fruity Rosé, buoyant MERLOT.

Matua Auck r w ★›★★ Producer of NZ's 1st SAUV BL in 1974 (from AUCK grapes) long known as Matua Valley. Formerly an industry leader, but now declining in profile in NZ. Owned by TWE. Most wines offer pleasant, easy-drinking; some impressive single-v'yd wines.

Maude C Ot r w ★★ Scented PINOT GR 19'; outstanding RIES (dry 19', medium 19') from mature vines at Mt Maude V'yd, at Wanaka. Generous PINOT N 18', esp age-worthy Mt Maude V'yd.

Pinot N in C Ot competes with cherry orchards for space. Chekhov had the answer.

Mills Reef B of P r w ★★ ›★★★ Easy-drinking Estate range from GIMBLETT GRAVELS and other HAWK grapes. Top Elspeth range incl age-worthy CHARD 17'; fine-textured B'x-style reds and SYRAH. Mid-tier Res range (r w) typically gd value. 2018 release of two prestige reds, labelled Arthur Edmund, at $350 each: fragrant, Syrah 13' and dark CAB/MERLOT 13'.

Millton Gis r (p) w ★★ ›★★★★ Region's top wines from NZ's 1st organic producer, despite warm, moist climate. Arresting, hill-grown, single v'yd Clos de Ste Anne range: CHARD, CHENIN BL, VIOGNIER, SYRAH, PINOT N in favourable seasons. Long-lived Chenin Bl 17 (honeyed in wetter vintages) one of NZ's finest. Drink-young range: Crazy by Nature (gd value). Classy La Cote Pinot N 16'.

Misha's Vineyard C Ot r w ★★ Large v'yd at Bendigo. Dryish PINOT GR 19'; RIES (dry Lyric 16', slightly sweet Limelight 16'). PINOT N: High Note (graceful 17'); Impromptu (drink-young 18').

Mission Hawk r (p) w ★★ NZ's oldest wine producer, 1st vines 1851; still owned by Catholic Society of Mary. Winemaker Paul Mooney celebrated his 40th vintage 2019. Wide range of gd-value regional varietals, with V'yd Selection next up the scale. Res range incl excellent MERLOT, CAB SAUV, SYRAH, MALBEC, CHARD, SAUV BL. Top label: Jewelstone. Also large v'yd in AWATERE VALLEY. Owns Ngatarawa.

Mondillo C Ot r w ★★ Rising star with weighty RIES 19. Late-harvest Nina Ries 19'. Powerful PINOT N 17, dense Bella Res Pinot N 17'.

Mount Edward C Ot r w ★★ Small, respected, organic producer. CHARD 17; racy RIES 17'; dry CHENIN BL 17; complex PINOT N 16'.

Mount Riley Marl r (p) w ★★ Medium-sized, top-value family firm. Top range is Seventeen Valley.

Mt Beautiful Cant r w ★★ Large v'yd at Cheviot, n of WAIP. Fragrant, harmonious.

Mt Difficulty C Ot r (p) w ★★★ Quality producer with extensive v'yds at Bannockburn, now owned by US billionaire Bill Foley. Powerful PINOT N 17'. Roaring Meg is popular Cromwell Basin blend for early drinking. Expanding range of single v'yds, incl Chablis-like Packspur CHARD 18', fleshy Packspur RIES 17'; elegant Packspur Pinot N 17'. Consistently classy whites.

Mud House Cant r w ★★→★★★ Large, Australian-owned, MARL-based producer (also owns v'yds in WAIP, C OT). Brands incl Mud House, Waipara Hills, Hay Maker (lower tier). Gd-value regional blends. Excellent Single V'yd collection (Home Block Waipara PINOT GR 18') and Estate range (graceful Claim 431 C Ot PINOT N 16').

Nautilus Marl r w sp ★★→★★★ Medium-sized, rock-solid range, owned by S Smith & Sons (*see* Yalumba, Australia). Yeasty NV sparkler (min 3 yrs on lees), one of NZ's best. Excellent GRÜNER V 19; ALBARIÑO 19'.

Nelson Small region (3.0% planting) w of MARL; climate wetter (except in 19) but equally sunny. Clay soils of Upper Moutere hills (full-bodied wines) and silty WAIMEA plains (more aromatic). SAUV BL is most extensively planted, but also strength in aromatic whites, esp RIES, PINOT GR, GEWURZ; also gd (sometimes outstanding) CHARD, PINOT N.

Neudorf Nel r (p) w ★★★→★★★★ Smallish, long-est, big reputation. Refined, Moutere CHARD 15' 16' 17 one of NZ's greatest; lower-priced Rosie's Block Chard 18. Savoury Moutere PINOT N 17; lightly oaked SAUV BL; off-dry PINOT GR; RIES (dr s/sw) also top flight.

No. 1 Family Estate Marl sp ★★ Family-owned company of regional pioneer Daniel Le Brun, ex-Champagne. No longer controls Daniel Le Brun brand (owned by Lion). Specialist in v.gd fizz. Top end, vigorous Cuvée Virginie.

Nobilo Marl *See* CONSTELLATION NEW ZEALAND.

Oyster Bay Marl r w sp ★★ From DELEGAT. A marketing triumph: huge sales in UK, US, Australia. Vibrant, easy-drinking, mid-priced wines with touch of class from MARL, HAWK. Marl SAUV BL 19' is biggest seller, 1.5 million cases/yr.

Palliser Mart r w ★★→★★★ One of district's largest; multiple shareholders. Lower tier: Pencarrow (gd-value, majority of output). Top-end: new single-v'yd wines: Om Santi CHARD; perfumed Wharekauhau PINOT N.

Pegasus Bay Waip r w ★★★ Family firm, superb range: S Island's best MERLOT/ CAB SAUV 16'. Lovely dessert RIES, SAUV BL, MUSCAT. Second label: Main Divide is top value.

Peregrine C Ot r w sp ★★ Vibrant, flavoursome whites; gd NV (sp). Refined, vibrant PINOT N (grown mostly in Cromwell Basin). Charming organic Rosé. Second label: Saddleback.

Albariño is fastest-expanding new white variety in NZ.

Pernod Ricard NZ Auck r (p) w sp ★→★★★ One of NZ's largest producers, formerly Montana. Wineries in HAWK, MARL (closed Auck winery in 2020). Extensive co-owned v'yds for Marl whites, esp huge-selling BRANCOTT ESTATE SAUV BL. Major strength in fizz, esp big-selling DEUTZ Marl Cuvée. Wonderful value CHURCH ROAD reds and CHARD. Other key brands: STONELEIGH.

Prophet's Rock C Ot r w ★★→★★★ Small producer with gd PINOT GR, Dry RIES. Top-tier Cuvée Aux Antipodes PINOT N, 2nd-tier supple Rocky Point Pinot N.

Puriri Hills Auck r (p) ★★→★★★ Classy, complex, long-lived, distinctly B'x-like MERLOT-based reds from Clevedon 10' 13' 14'. Harmonie Du Soir (formerly Res) is impressive, with more new oak. Top label is dense, silky Pope.

Pyramid Valley Cant r w ★★→★★★ Tiny elevated limestone v'yd at Waikari, owned by US investor Brian Sheth and viticulturist Steve Smith (ex-CRAGGY RANGE).

Estate-grown wines with strong personality. New regional range: deep MARL CHARD 17'; savoury C OT PINOT N 17.

Quartz Reef C Ot r w sp ★★→★★★ Small, bio producer. Gd PINOT GR, PINOT N. Stylish, yeasty, lively fizz, esp Vintage Blanc de Blancs 13'.

Rapaura Springs Marl ★★ Skilfully crafted, gd value, esp Res whites. Graceful, supple Rohe Southern Valleys PINOT N 18.

Rippon Vineyard C Ot r w ★★→★★★ Pioneer v'yd on shores of Lake Wanaka; arresting view and wines. Fragrant, savoury style. Mature Vine PINOT N, from vines planted 1985–91, is "the farm voice"; Jeunesse Pinot N from younger vines. Tinker's Field Pinot N: oldest vines, age-worthy. Slowly evolving whites, esp outstanding Mature Vine RIES.

Legal marijuana in N America worries some NZ exporters: cd be more exciting.

Rockburn C Ot r (p) w ★★ Gd PINOT GR, RIES 16', PINOT N from Cromwell Basin (mostly) and GIBBSTON grapes 17'. Popular Stolen Kiss vibrant rosé. Devil's Staircase Pinot N: drink-young charmer.

Sacred Hill Hawk r w ★★→★★★ Mid-size producer. Acclaimed Riflemans CHARD 17' from mature vines at inland, elevated site. Wine Thief Chard from same v'yd, more toasty, upfront 18'. Long-lived Brokenstone MERLOT 15', Helmsman CAB/Merlot 15' and Deerstalkers SYRAH 16 from GIMBLETT GRAVELS. Halo and Res: mid-tier. New single v'yd range variable quality.

Saint Clair Marl r (p) w ★★→★★★ Largest family-owned producer in region. SAUV BL from relatively cool sites at Dillons Point, in lower WAIRAU VALLEY – esp punchy Wairau Res 19'. Res is top selection; then wide array of impressive, 2nd tier Pioneer Block (single v'yds), incl several classy PINOT N; then gd-value, large-volume regional blends. Now owns Delta, Lake Chalice.

Seifried Estate Nel (r) w ★★ Region's biggest winery, family-owned and best-known for medium-dry RIES 19', GEWURZ 19'. Gd-value, often excellent SAUV BL 19'. Best: Winemakers Collection. Old Coach Road: 3rd tier, gd value. Whites better than reds.

Selaks Marl r w ★→★★ Old producer of Croatian origin, now a brand of CONSTELLATION NEW ZEALAND. Solid, easy-drinking wines. Recently revived top Founders range, esp complex CHARD 16'. The Taste Collection range: Buttery HAWK Chard, Velvety MARL PINOT N.

Seresin Marl r w ★★→★★★ Quality organic producer. Winery building and adjacent v'yd (but not brand or other v'yds) sold 2018 (Seresin wines still made there.) Sophisticated SAUV BL one of NZ's finest; generous CHARD. Savoury PINOT N, partly wood-aged PINOT GR. Bone-dry sparklings. Res: top range. 3rd-tier Momo (v.gd quality/value). Complex, fine-textured, distinctive wines.

Sileni Hawk r (p) w ★★ Large producer, owned by investment company. Strong recent focus on HAWK PINOT N. Top: Exceptional Vintage SYRAH 13', MERLOT 14. Strong, mid-range Grand Res, incl complex Lodge CHARD 18; Advocate ALBARIÑO; generous Triangle Merlot 18; Plateau Pinot N. Cellar Selection: 3rd tier.

Smith & Sheth Cru Hawk r w ★★→★★★ Partnership of Steve Smith (ex-CRAGGY RANGE) and billionaire Brian Sheth. Single-v'yd ALBARIÑO, CHARD, SAUV BL, SYRAH.

Spy Valley Marl r (p) w ★★→★★★ High achievers, extensive v'yds. Flavoury whites superb value; impressive CHARD, PINOT N, SAUV BL. Classy Envoy top selection.

Staete Landt Marl r w ★★ V'yd at Rapaura (WAIRAU VALLEY). Second label: Map Maker (gd value).

Starborough Family Estates Marl (r) w ★★ Family-owned v'yds in AWATERE and WAIRAU VALLEYS. SAUV BL, CHARD, PINOT GR, PINOT N.

Stonecroft Hawk, Marl r w ★★ Small organic winery. NZ's 1st serious SYRAH (1989), and still v.gd. NZ's only ZIN, fresh, spicy 17.

Stoneleigh Marl r (p) w ★★ Owned by PERNOD RICARD NZ. Based on relatively warm Rapaura v'yds. Popular MARL whites. Top wines: Rapaura Series (esp tropical SAUV BL; smoky CHARD; rich PINOT GR; flavourful PINOT N). Wild Valley range: indigenous yeasts.

Stonyridge Waih r w ★★★→★★★★ Boutique winery known since mid-80s for exceptional CAB SAUV-based blend, *Larose*, one of NZ's greatest (sold mostly en primeur). Airfield, little brother of Larose. Dense, Rhône-style, SYRAH-based blend, Pilgrim. Super-charged Luna Negra MALBEC.

Te Awa Hawk r w ★★→★★★ GIMBLETT GRAVELS v'yd and site of key new VILLA MARIA winery. Smoky CHARD 18; refined MERLOT/CAB; strong SYRAH 17. Fruity TEMPRANILLO. Left Field range: easy-drinking, gd value.

Te Kairanga Mart r w ★★ One of district's oldest, largest wineries, recently rejuvenated. Runholder is mid-tier (graceful PINOT N 18'). Top tier is John Martin: complex CHARD 18'; savoury, age-worthy Pinot N 17'.

Te Mata Hawk r w ★★★→★★★★ Winery of high repute (1st vintage 1895) run by Buck family since 1974. Coleraine (CAB SAUV/MERLOT/CAB FR blend) 13' 14' 15' 16 17 is Bx-like, great longevity). Much lower-priced *Awatea* Cabs/Merlot also classy, more forward 17. *Bullnose Syrah* 18' among NZ's finest. New powerful Alma Hawk PINOT N 18'. Elegant Elston CHARD 17. Estate V'yds range for early drinking incl new Pinot N 18'.

Te Whare Ra Marl r w ★★ Label: TWR. Small WAIRAU VALLEY producer, some of region's oldest vines, planted 1979. Known for highly perfumed, organic GEWURZ; vibrant SAUV BL, RIES (dry "D", medium "M").

Terra Sancta C Ot r (p) (w) ★★ Bannockburn's 1st v'yd, founded 1991 as Olssens. V.gd-value, drink-young Mysterious Diggings PINOT N 18; mid-tier savoury Bannockburn Pinot N 17'; lovely Slapjack Block Pinot N (from district's oldest vines) 17'. Pinot N Rosé arguably NZ's finest **19**.

Tiki Marl (r) w ★★ McKean family own extensive v'yds in MARL, WAIP. Punchy Marl SAUV BL **19'**; fleshy Waip PINOT GR 19; gentle Waip PINOT N 18. Top tier: Koro, incl elegant HAWK CHARD 18'. Mid-range: single v'yd. Second label: Maui.

Tohu Marl, Nel r w ★★ Maori-owned venture, extensive v'yds in MARL, NEL. AWATERE VALLEY SAUV BL, Whenua Awa CHARD. Strong, dryish RIES 17; refined Blanc de Blancs fizz **15'**; new savoury Whenua Awa PINOT N 17'.

Trinity Hill Hawk r (p) w ★★ →★★★ Highly regarded producer, US-owned. Refined B'x-style blend The Gimblett 17. Stylish GIMBLETT GRAVELS CHARD **17'**. Prestigious Homage SYRAH **14'** 15' 16 17, Rhôney MARSANNE/VIOGNIER **16'**, Impressive TEMPRANILLO 17'. Lower-tier "white label" range gd value, esp drink-young MERLOT.

Two Paddocks C Ot r (w) ★★ Actor Sam Neill makes several PINOT NS. Main label is estate-grown, multi-site. Single-v'yd Prop Res range: First Paddock (more herbal, from cool Gibbston district 17'), Last Chance (riper, from warmer Alexandra 17'). Latest is earthy The Fusilier, grown at Bannockburn 17'. Picnic: drink-young 18'.

Two Rivers Marl r (p) w ★★ Herbal SAUV BL **19'**. Vibrant CHARD 18'; rich Rosé. Fragrant PINOT N 17'. Second label: Black Cottage (v.gd value).

Urlar Wair r w ★★ Small organic producer. Complex, PINOT GR 17; scented RIES 17; vigorous SAUV BL 18; deep PINOT N 17', esp youthful Select Parcels 17'.

Valli C Ot ★★→★★★ Superb range of single-v'yd PINOT N. Also piercing Waitaki RIES 18', weighty Gibbston PINOT GR 18'.

Vavasour Marl r w ★★→★★★ US-owned by Foley Family Wines. Rich CHARD, intense SAUV BL **19'**. Also lovely dry Rosé **19'**; generous PINOT N 18.

Vidal Hawk r w ★★→★★★ Owned by VILLA MARIA. Top Legacy range: smoky, CHARD 18'; floral SYRAH 16; superb CAB SAUV/MERLOT 16'. Soler range is 2nd tier. Great-value 3rd-tier Res range (esp Chard 18').

> **Why Oh Wai?**
> WAIPARA, WAIRARAPA, WAITAKI, WAIRAU, WAIMEA, WAIHEKE – why Wai? NZ's
> coastline is 9th-longest in world and "wai" is the Maori word for water.
> A bit like every town being called Something-on-Sea.

Villa Maria Auck r (p) w ★★ →★★★ NZ's largest fully family-owned winery; president Sir George Fistonich (a lean, fit, hard-driving 80-yr-old); daughter Karen chairs board. Also owns ESK VALLEY, TE AWA, VIDAL. Wine-show focus, with glowing success. Distinguished top ranges: Res (regional character) and Single V'yd (individual sites). New Platinum Selection (some organic, much use of lees ageing, elegant, tight Sur Lie CHARD 18'). Cellar Selection: 3rd tier (less oak), excellent, superb value. 4th tier, volume Private Bin wines can also be v.gd. Small volumes of v.gd ALBARIÑO, VERDELHO, GRENACHE, MALBEC. New icon red, Ngakirikiri The Gravels 14': CAB SAUV-based, powerful, lush.

Waiheke Island r (p) w (w) Lovely, sprawling island in AUCK's Hauraki Gulf (temperatures moderated by sea). Early acclaim for stylish CAB SAUV/MERLOT blends, esp from Onetangi district; more recently for dark, bold SYRAH. Popular tourist destination; many helipads.

Waimea Nel r (p) w ★★ One of region's largest and best-value producers, owned by an investment fund. Punchy SAUV BL 19', spicy PINOT GR 18 and generous GEWURZ 19'. V.gd RIES 17 and punchy GRÜNER V 18. Full-bodied PINOT N 17. Second label: Spinyback.

Waipara Valley Cant CANT's key subregion, n of Christchurch (89% plantings). High profile for PINOT N, RIES (also heavy plantings of SAUV BL). Currently repositioning itself as "North Canterbury", after name confusion in export markets.

Wairarapa NZ's 7th-largest wine region (not to be confused with WAIP). *See* MART. Also incl Gladstone subregion in n (slightly higher, cooler, wetter). Driest, coolest region in N Island, but exposed to cold s winds; v. little expansion in past decade. Strength in whites (SAUV BL, PINOT GR most widely planted, also gd CHARD, GEWURZ, RIES) and esp PINOT N (weighty, ripe, savoury, from relatively mature vines). Starting to promote itself as "Wellington Wine Country".

Wairau River Marl r (p) w ★★ Gd whites. Res is top label: single-v'yd CHARD full, harmonious 17'; weighty, VIOGNIER 18; fragrant PINOT N 17.

Wairau Valley MARL's largest subregion (1st v'yd planted 1873; modern era since 1973). Vast majority of region's cellar doors. Three important side valleys to s: Brancott, Omaka, Waihopai (known collectively as Southern Valleys). SAUV BL thrives on stony, silty plains; PINOT N on clay-based, n-facing slopes. Much recent planting in wetter, more frost-prone upper Wairau Valley.

Waitaki Valley C Ot Slowly expanding subregion in N Ot (54 ha), with cool, frost-prone climate. Handful of producers. V. promising PINOT N (but can be leafy); racy PINOT GR, RIES (superb in top vintages).

Whitehaven Marl r (p) w ★★ Medium-sized producer. Flavour-packed SAUV BL 19' big seller in US. Top range: Greg (savoury PINOT N 17').

Wither Hills Marl r w ★★ Big producer, owned by Lion brewery. Popular, gd-value.

Wooing Tree C Ot r (p) w ★★ Single v'yd, mostly reds. Bold PINOT N 17' (Beetle Juice is delicious, drink-young style 17'). Powerful Sandstorm Res Pinot N 15. Weighty CHARD. Less "serious" wines, all from Pinot N, incl Blondie, Tickled Pink.

Yealands Marl r (p) w ★★ NZ's biggest "single v'yd", at lower AWATERE VALLEY site, now owned by utility company, Marlborough Lines. Partly estate-grown, mostly MARL wines. Past high profile for sustainability, but most wines not certified organic. Punchy, gd-value Marl SAUV BL 19': regional blend. Weighty, high-impact Res AWATERE Sauv Bl 19'. Refined, pure Single Block S1 Sauv Bl 19'. Other key brands: Babydoll, Crossroads, The Crossings.

South Africa

Abbreviations used in the text:

Bre	Breedekloof	**Rdg/Up/V**	Ridge/Upper/Valley
C'dorp	Calitzdorp	**Oli R**	Olifants River
Cape SC	Cape South Coast	**Pie**	Piekenierskloof
Ced	Cederberg	**Rob**	Robertson
Coast	Coastal Region	**Sla**	Slanghoek
Const	Constantia	**Stell**	Stellenbosch
Ela	Elandskloof	**Swa**	Swartland
Elg	Elgin	**Tul**	Tulbagh
Fran	Franschhoek	**V Pa**	Voor Paardeberg
Hem	Hemel-en-Aarde	**Wlk B**	Walker Bay
		Well	Wellington

What people are talking about is water, and what to do when there's not enough of it. Will the current mix of grapes – dominated by Cabernet Sauvignon, Shiraz, Pinotage, Merlot, Chenin Blanc, Sauvignon Blanc and Chardonnay – still work in a warming world? Will it be a question of different rootstocks and different varieties, or of different sites, perhaps planting virgin land in new, cooler areas? All of the above, perhaps – and more, given that the effects of warming reach far beyond the vineyard. Wine-growers will take it in their stride. They are already responding to evolving global tastes by using alternatives like clay and concrete to reduce wood character in their reds, dialling down sugar in their pinks, harvesting early for zingier whites, producing lower-alcohol, vegan-friendly and no-added-sulphur wines. They are also seeking out old vines and special parcels with compelling stories, working with exotic and resurrected varieties, finding better ways to express grape, site and region. It is safe to assume the same can-do attitude will underpin the response to the evolving climate.

Recent vintages

2019 4th successive drought crop. Gd intensity with freshness, thanks to milder temps. Potentially stellar whites.

2018 Concentrated, flavourful wines though probably not for long cellaring.

2017 Quality, character comparable to 15. Accessible young, possibly peaking earlier too.

2016 Extreme conditions favoured later-ripening varieties, cooler areas. Many excellent wines to drink while waiting for 15.

2015 Among greats: exceptional flavour, balance, intensity. Slow-starting, structure for long ageing.

AA Badenhorst Family Wines Coast r (p) w (sp) ★★→★★★★ Heirloom varieties, old vines, wild yeasts, orange wine, every trend. Badenhorst cousins into new-wave wine-growing on Paardeberg Mtn. Mostly CHENIN BL, Med blends (r w), Tinta Barroca, PALOMINO. Gd-value range Secateurs artful blend of fun and gravity.

Alheit Vineyards W Cape w ★★★★ Fine, pure expressions of eg. old vines by Chris and Suzaan Alheit. Multi-region CHENIN BL/SEM Cartology, site-specific Chenin Bl (eg. Magnetic), Sem La Colline, Vine Garden field blend (w) from HEM home farm. Winemaker Franco Lourens' own boutique label right up there.

Anthonij Rupert Wyne W Cape r (p) w (br) sp ★→★★★ Extensive, impressive portfolio honours owner Johann Rupert's late brother. From own v'yds in DARLING, SWA, Overberg, elegant home farm L'Ormarins nr FRAN. Best: flagship Anthonij Rupert, site-specific Cape of Gd Hope, premium Jean Roi (p).

Babylonstoren W Cape r (p) w sp ★★→★★★★ C17 Cape Dutch farm nr PAARL (name means Tower of Babel) stylishly restored by ex-*Elle Decoration* ed Karen Roos and media-giant husband Koos Bekker. B'x red Nebukadnesar 15' 17' heads impressive line-up.

Bartinney Private Cellar Stell r (p) w (sp) ★★→★★★ Rising star on steep Banhoek Valley sides, family-owned; original CAB SAUV, CHARD since joined by Res versions. "Lifestyle" Noble Savage and newer upscale sibling brand Montegray.

Beaumont Family Wines Cape SC r w br (sw) ★★→★★★ Excellent wines from C18 estate in Bot River. Rare solo-bottled MOURVÈDRE 10' 15' 16 17', elegant Hope Marguerite CHENIN BL 12' 15' 16' 17' 18' 19, Chenin Bl-based New Baby 15' 17' 19.

BEE (Black Economic Empowerment) Initiative aimed at increasing wine-industry ownership and participation by previously disadvantaged groups.

Beeslaar Wines Stell r ★★★★ KANONKOP winemaker's personal take on PINOTAGE 13' 14' 16' 17' 19. Refined, rather special.

Bellingham Coast r w ★★→★★★ Enduring DGB brand with low-volume, high-class The Bernard Series (incl scarce monovarietal ROUSSANNE 15' 17' 19), Homestead Series with v.gd old-vine CHENIN BL.

Beyerskloof W Cape r (p) (w) (br) ★→★★★★ SA's PINOTAGE champion: 11 versions of grape (12, incl spirit to fortify Lagare Cape Vintage "Port"). Powerful varietal Diesel 13' 16' 17' 18, clutch of CAPE BLENDS. Even Pinotage burgers. Also classic CAB SAUV/MERLOT Field Blend 09' 11 14' 15' 16 17.

Boekenhoutskloof Winery Coast r (p) w sw ★→★★★★ Top FRAN winery, exemplary quality, consistency with SWA SYRAH 09' 10 12' 13 15' 16 17' 18 19; Fran CAB SAUV 10 11' 13 15' 16 17' 18 19 (also newer STELL version); *old-vines Sem*; dense Chocolate Block (r) 13 15' 16 17' 18 19; simpler Porcupine Ridge, Wolftrap, new Vinologist

---

Not growlers, surely?

Glass containers account for 44% of wine sold in S Africa (well over 50% in 75cl bottles). Almost 40% is non-traditional bag-in-box, the party-size 5-litre pack the most popular in this category. Recyclable aluminium cans are having a moment too, and spill-proof "pouches" and, if you're also a beer drinker, you'll recognize the 1 or 2-litre glass "growlers" from pub takeouts.

lines. Major development, Cap Maritime, underway in HEM; PINOT N and CHARD. *See* PORSELEINBERG.

Bon Courage Estate W Cape r (p) w (br) sw (s/sw) sp ★ →★★★ Bruwer family in ROB with broad range. Stylish Brut MCC, aromatic desserts (RIES, MUSCAT), delightful COLOMBARD (DYA).

Boplaas Family Vineyards W Cape r w br (sp) (sw) ★ →★★★ Wine-growers Carel Nel and daughter Margaux at C'DORP, known for Port styles, esp Cape Vintage Res 09' 12' 15' 16 17' and Tawny (mostly NV). Table wines made from Portuguese grapes.

Boschendal Wines W Cape r (p) w sp ★ →★★★ Much-loved DGB brand on lovely C17 estate nr FRAN. Notable SHIRAZ, SAUV BL, CHARD, MCC in various tiers, some ELG. Newer B'x/Rhône-red blends (Black Angus, Nicolas) designed to impress.

Boschkloof Wines Stell r w ★★★ Young gun Reenen Borman in soon-to-be-gazetted Vlottenburg WARD with stellar SYRAH (varietals, blends) and CHENIN BL, inter alia, under Boschkloof (the home farm's name) and Kottabos labels. Also, with partners, produces standout Syrah and Chenin Bl (single-v'yd and blend) in COLOMBARD in Patatsfontein range.

Botanica Wines W Cape r (p) w (sw) (sp) ★★ →★★★★ American Ginny Povall on flower- and wine-farm Protea Heights nr STELL. Superlative CHENIN BL Mary Delany Collection from old w-coast bush-vines; PINOT N, SEM both partly ex-HEM. New Flower Girl from CAB FR part of mini-boom in *méthode ancestrale* bubbly.

Bouchard Finlayson Cape SC r w ★★ →★★★★ HEM pioneer with fine versions of area specialities PINOT N (Galpin Peak, occasional barrel-selected Tête de Cuvée 13' 17') and CHARD (Missionvale, Sans Barrique; also ex-Ela Crocodile's Lair).

Breedekloof Large (c.12,600 ha) inland area known for bulk- and entry-level wine. Fine winemaking in small/family cellars eg. Bergsig, Deetlefs, Le Belle Rebelle (previously Stofberg Family), OLIFANTSBERG, OPSTAL. But recent initiative, Bre Makers, prompts large/co-op ventures to step up by offering special/old parcels of mostly CHENIN BL. Some thrilling results.

Bruce Jack Wines W Cape r p w (br) (sp) ★★ →★★★ Never a dull bottle at FLAGSTONE founder Bruce Jack's solo venture at Appelsdrift farm in cool S Cape. Endlessly creative blends, varietals, rosé and bubbly. New CHARD "Sherry" Vloermoer. Delightful labels by artist wife, Penny.

Buitenverwachting W Cape r (p) w sw (sp) ★★ →★★★ Classy family winery in CONST. Standout CHARD 14' 15 16 17, Husseys Vlei SAUV BL 13' 15 17', B'x red Christine 09' 10 11 13. Labelled Bayten for export.

Calitzdorp KLEIN KAROO DISTRICT (c.320 ha) climatically similar to the Douro, known for Port styles and latterly unfortified Port-grape blends (r w) and varietals.

Cape Blend Usually red with significant PINOTAGE component; occasionally CHENIN BL blend, or simply wine with "Cape character".

Cape Chamonix Wine Farm Fran r w (sp) ★★ →★★★ Excellent winemaker-run mtn property. Distinctive PINOT N, Ripasso-style PINOTAGE, CHARD, SAUV BL, B'x blends (r w), CAB FR, all worth keeping.

Cape Coast New umbrella for COAST (w and central) and CAPE SC REGIONS.

Cape Point Vineyards Cape T (r) w (sw) ★ →★★★ Most s winery nr tip of the CAPE TOWN peninsula. Complex, age-worthy SAUV BL/SEM Isliedh, CHARD, Sauv Bl. Gd-value label Splattered Toad. Recent sibling venture Cape Town Wine Company (r w sp).

Cape Rock Wines W Cape r (p) w ★★→★★★ OLI R's leading boutique grower. Characterful, strikingly packaged Rhône and Port-grape blends (r w).

Cape South Coast Cool-climate REGION (c.2630 ha) comprising DISTRICTS of Cape Agulhas, ELG, Lower Duivenhoks River (previously a WARD), Overberg, Plettenberg Bay, Swellendam and WLK B, plus standalone wards Herbertsdale, Napier, Stilbaai East. *See* CAPE COAST.

Cape Town Maritime DISTRICT (c.2710 ha) covering Cape Town city, its peninsula WARDS, CONST and Hout Bay, plus neighbour wards DUR and Philadelphia.

Cape Winemakers Guild (CWG) Independent, invitation-only association of 41 top growers. Stages benchmarking annual auction of limited premium bottlings.

Capensis W Cape w ★★★ S Africa-US venture, GRAHAM BECK's Antony Beck and Jackson Family's Barbara Banke, specializing in CHARD. Multi-region Capensis 15' 16; dual-v'yd Silene 17' 18 and single-site Fijnbosch 16, both new.

Catherine Marshall Wines W Cape r w ★★★ Cool-climate (chiefly ELG) specialist focuses mostly on PINOT N, MERLOT, SAUV BL, CHENIN BL; delightful dry, mineral RIES.

Cederberg Tiny (c.100 ha) high-altitude standalone WARD in Cederberg mtns. Mostly SHIRAZ, SAUV BL. Driehoek and CEDERBERG PRIVATE CELLAR sole producers.

Cederberg Private Cellar Ced, Elim r (p) w sp ★★→★★★★ Nieuwoudt family cellar, among S Africa's highest (CED) and most s (ELIM) v'yds. Elegant intensity in CAB SAUV, PINOT N, rare Bukettraube, CHENIN BL, SAUV BL, SEM, MCC, SHIRAZ (incl exceptional CWG Teen die Hoog).

Central Orange River Standalone N CAPE "mega WARD" (c.9300 ha, more than half is Sultana for dried/table grape market). Hot, dry, irrigated; traditionally white and fortified but major producer ORANGE RIVER CELLARS pushing boundaries.

Charles Fox Cap Classique Wines Elg sp ★★★ Traditional-method bubbly house, French consultant. Six classic, delicious Bruts incl newer pair of NV Res (p w).

Coastal Largest REGION (c.42,700 ha). Incl sea-influenced DISTRICTS of CAPE TOWN, DARLING, STELL, SWA and Lamberts Bay WARD. Plus Lutzville Valley district and Bamboes Bay ward, both previously part of OLI R. Confusingly, Coastal also incl non-maritime FRAN, PAARL, TUL, WELL.

Colmant Cap Classique W Cape sp ★★★→★★★★ Exceptional *méthode traditionnelle* house at FRAN, Belgian family-owned. Brut and Sec Res, Rosé, CHARD, Absolu Zero Dosage; all MCC, NV and excellent.

Constantia Scenic CAPE TOWN WARD (c.430 ha) on cool Constantiaberg slopes, S Africa's 1st and most historically famous area, revitalized in recent yrs by GROOT CONST, KLEIN CONST et al.

Constantia Glen Const r w ★★★ Waibel-family owned gem on upper reaches of Constantiaberg. Superb B'x blends (r w), varietal SAUV BL.

Constantia Uitsig Const r w br sp ★★★ Premium v'yds, cellar producing mostly MCC and still white. Consistent, striking, individual SEM.

Creation Wines Cape SC r w ★★→★★★ Family-owned/vinified range of B'x, Rhône, Burgundy and newer Loire varietals/blends. Among buzziest cellar doors Gerhard Smith's newer project, Die Kat se Snor ("The Cat's Whiskers") more than a joke.

Darling DISTRICT (c.2750 ha) around this w-coast town. Best v'yds in hilly Groenekloof WARD. Cloof, Darling Cellars, Groote Post/Aurelia, Mount Pleasant, Ormonde, Withington bottle under own labels; most other fruit goes into 3rd-party bottles, some spectacular.

David & Nadia Swa r w ★★★★ Sadie husband and wife follow natural-wine principles of SWA Independent Producers. Exquisite Rhône red Elpidios, GRENACHE Noir, CHENIN BL (varietals incl new Plat'bos and blend Aristargos), SEM, PINOTAGE, mostly from old vines. Assistant André Bruyns' own City on a Hill also fine.

De Krans Wines W Cape r (p) w br (sp) ★→★★★ Nel family at C'DORP noted for Port styles (esp Vintage Res 10' 11' 12' 13' 16' 17' 18) and fortified MUSCAT. Latterly success with unfortified Port grapes.

De Toren Private Cellar Stell r ★★★ Majority Swiss-owned, on ocean-facing Polkadraai Hills. Consistently flavourful B'x Fusion V, earlier-maturing MERLOT-based Z; light-styled Délicate (DYI).

De Trafford Wines Stell r w sw ★★★→★★★★ Boutique grower David Trafford with track record for bold yet harmonious wines. B'x/SHIRAZ Elevation 393, CAB SAUV, SYRAH Blueprint, CHENIN BL (dry and *vin de paille*). *See* SIJNN.

De Wetshof Estate Rob r w sw sp ★★★ Famed CHARD pioneer and exponent; five versions (oaked/unwooded, still/sparkling) topped by single-v'yd The Site.

Delaire Graff Estate W Cape r (p) w (br) (sw) (sp) ★→★★★★ UK diamond merchant Laurence Graff's eyrie v'yds, winery and visitor venue nr STELL. Glittering portfolio headed by age-worthy CAB SAUV Laurence Graff Res.

Shrub that gives rooibos tea has antioxidant-rich tannin to replace sulphur in wine.

Delheim Wines Coast r (p) w sw (s/sw) (sp) ★→★★★ Eco-minded family winery nr STELL. Vera Cruz SHIRAZ, PINOTAGE; cellar-worthy, best-yrs CAB SAUV-driven Grand Res, scintillating botrytis RIES Edelspatz.

DeMorgenzon Stell r (p) w (sw) (sp) ★★→★★★★ Appelbaums' manicured property hits high notes with B'x and Rhône varietals/blends (r w), CHARD, CHENIN BL. Occasional Chenin Bl The Divas 13' 17' is spectacular. Classical music played in the v'yds/cellar 24x7.

DGB W Cape Long-est, WELL-based producer/wholesaler, owner of high-end brands The Bernard Series, BOSCHENDAL and newer Old Road Wine Company; easy-drinking BELLINGHAM, Brampton, Douglas Green and others.

Diemersdal Estate W Cape r (p) w ★→★★★ DUR family farm excelling with various site/row-specific SAUV BL (incl Winter Ferment using frozen grape must), red blends, PINOTAGE, CHARD, S Africa's 1st/only commercial GRÜNER V.

Diemersfontein Wines W Cape r (p) w ★→★★★ Family estate, restaurant and guest lodge at WELL, esp noted for full-throttle Carpe Diem Res range. Entry-level PINOTAGE created emulated "coffee style". BEE brand Thokozani.

Distell W Cape S Africa's biggest drinks company, in STELL. Owns or has interests in many brands, spanning styles/quality scales. New premium- and fine-wine arm, Libertas V'yds & Estates, inter alia oversees NEDERBURG Auction, now re-imagined/-branded "Cape Fine & Rare Wine Auction". *See* DURBANVILLE HILLS, FLEUR DU CAP, JC LE ROUX, NEDERBURG WINES.

District *See* GEOGRAPHICAL UNIT.

Dorrance Wines W Cape r (p) w ★→★★★ French-toned, family-owned, with cellar in Cape Town city (reason to visit). Gorgeous SYRAH, CHARD, CHENIN BL.

Durbanville Cool, hilly WARD (c.1420 ha) in CAPE TOWN DISTRICT, best-known for pungent SAUV BL; also MERLOT, w blends. Corporate co-owned DURBANVILLE HILLS and many family ventures.

Durbanville Hills Dur r (p) w (sw) (sp) ★→★★★ Owned by DISTELL, local growers and staff trust; awarded PINOTAGE, CHARD, SAUV BL. V.gd newer B'x blend Tangram (r w).

Eagles' Nest Const r (p) w ★→★★★ CONST family winery with reliably superior MERLOT, SHIRAZ, VIOGNIER. Also vibrant SAUV BL, cellar-door-only Little Eagle (p).

Edgebaston W Cape r w ★★→★★★ V'yds/cellar nr STELL owned by David Finlayson, of esteemed Cape wine family. V.gd CAB SAUV GS, old-vine Camino Africana series, classy early drinkers. With resident winemaker, fine CINSAULT and GRENACHE Noir under Van der Merwe & Finlayson.

Eikendal Vineyards W Cape r w ★★★ Swiss-owned high performer nr STELL. B'x red Classique, MERLOT, vintage-blend Charisma (r). Excellent wooded CHARD trio: multi-site, bush-vine, single-clone; unoaked version no slouch.

Elgin Cool-climate DISTRICT (c.730 ha) recognized for SAUV BL, CHARD, PINOT N; also exciting SHIRAZ, RIES, CHENIN BL, MCC. Mostly family boutiques, incl one of only two certified-bio wineries in S Africa, Elg Ridge (other is REYNEKE in STELL).

Elim Maritime WARD (c.140 ha) in most s DISTRICT, Cape Agulhas, producing aromatic SAUV BL, white blends, SHIRAZ. Grape source for majors like CEDERBERG and boutiques eg. TRIZANNE. Area pioneer Zoetendal back in production.

Ernie Els Wines W Cape r (p) (w) ★→★★★★ Star golfer's wine venture nr STELL; long-lived varietal/blended CAB SAUV under Signature and Proprietor's labels, earlier-ready Big Easy range. Co-proprietor Baron Hans von Staff-Reitzenstein also owns, and is revitalizing, nearby Stellenzicht.

Estate Wine Grown, made and bottled on "units registered for the production of estate wine". Not a quality designation.

Fable Mountain Vineyards W Cape r p w ★→★★★★ US-owned TUL grower with exceptional SHIRAZ (varietal/blend) and white blend. Also Rhône rosé, special-site/-vintage Small Batch Series, easy-drinking Raptor Post range. Winemaker Tremayne Smith moonlights as The Horsemen with college mate Jaco Engelbrecht for exceptional parcels; solo as The Blacksmith, out-there wines with names eg. Hell Yeah, label art by tattooist.

Fairview W Cape r w (br) sw (sp) ★→★★★ Charles Back's imagination creates a smorgasbord of varietal, blended and single-v'yd bottlings under Fairview, Goats do Roam, La Capra labels. Recent addition Bloemcool's labels inspired by PAARL home farm's name, Bloemkoolfontein ("Cauliflower Spring"); handmade labels have vcg's seeds embedded. SWA-based Spice Route now features qvevri-vinified Obscura (r w).

FirstCape Vineyards W Cape r p w sp ★→★★ DYA Huge export joint venture of four local co-ops and UK's Brand Phoenix. Mostly entry-level wines in over dozen ranges, some sourced outside S Africa.

Flagstone Winery W Cape r (p) w (br) (sw) ★→★★★ Accolade Wines' high-end venture in former dynamite factory at Somerset W. Sources widely for impressive PINOTAGE, SAUV BL, B'x white, new vine-dried Sauv Bl dessert named Ice. Sibling brands mid-tier Fish Hoek, entry-level KUMALA.

Fleur du Cap W Cape r (p) w sw ★→★★★ DISTELL premium label, incl v.gd Series Privée Unfiltered, always stellar botrytis dessert and B'x red Laszlo.

Foundry, The Stell, V Pa r w ★★★→★★★★ Winemaker Chris Williams and partner James Reid specialize in varietal Rhône bottlings since 2000, now based in PAARL's Voor Paardeberg. Sensational GRENACHE BL; rare solo ROUSSANNE. Geographica is new showcase for special parcels, varieties, techniques etc.

Franschhoek Huguenot-founded DISTRICT (c.1230 ha) known for CAB SAUV, CHARD, SEM, MCC. Home to some of S Africa's oldest wine farms and vines, eg. retired industrialist Basil Landau's La Brie property, founded 1689 and home to 1905 parcel producing Landau du Val Sem.

Free State Province and GEOGRAPHICAL UNIT. Mile High V'yds (previously The Bald Ibis) sole producer in viticulturally challenging e highlands.

Gabriëlskloof W Cape r (p) w (sw) ★→★★★ Peter-Allan Finlayson vinifies this expanding and improving line-up, headed by Landscape Series (CAB FR, SYRAH (esp *Syrah on Shale*), old-vine CHENIN BL, B'x white), plus own excellent Crystallum PINOT N, CHARD, in family cellar nr Bot River.

Geographical Unit (GU) Largest of four main WINE OF ORIGIN demarcations (others, in descending size, are REGION, DISTRICT, WARD): FREE STATE, Greater Cape (new umbrella appellation for E, N, W CAPE), KWAZULU-NATAL and LIMPOPO.

Hunt for terroir style leads back 8000 yrs to Georgian qvevri. (Georgia, not George I).

Glenelly Estate Stell r w ★★★ Former Ch Pichon-Lalande (*see* B'x chapter) owner May-Eliane de Lencquesaing's "retirement" venture in STELL. Impressive flagships Lady May (B'x red) and Estate Res duo (B'x/SHIRAZ, CHARD). Superior-value Glass Collection.

Graham Beck W Cape sp ★★★ Front-rank MCC house nr ROB. Seven labels (vintage/NV; Brut, Brut Nature, Demi-Sec) led by *Chard Cuvée Clive*. Long-time cellarmaster Pieter Ferreira's new own-brand CHARD Blanc de Blancs 12' arguably

---

Fumé rises

After a relatively brief flowering in the late 70s and 80s, varietal wooded SAUV BL (then often labelled Blanc Fumé) all but wilted on the vine. There's since been a 2nd bloom, and now quadruple the number of bottlings vs. 10 yrs ago. Best show succulent fruit and ample charm with modest oak influence. Volumes still limited though. Try De Grendel, Highlands Road, Le Grand Domaine, Lismore, NEIL ELLIS, Rousseau, The Fledge & Co.

S Africa's finest MCC. His ex-colleague Erika Obermeyer also has her own new label (r w) topped by v.gd STELL CAB SAUV.

Groot Constantia Estate Const r (p) w (br) sw (sp) ★★→★★★★ Historic property and tourist mecca in S Africa's original fine-wine area. Suitably serious wines, esp MUSCAT de Frontignan Grand Constance 12 13 14' 15' 16 helping restore CONST dessert to C18 glory.

Hamilton Russell Vineyards Swa, Wlk B r (p) w ★★ →★★★★ 1st to succeed with Burgundy grapes in cool S Cape in 1979 at Hermanus. Elegant PINOT N, long-lived CHARD under HRV label. Super SAUV BL, PINOTAGE (varietal, blend), Sauv Bl/Chard in Southern Right, Ashbourne ranges. Newer Pinot N venture in Oregon.

Hartenberg Estate Stell r w (sw) ★→★★★ Welcoming STELL family farm never disappoints with SHIRAZ (several varietals and newer blend, The Megan); B'x red, CHARD, RIES (dry, botrytis).

Hemel-en-Aarde Trio of cool-climate WARDS (Hem V, Up Hem, Hem Rdg) in WLK B DISTRICT, producing outstanding PINOT N, CHARD, SAUV BL.

Hermanuspietersfontein Wynkelder W Cape r p w ★★→★★★ Leading SAUV BL and B'x/Rhône blend (r w) specialist; creatively markets connections with seaside resort Hermanus; sources (esp fresh CAB FR) mostly from cool Sunday's Glen close by.

Iona Vineyards Elg r (p) w ★★→★★★ Co-owned by staff, with high-altitude vines. Excellent One Man Band blends (r w), CHARD (incl new single-site pair), SAUV BL, PINOT N, Solace SYRAH from lower-lying Brocha farm. Cut-above lifestyle brand Sophie & Mr P.

JC le Roux, The House of W Cape sp ★→★★ S Africa's largest specialist bubbly producer at STELL, DISTELL-owned. Best: PINOT N, Scintilla and Brut NV, all MCC.

Jean Daneel Wines W Cape r w ★★→★★★ Family winery at Napier producing high-octane Director's Signature Series, esp B'x red, CHENIN BL.

Joostenberg Wines Paarl r w sw ★→★★★ Family's organic label for SYRAH, CHENIN BL incl new orange version in Small Batch Collection Kaalgat ("Buck Naked"). Owners Tyrell and Philip Myburgh honour famous forebears via newer Myburgh Bros, also partner with STARK-CONDÉ in STELL estate Lievland, and in value brand MAN Family Wines.

Jordan Wine Estate W Cape r (p) w (sw) (sp) ★→★★★★ Admired family venture nr STELL, consistency, quality, value from entry Chameleon to immaculate CWG bottlings. Flagship CHARD Nine Yards and B'x red Cobblers Hill. Planting S Africa's 1st ASSYRTIKO (Santorini-inspired basketry for wind protection); setting up to produce bubbly in England.

Julien Schaal Cape SC r w ★★★ Alsace vigneron Julien Schaal and wife Sophie handcraft thrilling SYRAH, CHARD from cool-grown ELG, HEM and Bot River parcels.

Kaapzicht Wine Estate Stell r (p) w ★★→★★★ Family winery in STELL; widely praised top range Steytler (best-yrs CAPE BLEND Vision, PINOTAGE, B'x-red Pentagon, old-vines CHENIN BL The 1947). Newer CINSAULT Skuinsberg.

Kanonkop Estate Coast r (p) ★★→★★★★ Decades-long undisputed "First Growth", mainly with PINOTAGE ("regular", and old-vines Black Label, B'x red Paul Sauer and CAB SAUV). Insatiable demand for 2nd-tier Kadette (featuring new Cab Sauv).

Keermont Vineyards Stell r w sw ★★★ Low-key but high-performing family boutique, neighbour and a grape-supplier to top-ranked DE TRAFFORD, on steep STELL Mtn slopes. SHIRAZ and CHENIN BL (single-v'yd and blended), newer MERLOT.

Ken Forrester Wines W Cape r (p) w sw (s/sw) (sp) ★ ⇢★★★ With international wine specialist AdVini as partner, STELL vintner/restaurateur Ken Forrester concentrates on Med varieties and CHENIN BL (dry, off-dry, botrytis). Unputdownable budget line-up, Petit.

Klein Constantia Estate W Cape r (p) w sw sp ★★ ⇢★★★★ Iconic property re-energized. Focused on SAUV BL, with nine different varietal bottlings, and on luscious, cellar-worthy non-botrytis MUSCAT de Frontignan *Vin de Constance* 12' 13' 14' 15' 16 18, convincing re-creation of legendary C18 CONST.

Kleine Zalze Wines W Cape r (p) w sp ★ ⇢★★★★ STELL-based star with brilliant CAB SAUV, SHIRAZ, CHENIN BL, SAUV BL in Family Res and V'yd Selection ranges. Exceptional value in Cellar Selection series.

Klein Karoo Mainly semi-arid REGION (c.2150 ha) known for fortified, esp Port style in C'DORP. Revived old vines beginning to feature in young-buck bottlings, eg. Patatsfontein Steen, Le Sueur Wines' Kluisenaar (both CHENIN BL), Radicales Libres CHARD by LEEU PASSANT.

Krone W Cape (p) sp ★ ⇢★★★ Mostly MCC, refined, classic, incl RD 02' 06' 09 and new Blanc de Blancs (Extra Brut, Brut Nature, latter's 1st ferment in amphorae) ex-famed Kaaimansgat/Crocodile's Lair CHARD v'yd in Ela. Made at revitalized Twee Jonge Gezellen in TUL.

Kumala W Cape r p w s/sw (sp) ★ DYA Major entry-level export label, and sibling to premium FLAGSTONE and mid-tier Fish Hoek. All owned by Accolade Wines.

KwaZulu-Natal Province and GEOGRAPHICAL UNIT on e coast; summer rain; sub tropical/tropical climate nr ocean; cooler in central Midlands plateau, home to Abingdon, Highgate wineries, and, further n, Cathedral Peak Estate in central Drakensberg mtn area.

KWV W Cape r (p) w br (sw) (s/sw) sp ★ ⇢★★★ Formerly national wine co-op and controlling body, today one of S Africa's biggest producers and exporters, based in PAARL. Reds, whites, sparkling, Port styles and other fortifieds under more than a dozen labels, headed by serially decorated boutique range The Mentors, with new, rarely seen varietal CARMENÈRE.

La Motte W Cape r w sw sp ★★ ⇢★★★ Graceful winery and cellar door at FRAN owned by Koegelenberg-Rupert family. Old-World-styled B'x, Rhône varietals, blends, CHARD, SAUV BL, MCC, VIOGNIER *vin de paille*.

Lammershoek Winery Coast r (p) w ★ ⇢★★★ German-owned, influential in recent SWA/S African evolution, early emphasis on old vines, organic cultivation, "natural" practices, etc. Plenty to discover in The Mysteries range, eg. new Ounooi from S Africa's original CHARD (c.1980).

Le Lude Méthode Cap Classique W Cape sp ★★★ ⇢★★★★ Celebrated, innovative MCC house, family-owned. Premium-priced offering incl CHARD/PINOT N Agrafe, 1st locally to undergo 2nd fermentation under cork.

Le Riche Wines Stell r (w) ★★★ Fine, modern-classic boutique CAB SAUV (varietal, incl Res, and blend Richesse) by Christo le Riche and family.

Leeu Passant *See* MULLINEUX.

Limpopo Most n province and GEOGRAPHICAL UNIT in WINE OF ORIGIN system. Currently single grower, Camel Thorn Estate.

Lowerland N Cape r w sp ★★ ⇢★★★ Literally "Verdant Land", area's most exciting winery and part of extensive Coetzee family agribusiness in Prieska WARD beside Orange River. Just 9 ha, vinified in W CAPE by top names. TANNAT, VIOGNIER and two outside-the-box versions of heirloom COLOMBARD, wooded and Brut Nature MCC.

> **Lighter, brighter Pinotage**
> Ironic that PINOTAGE, the 1925 cross of two of the least macho
> varieties (in S Africa, at least), PINOT N and CINSAULT, tended to be used
> for big, muscular wines emphasizing weight and power. Now a new
> style is challenging this, aiming for freshness, texture and palate
> appeal (hopefully without sacrificing longevity). Avant-garde
> Pinotage heroes: The Blacksmith, DAVID & NADIA, The Giant Periwinkle,
> Holder V'yd, OLIFANTSBERG, Reverie, Scions Of Sinai, SPIOENKOP,
> Swartberg, Wolf & Woman.

Agulhas Wine Triangle, new body for Cape's most s v'yds. Don't disappear.

MCC (Méthode Cap Classique) EU-friendly name for bottle-fermented sparkling, one of S Africa's major success stories; c.360 labels and counting.

Meerlust Estate Stell r w ★★★★ Myburgh-family-owned v'yds, cellar since 1756. Elegance, restraint in flagship Rubicon 09' 10' 15' 16 17, was among S Africa's 1st B'x reds; excellent MERLOT, CAB SAUV, CHARD, PINOT N.

Miles Mossop Wines Coast r w sw ★★★ →★★★★ Ex-TOKARA wine chief and CWG member Miles Mossop's own brand; elegantly styled, consistently excellent. Red and white blend, botrytized CHENIN BL, mostly ex-STELL; recently more labels, wider sourcing, eg. The Chapters range with CINSAULT from SWA.

Morgenster Estate W Cape r (p) w (sp) ★ →★★★ Prime Italian-owned wine/olive farm nr Somerset W, advised by Pierre Lurton of B'x (Cheval Blanc). Elegant Morgenster Res 11' 14 15' and second label Lourens River Valley (both B'x red). Old-country varieties incl one of only two S African VERMENTINOS.

Mount Abora Vineyards Bot R r w ★★★ High-scoring, naturally fermented old-vine CINSAULT, Rhône-red blend, bush-vine CHENIN BL Koggelbos 14' 17' 18, all ex-SWA. Owners Pieter de Waal and (winemaker) Krige Visser added Bot R CAB FR string to bow with 1st tiny bottling of 18.

Mulderbosch Vineyards Coast r p w (sp) ★★ →★★★ Highly regarded, US-owned STELL winery, sibling to FABLE MTN V'YDS, with single-block CHENIN BL trio, B'x-red Faithful Hound, huge-selling CAB SAUV rosé. Both winemakers have noteworthy parallel ventures: Yardstick Wines/Raised By Wolves (r w) by Adam Mason, terroir, old-vine and rare-grape bottlings; and Craven Wines, Aussie Mick Craven and S African wife/co-winemaker Jeanine, vivacious feel-good wines (r w).

Mullineux & Leeu Passant W Cape r w sw ★★★ →★★★★ Chris M and US-born wife Andrea with star viticulturist Rosa Kruger transform SWA SHIRAZ, CHENIN BL and handful of compatible varieties into ambrosial monovarietals and blends based on soil type (granite, quartz, schist), CWG bottlings and *vin de paille*. Wider-sourced Leeu Passant portfolio, Dry Red and two CHARD (ex-STELL, KLEIN KAROO), as sublime. Assistant Wade Sander's family brand, Brunia Wines, one to watch.

Mvemve Raats Stell r ★★★★ Mzokhona Mvemve, S Africa's 1st qualified black winemaker, and Bruwer Raats (RAATS FAMILY): stellar best-of-vintage B'x blend, MR de Compostella.

Nederburg Wines W Cape r (p) w sw s/sw (sp) ★ →★★★★ Among S Africa's biggest (two million cases) and best-known brands, PAARL-based, DISTELL-owned. Excellent Two Centuries flagship CAB SAUV, Heritage Heroes, Ingenuity ranges, Manor House. Delicious, long-lived desserts CHENIN BL botrytis Edelkeur, MUSCAT Eminence. Many value quaffers.

Neil Ellis Wines W Cape r w (sw) ★★ →★★★★ Pioneer STELL-based négociant sourcing mostly cooler-climate parcels for site expression. Masterly Terrain Specific range, esp Jonkershoek Valley CAB SAUV and Pie GRENACHE. PALOMINO blend Op Sy Moer is 1st foray into orange wine. Newer No Added Sulphite line-up (r p w).

Newton Johnson Vineyards Cape SC r (p) w (sw) ★→★★★★ Acclaimed family winery in UP HEM. Top Family V'yds PINOT N, CHARD, SAUV BL, SYRAH/MOURVÈDRE Granum, from own and partner v'yds; S Africa's 1st commercial ALBARIÑO. Lovely botrytis CHENIN BL L'illa, ex-ROB; entry-level brand Felicité.

Northern Cape Largest province and GEOGRAPHICAL UNIT in WINE OF ORIGIN scheme. Semi-arid to arid, with temperature extremes. Handful of producers: ORANGE RIVER CELLARS, LOWERLAND in recent Prieska WARD. *See* SUTHERLAND-KAROO.

Olifantsberg Family Vineyards Bre r (p) w ★★→★★★ Dutch-owned rising star, focus on Rhône grapes (r w), new-wave PINOTAGE, CHENIN BL on mtn slopes. SHIRAZ-based Silhouette 14' 15' 16, sophisticated Blanc (mostly ROUSSANNE/GRENACHE BL).

Olifants River W-coast REGION (c.9530 ha). Warm valley floors, conducive to organics; cooler, fine-wine-favouring sites in Citrusdal Mtn DISTRICT and its vaunted WARD, PIE. Bamboes Bay and Koekenaap wards nr Atlantic now part of COAST region.

Opstal Estate Sla r p w (sw) ★→★★★ One of BRE's quality leaders, family-owned, in mtn amphitheatre. V. fine CAPE BLEND (r w), old-vines CHENIN BL, SEM. New CINSAULT.

Orange River Cellars N Cape (r) w s s/sw (sp) ★→★★★ Vast operation with 750 grower-owners, 3700 ha under vine and five cellars on Orange River banks. Increasingly impressive/interesting Res portfolio incl new bubbly from obscure Hungarian grape IRSAI OLIVÉR.

Paarl DISTRICT (c.8740 ha) around historic namesake town with WARDS Agter Paarl, Simonsberg-Paarl, Voor Paardeberg. Diverse styles, approaches; best results from Med vines (r w), CAB SAUV, PINOTAGE, CHENIN BL. Major kosher houses (Backsberg, Zandwijk/Kleine Draken) based here.

Paul Cluver Estate Wines Elg r w sw s/sw ★★→★★★ Area's pioneer, Cluver family-owned/run; convincing PINOT N (incl nr-gluggable Village bottling), elegant CHARD, knockout RIES (botrytis and partly *foudre*-fermented semi-dry).

Porseleinberg Swa r ★★★★ BOEKENHOUTSKLOOF's organically farmed v'yds and cellar with expressive SYRAH. Handcrafted, incl ethereal front label printed on-site by the winemaker.

Raats Family Wines Stell r w ★★★→★★★★ Pure-fruited CAB FR and CHENIN BL, oaked and unwooded, esp Eden single-v'yd bottlings. Bruwer Raats and cousin Gavin Bruwer Slabbert are also partners in B Vintners, unearthing vinous gems. *See* MVEMVE RAATS.

Radford Dale W Cape r w (sw) ★→★★★ STELL venture with Australian-French-S African-UK owners. Creative and compatible blend of styles, influences, varieties and terroirs. Winemaker Jacques de Klerk's Reverie CHENIN BL and new PINOTAGE subtle beauties.

Rall Wines Coast r w ★★★→★★★★ Owner/winemaker and consultant Donovan R has phenomenal track record since debut 08 vintage. Original SWA blends (r w) since joined by eg. S Africa's only Cinsault Blanc, from minuscule old WELL parcel, showing house's flavour-filled understatement.

Lapel pin for aficionados: "Fan de Chenin". What about Que Syrah Syrah? À la Recherche du Tempranillo Perdu? Ours not to Riesling why?

Region *See* GEOGRAPHICAL UNIT.

Reyneke Wines W Cape r w ★→★★★ Leading certified-bio producer nr STELL; apt Twitter handle (and wine brand) "Vine Hugger". Luminous SHIRAZ, CHENIN BL, SAUV BL, newer limited-release, barrel-selected CAB SAUV.

Richard Kershaw Wines W Cape r w ★★★→★★★★ UK-born MW Richard Kershaw's refined PINOT N, SYRAH, CHARD from ELG; consituent sites showcased in separate Deconstructed bottlings. Newer GPS and Smuggler's Boot ranges spotlight other areas and new techniques, respectively.

Robertson Valley Low-rainfall inland DISTRICT with record 14 WARDS; c.12,790 ha; lime soils; historically gd CHARD, desserts; more recently SAUV BL, SHIRAZ, CAB SAUV. Major cellars: GRAHAM BECK, ROBERTSON WINERY. Family boutiques: recently reborn Mont Blois (r w) making waves with single-v'yd Chard and CHENIN BL.

Robertson Winery Rob r p w sw s/sw sp ★→★★ (Mostly DYA) Consistency, value throughout extended portfolio. Best: Constitution Rd (SHIRAZ, CHARD).

Rupert & Rothschild Vignerons W Cape r w ★★★ Top v'yds and cellar nr PAARL owned by Rupert family and Baron Benjamin de Rothschild. Baron Edmond and Classique (both r blends), CHARD Baroness Nadine.

Rustenberg Wines W Cape r (p) w (br) (sw) ★→★★★★ Venerable Barlow family cellar and v'yds nr STELL. Beautiful site, gardens. Flagship CAB SAUV Peter Barlow. Outstanding red-blend John X Merriman, savoury SYRAH, single-v'yd CHARD *Five Soldiers*. New wooded small-batch SAUV BL Wild Ferment.

Rust en Vrede Wine Estate Stell r (w) sw ★→★★★★ Jean Engelbrecht's powerful, pricey offering, incl Rust en Vrede red varietals, blends; Cirrus SYRAH joint venture with California's Silver Oak; Res range; Donkiesbaai PINOT N, CHENIN BL (dry and *vin de paille*); v.gd Guardian Peak.

Sadie Family Wines Stell, Swa, Oli R, Pie r w ★★★★ Revered SWA-based Eben Sadie's traditionally made portfolio. SHIRAZ/MOURVÈDRE Columella, multivariety Palladius (w) Cape benchmarks. Seminal Old Vines series celebrates heritage: profound, pure-fruited. Similar linear styling for co-winemaker Paul Jordaan and partner Pauline's CHENIN BL, Bosberaad, under new brand named (what else?) Paulus.

Saronsberg Cellar W Cape r (p) w (sw) sp ★→★★★ Art-adorned TUL family estate with awarded B'x blends (r), Rhône varieties, blends (r w), bracing CHARD MCC.

Savage Wines W Cape r w ★★★ Duncan S ranges far and wide from Cape Town city base for thrilling, understated wines from mostly Med varieties, CHENIN BL. Talent for catchy names in irresistible new dessert Not Tonight Josephine. Consults to promising UK-owned Brookdale v'yds/cellar overlooking PAARL.

Shannon Vineyards Elg r w (sw) ★★★→★★★★ Arguably S Africa's top MERLOT, also cracking PINOT N, SAUV BL, SEM, newer B'x w Capall Bán, grown by Downes brothers, James and Stuart, vinified at/by NEWTON JOHNSON.

Sijnn r p w ★★★ DE TRAFFORD co-owner David Trafford and partners' pioneer venture on CAPE SC. Pronounced "Sane". Stony soils, maritime climate, distinctive varietals and blends, brilliant rosé. Winemaker Charla Haasbroek's eponymous brand also compelling.

Simonsig Estate W Cape r w (br) (sw) (s/sw) sp ★→★★★ Malan family venture nr STELL admired for consistency and lofty standards. Pinnacle wine is powerful mtn CAB SAUV The Garland; v. fine SYRAH Merindol, PINOTAGE Red Hill; S Africa's original MCC, Kaapse Vonkel, still a delicious celebrator.

Spice Route *See* FAIRVIEW.

You could buy a boutique wine farm in the beautiful Cape from around $700,000.

Spier W Cape r p w (sw) sp ★→★★★★ Large, multi-awarded winery and tourist magnet nr STELL. Highlights incl Frans K Smit flagship, Creative Block, 21 Gables and new Seaward ranges, now also certified vegan.

Spioenkop Wines Elg, Stell r w ★★★ Belgian Koen Roose and family on ELG estate with Second Boer War-themed ranges ("1900", Spioenkop, new Tugela R). V.gd and individual PINOTAGE, CHENIN BL, RIES.

Springfield Estate Rob r w ★→★★★ Cult grower Abrie Bruwer, traditionally vinified CAB SAUV, CHARD, SAUV BL, B'x red, PINOT N. Newer ALBARIÑO sells out instantly.

Stark-Condé Wines Stell r w ★★★→★★★★ US-born boutique vigneron José Conde in STELL's Jonkershoek. Exceptional CAB SAUV, SYRAH and Field Blend (w) in Three

> ### The 18,000-ha Petri dish
> S Africa's CHENIN BL is a vast and continuing terroir experiment. It's omnipresent: c.18,000 ha planted, and until relatively recently most were overlooked workhorses, labelled Steen or going into fortifieds. Now old parcels are revered, and neglected and forgotten ones revived and cosseted; the gamut of techniques and equipment are harnessed to express grape and site. Stylistically, there are two main styles: fresh and fruity, and ripe and rich (latter often with oak element). Between lies delicious diversity, from linear and cerebral (BOTANICA Mary Delany, SADIE Skurfberg) through opulent and curvy (MULLINEUX Straw Wine, KEN FORRESTER The FMC) to effortless and gluggable (SIMONSIG Cultivar, SPIER Signature). It's too early definitively to map particular aroma/flavour profiles to specific regions. But getting there will be half the fun.

Pines, Stark-Condé ranges. Winemaker Rüdger van Wyk's fledgling PINOT N Kara-Tara looks gd too.

Steenberg Vineyards W Cape r (p) w sp ★★→★★★★ Top CONST winery, v'yds and chic cellar door, GRAHAM BECK-owned; SAUV BL, Sauv Bl/SEM blend, MCC, polished reds incl rare varietal NEBBIOLO.

Stellenbosch University town, demarcated wine DISTRICT (c.12,330 ha) and heart of wine industry – the Napa of S Africa. Many top estates, esp for reds, tucked into postcard mtn valleys and foothills. All tourist facilities.

Stellenbosch Vineyards Coast r (p) w (sw) ★→★★★ Big-volume winery with impressive Flagship range. Limited Release line-up has new curvaceous single-v'yd CHENIN BL. Budget range Welmoed.

Storm Wines U Hem, Hem V, Hem Rdg r w ★★★ PINOT N, CHARD specialist Hannes Storm expresses favoured HEM sites with precision and sensitivity.

Sutherland-Karoo DISTRICT in challenging N CAPE, not to be confused with separate, distant KLEIN KAROO. 5 ha, chiefly PINOT N, SHIRAZ, CHARD nr S Africa's coldest town, Sutherland. Scintillating SYRAH by Super Single V'yds, vinified offsite in STELL. NEBBIOLO, TEMPRANILLO also v. fine.

Swartland Coastal district, many fans abroad. C.10,000 ha of mostly shy-bearing, unirrigated bush-vines produce concentrated, distinctive, fresh wines. Home to heavies like SADIE FAMILY, MULLINEUX and KALL, source for lengthening list of others.

Testalonga Swa r w ★★→★★★ Range name El Bandito says it all: Craig Hawkins' vinifications of Med/heritage varieties and fizz defy convention; much-respected, loved and Instagrammed nonetheless. Earlier/easier-approachable Baby Bandito label for iconoclasts-in-training.

Thelema Mountain Vineyards W Cape r w (br) (s/sw) (sp) ★→★★★★ STELL-based pioneer of S Africa's modern wine revival, still beacon of quality, consistency; CAB SAUV, MERLOT Res, B'x red Rabelais, new premium fortified MUSCAT Gargantua. Extensive Sutherland v'yds in ELG broaden repertoire.

Thorne & Daughters Wines W Cape r w ★★★ John Thorne Seccombe and wife Tasha's wines, some from v. old vines, marvels of purity and refinement. White blend (with CLAIRETTE Blanche) Rocking Horse epitomizes Bot River vintners' cerebral-yet-sensual style and love of heirloom varieties.

Tokara W Cape r (p) w (sw) (sp) ★★→★★★★★ Wine, food, art showcase nr STELL. V'yds also in ELG. Gorgeous Director's Res blends (r w); elegant CHARD, SAUV BL. Newer CAB SAUV Res, CAB FR, Chard MCC.

Trizanne Signature Wines W Cape r w ★★★ Number of women in S Africa's cellar teams rising rapidly, but female soloists like Trizanne Barnard still v. unusual. Avid surfer sources from mostly COAST v'yds for seriously gd boutique SYRAH, varietal/blended SAUV BL, local rarity BARBERA. Sister lone rangers: Christa von

La Chevallerie (Huis van Chevallerie), Lucinda Heyns (Illimis Wines), Jocelyn Hogan Wilson (Hogan Wines), Marelise Niemann (Momento Wines).

Tulbagh Inland DISTRICT (c.1000 ha) historically associated with MCC and still white, latterly also red (esp PINOTAGE, SHIRAZ), some sweet styles. Look for FABLE MTN, KRONE, Rijks, SARONSBERG.

Uva Mira Mountain Vineyards Stell r w ★★★ Helderberg Mtn eyrie v'yds and cellar reaching new heights under owner Toby Venter, CEO of Porsche SA. Brilliant line-up incl newer SYRAH plus longtime performers CHARD, SAUV BL.

Vergelegen Wines Stell r w (sw) ★★★→★★★★ Historic mansion and gardens (incl c.1700 camphor trees) owned by Anglo-American plc mining company. Immaculate v'yds and wines, stylish cellar door at Somerset W. Powerful CAB SAUV "V", sumptuous and perfumed B'x blends (r w) GVB, new V'yd range with single-site Cab Sauv and MERLOT.

Vilafonté Paarl r ★★★ California's Zelma Long (ex-Simi) and Phil Freese (ex-Mondavi viticulturist) partnering ex-WARWICK Mike Ratcliffe. Trio of deep-flavoured B'x blends given distinctiveness by CAB SAUV, MERLOT or MALBEC predominance.

Villiera Wines Elg, Stell, Hem Rdg r w (br) (sw) (s/sw) sp ★→★★★ Grier family nr STELL with exceptional quality/value ratio, esp brut MCC bubbly quintet. New fortified dessert Dakwijn aged alfresco in demijohns.

Vondeling V Pa r (p) w sw sp ★→★★★ UK-owned, sustainability focused estate in PAARL'S Voor Paardeberg. Eclectic offering incl S Africa's 1st certified *méthode ancestrale* bubbly.

Provence is the rosé template: pinks get paler and drier, shellfish-friendlier and prawnier, not brawnier.

Walker Bay Highly regarded maritime DISTRICT (c.1000 ha); WARDS HEM, Bot River, Sunday's Glen, Stanford Foothills. PINOT N, SHIRAZ, CHARD, SAUV BL standout. German-owned PINOTAGE and CHENIN BL specialist Springfontein Estate in new Springfontein Rim ward well worth tasting/visiting.

Ward Smallest of the WO demarcations. *See* GEOGRAPHICAL UNIT.

Warwick Estate W Cape r (p) w (sp) ★★→★★★★ Tourist drawcard on STELL outskirts, prime v'yds extended and redeveloped under recent US owners. V.-fine full-flavoured CAB SAUV, CAB FR, CHARD, B'x Professor Black (w).

Waterford Estate W Cape r (p) w (sw) (sp) ★→★★★ Stylish family winery nr STELL; great cellar door. Savoury Kevin Arnold SHIRAZ, elegant CAB SAUV, intricate Cab Sauv-based flagship The Jem, new CHENIN BL ex-1966 vines.

Waterkloof W Cape r (p) w sp ★→★★★ British wine merchant Paul Boutinot's organic v'yds, winery and glass-curtained cellar door nr Somerset W. Top tiers: Waterkloof, Circle of Life, Seriously Cool and Astraeus MCC. Quality easy-drinkers False Bay, Peacock Wild Ferment.

Wellington Warm-climate DISTRICT (c.3870 ha) bordering PAARL and SWA. Growing reputation for PINOTAGE, SHIRAZ, red blends, CHENIN BL. BEE exemplar Bosman Family (winemaker Natasha Williams' debut own-range, Lelie van Saron, excellent too), Fairtrade-accredited Imbuko, organic Upland.

Western Cape Most s province and most important GEOGRAPHICAL UNIT in WINE OF ORIGIN system, with 113 of 127 official appellations.

Wine of Origin (WO) S Africa's "AOC" but without French restrictions. Certifies vintage, variety, area of origin. Opt-in sustainability certification additionally aims to guarantee eco-sensitive production. *See* GEOGRAPHICAL UNIT.

Worcester Sibling DISTRICT (c.6380 ha) to ROB, BRE in Breede River basin. Mostly bulk produce for export but bottled wines of Alvi's Drift, Arendskloof, Conradie, Leipzig, Stettyn, Survivor (by Overhex) and Tanzanite are taste-worthy, mostly family-made exceptions.

What is this thing called terroir?

It's one of those wine words that get bandied about, often wildly. But what is it, actually?

The truth is that the more we know about it, the more subtle it gets. The classic definition of terroir is that it is the combination of soil, climate and exposure to the sun – aspect, in other words – that makes each vineyard unique. That's an easy enough idea to grasp: anyone with a garden knows that one part is shadier, one part better drained, and one part gets the evening sun. That's terroir: it's as simple as that. And just as a sun-loving plant won't thrive in the shady bits of your garden, different grape varieties like different terroirs. The skill of the viticulturalist comes into play here: which, of the (usually) many varieties to choose from, will give the best, most balanced results on a given site? And is there anything to be done to help things along – put in a windbreak, perhaps, or more drainage? Most definitions for terroir also include the hand of man in the elements of terroir, for the simple reason that people have been interfering with raw nature for the last few thousand years.

Climate, in all its detail, is vital. And that's not just sun, rain and frost. Cloud cover, whether the sunlight is clear or filtered, what time of the year the rain arrives, whether nights are cool or warm, whether there's a large river or lake close by to moderate extremes of temperature: all these things have an effect.

Then there's soil structure and type. Gravel is different to clay, sure. Sand is different again. Better-drained soils are warmer, so encourage ripening. These things have a clear effect on flavour. Altitude affects acidity, because it's cooler higher up. This is all straightforward.

What is not straightforward is to say that one terroir is intrinsically better than another. What's hopeless for one grape might be brilliant for another. Or not. This is where the judgement of the grower comes into play, and it's why the great terroirs of the world can look so very different: the almost flat Médoc, even flatter Coonawarra, almost vertical Mosel.

Wine is about place. The influence of place is what we'll take a look at in the next few pages.

What difference does terroir make?

You can grow tomatoes hydroponically, and many people do. That's the antithesis of terroir. And what do they taste like?

Quite. A tomato more different to one grown in the sunshine of an Italian garden could hardly be imagined.

Let's define the ideal terroir for wine. It would give you sugar ripeness, flavour ripeness and phenolic ripeness all at the same time, assuming that you've chosen vines that are suitable in the first place.

What are these things, you ask? A ripening grape develops in several directions at once. Sugar ripeness is sufficient sugar to give you the alcohol level you want – let's say 13–14% alcohol after fermentation. Phenolic ripeness is about tannins in skins and pips: unripe, green tannins taste harsh and raw; ripe ones are softer, silkier. Flavour ripeness is just that: actual flavour, different to sweetness and acidity.

To get all these simultaneously doesn't often happen. Sugar ripeness increases with warmth, whereas tannin (phenolic) ripeness needs time. So, summer and autumn heat can send sugar levels through the roof while the grower is waiting for the tannins to ripen. That's what gives wines with jammy fruit and high alcohol: in hot climates, if you're determined to wait for really soft, supple tannins, you might have to put up with whopping sugar levels and pruney fruit flavours.

Climate change is tending to drive those different ripening times further apart; what was a great terroir in cooler decades might be less good now – and vice versa. Marginal vineyards in the Mosel, which had been abandoned as too cold, are now being cultivated again. Warmer summers have brought Champagne further into its climatic comfort zone (*see also* box, p.296).

Spiced Lamm: the terraced Lamm vineyard gives a particular flavour.

When the terroir is right, the grape variety will shine. Some grapes reflect their terroir more transparently than others: Riesling, Garnacha, Chenin Blanc and Pinot Noir are among those able to mirror every detail. Sometimes it's easy to see the link between flavour and terroir: water-retaining clay might give better balance in a dry year. But clay varies a lot; whole research papers have been written on the internal structures of clay molecules and what that means for vine roots. Sometimes all you can say for certain is that this wine is what that terroir produced in that year. It's about the influence a warm April had on this site compared to that: what effect a dry August had on riper and on less ripe grapes.

For example: there's an acid called abscisic acid. It is formed by the vine it's not present in the soil – and it's formed at the moment when soil begins to dry out. Its formation seems to tell the vine to stop putting nutrients into the leaves and start putting them into the grapes. So, the moment at which the vine starts making it has a big effect on ripening and final flavour. That is pure terroir effect, and it's far more precise than the simple association of warmth with ripeness, or coolness with acidity.

Terroir is the reason that different vineyards taste different. Look at Austrian Grüner Veltliner, from three vineyards in Kamptal: Lamm gives spicy wines; Grub is rich; Renner is juicy. These are terroir differences. And while it's true that producers might adjust their winemaking to emphasize the character of the vineyard, the differences are there. In the Kremstal region the grapes ripen about ten days earlier than they do in the Wachau. That's terroir – not all soil, either. Climate has a big part to play.

Does terroir have to be remarkable to qualify as terroir, or can anyone play? That's what we'll look at next.

Can you find terroir anywhere? And can you create it?

Let me start with a story. I was in Bordeaux, looking at a mid-range classed growth: a good château, in other words. Gravel was piled high beside the vineyard tracks.

"Oh," I said, half joking, "are you going to put that on the vineyards to improve the terroir?" The viticultural consultant looked alarmed. "You can't do that," he told me solemnly. "A terroir takes thousands of years to form. You can't just create one." Afterwards, the château-owner took me aside. Yes, the gravel was intended for the vineyards, to improve the drainage. That's exactly what they were going to do.

Yes, you can improve a terroir; every time a grower puts porous drainage pipes into his vineyard he is adjusting the terroir. Ditto, every time he plants a row of trees as a windbreak. Ditto, every time he puts in frost protection. Terracing a steep hillside in Portugal's Douro Valley? Yes, ditto. Re-landscaping entire areas, as some producers have done in the New World? Ditto. Some hillsides in the Douro are so intractably rocky that you need dynamite to make them plantable. That makes adding a bit of gravel look rather small-scale.

There wouldn't be nearly so many vines in the Médoc if Dutch engineers hadn't drained this damp, low-lying area back in the 17th century. The chalky-clay slopes of southern England wouldn't be so suited to vines if they hadn't been deforested from the Bronze Age onwards. Man has always interfered with landscape. If you want a pristine terroir to plant with vines you might still find one

Row direction, section of slope and grass vs bare earth all help to adapt terroir to vine.

Terraces in the Douro: an ingenious solution to an intractable problem.

in Australia – might. But chances are you'll have to install drip irrigation. And if that isn't altering a terroir I'd like to know what is.

What is less easy to change is the bedrock (*see* box, p.205). There will be layers of soil on top (perhaps some alluvial, some wind-blown) but what that rock is, how it is structured and how near the surface it is will affect how vines grow. If it is fractured, or perhaps with vertical strata (as with the schist of the Douro), then vine roots will be able to go deep. If those schist strata in the Douro were impenetrably horizontal, there would probably never have been a wine industry there.

The mountains of the Douro, the hills of southern England and the flat Médoc are all so different that one has to say, yes, great terroir can be found anywhere. A site might produce mediocre wine until someone has a brainwave and tries a different grape variety; and suddenly it gives terrific wine. The winemaker at one estate in Tuscany talks of her Cabernet Franc as being happy in its terroir, poised and at ease with itself, whereas the Merlot struggles every year and is never quite happy. Finding the perfect match to make a terroir sing is where great growers come into the picture.

For example, Australian Brian Croser says that he has spent his life trying to prove that his Tiers vineyard is a great site. "But," he says, "what if it isn't? And what if it is? And what if I've really stuffed it up?"

Yes, you can find terroir anywhere, which is not the same as saying that all terroir is potentially great. You can also misunderstand it, anywhere. And you don't know until much, much later.

Terroir in chunks: soil and geology

Tearing chunks out of terroir to examine them separately is never easy, because everything is so interlinked, but let's try.

Most winemakers agree that soil is the most important part of terroir. Obviously, it's the stuff that holds the vines up, but it's also vital in determining ripeness. Warm, well-drained soils encourage vines to bud earlier and grapes to ripen earlier; cooler, damper soils do the opposite. Some soils, like the chalk of Champagne, naturally give wines with a lower pH.

Can one associate particular flavours with particular soils? Volcanic soils seem to give a particular sparkiness; chalk gives a particular finesse to acidity. What about the *licorella* soils of Spain's Priorat? The slate of the Mosel? Surely we can taste such soils (*see* boxes, pp. 180, 156)?

Think of Australia's Coonawarra, one of the world's most famous terroirs. Its *terra rossa* – a long narrow strip of red soil over limestone – provides one of the very few spots in the world where Cabernet Sauvignon makes great varietal wines, without help from Merlot or anything else. The *terra rossa* of Coonawarra has been the subject of great argument, mostly from those growers with land off the strip. "It all looks flat, but it tastes different," says Sue Hodder, senior winemaker at Wynns. "The black soil [outside the *terra rossa*] is great for potatoes, onions; the transitional soil is good for whites." But for reds, only the *terra rossa* gives great results.

Or think of Burgundy's Côte d'Or, where one wine might cost €20 and another, from perhaps 100m (328ft) away, costs €200. It's not the climate that varies; it's the soil. The Côte d'Or, being on a geological fault, has a crazily complex pattern of soils, reflected (though not perfectly) in its vineyard boundaries.

The Mosel's slate...

...Chablis' limestone and...

In Alsace, where the geology is just as complex, the vineyard tracks mark the position of the geological faults. Such faults, where the soils change frantically within small areas, are catnip to terroir-focused growers.

Soil affects acidity, and it affects tannins: how they ripen, how chunky they are. Clay gives more solid tannins (not a flavour, I agree, but noticeable in the mouth), while sand gives lighter ones. Carl Schultz, winemaker at Hartenberg in S Africa, says that if you add stones to any soil it will increase the sensation of tannin.

What we are not tasting, however, is the soil itself. Growers are fond of telling us that their wines gain "minerality" from the soil and bedrock by directly sucking it up, but there is no mechanism for this. Soil minerals are not the same as the nutrient minerals found in trace quantities in wine; vines do not swallow slate or schist and eject them into the grapes. "Minerality" is a descriptor to be used in the same way as we use "peachy" or "grassy"; in other words, not literally. When they tell you the opposite, take it with a pinch of (mineral) salt.

The link between vine and soil is far more complicated, and is made by micro-fungi that live symbiotically on the vine roots and make nutrients in the soil available to the vine. Vines get very little from the bedrock, no matter how far down their roots go; 95% of their nutrients come from the topsoil. Terroir differences in wine depend on a great many different processes and reactions; destroy the life of the soil and you destroy the soil, as far as the vine is concerned. It's why terroir-obsessed growers around the world have turned to biodynamics to restore the life of their soils, and thus their wines. And it works: dead soil cannot claim to be great terroir.

...Coonawarra's *terra rossa* over limestone – just beware the word "minerality".

Terroir in chunks: climate

Soil is the most important part of terroir, as we've said, but it's how climate interacts with soil that makes a terroir suitable for this grape rather than that.

For example: cold, water-retaining soil can be a problem in a damp climate like Bordeaux, when autumn rains can spoil grapes still laboriously ripening in October. That's why earlier-ripening Merlot gets the cooler clay soils there. But in a dry climate like Chile, it doesn't matter. Almaviva's vines near Santiago, cooled by cold winds from the Andes, have all the time in the world to ripen. In Bordeaux you need a warm soil to balance a cool, damp climate; near Santiago you need cold winds to hold things back.

Bordeaux and Chile are a good comparison, because several top-flight Bordelais producers have joint ventures in Chile. It looked like a no-brainer if you wanted to make great Cabernet blends. Common sense would tell you that all you had to do was replicate Bordeaux's rainfall patterns with irrigation, and voilà!

Except that it didn't work like that. The vines, squeezed into an alien pattern, produced wines that weren't as good as they were supposed to be – not as graceful, not as poised. The growers realized they had to work with the climate, not against it, and adjust their ideas of what top Chilean wine should be. Once they started thinking Chilean, with different views on ripeness, grape blends and yes, irrigation, things improved rapidly.

The usual definition of a great terroir is one that is marginal: where grapes will ripen perfectly, but only just. The classic regions of Europe have endured many swings of climate and have shrunk and expanded accordingly, and changed their vine varieties too. The point of growing several varieties is that it gives you more chance of ripening something no matter what the year brings: it's why Bordeaux, Champagne, Tuscany, Piedmont, Vienna and others tend not to focus on one grape. Before phylloxera, most grew even more.

The cold middle decades of the 20th century were particularly difficult for northern Europe: most places struggled to ripen grapes, most years. Warmer summers began to be noticed from about 1988, and climate change – for all the problems it brings in other ways – has brought northern Europe back to its vinous comfort zone. Better clones and better viticulture have helped, and the wines are better, riper, better balanced than they have been in living memory. Chaptalization – that remedy for underripeness, adding sugar at fermentation to increase alcohol – is far less used; there has even been some need to acidify, which is the usual remedy for overripeness.

In the southern hemisphere the problems of climate are different. That large expanse of ocean around Australia helps to mitigate climate change, but many Australian growers wish their continent

Cacti and vines in Chile's Limarí Valley: a cool climate, believe it or not.

extended a bit further south, to give them more cool-climate options. Tasmania ticks that box, but has had to learn to deal with damaging winds. Chile's vineyards now extend very far south indeed; but counter-intuitively, some of its best vineyards are proving to be right up north, in the Atacama Desert, rubbing shoulders with large cacti. Weirdly, they're thought of as cool-climate sites.

How so? Because oceans, indeed all large bodies of water, and small ones too, are key to climate. Chile's climatic regions for wine are divided east-west: the coastal regions, within reach of cold winds from the ocean, are the coolest, even if cacti flourish there. The hottest are close to the Andes – though here again, if you can catch cold night winds from the Andes, as Almaviva does, you'll be cooler than otherwise. California's Napa Valley is the classic example of ocean fog blowing through gaps in the coastal range to cool the vines.

Big rivers, like the Douro or Mosel, can reflect heat and light and help to ripen grapes at those two extremes, Port and Mosel Riesling. (Douro table wine prefers cooler conditions, usually higher up.) And in France's Sauternes and Hungary's Tokaj, the presence of a river brings about the morning mists that enable *Botrytis cinerea* to flourish – which on the wrong grapes would be a disaster. The ingenuity of growers in exploiting quirks of climate is endless.

Terroir in chunks: exposure and altitude

If climate is a way of balancing soil, exposure and altitude are ways of balancing climate.

The obvious example of exposure is that in northern European vineyards the classic exposure is south-facing, vice-versa in the southern hemisphere. East-facing gets the morning sun; west-facing gets the hotter afternoon sun. In hotter, more southerly European regions there is the option of north-facing sites for less heat-loving vines. In Barolo, for example, Dolcetto vines get the north-facing slopes, Barbera the east- or west-facing ones, and Nebbiolo the warmest ones. (I'm talking generally: the vineyards there are so convoluted that if you want to throw contradictory examples at me, you'll find plenty of ammunition.) In chilly Champagne the north-facing slopes might be used for Pinot Noir, because they give a particular character that is useful in a blend. But in southern England, where ripening is just that bit more difficult than in Champagne, north-facing slopes are not in demand.

But why would you want a slope in the first place? Why not just plant on flat ground, which is easier for tractors and easier for people? The original reason for choosing slopes was probably because the poorer soil there meant that other crops wouldn't flourish; and then it turned out that you got greater concentration and intensity from slope-grown vines, so the idea stuck. If you compare Italy's Soave grown down on the plain with Soave Classico from the hills you'll get the idea (*see also* box, p.255).

Slopes are a good idea in a climate prone to late frosts, because the cold air keeps sliding downwards rather than settling among your vines.

Not as high but scarily steep, in Cinque Terre, Liguria, Italy.

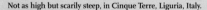

Way up high in Salta province, Argentina.

The traditionally greatest spots of the Médoc, like the Carruades plateau, are slightly raised: even a couple of metres can make a difference.

But when you head for the hills, as well as the benefits of air drainage and water drainage, you find new problems. The soil, which is always deeper and richer on the valley floor, can be just too thin on the slopes, and the higher you go the thinner it gets. Plus the higher you go, the cooler it gets. That's why you'll often find forest on the very tops of the hills. Even the Corton hill in the Côte d'Or has its little tonsure of forest. And in the Côte d'Or the best sites – the Grands Crus – are on the mid-slopes. This is the Goldilocks point of both soil and temperature.

The usual rule of thumb is that for every 100m (330ft) of height you lose 1°C of average temperature. In Argentina's Salta province they take altitude to extremes, and plant at over 3000m (9840ft) in search of freshness and acidity and a long ripening season. In most places, even if that altitude existed, it would be far too cold.

Some slopes can just be too steep – too steep to work, too steep for tractors. Terraces are the traditional answer here: shallow ones with just a couple of rows of vines, or broader, gently sloping ones. You'll find both in the Douro Valley, for example, and both are labour-intensive ways of reshaping an unpromising landscape. Much of Germany's vineyard was terraced until a replanting programme from the 80s onwards smoothed most away. Austria's Wachau is still terraced, and if you climb to the top of the Achleiten vineyard you'll see old abandoned ones in what is now forest. They were cultivated in warmer times, and now some of them are being cultivated again. It's a long walk up there, and a long walk back down again. But the best growers don't believe in making life easy.

Terroir in chunks:
the role of the grower

So: you've found your hillside, you like the climate and you like the soil. How are you going to plant it, and with what?

This is the first of many interventions made by growers – when they talk about low-intervention wines they know that wine-growing is one big intervention from soil to bottle. Every move they make is a decision and an intervention, and is directed to what the winemaker wants that vineyard to express. That might be the terroir, or it might be the balance sheet. Probably it will be both, given that growers have to live, and that a pure expression of terroir in a year of rain and rot is not what people want to drink.

The point is that terroir expression is what the grower says it is. More ripeness or more freshness? More tannin or less? The grower decides. Even a producer focused on pure terroir expression first selects certain terroirs, and then selects those barrels or vats with a less clear expression – a more dissident expression, one might say.

The decisions begin with bare earth. Does it need additional drainage? Different nutrients? Now is the time. Growers in Spain's Rías Baixas traditionally added shellfish shells to the vineyards to change the soil pH. They eat a lot of shellfish round there.

Then: which grape variety? At what density of planting? On which rootstock? Which clones? All these factors affect flavour. How will you train the vines and orient the rows? North-south means the vines will get full morning and afternoon sun. You'll need to take wind direction, and the slope, into account here too. Rows across the slope may trap cold air, but hinder erosion. Brisk winds along the rows keep away disease but might be too cold. Take your pick. The height of the trellis? Higher can mean more ripeness, but is that what you want? Letting grass and weeds grow between the rows can retain soil humidity or increase competition for water, depending on your point of view. Will you irrigate? If so, how much?

The pendulum is swinging towards sustainability and biodiversity, which can only be a good thing. Pheromone traps are better than sprays for controlling pests. In S Africa, leafroll virus and "dead arm" disease are now being conquered by re-introducing native insects that control the vectors – insects like ladybirds and a couple of parasitic wasps that were indigenous to vineyards until chemical sprays knocked them out in the 70s.

But these are things we tend to take for granted. What is more noticeable now is the way growers are dealing with climate change.

The more excitable predictions for how wine regions will react involves mass replanting: Cabernet on the Côte d'Or, Syrah in Bordeaux (*see* box, p.119). Already we're seeing grape varieties

Biodiversity at Bordeaux's Château Guiraud: we love insects now.

grown where they would never have ripened in the past: Pinot Noir in the Mosel, for example. If growers are looking for other varieties, it's usually for grapes that will give lower alcohol and more acidity, like Manseng Noir in southwest France, being re-introduced by Producteurs de Plaimont.

Otherwise it's about managing heat. In Europe, the 2003 vintage taught them an awful lot. You can leave on more leaves, for shading: sunburn on grapes kills acidity. You can also stop using potassium-based fertilizers: they're acidity-killers. Details like removing or keeping the secondary clusters, those tiny clusters that come from a smaller, second flowering: leave them on and you extend the ripening season. In a dry year, you might plough a bit later, which keeps the vines working until the grapes are ripe: that can make the difference between ripe tannins and green ones.

These are details: finicky, unglamorous and needing detailed knowledge of your vines and terroirs. They're the reason why the hand of the grower can never be dissociated from the terroir; people are part of terroir, understanding it, manipulating it and deciding how it should speak. If that sounds authoritarian, remember that without growers, there wouldn't be any wine, ever.

How to kill your terroir

There are two ways to do this: in the vineyard, and in the winery.

In the vineyard, it's simple. If you pour chemicals on for long enough, and in sufficient quantity, you'll kill everything. With no insect life, except those grape pests that manage to dodge the poison, and no life in the soil – no earthworms, no micro-fungi, nothing – your soil will be dead. At that point, you can still make wine; you will still have sun and rain, and the vine will ripen its grapes. You can still point to your slopes, and where the wind blows in the evening, and can talk about your terroir – except that you will have killed it. The wine will not taste alive.

It was fear of this – and often the realization that it was happening – that inspired one Burgundy grower after another to switch to biodynamics. Pinot Noir reflects its terroir particularly faithfully, and one of the world's greatest wine regions was being killed by chemicals. Some Côte d'Or soils had less life in them than the Sahara Desert. Biodynamics, they reasoned, might sound wacky, but it might work.

Not cigarette ends but cow horns filled with manure for biodynamic Preparation 500.

And it did, and it does. The first thing growers usually noticed was the acidity returning to their wines. The vines were healthier, the leaves bouncier, stronger. The soil was better, richer – and it had things wriggling in it. Not all growers on the Côte d'Or are biodynamic, but a lot are. Same in Alsace: the terroir here is as complicated as it is in Burgundy, but there's no point in having such great terroir if all your vines can express is potassium fertilizer.

You can't kill your terroir in the winery, but you can slaughter terroir expression. There are several ways you can do this. First of all, overripeness. We can argue for years about what constitutes ideal ripeness, but for me it can be summed up as what Toro estate Pintia describes as picking al dente. If the wine tastes of jam, things have gone too far.

Is chaptalization or acidification a negation of terroir? How long have you got? Technically, yes: you're contradicting what your terroir is saying, because you don't like the sound (or the taste) of it. They usually acidify in Coonawarra; even today, there will be wines in Champagne that get chaptalized. If we're going to forgive one (and are we?), we should forgive the other.

Overextraction is another killer: bashing up the grape skins during fermentation to give as much tannin, colour and flavour as possible. Some would say that gives maximum terroir expression: why would it not? Because, I would suggest, terroir should not need to shout. If it has a distinctive voice it will be heard without bellowing. A stage whisper can be more effective than a rant.

Overoaking is a subject in itself. How can you express terroir by putting wine into oak barrels, which give it their own flavour? Mercifully the days of obvious oak flavours are over, except for a few serial offenders. Older oak is in fashion: bigger barrels too, which make less noise. Even so, consultant Alberto Antonini asks why a grower's best wines should get more oak than their lesser wines. They're the best grapes, he argues; they need less help.

But an awful lot of wine is less about the vineyard than about the winemaker. It's easy to see why; the winemaker has a vision, and he has a competitive marketplace in which he needs to shine. Wines tend to bear the stamp of their maker: flamboyant, extrovert personalities make flamboyant, extrovert wines. It takes winemakers who are not only modest but also strong-minded to edit themselves out of the wine, if they ever can.

The traditionally great terroirs of the world have survived not only changes of climate and economics but many changes of winemaker. Winemakers come and go; terroir endures.

Terroir in small chunks: yeasts

A small section for something even smaller. Something unnoticeable to most of us, indeed.

You can't have wine without yeasts. But yeasts can be unpredictable. There are very, very many different kinds, and they do different things and give different flavours. A vineyard will be full of a mixed population of yeasts, and a winery likewise. That population mix will vary from year to year too, although it does seem that different vineyards maintain basically the same kinds of yeasts from year to year; they're one of the markers of terroir, one of the ways in which one vineyard differs from another. Yeasts will also come into a winery on oak barrels, and with people: wineries are busy with unseen life.

If you leave grape juice alone it will ferment spontaneously, from the yeasts in the winery and on the grape skins. Growers who welcome that say that it is a vital part of terroir expression, and wild ferments, indigenous yeasts – whatever term you want to use – give wines that taste different. They are often less clearly fruity. They are more complex, subtle: often more energetic and vital. They are also riskier to make by a multiple of goodness-knows-what. You don't know what yeasts are there, and you don't know what's going on. The wine might be spoiled, or the fermentation might stick, which is nearly the same thing. So, most winemakers kill everything with sulphur, and start again with a laboratory yeast chosen because either it is as neutral in flavour as possible, or because it's what's known as an "aromatic yeast", designed to enhance particular flavours.

All yeasts affect flavour: it's why wine doesn't taste like grape juice. Yeast doesn't create flavours from nothing, but it transforms what is there. If you're using a selected yeast you might choose to emphasize red-fruit or black-fruit flavours, emphasize the gooseberry or passion fruit flavours in Sauvignon Blanc. It's another manipulation.

The middle way is to select a wild yeast from your own vineyards and get a laboratory to produce it in quantity for you. Then it becomes your signature: always the same, always trustworthy.

Fermenting wine: first choose your yeast.

A little learning...

A few technical words

The jargon of laboratory analysis is often seen on back labels. It creeps menacingly into newspapers and magazines. What does it mean? This hard-edged wine talk is very briefly explained below.

Acidity is both fixed and volatile. Fixed is mostly tartaric, malic and citric, all from the grape, and lactic and succinic, from fermentation. Acidity may be natural or (in warm climates) added. **Volatile (VA)**, or acetic acid, is formed by bacteria in the presence of oxygen. A touch of VA is inevitable, and can add complexity. Too much = vinegar. Total acidity is fixed + VA combined.

Alcohol content (mainly ethyl alcohol) is expressed as per cent (%) by volume of the total liquid. (Also known as "degrees".) Table wines are usually 12.5–14.5%; controlling alcohol levels is big challenge of modern viticulture.

Amphora the fermentation vessel of the moment, and the last 7000 years. Remove lid, throw in grapes, replace lid, return in six months. Risky: can be wonderful or frankly horrible.

Barriques small (225-litre) oak barrels, as used in Bordeaux and across the world for fermentation and/or ageing. The newer the barrel, the stronger the smell and taste of oak influence; French oak is more subtle than American. The fashion for overpowering wine of all sorts with new oak has waned; oak use is now far more subtle across most of the globe.

Bio (Biodynamic) viticulture uses herbal, mineral and organic preparations in homeopathic quantities, in accordance with the phases of the moon and the movements of the planets. Sounds like voodoo, but some top growers swear by it. NB: "bio" in French means organic too, but here it means biodynamic.

Concrete eggs the fermentation vessel of the moment. Concrete generally has returned to fashion as part of move away from oak.

Malolactic fermentation occurs after the alcoholic fermentation, and changes tart malic acid into softer lactic acid. Can add complexity to red and white alike. Often avoided in hot climates where natural acidity is low and precious.

Micro-oxygenation is a widely used bubbling technique; it allows controlled contact with oxygen during maturation. Softens flavours and helps to stabilize wine.

Minerality a tasting term to be used with caution: fine as a descriptor of chalky/stony flavours; often wrongly used to imply transference of minerals from soil to wine, which is impossible.

Natural wines are undefined, but start by being organic or biodynamic, involve minimal intervention in the winery and as little sulphur as possible; sometimes none. Can be excellent and characterful, or oxidized. Often made in amphora or concrete eggs.

Old vines give deeper flavours. No legal definition: some "vieilles vignes" turn out to be c.30 years. Should be 50+ to be taken seriously.

Orange wines are tannic whites fermented on skins, perhaps in amphorae. Like natural wines, some good, some not, but good to excellent with food.

Organic viticulture prohibits most chemical products in the vineyard; organic wine prohibits added sulphur and must be made from organically grown grapes.

Pét-nat (pétillant naturel) bottled before end of fermentation, which continues in bottle. Slight residual sugar, quite low alcohol. Dead trendy.

pH is a measure of acidity: the lower the pH, the sharper the acidity. Wine is normally 2.8–3.8. High pH can be a problem in hot climates. Lower pH gives better colour, helps stop bacterial spoilage and allows more of the SO_2 to be free and active as a preservative. So low is good in general.

Residual sugar is that which is left after fermentation has ended or been stopped, measured in grams per litre (g/l). A dry wine has almost none.

Sulphur dioxide (SO_2) added to prevent oxidation and other accidents in winemaking. Some combines with sugars, etc. and is "bound". Only "free" SO_2 is effective as a preservative. Trend worldwide is to use less. To use none is brave.